Manmade Fibers: Their Origin and Development

Based on Proceedings of an International Symposium cosponsored by the History of Chemistry, Polymer, Material Science and Engineering, Polymer Chemistry and Cellulose, Paper and Textile Divisions of the American Chemical Society at its 201st National Meeting at Atlanta, GA, April 17–19, 1991.

Manmade Fibers: Their Origin and Development

Edited by

The late Raymond B. Seymour

Department of Polymer Science,
University of Southern Mississippi,
Hattiesburg, MS 39406, USA

and

Roger S. Porter

Department of Polymer Science and Engineering,
University of Massachusetts,
Amherst, MA 01003, USA

and

Department of Polymer Science,
University of Southern Mississippi,
Hattiesburg, MS 39406, USA

ELSEVIER APPLIED SCIENCE
LONDON and NEW YORK

ELSEVIER SCIENCE PUBLISHERS LTD
Crown House, Linton Road, Barking, Essex IG11 8JU, England

WITH 34 TABLES AND 163 ILLUSTRATIONS

British Library Cataloguing in Publication Data
Manmade Fibers: Their Origin and
Development
I. Seymour, Raymond B.
II. Porter, Roger S.
677

ISBN 1-85166-888-8

Library of Congress Cataloging-in-Publication Data
Manmade fibers: their origin and development / edited by Raymond B.
Seymour and Roger S. Porter
p. cm.
Includes bibliographical references and index.
ISBN 1-85166-888-8
1. Textile fibers, Synthetic. I. Seymour, Raymond Benedict,
1912–91. II. Porter, Roger Stephen, 1928– .
TS1548.5.M264 1992
677'.4—dc20

92-30058
CIP

Printed in Great Britain by Galliard (Printers) Ltd, Great Yarmouth

Dr Raymond B. Seymour, Prominent Polymer Scientist and Educator

Raymond B. Seymour, an inventor and researcher who dedicated much of his working life to polymer education, died on November 15, 1991. He was 79. Seymour died at his home in Hattiesburg. He had been Distinguished Professor of Polymer Science at the University of Southern Mississippi.

Ray Seymour had earned both a bachelor's and a master's degree in chemical engineering from the University of New Hampshire. In 1937 he earned a doctorate from the University of Iowa.

In 1989 the Society of Plastics Engineers presented Dr Seymour with its highest award, the International Gold Medal. During his career, Seymour received more than 40 other professional honors and awards, wrote more than 1500 scientific reports and helped develop products protected by 45 US patents and 100 patents in foreign countries.

He demonstrated plastics on television science shows. He wrote books ranging from technical works to an introductory text for grade school children. In the 1940s, Dr Seymour introduced the first undergraduate course in polymer science.

FOREWORD

Since the recording of the history of various phases of polymer science is of the utmost importance, it was reassuring to note the previous publications on the history of polymer science, edited by Professor Seymour and published by Marcel Dekker, Inc. in 1981; *Polymers: Their Origin and Growth as a Science*, authored by Dr H. Morawetz and published by Wiley in 1985; *Origin and Development of High Performance Polymers*, coedited by Drs Seymour and G. Kirschenbaum and published by Elsevier Science Publishing Co. in 1986; *History of Polyolefins*, coedited by Drs Seymour and T. Cheng and published by Reidel in 1986; and *Pioneers in Polymer Science*, coedited by Drs Seymour, H. Mark, L. Pauling and C. Marvel and published by Kluwer Academic in 1989.

It was also my pleasure to serve as coeditor of *Organic Coatings: Their Origin and Development*, which was published by Elsevier Science Publishing Co. in 1990. Fortunately, Drs Seymour and Porter have added to the history of polymer science literature by coediting this book on *Manmade Fibers: Their Origin and Development*. As was the case with other books in this series, many of the chapters are written by pioneers in polymer science who were responsible, to a large extent, for today's polymer industry.

H. MARK
Polytechnic University

PREFACE

Natural fibers, such as wool, silk and cotton, have been used for many thousands of years. However, the possibility that manmade fibers could be synthesized was overlooked until three centuries ago when Hooke in his *Micrographia* suggested that man could utilize artificial substances that could be wire drawn, like the silkworm's 'clew'. De Réaumur promoted Hooke's suggestion in 1774, but a useful filament was not produced by man until Swann and de Chardonnet extruded a solution of cellulose nitrate (collodion) through small holes (spinnerets) in the last part of the 19th century.

These pioneer manmade fibers were replaced by rayon fibers, which were spun from an alkaline cellulose xanthate solution (viscose) and were subsequently supplemented by cellulose acetate and many synthetic fibers, such as nylon, polyesters, polyolefins, and acrylic fibers.

Much has been written about the history of natural fibers but very little has been written about the history of manmade fibers. Chapters on the history of natural and synthetic fibers by Drs Charles Fisher and Paul Morgan, respectively, were published in *History of Polymer Science and Technology*, which was published by Marcel Dekker, Inc. in 1982.

According to Fisher, about 15 million tons of natural fibers are produced annually. Their history is not discussed in this book, but considerable information is presented on the history of manmade fibers which according to Morgan are produced at an annual rate of about 36 million tons.

RAYMOND B. SEYMOUR
ROGER S. PORTER

ACKNOWLEDGMENTS

The editors would like to thank the many authors who presented reports at the International Symposium on History of Manmade Fibers at the National Meeting of the American Chemical Society at Atlanta, GA on April 17–19, 1991. The editors also appreciate the assistance of Ms Paula E. Phillips and Mrs Mischa Thomas, who aided in the preparation of this book.

CONTENTS

AN UPDATE ON FIBER SCIENCE AND TECHNOLOGY

RAYMOND B. SEYMOUR
Department of Polymer Science, The University of Southern Mississippi, Southern Station Box 10076, Hattiesburg, MS 39406-0076

The original fibers were natural products, such as flax and cotton, which were obtained from plants and animal products, such as wool and silk. The advance of textile arts, such as spinning and weaving, paralleled other advances in civilization from the Stone Age to the Industrial Revolution and to the present era. Mechanization from the cotton gin to the spinning jenny catalyzed a new era in the textile industry and made cotton king.

Manmade fibers, such as regenerated cellulose (rayon) and cellulose acetate (acetate rayon), which were commercialized in the early 1900's, provided a new source of fibers, but cotton continued to be the principal fiber until after the advent of the synthetic fibers, viz, nylon, polyester and acrylic. Cotton accounted for less than 50 percent of the 3.8 million tons of fibers used in the U.S. in 1965, and this share continues to decrease. Twenty-five kg per capita of fibers were consumed in the U.S. in 1925 and manmade and synthetic fibers accounted for 16kg per capita or 69 percent.

The worldwide production of fibers in 1975 was 23.7 million tons and the product mix, in millions of tons, was as follows: synthetic fibers (7.4), rayon and acetate (3.0), cotton (11.9), wool (1.5), and silk (0.02). The growth of the fiber industry continues at a rate that is 50 percent greater than that of the population growth [1]. The production of synthetic fibers, worldwide, increased by about 6 percent in 1987 [2].

Western Europe accounted for 38 percent of the world's output of synthetic fibers in 1987, Japan accounted for 16 percent, and the U.S. accounted for 11 percent. Seven and one-half million tons of polyester fibers were produced, worldwide, in 1987, and the output by various countries, in percent, was as follows: China (9%), Japan (7%), other Asian nations (30%), U.S. (12%), other American nations (7%), Western Europe (13%), Eastern Europe (10%), and Middle East, Africa, and Oceania (2%). Nylon, which was the number 2 synthetic fiber, was produced at an annual rate of 4 million tons, and the worldwide production of acrylic fibers in 1987 was 3.5 million tons.

The production of noncellulosic and cellulosic manmade fibers, in thousands of tons, in European nations in 1987 was as follows:

Country	**Noncellulosic**	**Cellulosic**
West Germany	821	16
Italy	655	3
Spain	268	0.5
Turkey	260	1
U.K.	227	12
Benelux countries	176	5
France	173	2
Switzerland	95	0
Ireland	78	0

Country	Noncellulosic	Cellulosic
Portugal	63	0.1
Austria	27	120
Finland	1	54

The distribution between filament and staple synthetic fibers, worldwide, is 57 percent and 43 percent, respectively [3]. The production of synthetic fibers in the U.S. and the entire world, in thousands of tons, in 1988 was as follows:

Fiber	U.S.	World
Polyester	1,682	7,718
Nylon	1,261	3,718
Polyolefin, vinyl	784	
Acrylic	275	2,523
Total	**3,413**	**13,959**

The production of rayon and acetate rayon, in thousands of tons, in the U.S. in 1988 was 1.8 and 0.95, respectively. Autex closed its rayon plant at Front Royal, VA, but reopened after receiving a long-term contract from NASA [4,5]. Forty tons of aerospace grade rayon is required for lining a pair of shuttle booster nozzles in order to hold the temperature of the nozzle below 100°C. The lining is carbonized and coated with a phenolic resin coating before firing. Other U.S. producers of rayon are BASF, Courtaulds, Enka, and North American Rayon.

The production of fiberglass, in thousands in tons, in the U.S. in 1988 was as follows:

Owens Corning	182
PPG	155
Certain Teed	57
Manville, etc.	11
Total	**405**

Since Amoco was the only domestic source of carbon fibers and most of these fibers were imported from Japan, the U.S. Government requested that at least 50 percent of the domestic requirement of carbon fibers be produced in the U.S. [6]. The annual worldwide consumption of carbon fiber is 2.7 thousand tons and 2 thousand of this fiber is consumed in the U.S. The U.S. aerospace industry consumed one thousand tons of carbon fiber in 1988.

As shown in the following table, there will be an annual capacity for the production of 3.8 thousand tons of carbon fiber by U.S. firms in 1989.

Hercules	1.2	Zoltech	0.11
Amoco	0.9	Polycarbon	0.45
BASF (Celion)	0.45	Textron	0.45
Courtaulds	0.41	Witco	0.23
Ashland	0.14		

Most of this fiber will be wet spun from acrylic fibers but Zoltech will use rayon as the precursor. Ashland will use pitch and 25 percent of Amoco's production will be from pitch at Greenville, SC. Celion will produce its carbon fiber by melt spinning [6] at Rock Hill, SC, and Williamsburg, VA [7]. Courtaulds will produce its carbon fiber in a joint venture with Dexter. Amoco will produce its fiber at Sacramento, CA, and Ridgefield, CT, and Hercules will produce carbon fiber at Magna, UT, and at Decatur, AL, in a joint venture with Sumitano [8].

Enimont produces polyester, nylon-66, and rayon, and is attempting to solve Italy's troubled fiber industry. The polyester facilities at Fayetteville, NC, and Darlington, SC, which Hoesch was forced to sell after its acquisition of Celanese, were sold to Fiber Industries which also acquired the tradename Fortrel [9]. Fiber Industries' sales in 1988 exceeded $350 million. Rhone Poulenc is investing $46 million to upgrade its polyester and nylon production facilities in several European nations.

Mann has acquired the acrylic fiber business at Williamsburg, VA, from BASF. Allied Signal is expanding its fiber production facilities at Moncure, NC, to 63 thousand tons. Hoechst is also expanding its fiber facilities at Bodingen, West Germany, and Sao Paulo, Brazil [10]. Monsanto has sold its Astroturf business to Balsam of West Germany for $15 million. Eastman has phased out its 110 thousand ton polyester business at Columbia, SC, but has retained its 73 thousand ton facility at Kingsport, TN. Alavere is spending $240 million to expand its polyester staple fiber capacity by 50 thousand tons at Shelby, NC, but has phased out its polyester yarn facility at Spartenburg, SC.

The mechanical properties of wood fibers have been improved by grafting with acrylic or ethylene monomers [11]. In a joint effort, Arco and DuPont are attempting to improve the water absorbency of textiles for the disposable diaper market which consumed 90 thousand tons of material in 1984 [12]. Formaldehyde-free wrinkle-free cotton fabrics are being made by crosslinking cellulose with 1,2,3,4-butane tetracarboxylic acid [13].

Investigations at the University of Tokyo have improved the properties of PET by swelling the imperfect crystals in acetone and drawing to produce more perfect crystals [14]. Monsanto has improved its synthetic fiber operations at Pensacola, FL, by balancing its energy requirements, etc. [15]. Monsanto is also producing nylon filament carpet with anion sulfonate surface groups [16].

Because of a decrease in demand for knitted textiles, the sale of acrylic fibers has decreased. Toray is promoting the use of acrylic staple fibers as replacement for asbestos fibers in Portland cement composite sheets. In addition to improving the quality of its acrylic fibers in the U.S., DuPont is aiding comparable improvements at Celbras Qinica at Aratu, SA [17]. Facilities for an annual production of 120 thousand tons of polypropylene (PP) fiber are being built at Dunkirk, using Union Carbide technology [18]. Amoco is expanding its annual PP yarn production by 10 thousand tons at Bainbridge, GA. Exxon, which is the leading supplier of spun bonded fabrics is using Reicofil technology in which PP filaments are kept separate until they are bonded in the mat [19].

Exxon has donated its melt flow non-woven fiber unit to the University of Tennessee at Knoxville. DuPont, which is the leader in nylon carpet face yarns, has purchased Hercules $50 million carpet fiber business and is now promoting PP face yarn under the tradename of "Pro Select" [20].

Exxon has licensed the heat bonded, non-woven "Teram" PP process from ICI.

DuPont is investing $150 million in its Tyvex spun bonded HDPE facility at Luxembourg. Geotextiles, which are superior to clay for stabilizing soil, are being used for road building and for increasing the capacity of landfills [21]. HDPE grid materials (Geogrid) are being used in several projects to control flooding [22]. The use of various polymers for geotextiles has been reviewed [23].

The use of vinyl-coated textiles is increasing. The name Ebb Tide is now being used instead of Naugahide for coated fabrics. The use of acrylic and nylon fiber-flocked textiles is also increasing [24]. The relationship of strength and length of carbon fibers has been investigated [25]. Melt-spun carbon fibers have non-circular cross sections. A new book on carbon fibers is available [26]. Fiber Materials (FM) is producing metal-filled and metal-coated quartz fibers at Columbus, OH [27].

Polybenzamide fibers (H-Cel), which are being added to aluminum to decrease cracking tendencies, are being considered as substitutes for asbestos for brake linings. Spider silk, consisting of crystalline and amorphous regions, which is being produced by bioengineering, is being considered for dragline silk [28]. Rovings are being admixed with resins and pultruded to provide high performance composites [29].

The court in U.K. upheld Akzo's patented infringement case on Twaron vs. DuPont's Kevlar. The two firms have agreed to cross license but Akzo is prohibited from selling its *p*-phenylene terephthalamide in the U.S. [30]. Tejin continues to produce this filament under a process that does not conflict with DuPont's patent claims. The present market for these aramid fibers is 10 thousand tons in the U.S., 5 thousand tons in Europe, and 1 thousand tons in Japan. A new aramid, Kevlar 149, which is stiffer and more moisture resistant than conventional Kevlar, is being used as a reinforcing fiber for composites in helicopter blades. GE has developed a hollow liquid crystal polymer (LCP) filament, which provides composites analogous to wood.

Allied Signal continues to produce extended chain HDPE (Spectra) using a DSM process but has discontinued R & D efforts on LCP filaments [31]. Fiberglass continues to account for 93 percent of reinforcing fibers but carbon, aramid, polyester, boron nitride, and phosphate (Fiberfax) fibers are also being used in resin composites. Continuous filaments with low dielectric constants, are being produced from boron, boron carbide, silicon nitride, and silicon carbide. Silica, alumina, metal oxide fibers, and whiskers also being used for the production of composites [32].

The properties of ceramic fiber-polymer composites with alumina (Fiber FP), alumina/silica (Nextil, Sumica), silica/silicon carbide/carbon (Nicalon) and amorphous silica have compared with those made from E glass, carbon, aramid, and quartz [33]. The annual production, in thousands of tons, of high performance fibers in 1987 and for 1992 are shown below [34]:

Fiber	**1987**	**1992 (est)**
Carbon	2.0	4.3
Aramid	10.0	16.0
S Glass	2.5	4.4
Ceramics	0.16	0.7
Boron	0.16	0.2
Quartz	0.16	0.2
HDPE (Spectra)	0.4	1.0
Total	**14.5**	**26.0**

REFERENCES

1. R. B. Seymour, Polymer News, 14, 89 (1989).
2. J. F. Dunphy, Chem Week, 143(5), 8, (1988).
3. D. O'Sullivan, Chem Eng News, 66(10), 16 (1988).
4. P. Kenezis, Chem Week, 143(20), 17 (1988).
5. S. Stinson, Chem Eng News, 66(38), 16 (1988).
6. G. Parkinson, Modern Plastics, 65(6), 24 (1988).
7. G. Parkinson, Chem Week, 142(16), 6 (1988).
8. M. Reisch, Chem Eng News, 66(8), 17 (1988).
9. M. Reisch, Chem Eng News, 66(10), 13 (1988).
10. E. Goldbaum, Chem Week, 144(6), 13 (1988).
11. B. V. Kobta, Polymer News, 13, 331 (1988).
12. L. Brockinton, Chem Week, 143(18), 13 (1988).
13. R. J. Spaulding, Chem Week, 143(10), 38 (1988).
14. R. S. Porter, A. E. Zachariades, Polymer News, 14, 111 (1989).
15. R. Boyle, Chem Week, 143(2), 31 (1988).
16. S. Stinson, Chem Eng News, 66(39), 29 (1988).
17. J. F. Dunphy, Chem Week, 143(5), 8 (1988).
18. P. L. Layman, Chem Eng News, 66(5), 15 (1988).
19. P. Kenzis, Chem Week, 144(15), 18 (1988).
20. P. Kenzis, G. D. L. Morris, Chem Week, 144(4), 24 (1989).
21. R. D. Leversuch, Modern Plastics, 16(3), 77 (1989).
22. P. Kenzis, Chem Week, 143(23), 20 (1988).
23. C. E. Carraher, Polymer News, 13, 112 (1988).
24. W. Ward, Plast Eng, 43(7), 31 (1987).
25. K. K. Phani, Compos Sci Technol, 30, 59 (1987).
26. E. Fitzer, "Carbon Fiber and Their Composites," Springer-Verlag Berlin, 1986.
27. L. Nazarenko, Modern Plastics, 65(5), 214 (1988).
28. D. O'Sullivan, Chem Eng News, 66(30), 24 (1988).
29. H. I. Glasson, J. N. Gaushel, Modern Plastics, 65(8), 114 (1988).
30. P. L. Layman, M. Reisch, Chem Eng News, 66(34), 10 (1988).
31. A. S. Wood, Modern Plastics, 65(1), 30 (1988).
32. F. Cooke, G. C. Tesoro, Polymer News, 13, 302 (1988).
33. J. D. Bolt, R. H. French, Adv Mat & Proc, 139(1), 32 (1988).
34. G. S. Kirshenbaum, Polymer News, 13, 150 (1988).

NEW HORIZONS IN THE FIBER INDUSTRY

RAYMOND B. SEYMOUR
Department of Polymer Science, The University of Southern Mississippi, Southern Station Box 10076, Hattiesburg, MS 39406-0076

Introduction

The growth and harvesting of natural fibers and the production of manmade fibers is one of man's most important activities, worldwide. Nature has been particularly generous in supplying cellulose fibers, such as cotton (Gossypium hirsutium), linen (Linum usitatissimum), hemp (Cannabis sativa), jute (Boehmeria nivea), sisal (Agave sisallina), kapok (Ceiba pentandra), and coir (Cocos Nocitera). Proteinaceous fibers such as wool, mohair, cashmere, alpaca, vicuna and silk as well as inorganic fibers, such as asbestos, which is subject to much scrutiny because of toxicity.

The first man made fibers were regenerated cellulose (rayon), cellulose di and triacetate and regenerated protein (azlon). The first synthetic fibers were nylon, polyesters and PVC but these have been supplemented by acrylics, polyolefins, polyurethanes, and aromatic polyamide (aramids). Inorganic fibers, such as glass, metal fibers and ceramic fibers are also available [1].

Fiber Production

The fiber industry is a mature industry whose increase in production should parallel increases in population. However, as noted by Elias [2], the growth of the fiber industry, during the past forty years, has been 50 percent greater than the population growth [2].

The annual world wide production of natural fibers and that of manmade fibers is approximately 16 million tons. Rayon and acetate rayon account for 3 million tons. The annual consumption of synthetic fibers in millions of tons in the U.S., ECC, Japan and other countries is 3.2, 2.6, 1.4, and 6.3, respectively [3].

Polyester is the major synthetic fiber, but as shown in the following table, sizable quantities of other synthetic fibers are also consumed in the U.S., Western Europe, and Japan.

ANNUAL CONSUMPTION OF SYNTHETIC FIBERS (Million Tons)			
	U.S.	ECC	Japan[4]
Polyester	1.5	0.85	40.62
Polyamides (nylons)	1.1	0.64	0.26
Acrylics	0.29	0.8	0.4
Polypropylene etc.	0.8	-	-

American mills used 1.3 million tons of cotton in 1985 and this accounted for 30% of all fibers processed. The U.S. Government compensates cotton growers so that they are able to sell cotton competitively in world markets [5]. A new multifiber arrangement (MFA-4) effective until 1999 has been made with 54 countries which export cotton, abaca, coir, henequen, jute, ramie and sisal.

The rate of imports of synthetic fibers by U.S.A. has increased by 14 percent and the rate of exports has decreased by 13 percent [6].

Cellulosics

A process for the preparation of amorphous dietary cellulose from nonwoody plants has been developed by the U.S. Department of Agriculture [7]. 75 million tons of pulp and paper board were produced in the U.S. in 1986. This volume included 6 million tons of coated paper.

Because of the waste disposal problem, attempts have been made to replace or modify the kraft or sulfite pulping process. In one of these alternatives, developed accidentally by the son of the inventor, Dr. Young, ethyl acetate is added to a dilute solution of acetic acid and this "ester process" removes lignin from the wood and leaves a 60-70% yield of high quality pulp [8]. In another alternative process, oxygen or ozone is added to a caustic soda-anthraquinone pulping process at 160°C and the wood is defiberized in this pulping process in the presence of magnesium salts which inhibit the degradation of the cellulose.

There are more than 10 lignin degrading enzymes in the white rot fungus but these are not very active and not available in large quantities. The Institut National-de la Recherche Agronemique (INRA) uses an improved ligninase (Phanerochate chrysisperium) with hydrogen peroxide to obtain purified cellulose. The USDA has funded a program to investigate ligninase degradation of wood at its Forest Products Laboratories in Madison, WI and is seeking additional financial support from paper companies [9].

Vigo of USDA has produced a fiber which absorbs thermal energy when heated and releases this energy when cooled by crosslinking cellulose fibers with polyethylene glycol and urea resins in the presence of acids. This reversible thermal effect is based on a phase change in crystalline structure [10,11]. Over 60 thousand tons of soluble cellulosics are used annually by the U.S. pulp and paper industry [12].

Asbestos

Over 225 thousand tons of chrysotile asbestos was consumed in the U.S. in 1984 [13]. Because asbestos has been identified as a carcinogen, its use is restricted to applications where no economical substitute is available. Polytetrafluorethylene (PTFE) and flexible graphite have been used in place of asbestos in gaskets. The graphite is suitable for high temperature applications (2750°C) in the absence of oxygen but only at 425°C in the presence of oxygen. PTFE has poor resistance to creep but this deficiency can be overcome by the incorporation of fillers [14]. Aramids, ceramics and steel have also been considered as replacements for asbestos [15].

A sharp reduction in the toxicity of asbestos has been observed when it is heated at 120°C in the presence of phosphorus oxychloride vapors. No increase in cancer, tumors or lesions was observed in a $2.5 million study at Los Alamos National Laboratory when animals were exposed to extremely high concentrations of asbestos. According to Dr. M. F. Stark, the size and shape of the asbestos particles and not the asbestos are responsible for adverse health effects in asbestos environments.

Inorganic Fibers

Fiber glass, which is produced by the melt spinning of small glass beads is the only widely used inorganic fiber. Both blown glass fibers (staple) and continuous filament glass yarns, treated with coupling agents, are used as reinforcement for plastics. Glass yarns are also used as reinforcements in pneumatic tires. Fibers with high silica content are made by the acid leaching of glass fibers.

Inorganic fibers also being made by chemical vapor deposition (CVD) from boron, and silicon carbide. These fibers have been developed by AVCO and GTE. A matrix of silver nitride and silicon carbide whiskers is being injection molded for internation combustion engine parts. In some cases, the inorganic fibers are produced by the ignition and sintering of an extruded viscous-inorganic filament. Glass fibers are being rubber coated when used in alkaline environments [16].

Nylon-Aramids

Aliphatic-didactic nylons, which were the first synthetic fibers, monadic nylons and aromatic polyamides (aramids) are being used both as textiles, reinforcements, and molded plastics. Nylon-66 is made by the condensation of hexamethylenediamine and adipic acid and nylon-6 is produced by the self condensation of caprolactam.

The first aramid (Nomex) was polyphenyleneisophthalamide. Another aramid (Kevlar) which is poly-*p*-benzamide, is a nematic liquid crystal in sulfuric acid. Over 10,000 tons of Kevlar was produced in 1986.

The rate of spinning of nylon-66 has been increased by extrusion in an alkaline solution of cellulose. The strength and toughness of nylon-66 reinforced plastics has been improved by pultrusion of resin-impregnated continuous filaments [17]. Woven-aramid-epoxy resin laminates have less tendency to fail by delamination fracture [18].

The rapid growth of Nylon-66 fibers has been based on an increase in use of nylon carpets. Sales of intermediates for fiber production have remained unchanged in recent years [19]. A 750Km trip to the North Pole on May 1, 1986 by six member of the Steger International Polar Expedition would have been impossible without the use of nylon clothing.

Allied is promoting the use of nylon-66 (Stanyl) produced by DMS in the Netherlands. The aliphatic polyester fibers produced by Carothers in the early 1930's lacked thermal resistance and were replaced by melt-spinnable polyethylene terephthalate (PET). PET fibers are hydrophobic but blends with cotton and wool are hydrophilic and more "breathable" than cotton. Both nylon and polyester knitted or tufted fibers have been used as artificial playing surfaces [20]. Hoechst acquired Celanese in a $2.85 billion acquision but the Federal Trade Commission (FTC) has ruled that the new company (Hoechst-Celanese) must divest much of the Celanese fiber production facilities but can retain its 70 thousand ton industrial polyester filament business. The spinning of a copolyester of ethylene terephthalate and *p*-oxybenzoate from a nematic melt has been described [21]. A new book on polyvinyl alcohol fibers has been published [23,24]. Information has also been provided on the relationship of tensile strength to the molecular weight of fibers [25].

Graphite Fibers

Carbon filaments, which were used by Swan and Edison in the early 1900's, combined high strength and low specific gravity and hence are being used as reinforcements for plastics. In addition to their production by the pyrolysis of polyacrylonitrile filaments, carbon (graphite) fibers are also being made from mesotropic pitch. New data have shown that composites made by reinforcing epoxy resins with graphite fibers have an excellent wearing surface [26]. A new book on carbon fibers and composites has been published [27].

The world wide production capacity for carbon fibers is 4.5 thousand tons. This fiber is being produced in the U.S. by five American firms under the trade names shown below:

Hercules:	Magnamite	BASF:	Celion
Hysol:	Grafil	AMOCO:	Thornel
Ashland:	Carboflex		

The importance of plastics reinforced by carbon and aramid fibers was demonstrated by the round-the-world flight, without refueling, of the reinforced plastic Voyager in December 1986.

Polyolefin Fibers

Woven and nonwoven polyolefin textiles (Geotextiles) which have been used for 15 years, continue to be used in road construction, for erosion control, soil reinforcement, irrigation and drainage applications. The interlaminar shear strength of polyolefin fiber-reinforced epoxy resin composites has been improved by exposing the fiber to ionized plasma in the presence of oxygen. The increase in impact resistance and resistance to spalling of this composite was attributed to good adhesion of the resin and fiber resulting from the formation of many small pits in the fiber surface [28]. Epoxy resin composites, with low thermal expansion and good impact resistance, have been provided by reinforcing a blend of chain-extended ultra high molecular weight polyethylene and graphite fibers [29].

The cost of producing fibers has been reduced at Celanese by the use of a multistage melt spinning and drawing process in place of the gel spinning process. Other filaments, including polypropylene (Spectra) can be produced by comparable techniques [30].

Other Organic Polymeric Fibers

An impact strength improvement of 5 to 7 fold has been reported for polyphenylene sulfide fiber (Sulfat)-reinforced plastics [31]. High strength composites have also been produced by reinforcing thermoset resins with polybenzothiazole or polybenzoxazole fibers [32]. Polyimide fibers (Lenzig) were used to tie down elements on NASA's space shuttle. The present use and potential use of high strength fibers for reinforcing resins has been reviewed [33].

References

1. L. Rebenfield in "Encyclopedia of Polymer Science and Engineering," J.I. Kroschwitz, ed., Vol. 6, Wiley-Interscience, NY, 1986.
2. H.G. Elias, Polymer News, 12, 69 (1986).
3. D. O'Sullivan, Chem Eng News, 65(8), 20 (1987).
4. P.P. McCurdy, Chem Week, 136(15), 31 (1985).
5. A. Agoos, Chem Week, 137(3), 18 (1986).
6. W. Storch, Chem Eng News, 64(33), 9 (1986).
7. P.P. McCurdy, Chem Eng News, 64(33), 9 (1986).
8. J. Alper, Chem Week, 138(11), 35 (1986).
9. J. Tremkitt, Chem Week, 139(19), 12 (1986).
10. K. Portnoy, Chem Week, 139(2), 21 (1986).
11. R. Seltzer, Chem Eng News, 64(42), 15 (1986).
12. P.P. McCurdy, Chem Week, 136(15), 31 (1985).
13. G. Parkinson, Chem Eng, 93(20), 9 (1986).
14. G. Parkinson, Chem Eng, 93(20), 23 (1986).
15. J.J. Wilson, Chem Week, 140(2), 12 (1987).
16. P. Dreyfus, et al, "ACS Symp Ser," 8(322), 349 (1986).
17. J.M. Crosby, T.R. Drye, Modern Plastics, 63(11), 74 (1986).
18. C.K.H. Dharan, J. Eng Mater Technol, 108, 290 (1986).
19. B.F. Greek, Chem Eng News, 64(40), 11 (1986).
20. T.A. Orofino, Polymer News, 10, 284 (1985).
21. H. Muramatsu, W.R. Krigbaum, J. Polym Sci, Polym Phys Ed, 24, 1685 (1986).
22. I. Sakurda, "PVA Fibers," Marcel Dekker, New York, 1986.
23. G. Tesora, Polymer News, 10, 202 (1985).
24. R.B. Seymour, Polymer News, 11, 72 (1985).
25. Y. Termonia, P. Meakin, P. Smith, Macromolecules, 18, 2246 (1985).
26. C. Subramanian, et al, J. Reinf Plast Comp, 5(7), 200 (1986).
27. E. Fitzer, ed, "Carbon Fibers and Their Composites," Springer Verlag, New York, 1985.
28. N.H. Ladizosky, I.M. Ward, Compos Sci Technol, 26(6), 129 (1986).
29. D.F. Adams, R.S. Zimmerman, Sampe J, 22(11/12), 10 (1986).
30. Allied Corp., U.S. Pat 4,619,988 (1986).
31. J. Chowdhury, Chem Eng, 91(17), 37 (1984).
32. S.R. Jones, Chem Week, 139(24), 11 (1986).
33. M.C. Reisch, Chem Eng News, 65(5), 9 (1987).

LOUIS ATWELL OLNEY: PIONEERING TEXTILE EDUCATOR AND FOUNDER OF THE AATCC

RUDOLPH D. DEANIN
University of Lowell, Lowell, Massachusetts 01854

In The Beginning

Dr. Louis A. Olney's greatest achievements were the founding and leadership of textile chemistry at Lowell Textile School/Institute and of the American Association of Textile Chemists and Colorists. In addition, his great ability, energy, and dedication made many other major contributions to the textile industry worldwide and to the welfare of his local community in particular [1-6].

The Olney family name appeared in early ninth-century England, as a Saxon place-name meaning Holly + Water. In old Anglo-Saxon and Norman records it was spelled at least 23 different ways. Thomas Olney of Hertford emigrated to Salem, Massachusetts in 1635, then joined Roger Williams as a member of the "Original Thirteen Proprietors of Providence," Rhode Island in 1638, where he served as treasurer, commissioner, judge, pastor, minister, surveyor, and many other duties in the Colony [7].

Louis Atwell Olney was born April 21, 1874, son of Albert H. and Frances E. Olney, in East Providence, RI [6]. He attended public school, high school, and Bryant and Stratton Business College in Providence, then studied chemistry at Lehigh University, where he received a B.S. in 1896. Lehigh later awarded him an M.S. in 1908 and an Sc.D. in 1926.

Growing up in a major center of the textile industry, he naturally developed an early and lasting interest in the field. As he later recalled, "During my senior year at Lehigh University, we went on a sol-called 'shop inspection trip' to Philadelphia. Among other places, we visited the Eddystone Print Works. I was, to say the least, thrilled to see large rolls of plain white cloth put into one end of a machine and then to see the same cloth come out at the other end in a few seconds, covered with a multi-colored design of great beauty and brilliancy. It was my desire to know the whys and wherefores of these marvelous results, probably more than anything else, that headed me in the direction which I have followed for nearly 50 years" [5].

Finishing his studies at Lehigh, he spent a year as Instructor in Chemistry at Brown University, 1896-1897, where he developed a course in textile and dyestuff chemistry.

Textile Chemistry at Lowell

Lowell Textile School was founded in 1895 as a private school to serve the local textile industry; it later became a state school, Lowell Textile Institute, and ultimately the University of Lowell. In 1897 they hired Louis Olney as Professor of Chemistry and Dyeing and Director of the Department of Chemistry and Textile Coloring, a position he held for 47 years until he became Professor Emeritus in 1944. His original contract, for

Figure 1

Figure 2

$1250/year, included a four-week vacation, plus sick leave at the discretion of the Board of Trustees [8].

According to his own account, "When I came to Lowell in June of 1897, I was ushered into an empty room of 5000 square feet. The summer was a whirl of carpenters, plumbers, electricians, painters, and packing cases, but we were ready for business in the fall with double the number of students we had planned upon. The first year I had only the freshman class, largely straight chemistry. This gave a whole year to prepare for the sophomore class, and so I was able to keep well ahead of the students. Furthermore in my first evening class there were dyers from nearly every mill in the vicinity and the superintendents of two large print works and a bleachery. For several years I spent most of my spare time in their plants" [5].

He soon realized that the school sorely needed a library. He offered to found the library in memory of this father-in-law, Edward Daniel Holden [9], but encountered so much squabbling among the Board of Trustees that he gave up the attempt [10]. Instead he wrote, "As it is absolutely necessary for the students to have more books of reference, I shall immediately purchase $300 worth of books and place them in the school as a loan, during my connection with the school" [11].

In 1906 the great British textile chemist Sir William Henry Perkin visited the United States. At Lowell, he "showed great interest and was delighted with the quality and magnitude of the work and the students" [4,5]. As Olney later recalled, "The whole affair wound up with a grand banquet at the Algonquin Club. It was rather difficult to entertain Sir William, as he was a vegetarian and a teetotaler and we did the inappropriate thing, of presenting him with a magnificent silver punch bowl" [5].

Starting in rented quarters in downtown Lowell, the school moved to "its imposing location on the Merrimack River during 1902-1903" [5], and in 1943 he opened Louis Pasteur Hall as a new building for the Department of Chemistry and Dyeing [2]. In 1956, a new building was dedicated with the name Olney Hall; and in 1974, on his centenary, the new Louis A. Olney Science Center was dedicated and became the home of the College of Science at the University of Lowell.

"Of the long succession of students who have passed through the portals into Dr. Olney's lecture classes, there is probably at least one representative in every dyeing and finishing plant in the country, as well as a long list of those who have gone with dyestuff and chemical manufacturers, testing and service laboratories, and synthetic yarn manufacturers" [2].

"Olney devoted his lifetime to teaching at Lowell, and was affectionately known as 'the father of the institution'" [6]. "His department grew in importance and stature through the years due to his untiring efforts and the institute attained an international reputation. Although he was continually tempted by offers, he remained firm in the belief that he could serve the textile industry best by remaining in his educational capacity" [3].

AATCC

Louis A. Olney is also "truly the father of AATCC" [2,3]. As early as 1910 American textile chemists discussed the possibility of forming an American section of the

British Society of Dyers and Colourists. Olney early expressed the hope for an American association. As editor of the American Dyestuff Reporter, he launched an editorial campaign toward this goal. Following discussions at the Engineers Club in Boston August 26, 1921, and the Chemists Club in New York September 13,, 1921, he held the inaugural meeting of the American Association of Textile Chemists and Colorists in Boston November 3, 1921. Against his recommendation, they elected him first president and reelected him seven times [12]. Starting with a membership of 271, he saw it grow to 6350.

He also organized and presided over the Research Committee and its 30 subcommittees, 1921-1947. In 1944 he asked for Corporation Memberships to provide the modest sum of $35,000/year to support his work, but dreamed of much more in the future [5]. "He never for one moment released the essential pressure to keep the machine going" [13].

A. P. Howes, of Howes Publishing Company, established the Olney Medal in 1943, and awarded it first to Louis A. Olney on October 14, 1944, in recognition of all he had done for the AATCC and the textile industry [14]. Since then the medal has been awarded to a long line of leaders in the field.

As was only fitting, Olney organized the first student chapter at Lowell Textile School in 1923.

Writing

While teaching Textile Chemistry and Dyeing, and Chemical Technology of Fibers, Olney developed textbooks for these courses, starting with slim paperbacks in 1903 and going through 10 editions by 1947. Always alert to new developments, he included chapters on rayon in 1933; rayon, casein, and gelatin in 1937; and rayon, nylon, and vinyon in 1945. To prepare his students for these courses, he also taught and wrote his own Elementary Organic Chemistry and brought it through several editions from 1927 to 1941.

As part of his organizing activity in AATCC, he started and produced their Yearbook, which became the standard in its field. He co-authored Roger's Industrial Chemistry, and contributed chapters or sections to other books on Colloidal Chemistry, Chemistry In Industry, Municipal Chemistry, Textile Research, Acetate Silk and Its Dyes, Review of American Chemical Research, and Casein.

While he also write numerous technical articles, "you will find paper after paper that owes its originality to Dr. Olney, but will not carry Dr. Olney's name. He thinks it better to inspire than to claim credit. They will be published in the name of the younger man to whom it meant so much" [13].

His editorial activities were legion. For many years he was editor of American Dyestuff Reporter, Assistant Editor of Chemical Abstracts, Associate Editor of Century Dictionary, and ultimately Director and President of Howes Publishing Company.

In the words of one of his proteges, "We lacked a language to communicate knowledge to each other and an organization to use the language. His indomitable energy created an organization (AATCC) to communicate with each other through the American Dyestuff Reporter" [13].

Textile Industry

Even while teaching at Lowell, organizing AATCC, writing and editing, Louis Olney exercised many positions of leadership in the textile industry. He was President and Treasurer of Stirling Mills, President and Director of Wannalancit Textile Company, and President and Director of Lowell Lingerie Company. At the U. S. Institute for Textile Research, later the Textile Research Institute, he was a member of the Research Committee, Director of Research, Vice President, and Acting President. He consulted for a number of companies and testified in a number of court cases. In addition to leading AATCC, he was also a member of the National Association of Wool Manufacturers and the British Society of Dyers and Colourists.

Chemical Industry

As a graduate chemist, Olney consulted for the Lowell Machine Shop, 1902-03, Lowell Board of Health, 1904, and Lowell Gas Light Company, 1904-1905. He maintained long-standing membership in the American Chemical Society, Chemists Club of New York, British Society of Chemical Industry, Lehigh University Chapter of Sigma Xi, and the Engineers Club of Boston. He was a charter member of the American Institute of Chemical Engineers, and a Fellow of the American Institute of Chemists and the American Association for the Advancement of Science. He was Chairman of the Northeast Section of the American Chemical Society.

The Financial World

Because of his experience, energy, and acuity in business and finance, Olney also held positions of major responsibility in the financial world. He was Trustee, Chairman of the Board of Investment, and President of the Lowell Institution for Savings. He was also President and Director of the Lowell Morris Plan Company.

Community Service

Louis A. Olney was a vigorous contributing member of his community. He was a Republican, a Director of the Lowell Chamber of Commerce, and a member of Draft Board #87 in Lowell. In the YMCA he was Trustee, Director and President.

He was an active Congregationalist, Deacon of the Eliot Union Congregational Church, and President of the Lowell Congregational Club. He was also Director of the Isle of Shoals Congregational Corporation, Vice President of the Lowell Ministry at Large, and Treasurer of the Northfield Conference of Religious Education.

In a lighter vein, he was active in the Society of Philatelic Americans, Boston Philatelic Society, and Lowell Philatelic Society; and the Lowell Historical Society.

His club memberships included the Masons (Knight Templar, 32 Degree), President of the Lehigh University Club, Temple Club, Vester Country Club, and Yorick Club of Lowell.

In one comprehensive listing, he was counted as officer and director of 25 organizations [4].

Personal Life

When Dr. R. E. Rose reviewed Louis Olney's accomplishments to present him with the first Olney Medal, he described him as a "a true scholar, one who constantly wanted to learn and to pass his knowledge on to others" [13]. "Wherever you go in the world, you will find those who, when you mention Dr. Olney, will greet his name with an obvious expression of happiness at recalling that time at Lowell when they learned to be textile chemists and colorists." At the same ceremony, A. P. Howes rejoiced that "I know from frequent contact with others whose paths have crossed yours that your influence upon them has been equally helpful and inspirational." [14]

Louis Atwell Olney married Bertha Haynes Holden on June 24, 1903, and they were the proud parents of Margaret Lucia (Mrs. Edward Alan Larter), Edna Elizabeth (Mrs. Dexter Nichols Shaw), and Richard Holden Olney. They enjoyed 46 years of happy family life together. During this time, Dr. Olney took up the hobby of photography, primarily to record the joys his family brought him.

Their home was at 118 Riverside Street in Lowell. they owned a summer home on Lake Penacook, near Concord, NH, from 1904 to 1949. In the early days there were no cars and no roads. Dr. Olney would commute to school during the summer, taking the train between Lowell and Concord, and Mrs. Olney would chauffeur him in their motor boat between West Concord and Lake Penacook.

The reason he commuted to work was that their summer home was strictly for pleasure and relaxation. Once when A. P. Howes of Howes Publishing Company visited to invite him to become editor of American Dyestuff Reporter, Mrs. Olney sternly warned Howes not to talk business; he had to wait until Dr. Olney was taking him back to the train before he could ask him to volunteer [2].

The personality traits his associates admired most were his 'carelessly correct attire, distinguished appearance, calm unruffled manner, persuasive nature, and keen intelligence. He was always ready to help others, and in return they expressed their good will, affection, and indebtedness toward him" [2,6,14].

One biographer noted that his time in business college was well-spent learning a flourishing signature which was "his mark" as a business executive, particularly in the world of finance [2].

His philatelic career gave rise to a number of apocryphal anecdotes. Her used to tell his class that, by giving up smoking, he saved enough money to become a successful stamp collector [15]; actually he never smoked very much, but he took the opportunity to give his students some fatherly advice. When his friends mailed him packages, which required multiple postage stamps, he always asked them to include the four stamps from the corners of the sheet [2].

Dr. and Mrs. Olney died tragically in an automobile accident in February 1949. We can only guess how many more fine contributions he would have made if he had more time.

References

1. Who Was Who in America, II, 1943-1950, A. N. Marquis Co., Chicago, 1950, Pg. 405.
2. N. A. Johnson, Chem. Eng. News, 21, 1620 (10/10/43).
3. Cenco News Chats, 73 (Fall, 1951).
4. Lowell Technological Institute, Special Convocation in Observance of the Perkin Centennial and Dedication of Olney Hall (9/27/56).
5. L. A. Olney, Am. Dye. Rep. 468-473 (11/6/44).
6. P. Perkins, Unpublished Speech to the Masons, Lowell, MA.
7. James H. Olney, "A Geneology of the Descendents of Thomal Olney," E. L. Freeman & Sons, Providence, RI, 1889.
8. Lowell Textile School contract 6/29/98.
9. L. A. Olney, letter to the Trustees, 7/6/06.
10. L. A. Olney, letter to President A. G. Cumnock, 10/15/06.
11. L. A. Olney, letter to President A. G. Cumnock, 10/17/06.
12. Text. Chem. Col., 21 (1), 20 (1/89).
13. R. E. Rose, Am. Dye. Rep., 467 (11/6/44).
14. A. P. Howes, Am. Dye. Rep. 468 (11/6/44).
15. B. Baum, Private Communication.

FULL CIRCLE IN CELLULOSE

BRUCE A. TOWNSEND
Courtaulds Research, P. O. Box 111, Lockhurst Lane, Coventry, England CV6 5RS

This is the story of a British company, Courtaulds, and the natural polymer, cellulose. They came together at the beginning of this century and became completely intertwined. In the second half of the century there was a period of partial estrangement but now they are very much back together again. This is the story of a hundred years which started with cellulose as the dominant theme and will end in a similar way.

Courtaulds to 1904

The name Courtaulds comes from a Huguenot family who lived near La Rochelle halfway down the west coast of France - a beautiful area which is now a very popular holiday resort for French families. (Illustration 1) They were Protestants and left the area because of religious intolerance. Augustin Courtauld moved to London, in England, at the end of the seventeenth century, and became a wine merchant. His sons were apprenticed to the trade of goldsmith and silversmith. The family was successful in this field during the eighteenth century and some of their work is now on show at Somerset House in London. (Illustration 2)

In the second half of the eighteenth century a Samuel Courtauld married a Louisa Perina Ogier who was the daughter of a silk weaver. At that time the silk business was booming and in 1775 their seventh son, George, was apprenticed to Peter Merzeau, silk throwster of Spitalfields, London. This was the starting point for Courtaulds in fibres: polypeptides, with cellulose to come much later.

At this time the three centres of the silk industry were the cities of London, Coventry and Norwich, with a total throughput of one million lb per year. (Illustration 3) A recession at the end of the century led to pressure on costs and silk businesses began to move into the country where labour was available more cheaply. George Courtauld ended up in a joint venture in Braintree, Essex, (Illustration 4) 40 miles north east of London, but this was after a very varied life including periods farming in New York State and in Kentucky. Unfortunately the partnership was dissolved in 1818 and George Courtauld was out of the silk business. However his eldest son Samuel (who had been born in North America) set up in the silk business independently, in Bocking, (Illustration 5) next to Braintree, in 1816 and was very successful. By 1828 the partnership of Courtauld, Taylor and Courtauld came into being and this lasted until 1849 when the partnership was dissolved and the company became Samuel Courtauld and Co. The company had a number of silk products of which crepe was one and this steadily increased at the expense of others. One use of crepe was as a black mourning fabric and Samuel Courtauld steadily increased their share of this market. This was certainly a good early example of niche marketing and by the time of Samuel Courtauld's death in 1881 mourning crepe totally dominated the business.

The use of mourning crepe increased during the nineteenth (Illustration 6) century and after a death in a family the general rules were as follows:

"A widow who wished also to be regarded as a lady had, in the 1870s, to wear black for a year and a day in the shape of a dress of paramatta entirely covered with crepe to within an inch or two of the waist; wear a mantle also of paramatta and crepe, a bonnet entirely of crepe with a widow's cap inside, and a crepe veil. After a year and a day the widow 'slighted her mourning' and propriety demanded a black silk dress heavily trimmed with crepe and a final three months in black without the crepe trimmings. Even after these two years of full mourning it was said to be 'in much better taste to wear half-mourning for at least six months more'. A daughter's mourning for a parent or a parent's for an offspring lasted twelve or fifteen months, for the first six of which crepe was to be worn; brothers, sisters, or grandparents rated six months (three with crepe), uncles and aunts three months, and first cousins six weeks. A wife was supposed to wear mourning for her husband's relatives as she would for her own. Household servants went into mourning along with their employers."

Although Queen Victoria was not the only cause of this increase, her protracted retirement into mourning, after the death of her husband Albert in 1861, must have had positive effects on sales.

The combination of Samuel Courtauld and mourning crepe led to a successful venture which had fixed assets valued at 190,000 and net profits of 80,000/year in 1881, the year of Samuel Courtauld's death. (Illustration 7)

In the UK 1873-1896 were years of economic depression. Samuel Courtauld started to feel the effect by 1886 and it was 1898 before recovery was on the way. (Illustration 8) The company had been very successful for many years and was not skilled in dealing with downturns. Volume kept up but prices dropped and they steadily moved into exports, into colour, and improved production methods. In 1893 Henry Tetley was brought in to oversee production and Thomas Latham sales. They saw silk prices rising and they saw new technology coming. In 1904 Tetley persuaded the company to buy a licence in a new artificial silk process.

This is what we now know as the viscose process and was based on wood pulp as the raw material. The Samuel Courtauld weaving business was almost completely in Essex but they decided to set up their artificial silk business in Coventry, in Warwickshire, in the centre of England. (Illustration 1) It is uncertain why they made this move but there certainly was a significant textile industry in Coventry, there was labour available familiar with textiles, and another artificial silk operation (Chardonnet) had started a few years earlier about 6 miles east of Coventry.

Early Cellulose Fibres

Looking at the other side of the story the major traditional fibre in Europe had been wool and this was the source of immense wealth for many centuries. The opening up of China and the establishment of trade routes introduced silk, and the use of cotton expanded as semi-tropical areas were developed. Silk was the luxury fibre but was very expensive. As early as 1664 the possibility of an artificial replacement was first suggested by Robert Hooke, who hoped that "some ingenious inquisitive person would try wire drawing with some transparent artificial glutinous composition to make silk as the worm does".

In 1734 Reaumur tried to spin silk from the gum of silk worms, in 1770 Dubet reported success and in 1842 Schwabe produced a fibre from glass. The target was an artificial replacement for silk but the first breakthrough was in a different direction. John Mercier, in 1843, treated cotton with caustic soda and obtained a silk look. The caustic soda made the cotton swell, it became translucent, it gained strength and the dyeability improved. The process became mercerisation and has prospered ever since.

By the middle of the nineteenth century various workers were involved in the problems of processing cellulose. The initial incentive was the search for a filament for an electric lamp and in 1883 Joseph Swan (Edison + Swan - Ediswan) took out a patent for dissolving nitrocellulose in acetic acid and extruding under pressure. Count HB de Chardonnet evolved a very similar process at about the same time and should be regarded as the father of the man-made fibre industry because his aim was artificial silk. The nitrocellulose produced was also the source of the first successful plastic material. Billiard balls were manufactured from a mixture of nitrocellulose and camphor, replacing ivory. These had the right characteristics but, unfortunately, there were a number of minor explosions on various billiard tables!

In 1857 ME Schweizer discovered that cellulose could be dissolved in a solution of copper oxide in ammonia. This led to a German electric light filament business and by 1899 M Fremery and J Urban launched VGF (Vereinigte Glanzstoff Fabriken) for the successful production of cuprammonium silk.

Eventually Courtaulds and VGF came together in a joint viscose operation. In the 1950s as a young chemical engineer with Courtaulds (Illustration 9) I visited the Glanzstoff Courtaulds factory in Cologne to be a part of a team looking into the possibility of manufacturing carbon disulphide from methane. The team travelled from Cologne to Hamburg by overnight train and I found myself sitting with a Dr Fremery. He was old and I was young and I found conversation difficult. However at one point I asked him if he had been in the rayon industry for very long and he pointed out that his father was the pioneer. I gave up small talk after that.

Later in the century Charles Cross and Ernest Bevan were working together on cellulose fibres at a laboratory in Kew Gardens in London. They first published a paper together, in the Chemical News, in 1881 entitled 'Cellulose and Coal'. Cross was also in contact with a Swede, CD Ekman, who had developed a calcium sulphite process for producing wood pulp and was interested in possible production in the UK. This led to Cross' interest in dissolving cellulose. In 1891 it was discovered that any mercerised cellulose, treated with carbon disulphide, became soluble in dilute caustic soda, producing a treacle-like solution whose viscosity depended upon the completeness of the reaction and the quantity of soda left in, or added. The solution was termed Viscose and the key patent was granted in 1892 to Cross, Bevan and Beadle. (Illustration 10) This patent was for a process and did not specifically mention fibre.

In 1893 Cross, Bevan and Beadle formed the Viscose Syndicate and in 1896 the British Viscoid Co Ltd with a capital of 12,000. They launched a factory in Erith in Kent to exploit the process. The products were 'plastic' objects such as cigarette holders, umbrella handles, etc and some early film. In 1902 these activities were taken over by the Viscose Development Company, who have exploited Viscose in most areas except fibres ever since. (Illustration 11)

It was apparent that this method of processing cellulose offered a simpler route to fibre than via nitrocellulose or cuprammonium. The Viscose Spinning Syndicate was established and opened another laboratory at Kew, with the object of developing a filament process. A CH Stearn joined and in 1898 had taken out a patent covering the production of filaments by extruding viscose (using the Cross, Bevan and Beadle patent) through an orifice and into a precipitating solution. Stearn had come from Northumberland where he had worked on filaments for electric lamps, with Edison and Swan. He had probably been involved with Cragside, (Illustrations 12 & 13) a large Victorian mansion which had been lit throughout with electric light in 1880. Stearn also brought with him a CF Topham, a skilled glass-blower and practical engineer, who played a crucial part in the development of hardware. As anyone who has been involved with the Viscose process will realise, there are a great many variables and even more pit-falls for going wrong.

Courtaulds and Cellulose Come Together

There were many visitors to Kew, Germans, Americans, French: lots of interest and much frustration. Samuel Courtauld appeared early in 1904 (Feb 19th) in the form of HG Tetley, A Guthrie and H Johnson. Things moved rapidly and on Aug 22nd that year Samuel Courtauld bought out the complete enterprise. A year later and it all moved to new premises in Coventry. (Illustrations 14 & 15)

Tetley (Illustration 16) had been the driving force behind the acquisition and his main arguments were:

a) The company was on a plateau and needed new developments for growth
b) The Chardonnet artificial silk process was now making good profits
c) Viscose fibre could be produced at a much lower cost than silk.

Initially the Courtaulds Board rejected the proposal to move into this new man-made fibre. However shortly afterwards the company went public and a slightly altered Board then approved the proposal. It is very difficult to know which was cause and which was effect, but all of these important events occurred over a very short time and must reflect Tetley's energy and determination.

The other processes for producing artificial silk were, at that time, ahead of viscose and in 1904, when Courtaulds first entered the arena, 4,000-5,000 tonnes/year of artificial silk were already being produced. The viscose rayon process effectively took a decade to really master but by that time most of the other processes for making artificial silk were falling away, with cuprammonium being the longest lived survivor. A search for simple solvents for cellulose went on all the time and failure was not due to lack of hard work. In reality the successful current solvents did not really exist at that time except possibly as laboratory curiosities.

Development work continued at Coventry and the major problem was the yarn defects arising from the clumsiness of the spinning process - coagulation in an alkaline spinning bath, treatment with iron sulphate and fixing in an acid bath. In 1906 experiments with sulphuric acid/soluble salts baths looked extremely encouraging but a German patent appeared to stand in the way. Patent litigation left the Germans (Dr Max Müller) in control and Courtaulds sought a way round. By 1907 a sulphuric acid/ammonium sulphate bath was in use in Coventry and in the same year it was discovered that the addition of glucose improved quality. Continued work led to a gradual reduction in the ammonium content and the introduction of zinc sulphate in 1910. The interpretation of the Dr Max Müller patent had been more flexible in the UK than in the remainder of Europe and this was certainly advantageous to Courtaulds' progress.

The Growth of Viscose Rayon

In 1907 Henry Johnson was appointed Works Manager in Coventry and his energy and determination had a major effect on quality, production, and productivity. (Illustrations 17, 18, 19 & 20) Together with the technical advances mentioned above, profits from Viscose soon exceeded those from silk weaving. The filament being produced was satisfactory but it was certainly not a replacement for silk. It established its own market at a lower price level and was particularly useful for blends with cotton and wool. A cotton warp/viscose weft went into shirtings and a silk warp/viscose weft into blouses. Lining fabrics from 100% viscose were satisfactory and this was a long enduring market. The additional factor in Courtaulds' success was its in-depth knowledge of fabric and its ability to use the Samuel Courtauld weaving as a real partner to yarn production.

Before Samuel Courtauld had purchased the Viscose Spinning Syndicate in 1904 the American rights had been sold and by 1907 we find Courtaulds in formal contact with Silas W Pettit who had a viscose enterprise based on cotton linters in Pennsylvania. Agreement was effectively reached on a commission to Pettit for import into the USA, but proposed duties of 45 c/lb led to plans to manufacture. The American Viscose Corporation was incorporated on February 14th 1910 and production at Marcus Hook, Pennsylvania started early in 1911. (Illustration 21) Results came rapidly with sales of 1.6 million lb and gross trading profits of $1.6 million by 1913. (Illustration 22)

The following year war broke out but the period up to 1920 continued the expansion of the company. (Illustration 23) In the UK Aber Works (North Wales) was bought from VGF and converted to Viscose whilst in the USA a second factory in Roanoke Virginia started up in 1917.

From 1920 until the start of the Second World War in 1939 growth was continuous. Courtaulds opened factories in Wolverhampton, Preston and Flint in the UK, in Calais, France, jointly with VGF in Cologne, Germany, jointly with Snia in Italy, (Illustration 1) and in Cornwall, Ontario, Canada. The American Viscose Corporation opened factories at Lewistown (Pennsylvania), and Parkersburg (West Virginia) and Front Royal (Virginia). (Illustration 24) The other two major developments were staple and Tyrecord.

From the earliest days the disposal of waste filament had led to its cutting and use as a cotton diluent. (Illustration 25) The First World War stimulated the deliberate production of staple viscose in Germany, to eke out supplies of cotton and wool. There was a reaction after the war but technical development continued and production was built up in Germany and Italy in the 1930s under the pressures of self-sufficiency. The need did not exist in the UK or the USA but Courtaulds links with VGF in Germany and Snia Viscosa in Italy enabled the company to obtain know-how and successfully start up a factory in Greenfield in North Wales in 1936, (Illustration 26) followed by an AVC unit at Nitro (West Virginia) in 1937. In the UK a cotton mill (Arrow) and a worsted mill (Westcroft) were purchased as an aid to getting staple accepted and these were very helpful in a rapid build-up in production. Over 1 million lb/week in the UK by 1939 and to this level in the USA by 1940.

Tyrecord started with samples to Dunlop in 1920 but no real progress was made until 1935 when JH Givens developed the 'hot stretch' process. It is interesting to note that this was really the first major technical advance emanating from Courtaulds Research Laboratories since the initial development of viscose thirty years earlier. In between most developments had been purchased or licensed. Even with the Givens breakthrough Tyrecord made little progress until the war started and cotton was in short supply. It then enjoyed 20 years of prosperity before the synthetics arrived and superseded rayon.

Another development resulted from the use of aeroplanes during World War One. Henry Dreyfus, in Switzerland, developed a coating for the fabric wings of aeroplanes based on acetylated cotton linters dissolved in acetone. He moved to the UK to set up in business, sponsored by the Government, and turned this into a yarn filament process after the war. There was a certain amount of acrimony surrounding this development as it was suspected that, effectively, Government money, had been used to develop an artificial silk process. Whatever the truth the product and the company formed - British Celanese - prospered and a brother moved to North America to found American Celanese. Although there were no financial links between these two companies the technical links existed nearly up to the present-day. Courtaulds saw the advantages of acetate fibre and started its own venture in the UK.

The Second World War

The Second World War brought a major traumatic event: the forced sale in 1941 of the American Viscose Corporation as a sign of goodwill, in order to ensure the passage of the Lend-Lease Act through Congress. AVC was regarded as a "$100 million business" but the proceeds of the sale totalled just over $62 million. After nearly $8 million selling expenses the UK Treasury received $54.5 million. Just prior to this Courtaulds had advised its shareholders that it valued AVC at $128 million. After a full assessment Courtaulds claimed $178 million from the UK Treasury plus interest at 3% to the date of payment. The Treasury offered $67 million plus interest at 2.5% and the issue went to arbitration. In July 1942 the case was settled at $109 million plus interest at 3%.

Looking back AVC was a very large and powerful company completely owned and directed from outside the USA. It is possible that partial US ownership would have led in a different direction.

At this time the Courtaulds Boardroom was still like a gentlemen's club and the links between the Board and the employees were very much at arms length. Even as late as 1952 the then chairman (Sir John Hanbury-Williams) wrote, in a memorandum relating to company organisation:-

"For a long time there has been no fundamental change in the constitution of the Board of Courtaulds Limited. The Executive Directors have directed and managed, and on the whole they have been successful. There has been a Gentlemen's Club atmosphere in the Board Room, and I believe it is true to say that over the years this has spread to all the Departments of our business. It is in fact part of the goodwill of the Company which we must safeguard. On the other hand great care must be taken to avoid inefficiency."

This obviously affected the way the Company operated and it is interesting that although Courtaulds' employees were involved in assessing the developments of the German rayon industry, after the end of the War, Courtaulds built a new viscose factory in Carrickfergus in Northern Ireland in a completely traditional way. It later transpired that during the thirties and forties the American Viscose Corporation had been developing their technology continuously, but very little of this was utilised by Courtaulds in the UK.

Know-how acquired by the German investigation was used in the Courtaulds Acetate factory but no doubt this was seen in a different light to the viscose operation where Courtaulds always knew best. When I joined the Company in the early 50s there was no doubt that it was a viscose rayon company and that acetate was an operation on the side looked after by people not quite up to viscose.

The New Team at Courtaulds

By the end of the War this technology deficit was beginning to be recognised and in 1945 Alan Wilson FRS, a mathematical physicist of distinction, was appointed to the Board. (Illustration 27) Wilson, in turn, attracted other young scientists including Frank Kearton (Illustration 28) and several of his fellow workers from ICI, in 1946. Wilson and Kearton were the key figures for the next thirty years.

Despite Sir John Hanbury-Williams' conservative view of the Board he was very keen to develop the company overseas and particularly keen to get back into the USA. In the latter he relied on Colonel FT Davies, a vigorous and assertive man whose interest was in getting things done. The unlikely combination of Davies and the new technical team led to a new viscose rayon staple factory in Mobile, Alabama, which started up in late 1952. (Illustration 29) To Courtaulds it was a revolutionary design, with viscose making almost continuous from beginning to end. Obviously there were start-up problems but the factory got going and is still operating very successfully. Courtaulds Mobile is now the leading viscose rayon manufacturer in North America and I suppose that there is a great deal of pride amongst the Courtaulds employees in having achieved this even after a big hiccup in the middle. Other overseas developments took place in France, Germany, Italy and Australia, and another new staple factory was built in Grimsby in the UK.

During the post-war years staple and tyre yarn grew steadily while textile filament gradually declined.

Tyrecord had struggled in the thirties but the war necessitated its use and it did not look back again until superseded by the synthetics. However within ten years the market was flattening out and people started to talk about viscose rayon as an end game. By the 1960s contraction was the name of the game and since then production in the western world has gradually declined. However production in third-world countries has increased largely as a cotton replacement to avoid dollar purchases, although the other incentive is that wood can be grown on worse land than cotton and so the better land can be kept for food production.

Wood pulp was becoming expensive and in 1949 prices had tripled since pre-war days, whereas rayon selling prices had risen by only 70 to 80 percent. In 1940 the Board had firmly rejected a move towards wood pulp but by 1947 Wilson emphasised the need to investigate the possibility of using pulp from such new sources as the tropical or sub-tropical forests of eucalyptus trees. (Illustrations 30 & 31) Courtaulds'links with Snia Viscosa were to be the key to progress. In the 1930s Mussolini had industrialised Italy and one of the projects was a pulp mill in the north-east of Italy not far from Venice. This was relatively old technology but using short fibre hardwood for making rayon pulp, which was novel. Many of the staff at the pulp mill had been brought up from Sicily and the southern part of Italy and the company was a subsidiary of Snia Viscosa. Between the First and Second World Wars eucalyptus had been brought into South Africa from Australia by the Rattray family and had become established in Zululand in North Natal. Conditions were ideal with a lot of rain-fall, a sandy well-drained soil, lots of heat, and no frost. The trees were growing to 80 ft and 8 inches diameter in 8 years and lots more land was available although at that time it was mostly covered with sugar cane. A three-cornered company was established and a pulp mill for 110 tonnes/day using the calcium-based sulphite process was established about 30 miles south of Durban on the Umkomaas river. (Illustration 32) The pulp produced was quite different from long fibre softwood and there were many problems involved in getting it into the Courtaulds' system.

Before the mill was properly commissioned Snia Viscosa became short of money and Courtaulds bought them out, so that it continued as a two-thirds Courtaulds venture, one-third effectively South African Government, with the management of the company in Courtaulds' hands. I was personally involved in both doubling and then trebling the production rate to 300 tonnes/day and this was done with comparatively little expenditure because of the over-design built into the factory. The Italians had been very conservative in the thirties as it was a government venture and this conservatism had ended up in South Africa as well. The Italian staff in the factory were very helpful in trying to debottleneck the process and for a young chemical engineer like myself it was a lot of fun carrying out irregular experiments in the middle of the night in order to find out where the limits were. No really major expenditure was required until above about 400 tonnes/day and the factory continued to expand up to its current rate of 1400 tonnes/day. At that point Courtaulds had a very low cost wood pulp source because the tree growth rate gave a low-cost tree and the quadrupling of the capacity had led to a low capital value/tonne. Other operating costs were probably similar to elsewhere. One interesting feature of the mill was the fact that notices were always in English, African, Zulu and Italian and this is still the case. There is still a substantial Italian community there and a lot of people who still only speak Italian. One of the interesting things about the Italians was their aptitude to stainless steel metal-work and there are quite a lot of companies now in the Durban area involved in stainless steel vessels etc owned and run by Italians who started with SAICCOR. Not all the Italians made it. Some of those who had come up from Sicily to the north-east of Italy and had then wanted to move to South Africa were refused work permits on the basis of their colour.

SAICCOR started up in 1955, three years after Frank Kearton went on to the Board and this was one of the first major projects that he drove forward and established. By 1960 he was running the acetate and acrylic businesses and had established Prinex. This was a company established to sell complete factories to Iron Curtain Countries and it brought in substantial profits over many years. Just before Kearton went on to the Board an existing Board member had written:

> "Very able, quick thinking and probably quick tempered. Immense ability for hard work, constructive thought, and is rapidly obtaining a very extensive knowledge of the Company's business. He thinks clearly and can express himself well on paper and in speech.

I don't think he will tolerate fools and may be over-ambitious. He is probably apt to jump to conclusions on subjects on which he is ill-informed without taking the trouble to ascertain the facts. His judgment is not therefore always sound.

I think he could be made a member of the Board in 1951, for I believe his worst failings will disappear when he finds his opinions subjected to critical analysis at Board level."

British Celanese Acquired

After SAICCOR the next important event was the acquisition of British Celanese at the end of 1957. Henry Dreyfus died in 1944 and the company had languished since then with profits steadily declining.

They had a very large factory in Derby (Illustration 33) and one in Wrexham in North Wales. The one in Derby was an enormous site and it was clear that Celanese had been run for many years by someone very technical and very secretive. The amount of research carried out into chemicals was phenomenal but it was very inefficient in that most of the work had never been exploited. Much of the profits of Celanese had clearly gone into this sort of research and development and then one man at the top had made a decision about what to exploit and what to file. All the buildings on the enormous site had numbers, and people in one building had no idea what went on in other buildings. The secrecy was incredible and even after 30 years the buildings are referred to by their original numbers. A naptha fraction was purchased and thermally cracked for ethylene and propylene through to ethanol and isopropanol and then through to acetic acid and acetone. Acetic anhydride was then manufactured from both acetic acid and acetone and acetone was used as a solvent for spinning. The rest of the cracked gases were used to some extent for chemicals for sale and this formed the basis of a chemical business which still exists. In fact Courtaulds now have a self-standing chemicals business which originated from acetate chemicals, from carbon disulphide and sulphur chemicals and to a smaller extent from acrylonitrile and acrylic fibres. There is no doubt that it is a chemicals business which originated in cellulose but is now self-standing and independent.

The thermal cracker was built in 1942, early in the war, with the support of the British Government and based on research which had gone on in the thirties. (Illustration 34) The trouble was that no further developments had taken place and the whole chemical complex had been overtaken by new technology. Eventually it was all closed down and acetone and acetic acid and acetic anhydride are now bought-in chemicals from much more efficient plants. The Celanese technical developments of the thirties were enormous and very exciting but it is a pity that they were never pulled through properly. Since that time acetate fibre has continued to flourish on a cyclical basis with fashion but in general becoming a more stable market as lustre becomes more consistently popular. The cellulose acetate cigarette tip replaced paper and established itself as a universal cigarette tip. This is now a steady but relatively static business. Obviously development work goes ahead all the time to improve the filtration characteristics but without removing the reasons why smokers smoke at all.

Carbon Disulphide

Another development was carbon disulphide and for many years Courtaulds had a plant on the outskirts of Manchester making carbon disulphide from charcoal. It was a strange mixture of technologies. They had tremendous competence in handling liquid sulphur and also a sophisticated system for the recovery of sulphur from hydrogen sulphide. However the cast iron retorts where charcoal and sulphur reacted were 18th century design but operated safely. Any leaks were immediately lit and so although there was a permanent smell of sulphur gases there were never any major accidents. At one point charcoal was in short supply and the price went through the roof. Courtaulds built a plant making briquettes from pulverised coal. The coal chosen was very low ash, ground to 300 mesh and then put together purely by pressure. It was a very difficult process indeed and I personally spent a year of my life getting it to work. Eventually it did and it certainly forced down the price of charcoal and saved us a lot of money. It was hard work as far as I was concerned but very rewarding. As you can imagine the carbon disulphide factory was very traditional and there was not a great deal of enthusiasm for this curious plant making imitation charcoal particularly when it was designed by young up-starts from the south and I had immense difficulty getting support from the other staff in the factory, particularly to get maintenance work done promptly. However the team and I had been there about 3 months and it was time for the traditional pre-Christmas hot-pot supper. The factory relented and we were invited along. It was a pleasant evening with a fair amount of beer disappearing but the five of us from the briquetting plant were very much on our own. However the time came round to telling stories, singing songs and telling jokes and one junior member of my team stood up and recited Eskimo Nell from beginning to end and without fault. (Illustration 35) The reception was unbelievable and the next morning the whole of the engineering team were on the briquetting plant doing maintenance. We never looked back. There is no doubt that human relations is the main feature involved in running a successful factory.

Soon after, briquettes went again, having served their purpose, and the CS_2 from methane plant appeared. Nearby Shell had a refinery and we licensed a process from FMC and built a pipeline from Shell. We did some of the detailed design and the erection but the start-up was under the control of FMC. At that time Kearton was Chairman and was pressing hard for a very fast start-up. The savings for the new process were significant and he wanted them as soon as possible. The FMC start-up man was adamant that he was not going to put sulphur on to the plant until he had checked everything and was absolutely confident that it would start up without trouble. I guess that 3 or 4 weeks went past with this increasing conflict but eventually sulphur was turned on, the plant ran and never gave any more trouble. I learnt a great deal over that start-up. I have no doubt if we had started up 3 or 4 weeks earlier we would have spent the first 6 months getting it right.

We also manufactured our own sulphuric acid but this was very straightforward and based on sulphur although we had a sortie into the use of iron pyrites on a fluidised bed. This was exciting, saved a little money but brought a fair amount of grey hair.

Looking back at the CS_2 from methane plant it was a great boon but it had bad effects in that the marginal price of carbon disulphide effectively became the price of sulphur and therefore the economic motivation to recover sulphur chemicals in the viscose process was absolutely minimal and I'm sure that this had a demotivating effect on environmental action and the processes involved. This is all in the past now but I'm sure that one or two sulphur recovery developments were never properly exploited for this reason.

Caustic soda was never manufactured as ICI was such a major supplier. However we did own some salt deposits and drew up threatening plans several times. I remember these particularly well and this was in an era when small presents from contractors were not frowned upon and I have an excellent full set of steak knives and forks which arrived from an electrolytic cell manufacturer.

Diversification

Up until the 60s cellulose had dominated Courtaulds completely although there was a self-developed acrylic fibre (Courtelle) which was launched in 1957. Technically, the slurry polymerisation process was unique and the Neochrome process enabled manufacture in colour. This latter was a real advantage and has been so ever since. For several years Courtelle factories were built continuously and profits came quickly.
With British Celanese absorbed, and with the other UK viscose companies acquired Courtaulds looked to diversification and in 1958 Group Developments Ltd came into being. Within a short period their thoughts were directed towards paints and packaging and both of these have survived to the present day.

It is interesting that although these two were diversifications they both had links with cellulose. After the early success of SAICCOR Courtaulds took the lead, in conjunction with the British Commonwealth Development Corporation, to design, build and operate a paper pulp mill (Usutu) in Swaziland, in Southern Africa. Separately, and many years before, Courtaulds had moved into cellophane manufacture (British Cellophane) in conjunction with the French company, Rhone-Poulenc. Both Usutu and British Cellophane led Courtaulds towards packaging. In the case of paints the link was the range of solvents produced in the British Celanese chemical complex.

In packaging several small companies were acquired and some of them were subsequently disposed of. The packaging business continued as a low-level activity but has expanded again over the last decade and is now a growth point again. The small paint company Cellon was acquired (0.8 million) in 1958 and then a larger company Pinchin Johnson (16.5 million) in 1960. This was followed by another substantial company International Paint. For many years this paint organisation gradually progressed having been left to itself and by the mid- to late-70s was a reasonable profit-earner. This situation has continued and improved every since. This particular diversification probably took 10 years to come to real fruition and it is interesting that as it progressed it became more separate from the rest of the Courtaulds operation. However by the 1980s its profit effectively allowed Courtaulds to overcome the very poor business days of most of the rest of the Group.

The ICI Bid

But back to the early sixties and another major traumatic event. In December 1961 ICI made a bid to acquire Courtaulds. At the time it was a sensation: "The Biggest Take-over Bid in Britain's History". The battle raged for three months and ICI ended up with 37% of Courtaulds ordinary share capital. It all really started late in 1939 when Dupont licensed ICI to produce nylon. A new company was set up - British Nylon Spinners (BNS) - which was equally owned by ICI and Courtaulds. This new company spun the nylon from ICI polymer, purchased at an agreed price. Not an easy arrangement when nylon started to boom and profits became significant: in the fifties the dividend Courtaulds received from BNS represented 5-10% of total profits.

In due course ICI moved into polyester and Courtaulds tried to persuade them to include this in BNS, but without success. ICI decided not to move into acrylics but by the late fifties were beginning to believe they had made a mistake. The Courtaulds acquisition of British Celanese gave them a nylon pilot plant and there were protracted discussions between Courtaulds and ICI on the exploitation of this unit. At one point both ICI and Courtaulds had announced plants to manufacture caprolactam - the basic material for nylon 6.

By 1960 synthetic fibres were advancing rapidly and were a key part of the chemical industry. ICI felt felt that a merger between them and Courtaulds would make the joint company the key player in man made fibres in Europe.
Courtaulds wanted its independence, although there were members of the Courtaulds Board who felt that the merger made sense and that the important issue was to get the best price for the shareholders. (Illustration 36)

A major defence of Courtaulds position was effectively led by Frank Kearton, with the underlying theme of utilising Courtaulds tied up assets partly to reward shareholders and partly to invest in expansion. (Illustration 37) Once again Kearton showed his stamina and strength and I seem to remember that during this period he had to have an operation and conducted the defence from his hospital bed. He certainly had the support of most of the company. Many of the younger technical people had looked at the possibility of joining ICI, after University and had rejected it. ICI technology had looked bureaucratic compared with the sense of freedom which had begun to flourish inside the Courtaulds Research organisation.

Courtaulds entered the 1961/62 financial year with an outside company owning 37% of its shares. It needed to keep this shareholder happy, and to live up to the promises made to its other shareholders. Both seemed successful and lengthy discussions eventually led to disengagement. In April 1964 the 37% shareholding was exchanged for the half share of BNS. Courtaulds was independent again and also received cash and agreement which allowed it to start is own nylon business. At a later date polyester was added but in the end neither contributed to the long term wealth of the company and both have now gone. Both were based on bought in technology and Courtaulds had very little that was unique: this is not a basis for a long term successful business. Cellulose and acrylic are both "home grown" and have enjoyed long term success.

Textiles

The major decision of the 1960's was to move downstream into textiles and over a period of ten years an enormous amount of the British textile industry was acquired and a lot of management time went into rationalising these textile businesses and integrating them with the fibre operations. With cellulosics, acrylic, nylon, polyester, and a wide range of textiles an efficient vertical chain was the objective. In some cases the verticality was complete: from a tree seed in South Africa, through pulp, fibre, yarn, fabric, right to a bed sheet wrapped and on a store counter. Unfortunately fabric and garment imports were a stronger factor and the objective was never really achieved. There were many closures, rationalisation, and a realisation that the textile business should be self standing. It should base its operations on design, customer requirements, rapid reaction, and be completely free to obtain its own raw materials.

During the 70's and 80's the Courtaulds textile businesses steadily moved in this direction and became a profitable and stable operation, gradually becoming more and more separate from the remainder of the company. Last year Courtaulds Textiles was demerged as a separate company and continues to operate very successfully. Looking back the rationalisation of the textile industry, precipitated by the Courtaulds acquisitions of the sixties, must be a major reason why a successful textile business now exists in the UK.

The New Courtaulds

The 30 year period between 1960 and 1990 - the post ICI years - is the new Courtaulds. These years have seen the growth of the new businesses, coatings, packaging, and more recently, performance materials; the rise and fall of nylon and polyester; the acquisition, rationalisation, and finally demerger of textiles; but in the background the continuity of cellulose, both viscose and acetate, and acrylic. Tyrecord had a successful life and other profitable viscose developments were the crimped staples Evlan and Sarille, for carpets and fashion; Vincel, a high wet modulus; Viloft, a hollow fibre with a very soft handle; and most recently a whole family of modified cross-section fibres with varying handle and absorbancy. Acetate developments kept the fibre where fashion dictated and steadily advanced the efficiency of filter tips. Acrylic developments allowed a wider range of use to avoid an over dependence on pullovers and knitting.

Back to Cellulose

Up to 1960 cellulose had dominated Courtaulds but it then started to decline as the company changed its objectives. In the last ten years the intrinsic values of cellulosic fibres have become more apparent - based on a renewable resource and biodegradable, but manufactured by processes which require very careful control to protect the environment. The value of a simple cellulose solvent has become more apparent. Courtaulds itself started to look seriously in this direction in 1980 and quite rapidly made the choice of n-methylmorpholine N-oxide (NMMO) oxide as the solvent.

Dee Lynn Johnson of Eastman Kodak, probably looking for surface activation, was the first to patent the use of amine oxides solvents to dissolve natural compounds - wool, feathers, hair, cellulose. The first major attempt to pursue this commercially was undertaken by American Enka who published their first patents on making and extruding amine oxide based dopes in early 1979. Enka worked on the process for 10 years and had a substantial pilot plant in operation. They did not take this through to a commercial operation and presumably the shift of the company away from the staple fibre business had an influence on this decision.

The first obvious technical challenge was to obtain a very high solvent recycle efficiency and Courtaulds had the confidence that this could be achieved, from its acrylic experience. A lot of work has gone into this area but to start with confidence of achievement was very important. Other major challenges were the dissolution itself, a very high viscosity dope, solution stability, and the spinning process. These have been overcome but are obviously dominated by the need to get good fibre properties, low operational costs, and minimum environmental and safety impact.

The timescale of the development, to the point of a full size production unit will have been:

Milestone	Date	Output
Project start	1980	Few grammes
First pilot plant	1983	10's kg/week
Second pilot plant	1987	100's kg/week
Semi-production scale plant	1988	25 tonnes/week
First full-size production plant	1992	350 tonnes/week

The first three steps were in the Research Laboratories in Coventry, England; the semi-commercial plant is in Grimsby, England, and the full-scale plant is being erected in Mobile, Alabama. Twelve years sounds a long time but each step of the scale-up felt risky at the time and looking back I cannot see any significant time savings that we could have made.

The properties of the fibres being produced in the laboratory in the early eighties were very good. Subsequent work has confirmed and, in some cases, improved these properties. Tencel has a high tenacity and is one of the strongest cellulose-based fibres. In its wet state it retains 85% of the dry tenacity and the wet modulus is high, which allows very straightforward wet processing and leads to a very good wash stability. Properties of a range of fibres are shown in Illustration 38 & 39. The fibre has a round cross-section and a smooth surface, and in yarn form the filament packing is close. This leads to a high strength conversion from fibre to yarn, with strengths approaching those of an equivalent polyester yarn, as shown in Illustration 40. The stress strain characteristics of Tencel make it ideal for blending with other fibres. At low blend levels it contributes significantly to the strength of cotton yarns, in contrast to polyester which, at low blend levels, leads to weaker yarns. This is shown in Illustration 41. Colour yield with reactive dyes is particularly high, resin application achieves better crease resistance than other cellulosics and enhanced fibrillation characteristics can be achieved for non-woven applications.

To summarise, the major features of Tencel are:-

High Fibre Strength
Excellent Wash Stability
High Colour Yield
Colour Yarn Uniformity
(Illustrations 42, 43, 44 & 45)

Tencel is environmentally friendly from all points of view. The raw material, cellulose, is based on wood pulp produced, in a very energy-efficient process, from renewable plantation-grown trees. The solvent (NMMO) has a very high recycle efficiency and is harmless over the range of concentrations used in the plant. Finally the fibre is biodegradable to carbon dioxide and water - the initial basic raw materials for the trees.

Courtaulds in the Future

The company has now been in existence for nearly two centuries. The first century was dominated by silk weaving and the second by cellulose fibres, with excursions into other fibres, and both backward and forward incursions into the textile chain. Cellulose fibres declined with the advent of synthetics but now look set for a major revival, with green issues becoming dominant.

Courtaulds is now a polymer company: a polymer company where the surface is dominant. Cellulose is a natural polymer (annual production 10^{15} kg/year) and has led Courtaulds into its other polymer and surface interests. It is now generating a revival through Tencel and the third century of Courtaulds will see the thread of cellulose continuing.

1. Europe showing Courtaulds sites

2. Examples of Courtaulds Silver

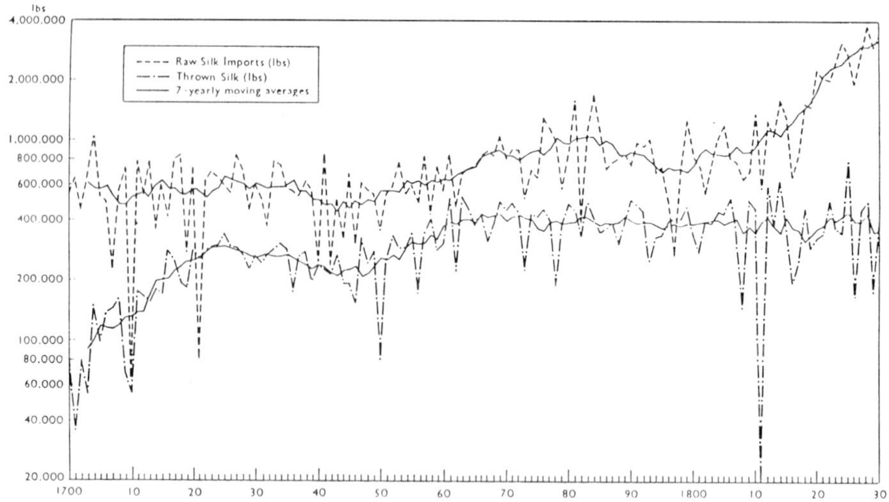

FIG. 2. *Imports: (retained) of raw and thrown silk, 1700–1830*

England & Wales to 1791; Great Britain, 1792–1830.

Sources: Mitchell & Deane; E. B. Schumpeter, *English Overseas Trade Statistics, 1697–1808* (Oxford, 1960); P.R.O.: Customs 3 and 5; *B.P.P.* 1821, vol. VII

3. Silk Imports to the UK 1700-1830

4. George Courtauld, Braintree, Essex, UK

5. Samuel Courtauld, Bocking, Essex, UK

6. Grand Maison de Noir, Paris, France

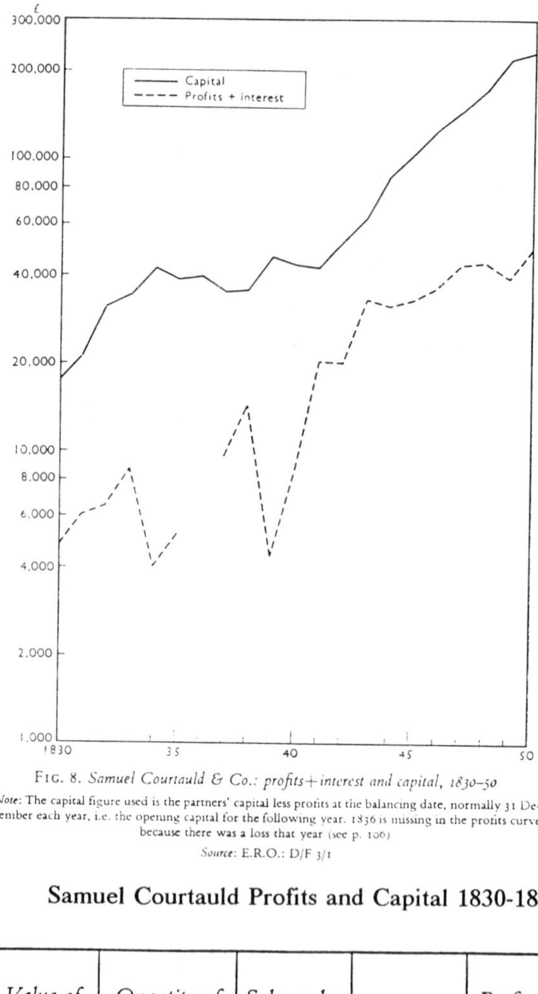

FIG. 8. *Samuel Courtauld & Co.: profits+interest and capital, 1830–50*

Note: The capital figure used is the partners' capital less profits at the balancing date, normally 31 December each year, i.e. the opening capital for the following year. 1836 is missing in the profits curve because there was a loss that year (see p. 106)

Source: E.R.O.: D/F 3/1

7. Samuel Courtauld Profits and Capital 1830-1850

Period	*Value of total sales*	*Quantity of total sales*	*Sales value per packet*	*Profits*★	*Profits per packet sold*	*Profit margins on sales*
	£'000	'000 *packets*	£	£'000	£	%
1880–2	337	46	7·3	78	1·7	23·3
1883–5	351	49	7·2	100	2·0	27·8
1886–8	266	42	6·4	50	1·2	18·7
1889–91	235	40	5·9	20	0·5	8·4
1892–4	172	30	5·7	11	0·4	7·0
1895–7	194	44	4·4	5	0·1	2·3
1898–1900	344	81	4·3	26	0·3	7·0
1901–3	447	93	4·8	39	0·4	8·3

★ Partners' profits (excluding interest) to 1890; net profits 1891–1903.

8. Samuel Courtauld Sales, Profits, Margins 1880-1903

9. Young Chemical Engineer 1954

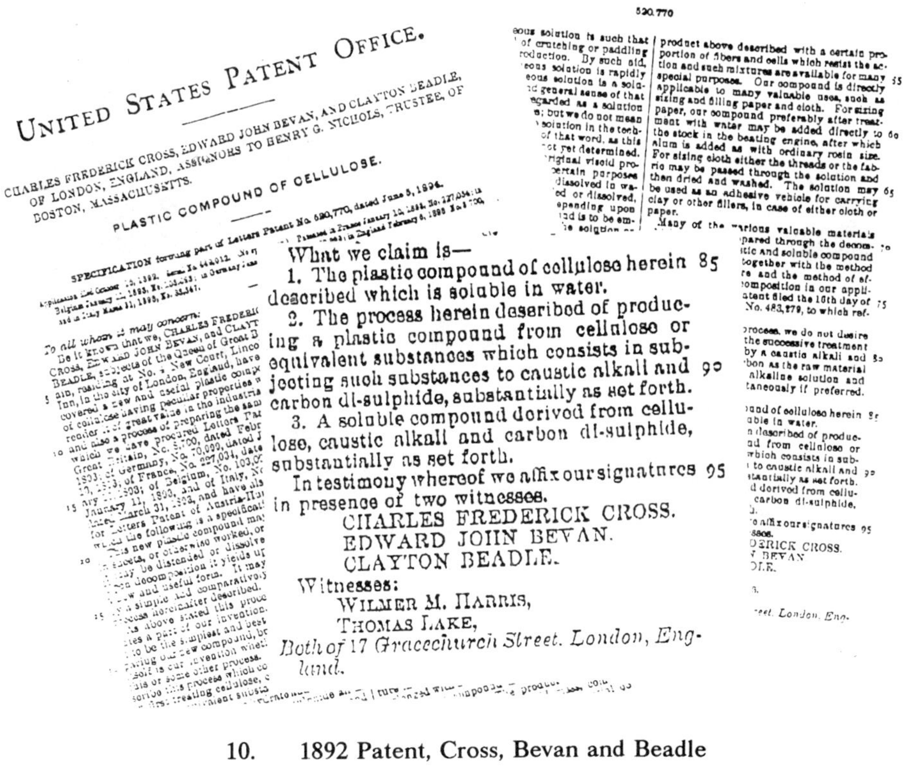

UNITED STATES PATENT OFFICE.

CHARLES FREDERICK CROSS, EDWARD JOHN BEVAN, AND CLAYTON BEADLE, OF LONDON, ENGLAND, ASSIGNORS TO HENRY G. NICHOLS, TRUSTEE, OF BOSTON, MASSACHUSETTS.

PLASTIC COMPOUND OF CELLULOSE.

SPECIFICATION forming part of Letters Patent No. 520,770, dated June 5, 1894.

520,770

product above described with a certain proportion of fibers and cells which resist the action and such mixtures are available for many special purposes. Our compound is directly applicable to many valuable uses, such as sizing and filling paper and cloth. For sizing paper, our compound preferably after treatment with water may be added directly to the stock in the beating engine, after which alum is added as with ordinary rosin size. For sizing cloth either the threads or the fabric may be passed through the solution and then dried and washed. The solution may be used as an adhesive vehicle for carrying clay or other fillers, in case of either cloth or paper.

What we claim is—

1. The plastic compound of cellulose herein described which is soluble in water.

2. The process herein described of producing a plastic compound from cellulose or equivalent substances which consists in subjecting such substances to caustic alkali and carbon di-sulphide, substantially as set forth.

3. A soluble compound derived from cellulose, caustic alkali and carbon di-sulphide, substantially as set forth.

In testimony whereof we affix our signatures in presence of two witnesses.

CHARLES FREDERICK CROSS.
EDWARD JOHN BEVAN.
CLAYTON BEADLE.

Witnesses:
WILMER M. HARRIS,
THOMAS LAKE,
Both of 17 Gracechurch Street, London, England.

10. 1892 Patent, Cross, Bevan and Beadle

11. Viscose Development Company 1960

12. Cragside, Northumberland, UK

13. Original Water Wheel, Cragside

14. 1901 Spinning Machine, Kew

15. 1904 Spinning Machine, Coventry

16. HG Tetley

17. Coventry Works 1908

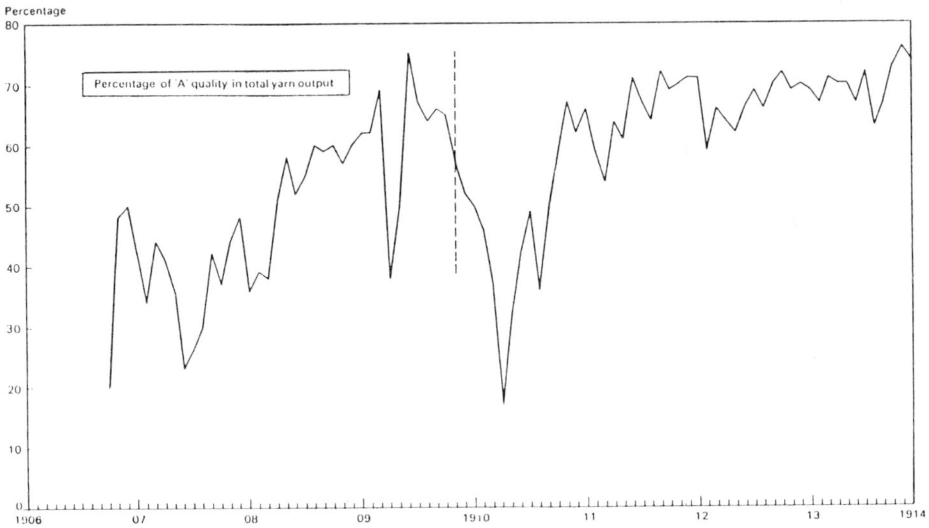

FIG. 4. *Coventry: quality of yarn, 1906–13*
(The upward revision of standards of classification in Nov. 1909 is shown by the vertical broken line.)

18. Yarn Quality 1906-1913

	1907	1908	1909	1910	1911	1912
lb. of yarn per spindle, per week	6·9	9·3	10·5	10·7	11·3	12·3
lb. of yarn per worker, per week	9·1	16·0	18·2	18·5	20·4	22·0
Output ('000 lb.) p.a.	157	554	1,216	1,852	2,176	2,547

19. Productivity 1907-1912

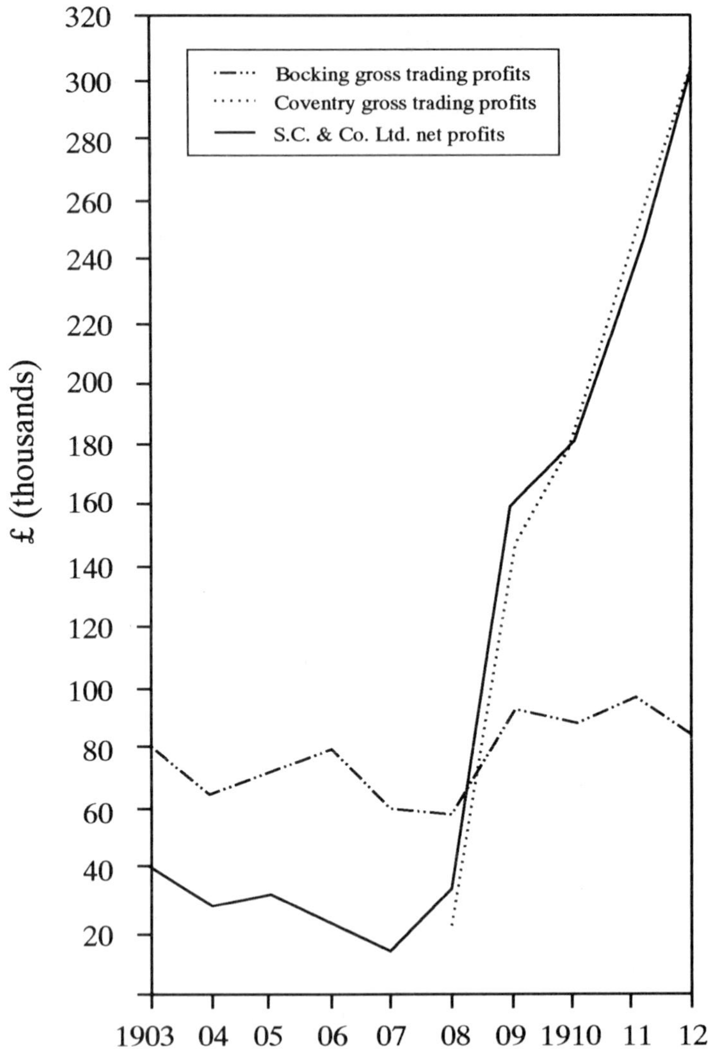

20. Samuel Courtaulds Profits 1903-1912

21. Marcus Hook, Pennsylvania 1911

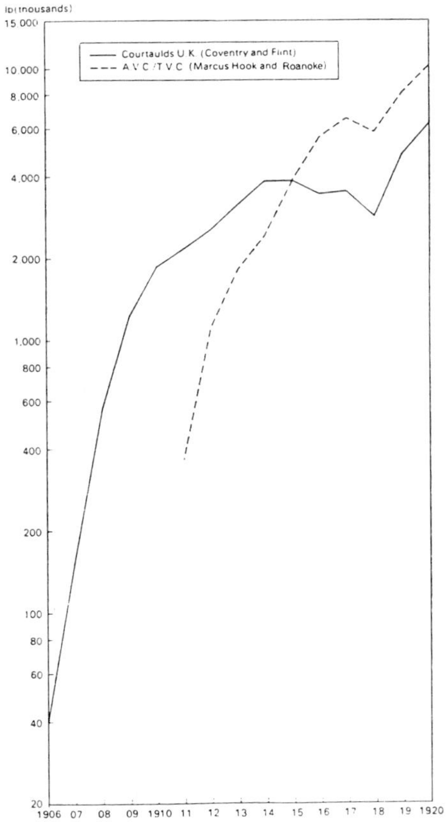

FIG. 8. *Courtaulds and A.V.C./T.V.C.: yarn output, and 1906–20*

22. Courtaulds Output 1906-1920

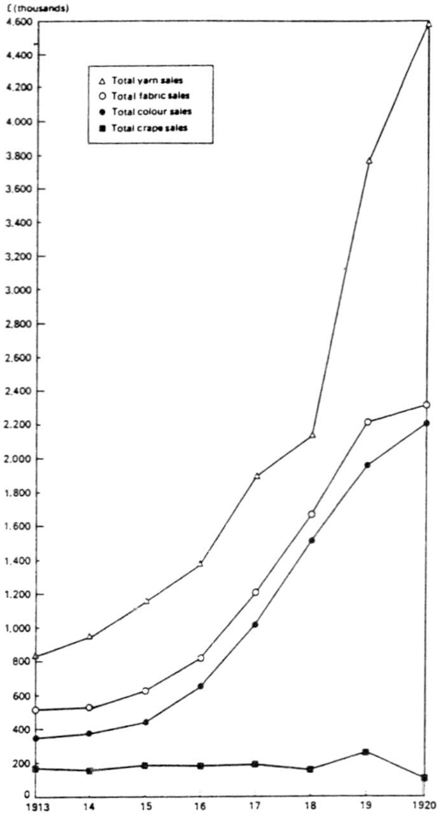

FIG. 7. *Yarn and fabric sales, 1913–20*

23. Courtaulds Yarn and Fabric Sales 1913-1920

24. Courtaulds Sites in USA

25. Doffers at Spinning Machinery 1930

26. Greenfield, North Wales

27. Sir Alan Wilson 1961

28. Sir Frank Kearton (now Lord Kearton)

29. Mobile Alabama 1955

30. Eucalyptus Plantation South Africa

31. Eucalyptus seedling

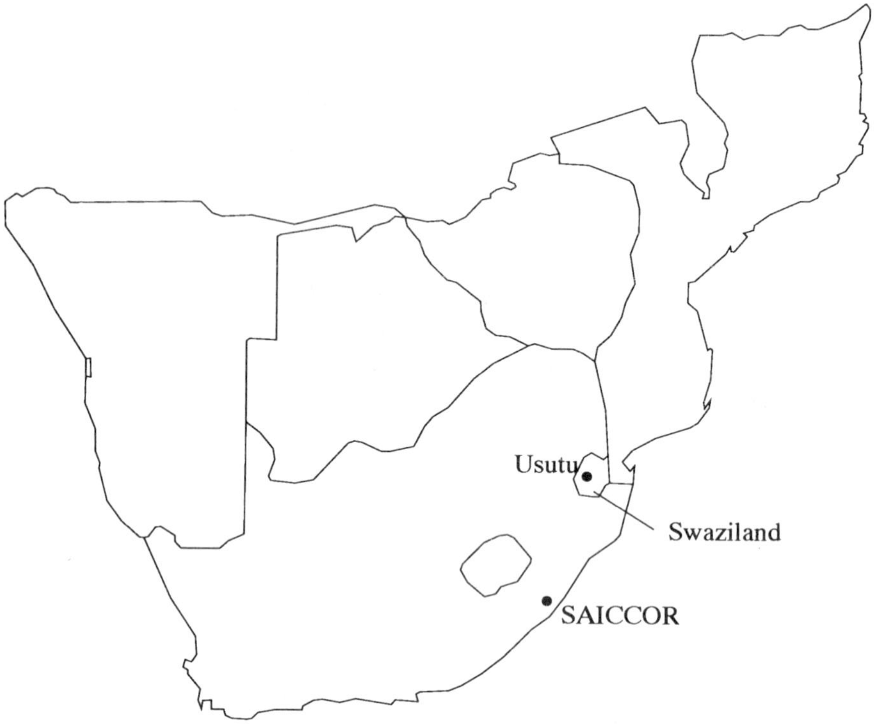

32. Map of South Africa

33. Spondon, Derby, England 1961

34. Thermal Cracker, Spondon

SO PULL UP A CHAIR, AND STAND ME A DRINK,
AND A TALE TO YOU I'LL TELL,
OF DEAD-EYE DICK AND MEXICAN PETE
AND A HARLOT CALLED ESKIMO NELL.

35. First Verse of Eskimo Nell

"Every shareholder should resist this iniquitous, invidious, immoral bid, until the offer goes up at least another five bob . . ."

36. The ICI Bid 1961

37. The ICI Bid 1961

	Tencel	Modal	Viscose	Cotton *	Polyester **
Titre (dtex)	1.7	1.7	1.7	N/A	1.7
Tenacity (cN/tex)	38-12	34-36	22-26	20-24	55-60
Elongation (%)	14-16	13-15	20-25	7-9	25-30
Wet Tenacity (cN/tex)	34-38	19-21	10-15	26-30	54-58
Wet Elongation (%)	16-18	13-15	25-30	12-14	25-30
Tenacity (@10% ext)	35	23	16	N/A	26
Wet Modulus (@5% ext)	270	110	50	100	210
Water Imibibition (%)	65	75	90	50	3

* US Middling

** High Tenacity

38. Comparison of Fibre Properties

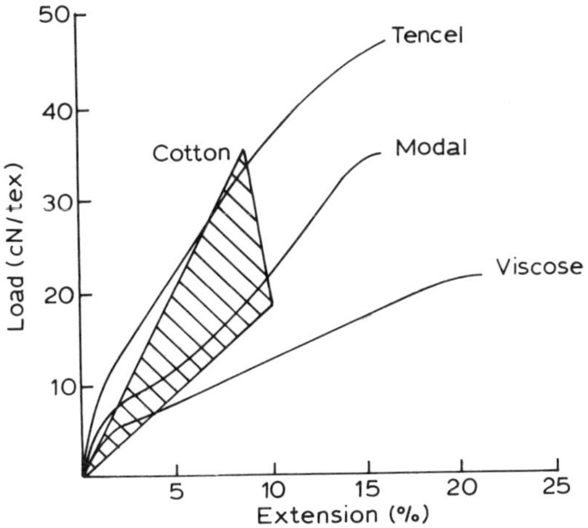

39. Comparison of Fibre Properties

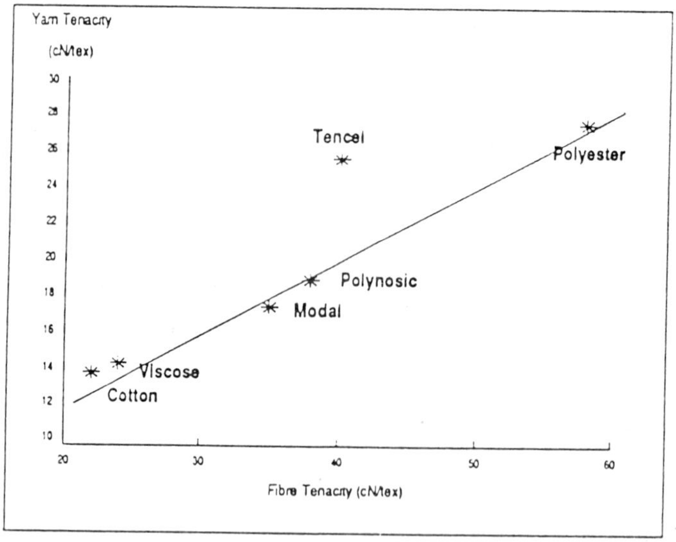

40. Yarn Formation

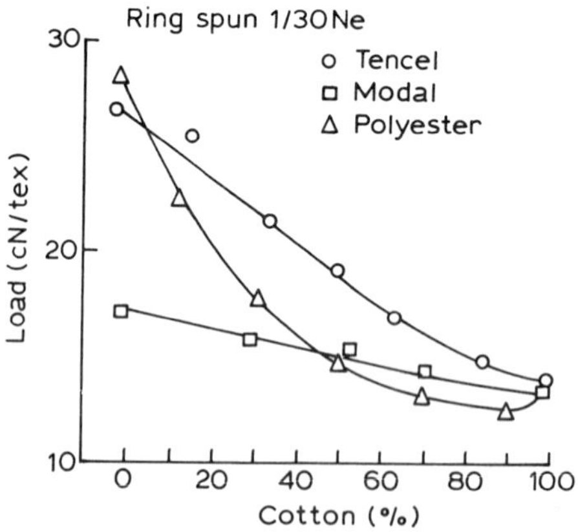

41. Strengths of Cotton Blend Yarns

- **HIGH FIBRE STRENGTH**
- **EXCELLENT WASH STABILITY**
- **HIGH COLOUR YIELD**
- **GOOD YARN UNIFORMITY**

42. Tencel Attributes

- **ADDS STRENGTH**
- **WIDE FABRIC RANGE**
- **COLOUR FLEXIBILITY**
- **EXCELLENT STABILITY**

43. Tencel in Apparel

- **GOOD PROCESSING**
- **WET PERFORMANCE**
- **BONDING CAPABILITY**
- **STRENGTH AND STABILITY**

44. Tencel in Non-Wovens

- **HIGH STRENGTH**
- **INCREASES TEAR STRENGTH**
- **HIGH UNIFORMITY YARNS**
- **NON-MELTING AND LOW CREEP**

45. Tencel in Industrials

46. Artists Impression of Mobile Tencel Factory

HISTORICAL APPLICATIONS OF SYNTHETIC RESINS FOR PASSIVE AND INTERVENTIVE CONSERVATION OF FRAGILE WOVEN VEGETABLE FIBERS AND COLLAGENEOUS SKINS

MAUREEN A ROBSON
Consultant Conservator, Studio, 29 Park Avenue, Birmingham, B18 5ND, England.

Former applications of early resins in the conservation of ethnographic and archaeological material present a number of problems to the conservator today. The use of synthetic resins in the late fifties enabled full consolidation of immersion under vacuum providing the material was mechanically sound and from which a sample had been removed for full future analysis. Often today, totally consolidated materials remain separated from their original samples lost in time. Current conservation methods more often require a passive approach and seek to ensure full documentation of both historical and modern methods; the retention, where appropriate, of old repairs during any re-restoration; adequate environmental monitoring in storage, transit, and on display; withholding where possible any interventive treatment which may cause an undesirable change in dimensional stability and colour of the antiquity and invalidate for the future any scientific examination of the original materials. More recently a new application of aqueous polymer emulsions whereby the dried film is reactivated by gentle heat, will ensure that no contact is made between the emulsion in the wet state and the dry materials undergoing interventive conservation and all particulate documentary evidence of indigenous origin is retained for future reference and research.

INTRODUCTION

Anthropologists both in the private and museum sector, acquire and study artifacts not merely as objects but to glean a better understanding of culture and technology from which the objects originated. The material, therefore, is a source of information rather than examples of exotica, particularly for societies lacking written documentation. They are, in fact, records in kind. Similar trends are now seen between ethnographic conservation and museum anthropology. During this century dramatic changes in direction have occurred in museums. The initial emphasis on massive acquisition programmes has been superseded by communication through didactic and interpretive displays and education programmes within the museum organization. The 1960's and 1970's saw the emergence of museum conservation as a professional discipline responding to the need for a physical preservation of objects. In the 1980's the importance of documentation was recognised as being an essential and important aspect of conservation work. By working with the artifact, conservators may learn much of anthropological importance.

This paper will outline the history of the use of synthetic resins in the conservation of fragile woven fibres and collagenous skins. Reference is made in the title to passive and interventive conservation. Passive conservation seeks to ensure that as possible the object is retained in its original state by controlling its environment. On the other hand, interventive conservation involves processes which treat the object both physically and chemically. By definition, conservation is a very long term process and the results and effectiveness of treatments cannot be assessed in a short period. Yet more new methods will supersede those in use today. To be both safe and effective the conservation treatment must meet a number of criteria:

1. The materials used in the conservation treatment must be compatible with the fabric of the restored object.

2. The conservation treatment must be reversible in the long term.

3. The final surface appearance of the restored object must be as close as possible to that of the original surface.

4. The consolidant should preferably form a colourless barrier against the atmosphere and give a penetration as deep as possible even when a vacuum cannot be used to assist impregnation.

A number of methods, the consequent problems they may present to today's conservator, and methods of solving these problems will be discussed. Up-to-date practices can ensure a very high standard of conservation for the future.

New discoveries and changing attitudes affect the field of conservation as much as any other, and since the turn of the century early advances in the conservation methods and materials may be demonstrated in the restoration and conservation of collections of fragile woven vegetable fibres and collageneous skins.

METHODS AND MATERIALS - INTERVENTIVE CONSERVATION

Cleaning

Current cleaning methods seek to ensure safe application, without loss or run of colour or fibre shrinkage or drying. Interventive conservation methods endeavor to guarantee the compatibility of the consolidant with the fabric of the artefact, their reversibility both in the short and long term and that although they structurally assist they are not conspicuous to the general observer.

The use of dry brushing, dental evacuator and photographic blow brush methods of cleaning when required (and depending on degree of the fragility of the fibre) may be recommended.

Wet cleaning methods of localized humidification and misting with deionised water to relax folds or unravel tangled vegetable fibres need to be carefully monitored to assess long term damage. Indiscriminate use may cause unforeseen damage in the future. It is often sufficient in ambient temperatures to use an inflated polythene bag wrapped in acid-free tissue to slowly relax folds resulting from bad storage methods without introducing humidification. Ethnographic textiles should never be washed. Very often on testing, dyes appear to be colourfast but may still run during prolonged immersion. (Robson, 1989)

STRENGTHENING THE WOVEN VEGETABLE FIBRE AND COLLAGENEOUS SKIN

Organic Conservation: Impregnation by Soft Resins

Polyethers. The use of carbowax [poly(ethylene glycol) PEG] was introduced around the 1950's. PEG is used primarily as a solution in water. Both the viscosity in water and the solidification temperature of water solutions are important in the choice of grade. The penetration of a PEG molecule depends on the size of the molecule and the size of the hole into which it must diffuse. The rate of diffusion of PEG through water is reduced by increasing the viscosity of the solution and the molecular weight, but increased by an increase in temperature and concentration gradient (Brown and Stilbs, 1982).

On excavation the wet organic material may be strengthened by direct immersion in a solution of the polymer in water. PEG is hygroscopic and there is hysteresis in response to variations in humidity. A disadvantage of this process is that the waxed surface would eventually attract atmospheric soiling as it is unable to sweat in high relative humidity. The composition of the wax solution may be varied according to the selected graded. Carbowax 300 to 400 were found suitable. PEG 300 - 400 is used as a humectant / lubricant for dry leather and other skin products (Werner, 1957).

Similar grades of PEG have been used for increasing flexibility in cellulose materials such as basketry and tapas (Schaffer, 1978, Bakker and Aarmo, 1981).

Organic Conservation: Impregnation by Hard Resins

Vinyl acetate derived polymers. Pure Poly(vinyl alcohol) PVAL has been shown to be very stable to ultraviolet/oxygen ageing. The first reports of PVAL use in conservation of paper and ethnographic textiles were in the early 1950's and it was used both as a size, and heat seal adhesive. By the early 1960's doubts were expressed about the solubility of PVAL which had been applied. It is wise to assume that these treatments were irreversible. PVAL continues to be used as a consolidant for textiles and as a heat seal adhesive for paper when strengthening without reversibility.

Poly (vinyl butyral) PVB. During the 1960's poly(vinyl butyral) PVB was widely used as both a consolidant for ethnographic textiles (Mowital B60HH) (Leene, 1972) and as an adhesive for bonding fractured organic fragments such as collageneous skins. (Butvar B98) A modelling material (AJK dough) originally based on poly (vinyl acetate) (Alvar) (Gednye, 1968) was later replaced by PVB (Butvar B98) (Patterson, 1978). This kneadable putty is prepared from an emulsion of water in a resin solution to which was added kaolin and jute flock. This models well and produces a strong light weight material. A latticework of strips of preshrunk AJK dough is used to bridge any gaps in the restoration. The putty consistency of AJK (Alvar 770) [poly(vinyl acetate)], jute flock, kaolin) dough was further worked across the lattice to match the required section thickness of the adjoining sherds. Unfortunately the dough undergoes short- and long-term shrinkage on drying, presumably through the loss of solvent and water.

Acrylic polymers. During the late fifties to early sixties a new approach was introduced with the idea of complete immersion in a synthetic resin. For the first time, account had to be taken of the possible effect of the conservation process on the external appearance of the organic material. Any testing or sample removal must be carried out prior to treatment or authenticity would be lost. The complete organic item or its associated fragments were immersed in a suitable solution under vacuum of bedacryl 122X [poly(butyl methacrylate)] and later Paraloid B72 (acrylic polymer) in xylene.

Bedacryl 122X (PBMA) has had a chequered career. During 1955-56 it was thought well of by the British Museum's conservation scientists, but fell from favour twenty years later when its long-term reversibility was in doubt. Currently, Bedacryl 122X (PBMA) is again in general use and can be purchased in two forms: either as a solid which can be dissolved to form the resin, or in a concentrated solution. Both forms may be diluted with toluene to the required concentration and when applied dries to a matt finish. These resins were effective as means of consolidation since they were used to strengthen whole items in both excavation and the laboratory workshop before poly vinyl alcohol emulsions were available. They were also used on individual fragments which could later be re-assembled and bonded with a suitable adhesive. However, the resin treatment did result in a certain amount of discolouration of the organic fabric. A vacuum desiccator is used for the purpose and a glass vessel of smaller dimensions is used as a container, in order to economise in the use of the solvent, and this vessel is placed inside the desiccator which is then evacuated. The object is allowed to remain under vacuum until all bubbling has ceased so that the consolidant is absorbed by the organic fabric. Air is then admitted and when the desiccator reaches atmospheric pressure, the item is lifted out of the container and allowed to drain: it can then be dried without acquiring a gloss by leaving it in an atmosphere of the solvent (Plenderleith, 1956).

DISPERSIONS

Polyvinyl Acetate (PVAC)

Dispersions of poly(vinyl acetate) PVAC became available in the late 1940's and by the early 1950's were widely used for ethnographic textile and paper conservation. Vinamul 6525 (Vinyl Products) was proposed as a heat set adhesive for attaching fragile textiles to supports (Beecher, 1959, 1963). It was probably too widely used judging from the difficulties encountered during the re-conservation of objects made necessary by severe changes in flexibility, colour and dirt pick up. [The polymer should be removable, if it remains soluble by soaking in an ethanol / water mixture] (Blum, 1983).

Acrylic Dispersions

Acrylic dispersions slowly increased in popularity during the 1950's because of the film's greater resistance to yellowing as compared with PVAC dispersion applied films. Acrylic dispersions have been widely used for heat set adhesives. More recently an application was found for supporting weakened vegetable fibres. Very fragile fibres require some support before they totally disintegrate during storage into minute piles of dust. The general practice of protecting the fibre with bandages during storage very often leads on removal to further disintegration and subsequent loss of the fragile woven fibre, and the problem still needs to be addressed before the item is displayed. Objects cannot be successfully displayed unless fibres are safely supported.

A safe method of bonding fragile fibres on to a support fabric of suitable weight has already been described (Robson, 1986). The use of aqueous polymer emulsions can be recommended (Robson, 1987). The support fabric is impregnated with the adhesive which is then allowed to dry to form a thin cohesive film. The film is then reactivated by the application of gentle heat from a spatula iron. (This method was discussed at Symposium '86 "The Care and Preservation of Ethnological Materials" at the Canadian Conservation Institute in Ottawa). Since the samples of adhesives varied in strength, they could be successfully used in supporting different strengths of fibre within the same object, ranging from the severely weakened to the more substantial.

EARLY ADVANCES IN RESIN BONDING

Organic fabric fragments of substantial strength, e.g., leather boot and saddle fragments were originally generously bonded with carpenters' glue made from animal bones, hooves, etc. This was warmed up with water and kept hot in a double boiler. During the 1920's, before the manufacture of epoxy resins, animal glue bought in sheet form from the oil show was broken up and reduced to a workable glue of gelatinous consistency by way of water and double boiler. A few drops of acetic acid were added to increase the working life of this strong glue, whereas the addition of dextrin speeds up the setting time. Another variant, French polish, was also added to animal glue in a 1 : 1 ratio. The animal glue was painted on and allowed to set. Any realignment of the assembly of sherds could be made by warming the glue with a naked flame. The use of Coaguline, a water soluble glue no longer available, manufactured by Kay Bros. of Stockport, UK, as a ready-made adhesive, has an advantage over the previous water soluble carpenters' glue. It was warmed in a similar fashion by immersing the bottle in a double boiler and produced a liquid of low viscosity which was easy to apply and strongly adhesive. (Robson, 1988)

During the 1950's the use of PEG, a water soluble wax for the consolidation of fragile organic fragments presented some problems in their repair work. Although this is preferable to assemble and bond the fragments before immersion, it is possible to remove the unwanted wax along the break edge of the wax-impregnated fragments with cotton wool swabs moistened in dilute ammonia thus ensuring subsequent bonding with a suitable adhesive.

During the late fifties and early sixties acrylic polymers Bedacryl 122X (PBMA) methacrylate and later Paraloid B72 were both used effectively as an adhesive. For combination repairs of fragmented organic materials fully consolidated with these resins.

Durofix and HMG (nitrocellulose adhesives) form satisfactory and compatible adhesives for the repair work. The two adhesives are easily reversible in the short term by application of acetone.

During the 1970's the use of soluble nylon was fashionable for bonding organic fabrics and both as a wet application to adhere a support fabric of a suitable weight and as a backing to fragile organic material when reinforcement was desirable. Although in the short term it is soluble in alcohol-based solvents, it cross-links and becomes insoluble after ageing. The use of N-methoxymethyl nylon in conservation has been strongly discouraged (De Witte, 1975; Sease, 1981).

A more suitable application as a heat set adhesive for attaching fragile fibres to supports is the use of acrylic dispersions Revacryl 1A and Mowilith DM131. the long term assessment of these aqueous polymer emulsions continues but in the 1980's satisfactory results were obtained both in the short term and in the long term accelerated ageing tests (Robson, 1986).

LONG-TERM REVERSIBILITY OF MATERIALS

In order to use an alternative and more suitable conservation method, the earlier materials used in the restoration have to be reversed and in each of the cases mentioned above this presents different problems.

The Carbowax process is perfectly reversible providing the organic fabric is physically sound and will withstand temperatures to a maximum of 60°C. It is possible and even desirable to remove any unsightly surface wax residues. 'There is a tendency for wax to creep out gradually from the inside and in a matter of time the surface will become sticky and tend to collect dirt. In addition, if a waxed object is exposed to the sun or exhibited in a glass case that is lit internally, the object may develop an objectionable gloss with rise of temperature' (Plenderleith, 1956).

The acrylic polymers Bedacryl 122X and Paraloid B72 are easily reversed in both the short and long term by either direct or atmospheric application of the solvent acetone. The removal of acrylic dispersion applied films is achieved by softening the polymer. A hydrocarbon solvent, e.g., toluene, xylene is wiped over to moisten impregnated support fabric which is then peeled off the fibres, with the rubbery solvent-swelled, polymer attached.

PASSIVE CONSERVATION

Environmental Monitoring

Various environmental controls according to specific requirements of the artefacts both on display, in storage, and during transit are inbuilt within any conservation system. The lighting units, built often into the tops of cases, are protected where desirable with ultraviolet absorbent sleeves and diffuse light through a sheet of opal perspex which provides a false ceiling to the exhibition cases. Storage facilities are monitored for suitable environmental controls.

Where necessary, portable humidifiers are installed in the galleries and stores, and hygrometers are displayed in the cases containing vulnerable material.

All sunlight whether direct or reflected is damaging. Light falling on an object must therefore be reduced. Structural problems such as skylights without sunblinds or ultraviolet protection may be blocked out by the assembly of a portable false ceiling which extends across the top of all showcases on exhibition, to maintain where necessary the correct light levels throughout the displays. It may be essential to reduce the time a particularly vulnerable object is exposed to light. Black cloths are draped over individual showcases containing such material outside viewing times.

Provided that all items were environmentally monitored and particularly vulnerable ethnographic textiles, etc., were adequately supported on suitable mounts and individually cased during transit, this passive conservation is sufficient to ensure the safe transportation to other venues where environmental monitoring may continue. The conservation of very fragile composite organic materials presents an often difficult choice between either leaving them in their natural state, which may result in subsequent damage or disintegration, or using interventive conservation methods which will structurally assist the long term preservation of the item but may lead to some damage in its structure and surface appearance.

DOCUMENTATION

The need to consider fully the ethics and consequences of conservation methods which may cause problems in the future reflects current advances in conservation. Certain treatments may destroy any cultural, historical and technological information associated with the artefact.

Thus full documentation without conservation, on selected antiquities or samples taken from them, will safeguard this information and provide a resource for future research. Provided the condition of the antiquities thus selected is continuously monitored under suitable environmental conditions on displays or in storage, it will no doubt pre-empt the necessity for interventive conservation.

The importance for the conservator to maintain adequate records of the variety of conservation and restoration materials and techniques used today is emphasized as a prerequisite for conservation on any formal training programme. On some occasions it may be necessary to document previous restoration which may be re-restored along with the antiquity as the possibility of greater damage resulting from its removal may not be ignored. Past restoration methods are of historical interest and very often their documentation no longer exists. Detailed records should be made of these early techniques to provide an invaluable resource for future research.

CONCLUSIONS

Passive conservation treatments include good housekeeping, safe packing, providing adequate support and appropriate environmental conditions in storage. transit and on display, with full documentation of objects and their past and present treatments. These measures may obviate to a large extent the need to introduce interventive conservation methods which will structurally assist in the long term preservation of the item but often lead to some undesirable change in the structure and surface appearance of the ethnographic object.

The initial aim of the conservator and the anthropologist is to prolong the life of materials which, in many cases, were never intended to last very long. Many objects were intended to alter their form during the course of their natural life, often changing their shape, colour, use and significance. But the moment an object enters a private or museums collection its original makers' and users' intentions are ignored or contradicted. Moreover, by extending the natural life expectancy of these items which on serious consideration have very limited worth, may be cost prohibitive.

Anthropologists and conservators may need to seriously question their professional judgements in justifying unnatural preservation of ephemeral items and during which stage their natural life cycle should be "frozen".

ACKNOWLEDGEMENT

The author should like to thank the late Distinquished Professor Raymond B. Seymour for inviting her to participate in the 201st National American Chemical Society Meeting in Atlanta Georgia, USA; and the following for their discussion, encouragement and support : David Bailey, Leo Biek, Debbie Boulton, Pat Brown, Andrew Carson, Eileen Collins and Christopher Radbourne, Anne Hudson, Hilde Smith, Pauline Timmins; and to Wendy Firmin, Editorial critic and Bunny Warren, and Kim Sumner Secretariat, Kevin and Robert Brown, Directors Advance Office Services, U.K for practical assistance and professional advice.

The paper is dedicated to the late Trudy Crilley, and Wilma Wells, UK, for their industrious assistance, sterling support and encouragement in progressing the authors research interests.

REFERENCES

Beecher, E.R. (1959) "Treatment of Weakened Fabrics," Museum Journal 58, 234-235.

Beecher, E.R. (1963) "Reinforcing Weakened Textiles with Synthetic Fibre Net," in Thompson (Ed) (1963) pp. 195-196.

Blum, D. (1983) "An Evaluation of Some Uses of Synthetic Resins in Textiles Conservation," in Tate et al. (Eds) (1983), Ch 8.

Brown and Stilbs, P. (1982) "On the Solution Conformation of Poly(ethylene oxide) an FT Pulsed Field Gradient NMR Self Diffusion Study," Polymer, 23, 1780-1784.

De Witte, E. (1975) "Soluble Nylon as Consolidation Agent for Stone," Studies in Conservation, 20, 30-34.

Plenderleith, H.J (1956) The Conservation of Antiquities & Works of Art, Oxford University Press.

Robson, M.A. (1986) "The Use of Aqueous Polymer Emulsions in the Conservation of Ethnographic Material: A New Method for Supporting Fragile Woven Fibre Decoration," in symposium 1986, The Care and Preservation of Ethnological Materials (Eds) R. Barclay, M. Gilberg, J.C. McCawley and T. Stone, Canadian Conservation Institute.

Robson, M.A. (1987) "A Note on the Use of Aqueous Polymer Emulsions in the Conservation of Archaeological Leather," Leather conservation News, Vol.4.

Robson, M.A. (1988) "Methods of Restoration Conservation of Bronze Age Pottery Urns in the British Museum," B.M. Occasional Paper.

Robson, M.A., (1989) "Passive and Interventive Conservation for a Travelling Exhibition," Museum Ethnographic Group UKIC.

Schaffer, E. (1978) "Water Soluble Plastics in the Preservation of Artifacts Made of Cellulose Materials," in ICOM (1978) 78/3/7.

Sease, C. (1981) "The Case Against Using Soluble Nylon in Conservation Work," Studies in Conservation, 26, 102-110.

Werner, A.E.A. (1957) "Synthetic Waxes," Museum Journal 57, 3-5.

A BRIEF HISTORY OF CUPRAMMONIUM RAYON

GEORGE B. KAUFFMAN
California State University, Fresno, Fresno, California 93740

INTRODUCTION

Rayon is a generic term for an artificial textile material composed of reconstituted, regenerated, and purified plant substance, viz., cellulose or cellulose compounds. Because its long-chain polymer structure is provided by nature and is only modified and partially degraded by man, it is referred to as a semi-man-made fiber in contrast to fully man-made fibers such as nylon (polyamide) or dacron (polyester), which the chemist synthesizes from low-molecular weight compounds by polymerization (chain-building) reactions. Although reactions for creating artificial fibers were discovered during the 19th century, such fibers did not become commercially significant until the 20th century, when processes for their manufacture were exploited, along with those for plastics and other polymers to which they are closely related. Developed as a substitute for silk, rayon is sometimes called "artificial silk" and "wood silk," but this is a misnomer because its chemical composition is entirely different from that of silk. In 1924 this first semi-man-made fiber was given the coined name rayon, particularly in the United States and Great Britain [1, 2a, 3]

CELLULOSE

The basic structural component of most plants, cellulose is widely distributed in nature, comprising more than one-third of all vegetable matter [4-8]. It is the world's most abundant organic compound and the principal constituent of cotton, kapok, flax, hemp, jute, wood, and an Asian urticaceous shrub, ramie. Each year land plants produce about 50 billion tons of this renewable resource by photosynthesis.

Cellulose is a polydisperse polymer, i.e., it consists of many different molecular homologues, varying from 1,750 to 18,000 repeating cellobiose units, each of which contains three reactive hydroxyl groups [9a] (Fig. 1). While celluloses are often largely linear polymers, they are insoluble in water because of their strong intermolecular hydrogen bonding and their small amount of cross-linking [10] (Fig. 2). Most linear celluloses can be dissolved in solvents capable of breaking the strong hydrogen bonds, e.g., aqueous inorganic acids, calcium thiocyanate, zinc or lithium chlorides, dimethyldibenzylammonium hydroxide, ammoniacal cadmium hydroxide, and ammoniacal copper hydroxide. The last name solvent (Schweizer's reagent) is used in manufacturing cuprammonium rayon. Only high-molecular weight native cellulose, which is insoluble in 17.5% aqueous sodium hydroxide and which is called α-cellulose, can be used in producing rayon; the only materials containing a high enough percentage of α-cellulose for making rayon are cotton (plants of the genus Gossypium) and some types of wood pulp [11a].

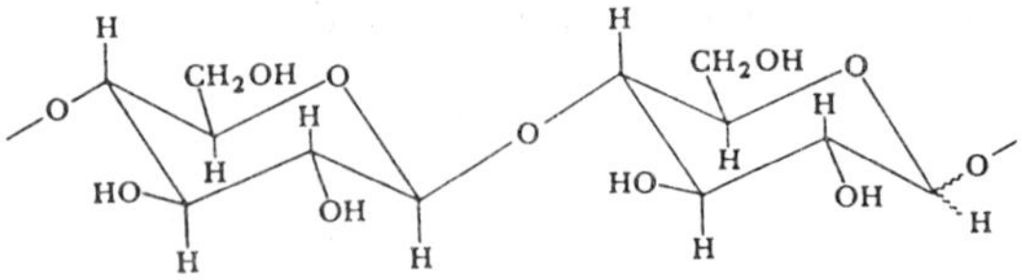

Figure 1. Chair Form of Cellobiose Unit in Cellulose [9a]

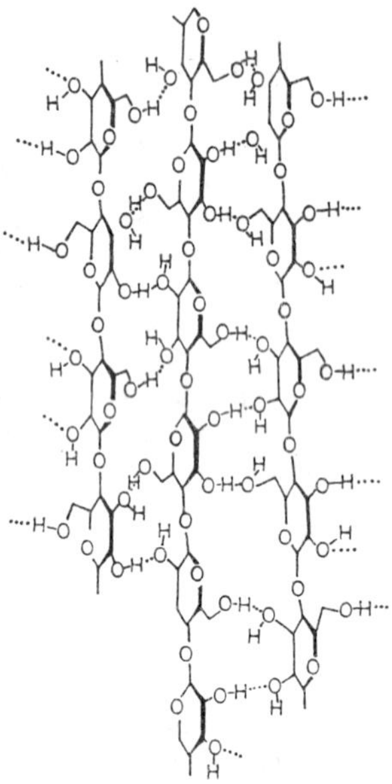

Figure 2. Puckered Sheet Structure of Cellulose Chains (The dots denote hydrogen bonding).[10]

Figure 3. Eduard Schweizer (1818-1860)[34]

ARTIFICIAL SILK

As early as 1665 [12] the English physicist Robert Hooke (35-1703) suggested that [2a, 13, 14]

> there might be a way found out to make an artificial glutinous composition much resembling, if not fully as good, maybe better, than that excrement or whatever other substance it be out of which the silk-worm [Bombyx mori] wire-draws his clew.

In 1734 the French naturalist and technologist Rene-Antoine de Réaumur (1683-1747) asked [2a, 11b, 15]

> Silk being nothing but dried-up liquid gum, why cannot we make silk with gum and resin?

Réaumur reported his attempt to produce silk-like fiber by forcing different kinds of varnish through tiny holes in the bottoms of sheet-iron cans or tin cans to form glutinous filaments that hardened in the air. Although his fibers were of no practical importance, his process formed the basis for all future attempts to produce artificial silk. In 1740 a Frenchman surnamed Bon reported to the Royal Society that he had woven stockings and gloves of yarn made from spider web filaments [2a], and in 1770 a Frenchman surnamed Dubet produced filaments from the silky gum extracted from dead silkworms [16]. At a meeting of the British Association for the Advancement of Science held at Manchester in 1842 a silk manufacturer named Schwabe suggested extruding solutions through fine holes as a method for producing fibers [2a].

The first practical progress toward producing a synthetic fiber was made in Basel, Switzerland by Christian Friedrich Schönbein (1799-1868), who in 1846 accidentally treated cellulose with a mixture of sulfuric and nitric acids and produced a cellulose derivative which he called guncotton (Schiesswolle) and proposed as an explosive [17-21]. Although guncotton was misnamed nitrocellulose, it is actually cellulose nitrate because the nitrogen is present as the nitrate ($-ONO_2$) rather than the nitro ($-NO_2$) group. Although cellulose is insoluble in water and most organic solvents, the bulky nitrate group in cellulose nitrate tends to separate adjacent polymer chains, reduce crystallinity, and promote solubility. In 1855 the Swiss chemist George Audémars (or Audhémar) of Lausanne received the first English patent for producing artificial silk by drawing out threads from an ethanol-ether solution of cellulose nitrate (collodion) with a steel pointer and reeling the threads [2a, 22].

However, it was not until 1883 that, while searching for a better carbon filament for his electric light bulb, which he invented almost simultaneously with, but independently of, Thomas Alva Edison, the English physicist and chemist Sir Joseph Wilson Swan (1828-1914) patented a process for squeezing an acetic acid solution of cellulose nitrate through holes to form fibers [2a, 11b, 23]. Swan realized that these filaments could be used to produce an artificial silk, and in 1884 and 1885 he exhibited fabrics made of this fiber in London [24].

The first practical fiber made from cellulose nitrate was produced by the French chemist and technologist Louis-Marie-Hilaire Bernigaud, Comte de Chardonnet (1839-1924), who is considered the founder of the rayon industry [25]. A pupil of Louis Pasteur, whose work on silkworm diseases led his student to being research on artificial silk in 1878, Chardonnet applied for his first patent and submitted a memoir titled "Une matière textile artificielle ressemblant à la soie" to the Académie des Sciences in 1884 [2a, 9b, 11b, 26, 27]. At the Exposition Universelle held in Paris in 1889 he was awarded the grand prize for the first gown made of cellulose nitrate fiber [28], and in 1891 he began to produce "soie

Chardonnet" in a factory that he had established in Besançon, his hometown. The new fiber, prepared by extruding collodion through fine glass capillaries to form continuous filaments that harden in the air as the solvent evaporates, had the disadvantage of being flammable and explosive. Chardonnet overcame this disadvantage by denitrification (removal of the nitrate groups by hydrolysis with nitric acid and later with sulfide solutions) [29]. Nevertheless, "Chardonnet silk" was largely superseded in the early 20th century by the more inexpensive viscose rayon.

"Viscose" was the name given by British chemist Charles F. Cross (1885-1935), Edward J. Bevan (1856-1921), and Clayton Beadle (1868-1917) to the syrupy, yellow liquid that they discovered in 1891 by the action of cellulose xanthate (a product of the reaction between carbon disulfide and alkaline cellulose) on dilute sodium hydroxide solution [2b, 11b, 30]. Although production of viscose rayon involved great technical difficulties, these were overcome gradually by research and development such as Charles F. Topham's invention in 1900 of the so-called Topham box, a device for centrifugally spinning twisted filaments from cellulose solution [31]. By 1905 Courtaulds, the British silk company, had built a plant for manufacturing viscose rayon in Coventry, and in 1910 the American Viscose Company, established by Courtaulds, at Marcus Hook, Pennsylvania, began production in the United States. Today the viscose process accounts for most of the world's total rayon production [11b].

A third type of rayon is based on the acetylation of cellulose by acetic anhydride achieved by French chemist Paul Schutzenberger (1829-1897) and Charles Naudin (1815-1899) in 1869 [32]. In 1894 viscose developers Cross and Bevan patented the first industrial process for the manufacture of cellulose acetate. In 1903 the American chemist George Miles (1868-1939) succeeded in dissolving partially hydrolyzed cellulose acetate in acetone. However, it was not used as the basis for a textile fiber until 1907, when it was so applied by Prince Guido Henkel von Donnersmark in Germany [11b]. It was only after World War I that a cellulose acetate fiber process was developed by the Swiss brothers Henri and Camille Dreyfus, who marketed their product in England under the name Celanese. By 1924 the Celanese Corporation was producing the fiber in the United States.

In addition to these three types of rayon--"Chardonnet silk" (nitrocellulose process), viscose, and cellulose acetate--there is a fourth type of rayon--cuprammonium rayon--prepared by a process that was first employed in 1890, the year after Chardonnet exhibited his product at the Exposition Universelle. This was based on a discovery by the Swiss chemist (Matthias) Eduard Schweizer.

SCHWEIZER AND SCHWEIZER'S REAGENT

Schweizer, the younger brother of astronomer Kaspar Gottfried Schweizer, was born on August 7, 1818 at Wyla in Canton Zürich [33, 34] (Fig. 3). Beginning with the winter semester of 1841 he was a Privat-Dozent (unsalaried lecturer) at the Universität Zürich, where he remained active until his death. An assistant to Karl Jakob Löwig (1803-1890), the first occupant of the chair of chemistry at the university, Schweizer had learned metallurgy and mining on trips to Germany, and he lectured on chemical technology. Beginning in 1842, he alternated with another Privot-Dozent, Friedrich Eichelberg, in lecturing on mineralogy. When Löwig was called to Breslau as Robert Bunsen's successor in 1852 and was succeeded by Georg Andreas Karl Städeler (1821-1871), an Extraordinariat (a position corresponding roughly to Associate Professor) for chemistry was created, and Schweitzer was appointed to the post. He also served as teacher of natural sciences (Lehrer der Naturwissenschaften) at the Obere Industrieschule in Zürich until 1855. Schweizer died in Zürich on October 23, 1860. In addition to his research articles, he wrote a textbook, a laboratory manual, and an account of chemical instruction at the Zürcher Kantonschule.

Schweizer's most important contribution to science and technology, which made his name known throughout the chemical world, was his discovery that cellulose dissolves in an ammoniacal solution of copper(II) hydroxide, now known universally as Schweizer's reagent or Schweizer's solution. His discovery, undertaken in the course of his investigations of metal-amines (ammoniakbasische Metalle), specifically those of copper, has great significance in the history of coordination chemistry. Schweizer is usually credited with discovering the solvent action of ammoniacal copper(II) solutions on cellulose. However, Edward Andrew Parnell claimed that John Mercer (1791-1866) had earlier observed this action and had recovered cellulose from these solutions by acidification with sulfuric acid [2a, 35].

Schweizer's study of copper(II)-amines began with his synthesis of unterschwefelsaure Kupferoxyd-Ammoniak (modern, tetraamminecopper(II) dithionate, $[Cu(NH_3)_4]S_2O_6$) by the metathesis of $[Cu(NH_3)_4]SO_4\ H_2O$ with BaS_2O_6 in warm aqueous solution [36a]. In his second paper on copper(II)-amines Schweizer attempted to prepare the free base Kupferoxyd-Ammoniak (modern, tetraamminecopper(II) hydroxide) by dissolving freshly prepared basic copper(II) dithionate in concentrated aqueous ammonia. Along with the main product ($[Cu(NH_3)_4]S_2O_6$), he obtained a dark blue solution ($[Cu(NH_3)_4](OH)_2$), the unusual properties of which are best described in his own words [36b, pp 110-111):

> This possesses to an extraordinary degree the power to dissolve plant fiber at ordinary temperature.
>
> If purified cotton is covered with the blue solution, is soon assumes a gelatinous slippery state, the fibers separate and disappear, and after some working up with a glass rod everything is converted into a slimy liquid...If an insufficient amount of solution is used, a portion of the fibers still remains visible; however, if an excess of the solution is added and shaken, an almost clear blue solution is obtained, which, after it has been diluted with water, can be filtered.
>
> If the filtered solution is supersaturated with hydrochloric acid, a voluminous white precipitate is formed, which, collected on a filter, has the exact appearance of moist aluminum hydroxide.
>
> This substance, unorganized, to be sure, but not changed essentially in its chemical nature, appears to be cellulose.
>
> If the gelatinous precipitate, freed completely from salts by washing, is dispersed in water, and potassium iodide and afterward some chlorine water is added, the substance is colored brown (not blue), proof that it is neither starch nor a starch-containing substance.
>
> On being dried on a water bath, the precipitate shrinks together strongly and leaves behind a horny, translucent, brittle mass, which is similar to dried paste but has no taste at all and does not stick to the teeth. Heated in the air, the substance burns without leaving a residue...
>
> If the fiber solution is painted on a glass plate and allowed to dry on it, a thin, bluish white coat remains, which adheres solidly to the glass...

Schweizer found that the Kupferoxyd-Ammoniak solution also dissolved paper and linen (somewhat more slowly than cotton), silk (more quickly than cotton), and wool (only on warming). Hair was gradually destroyed by the solution without complete solution occurring, while an animal bladder merely swelled up initially, but after some time also dissolved. He was surprised that the solution did not dissolve starch, so closely related to plant fibers. He concluded [36b, p. 111]:

> As anticipated, the solution obtained by dissolving basic copper sulfate in ammonia shows the same dissolving power toward the materials mentioned as that prepared from the dithionate salts, according to which the interesting solvent can be prepared more easily than in my first experiment [i.e., in ref. 36a].

In his third article on tetraamminecopper(II) [37c] Schweizer extended his study to other salts of the cation and reported that a saturated ammoniacal solution of basic copper carbonate was an excellent solvent for cellulose. He attributed the solvent power to pure, uncombined Kupferoxyd-Ammoniak (modern, $[Cu(NH_3)_4](OH)_2$) because the blue solution obtained by treating finely divided copper with aqueous ammonia in the presence of air possess this solvent power, an observation made the year before by Peligot [37, 38].

Schweizer's last publication on the subject was a letter to Otto Linné Erdmann (1804-1869), editor of the Journal für praktische Chemie, in which he disagreed with the view that "cellulose is not really dissolved in Kupferoxyd-Ammoniak but only strongly swollen." He asserted that the cellulose is actually dissolved for four reasons [39, p. 371]:

> (1) The fiber undeniably disappears on melting under the microscope... (2) No traces of structure or definite form appear any longer in the cellulose precipitated from the solution of cotton in Kupferoxyd-Ammoniak. (3) A solution of cotton in CuO,NH3, if it is completely saturated and diluted with water, can be completely filtered clear, even if indeed the filtration of greater amounts of the solution is linked with considerable difficulties. (4) The cotton solution permeates plant cell membrane as the endosmotic experiment made by Cramer (This journal J. prakt. Chem., 76, 6 (1858) illustrates).

Erdmann disagreed and "compared the cellulose solution with a hot filtered starch solution, from which, according to its behavior in the cold, etc., it must be assumed that it is no true solution" [40]. Reihlen later found that cellulose dissolved in Schweizer's reagent because of the formation of complex compounds between Cu^{2+} and $(C_6H_8O_5)^{2-}$ions [41]. The equilibrium reaction

$$nCu^{2+} + (C_6H_{10}O_5)_n + 2nOH^- \rightleftharpoons [CuC_6H_8O_5]_n + 2nH_2O$$

is pH dependent; cellulose ions are present in appreciable concentration only in a rather strongly alkaline solution. If the hydroxide ion concentration is lowered, the cellulose is reprecipitated. If it is increased by addition of NaOH solution, the cellulose redissolved [41]. The physical properties and chemical reactions of Schweizer's reagent have been extensively investigated [42].

CUPRAMMONIUM RAYON

Although a number of investigators recommended the use of Schweizer's reagent to dissolve cellulose, it was not until 1890 that the Frenchman Louis-Henri Despeissis developed and patented the second process of industrial importance for making textile fibers of artificial silk--using cellulose dissolved in this solution [2b, 11b, 28, 43-45]. Unfortunately, he died two years later without having utilized his patent on an industrial scale, and his process entered the public domain. However, Max Fremery, a German chemist from Cologne, and Johan Urban, an Austrian engineer, who had worked together in 1885 on the manufacture of carbon filaments in an electric light factory in Rotterdam, collaborated in improving Despeissis' process. In 1892 they founded the Rheinische Glühlampen Fabrik, Dr. Max Fremery & Co. to manufacture electric light bulbs. They established a factory in an old paper-making plant at Oberbruch near Aachen (Aix-la-Chapelle), and their enterprise prospered. In 1895 to develop their business further, they obtained a bank credit from the Bergische-Markische Bank, a condition of which, proposed by one of the bank's directors, a Doctor Jordan, was that they continue their research to produce textile filaments of artificial silk.

Fremery and Urban faced considerable difficulties in complying with this condition. An industry for manufacturing artificial silk from collodion already existed, especially in France, and its fabrication processes were well kept secrets, forcing them to carry out their own research and to solve numerous chemical and mechanical problems [44]. Among the problems involved in using Schweizer's reagent was the fabrication of machinery resistant to the high pressures required and yet allowing the necessary flow rate. After two years of experimenting, Fremery and Urbain filed their first patent under the name of Herman Pauly, director of a technical school at Munich Gladbach, a subterfuge intended to avoid both competition and attracting the attention of other researchers [46].

Although he is mentioned in historical accounts of rayon, Pauly made no contribution to this process or subsequent developments, and cuprammonium silk should not be called "Pauly silk" [43]. However, it is sometimes claimed that this first practical patent for cuprammonium silk was nothing but a repetition of Despeissis' patent of a dozen years earlier. However, unlike the earlier patent, it described a technical method for spinning the product and is truly the fundamental basis for the cuprammonium silk industry. Using their additional subsequent patents, as well as those of the German Emil Bronner, Fremery and Urban began to manufacture a cuprammonium artificial silk called "Glanzstoff" at their Oberbruch works, beginning in 1898. Their company eventually became the economically powerful Vereinigte Glanzstofffabriken.

Near the end of the 19th century the grandson of the founder of J. -P. Bemberg, a firm that held a prominent position in the German textile industry, followed Fremery and Urban's successful manufacturing efforts with considerable interest [44]. In 1900 he asked technician Emil Thiele to carry out research on this new fiber, for which he envisioned a promising future. Thiele began work in 1901 with the goal of producing very fine filaments comparable to those of natural silk. He returned to the ideas of the German chemist Lehner, who had taken out patents in 1889 and 1890 for the spinning of cellulose nitrate silk that employed slow-acting coagulation baths, which permitted drawing out the filament when it is still very malleable. Assisted by the chemist Emil Elsasser, Thiele required four years to put this process into practice by using a hopper in which the filament was drawn.

In 1908 Thiele perfected the process by using as starting material cotton linters (short fibers that stick to the cottonseeds after a first ginning), which were more economical and which imparted to the thread more resistance as well as a softer feel, enabling it to be used in the hosiery industry for making stockings. To indicate this improvement over the natural material Bemberg adopted the slogan "It is born of cotton and shines like silk" for what became known as "Bemberg silk" or "Eagle silk." Other variations of the cuprammonium process based on patents by Linkmeyer, Foltzer, Langhans, Dreaper, Friedrich, and Borzykowski were initially of lesser importance but later made inroads in the market.

The prosperity of the cuprammonium process was relatively short-lived; it was generally abandoned around 1910, except by Bemberg. However, cuprammonium rayon, prepared by dissolving cellulose in Schweizer's reagent, extruding the resulting solution through spinnerets, regenerating the cellulose in dilute sulfuric acid, and washing the fiber in acetic acid [47, 48], is still used to produce the most expensive rayon--one most resembling silk. Its extra fineness and strength, attractive hand, subdued luster, and excellent draping properties justify its added cost.

Cuprammonium rayon is used for satins and sheer fabrics such as chiffons, nets, and nylons. It is also popular for underwear, dress fabrics, and linings. Fine denier slub yarns are used for dresswear, sportswear, curtains, and fine drapery fabrics, while heavy denier slub yarns are used for suitings, upholstery fabrics, bedsheets, drapery, and other fabrics [1]. Currently, almost all artificial kidney dialysis units use membranes prepared from films and hollow fibers of cuprammonium rayon [49, 50].

REFERENCES

1. J. G. Cook, "Man-Made Fibres" in The Encyclopedia Britannica, 15th ed., Encyclopaedia Britannica: Chicago, IL, Vol. 7, pp 257-271.
2. E. Wheeler, "The Manufacture of Artificial Silk with Special Reference to the Viscose Process," D. Van Nostrand Co.: New York, 1928, (a) Chap. 1, (b) Chap. 2.
3. T. M. Letcher, N. S. Lutseke, J. Chem. Educ., 67, 361 (1990).
4. J. Bersch, "Cellulose, Cellulose Products and Artificial Rubber," Henry Carey Baird: Philadelphia, PA, 1904.
5. E. Ott, Ind. Eng. Chem., 32, 1641 (1940).
6. W. Haynes, "Cellulose, The Chemical That Grows," Doubleday: New York, 1953.
7. E. Ott, H. M. Spurlin, M. W. Graffin, "Cellulose and Cellulose Derivatives," Interscience: New York, 1954.
8. R. J. Brewer, R. T. Bogan, "Cellulose Esters, Inorganic", in Encyclopedia of Polymer Science and Engineering, Wiley: New York, 1985, Vol. 3, p. 139.
9. R. B. Seymour, C. E. Carraher, Jr., "Polymer Chemistry: An Introduction," 2nd ed., Marcel Dekker: New York, 1988, (a) pp. 180-182, (b) pp. 182-183.
10. R. B. Seymour, C. E. Carraher, Jr., "Giant Molecules: Essential Materials for Everyday Living and Problem Solving," Wiley: New York, 1990, pp. 84-85.
11. W. D. Darby, "Rayon and Other Synthetic Fibers," Dry Goods Economist: New York, 1929, (a) p. 34, (b) Chap. 1.
12. W. Bragg, "Old Trades and New Knowledge," six lectures before the Royal Institution, London, Christmas, 1925.
13. R. Hooke, "Micrographia: or Some Physiological Descriptions of Minute Bodies made by Magnifying Glasses: with Observations and Inquiries Thereupon," J. Martyn and J. Allestry: London, 1665.
14. Chem. and Ind., 45, 221 (1926), Times Artificial Silk Supplement, April 2, 1927.
15. R. -A. F. de Réaumur, "Mémoires pour servir à l'histoire des insectes," 6 vols, Imprimerie Royale, Paris, 1734-1742.
16. La Muriometrie, "Instruction nouvelle sur le lever de la soie," Paris, 1770.
17. C. F. Schönbein, Compt. rend., 23, 612, 678 (1846).
18. C. F. Schönbein, U. S. Patent, 4,874 (1846).
19. Plastics, 2(1), 9 (1926).
20. F. D. Miles, "Cellulose Nitrate," Oliver and Boyd: London, 1955.
21. E. C. Worden, "Nitrocellulose Industry," D. Van Nostrand: New York, 1911.

22. E. Audémars, English Patent 283 (1855).
23. J. W. Swan, English Patent 5,978 (1883).
24. J. Soc. Chem. Ind., 4, 34 (1885), Engineering, 40, 39 (July 10, 1885).
25. M. Kaurman, in "Dictionary of Scientific Biography," C. C. Gillispie ed., Scribners: New York, 1972, Vol. 3, p. 207 and references therein.
26. P. W. Morgan, in "History of Polymer Science and Technology," R. B. Seymour, ed., Marcel Dekker: New York, 1982, pp. 49-67.
27. L. -M. -H. -B. Chardonnet, Compte de. French Patent 165,349 (1885), German Patent 38,368 (1885).
28. F. Aftalion, "Histoire de la chimie," Masson: Paris, 1988, pp. 76-77.
29. P. J. T. Morris, "Polymer Pioneers: A Popular History of the Sciencc and Technology of Large Molecules," Center for History of Chemistry: Philadelphia, PA, 1986, pp. 28-29.
30. C. F. Cross, E. J. Bevan, C. Beadle, English Patent 8,700 (1892).
31. C. F. Topham, English Patents 23,157 and 23,158 (1900).
32. P. Schützenberger, Ann. Chim., 21, 235 (1870), Ann., 160, 74 (1871).
33. G. B. Kauffman, J. Chem. Educ., 62, 1095 (1984).
34. "75 Jahre chemischer Forschung an der Universität Zürich: Festschrift zur Eröffnung des neuen Chemischen Instituts, verfasst von den Studierenden der Chemie," n.d., n.p. [Zürich, 1909], pp. 14-16.
35. E. A. Parnell, "Life and Labours of John Mercer," Longmans, Green: London, 1886.
36. E. Schweizer, (a) J. prakt. Chem., 67, 430 (1856), (b) ibid., 72, 109 (1857), (c) ibid., 76, 344 (1859).
37. E. Péligot, Compt. rend., 47, 1034 (1858).
38. E. Péligot, Pogg. Ann., 58, 83 (1860), 63, 343 (1861).
39. E. Schweizer, J. prakt. Chem., 78, 370 (1859).
40. O. L. Erdmann, J. prakt. Chem., 76, 387 (1859).
41. H. Reihlen, Z. anorg. Chem., 257, 340 (1948).
42. "Gmelins Handbuch der anorganischen Chemie," 7th ed., System-Nr. 60, Kupfer, Teil B, Lieferung 1, Verlag Chemid: Weinheim/Bergstrasse, 1958, pp. 118-132 and references therein.
43. F. Reinthaler, (F. M. Rowe, Transl.), "Artificial Silk," Chapman & Hall: London, 1928, Introduction.
44. L. G. Fauquet, "Histoire de la rayonne et des textiles synthetiques," Librairie Armand Colin: Paris, 1960, Chap. 1.
45. L. -H. Despeissis, French Patent 203,741 (May 9, 1890).
46. H. Pauly, German Patent 98,642 (December 1, 1897).
47. G. Fowles, "Lecture Experiments in Chemistry," 4th rev. ed., G. Bell & Sons: London, 1957, pp. 232-233, 617.
48. G. B. Kauffman, M. Karbassi, J. Chem. Educ., 62, 878 (1984).
49. J. Lundberg, A. Turbak, "Rayon," in Kirk-Othmer Concise Encyclopedia of Chemical Technology, Wiley: New York, 1985, pp. 998-999.
50. A. Turbak, in "Concise Encyclopedia of Polymer Science and Engineering," J. I. Kroschwitz ed., Wiley: New York, 1990, pp. 960-962.

HISTORY OF VISCOSE RAYON

TERESA A. SUMMERS, BILLIE J. COLLIER, JOHN R. COLLIER*, JANICE L. HAYNES

*School of Human Ecology, and *Department of Chemical Engineering, Louisiana State University, Baton Rouge, LA 70803*

Viscose rayon was the third manufactured fiber process to be developed, following nitrocellulose rayon and cuprammonium rayon. Nevertheless, viscose rayon became the most commercially successful and is now the only type of rayon produced in the United States. In this process, cellulose is dissolved by forming a xanthate derivative, and the fiber is spun into a precipitating bath. The viscose process has proved to be versatile, allowing the production of fibers with controllable properties, and thus offering a wide range of end uses from formal apparel to tire cords.

BEGINNING DEVELOPMENTS

The first important step in the development of viscose rayon was the discovery by Cross, Bevan, and Beadle of the xanthate derivative of cellulose in 1892 [13-15]. In conducting studies to elucidate the chemical structure of cotton cellulose, they exposed alkaline-treated cotton to carbon disulfide (CS_2) vapor. The product was sodium cellulose xanthate, an ester of dithiocarbonic acid:

$$\text{Cell-ONa} + CS_2 \longrightarrow \text{Cell-O-}\overset{\overset{\displaystyle S}{\|}}{C}\text{-SNa}$$

Upon addition of water, the compound swelled and finally dissolved. The researchers purified and redissolved the product to study its properties. Treatment of the solutions with acids, acid salts, sulfites, or metallic oxides yielded a regenerated cellulose, or what the authors at the time called a "dissociation" of the compound into cellulose, CS_2, and NaOH. In those days before the discovery of polymers, Cross, Bevan, and Beadle attempted in their work to determine the size of the cellulose molecule and from their observations noted that "...the acting unit of cellulose is not constant, and may become a molecule of very considerable proportions, at least 4 $C_{12}H_{20}O_{10}$" [15, p. 842]. The cellulose xanthate derivative was important for dissolving cellulose, and the authors recognized the potential utilization of their discovery in films, fibers, and other products for which nitrocellulose was being used. The cellulose xanthate in dilute alkaline solution was termed viscose, a contraction of the terms originally used to describe the solution, "viscous cellulose" [23].

As with nitrocellulose, the impetus for production of fibers from the new viscose solutions was provided by the electric lamp industry. At the Incandescent Lamp Works in Zurich, C.H. Stearn and C.F. Topham, using the technology developed for spinning nitrocellulose fibers, studied the spinning properties of viscose in 1895, to determine the usefulness of the viscose fibers in light bulbs. Stearn patented a precipitation bath for spinning viscose fibers in 1895 [40], which employed ammonium chloride (NH_4Cl) to coagulate the filaments. Later baths used sulfates to precipitate the filaments and sulfuric acid to regenerate the cellulose [29]. Regeneration with acid will precipitate the cellulose, but the reaction is slow, and so the sulfate is added to enhance the coagulation [38].

The larger market of textile fibers for viscose was soon recognized, and a Viscose Spinning Syndicate was established. This firm began a development effort at a new experimental plant of the Zurich Lamp Company in Britain. It was here that the second definitive step in the development of viscose rayon occurred: the discovery of the Topham spinning box. This was a device that enabled the simultaneous collection and twisting of fibers into a yarn [43]. Topham had noticed that the take-up bobbins, which had been adopted from the nitrocellulose spinning processes, did not allow the filaments to be later rewound without entanglement [19]. He postulated that the addition of twist during spinning would eliminate this problem and determined to develop a twisting device. He soldered a blacking box (shoe polish can) to a spindle and added this to the spinning line, bringing the coagulated filaments from the spinning bath, up through a guide, and dropping them into the rotating can. The centrifugal force wound the fibers around the can surface and built up a cake of filaments from the outside in. Thus, the fibers could be collected and twist imparted by the rotation of the can. The blacking box was soon replaced by an acid resistant can, and a funnel was added above the spinning can to traverse up and down depositing the filaments evenly inside the can.

The patent rights to the viscose process developed by Cross and his partners, and subsequent improvements patented by Stearn and by Topham, were sold separately to different countries at the end of the 19th century, and production facilities for viscose rayon fibers were soon established in Germany, Britain, France, Italy, Holland, and the United States. The viscose process, being cheaper and safer than the previously discovered methods for spinning rayon fibers (nitrocellulose and cuprammonium rayon), soon replaced these in terms of production volume. As the viscose process developed, it became apparent that viscose rayon was a more versatile fiber with application in a variety of end use products.

FIRST VISCOSE PROCESSES

Viscose production attempts had been made in the United States even before the patent rights were sold. In 1900 in Boston, MA, A.D. Little, a friend of Cross and Bevan, made several attempts to spin viscose yarn but was unsuccessful and abandoned this venture. Little then became chemical director of the Cellulose Products Company organized by Daniel C. Spruance in Philadelphia, PA, and the company acquired the American rights to the viscose patents. Even with possession of the patent rights, the company went into bankruptcy [16]. The company's lawyer, Silas W. Pettit of Philadelphia, acquired the patent rights as payment for his legal fees and set up the General Artifical Silk Company (Genasco Silk Company) in Lansdowne, PA. The company produced approximately 500 pounds of viscose rayon a day for about four years but was ultimately unsuccessful and closed in 1909. Pettit died that same year and willed the patent rights to his son who sold the rights to Courtaulds for approximately 31000. Courtaulds established The American Viscose Company in Marcus Hook, PA, under the direction of Samuel Salvage in 1910 and produced its first shipments of viscose rayon in 1911 totalling 375,000 pounds selling at $1.85/pound. [1; 3, p. 137; 12]

Courtaulds acquired the British rights to the viscose patents in 1904 and set up a production site near Coventry. The facility steadily grew and increased its production capability and by 1913 was producing over 3 million pounds of viscose rayon annually.

This facility along with their wholly owned subsidiary in the United States made Courtaulds the largest and strongest of all firms that had purchased the viscose patent rights [12].

In these first commercial processes, production of viscose rayon fibers began with purified cellulose. The most commonly used raw material was sulfite wood pulp. Other forms of cellulose, wood pulp from sulfate processing or cotton linters, were not deemed as suitable because they gave inferior products or were too expensive [34, p. 56]. Cotton linters, however, were used by some firms in the United States.

As Cross, et al. [14,15] had determined, and others subsequently confirmed [31, p. 1684], it is necessary to react the cellulosic raw material with alkali before forming the xanthate. The first step, therefore, in commercial production of viscose fibers was soaking of the wood pulp, or other cellulosic material, in sodium hydroxide (NaOH). The concentration of the alkali is critical, and the optimum concentration was determined to be 18% [34; 21]. In practice, an excess of this concentration of alkali was used to assure uniform penetration of the solution into the cellulose sheets, and because the xanthate produced later in the process was dissolved more readily with an excess of alkali in the material. Originally the cellulose, in the form of pasteboard sheets, was placed in steeping vats with the alkali or in disintegrating machines that would break up the cellulosic material as it was sprayed with the alkaline liquor. After several hours, the excess hydroxide was squeezed out. Later processes used specially built vessels in which the pasteboard sheets could be placed vertically and separated by perforated iron plates. After steeping, the sheets were pressed to remove the excess alkali.

Because of the large volumes of alkaline solution used, it was important to recycle the alkaline steeping liquors. All but the final expressed solution could be regenerated by adding fresh NaOH. This more contaminated final solution, containing extracted hemicelluloses, was purified by oxidizing or precipitating the organic impurities or by dialysis.

The alkali cellulose from the above processes was shredded, and then held in covered iron vessels for 3 to 4 days at 20° to 25°C. This "ripening" or "ageing" resulted in a degradation or depolymerization of the cellulose. Attempts were made in some of these early production processes to increase the oxidative degradation of the cellulose with oxidizing agents such as peroxide and hypochlorite, but it was soon determined that oxidation with atmospheric oxygen was sufficient and easier to control [21, p. 269].

The next production step was treatment with CS_2 to form the cellulose xanthate. This reaction was carried out in heavy rotating vessels, or churns, around which cooling water was circulated because of the exothermic nature of the reaction. The crumbs of aged alkali cellulose were placed in the churns and the CS_2 added from an external measuring vessel by gravity flow. Cross, et al.,[15] in their original work used a 1:1 molar ratio of CS_2 to cellulose, and this was followed in many of the early production processes [21, p. 275]. Later, however, smaller quantities of CS_2 were used successfully. One of the by-reactions of xanthation is the formation of sodium trithiocarbonate, Na_2CS_3. This compound and other sulfur containing impurities impart to cellulose xanthate its characteristic orange-yellow color and its odor. After 3 to 4 hours, the excess CS_2 was removed from the churn by aspiration, and the xanthate dissolved with addition of NaOH and water.

The solution of cellulose xanthate was then filtered and transferred to ripening chambers. Some producers at this point purified the cellulose before the next ripening step by precipitation, washing, and redissolution in alkali. This was done to remove many of the sulfur containing impurities, but it was found that these steps were not economically beneficial in terms of the quality of the final product, and so these purification steps were eliminated.

The ripening step that followed was very important in achieving the optimum viscosity for spinning of the viscose solution. The xanthate is gradually decomposed during the ripening process with regeneration of cellulose, and subsequent liberation of CS_2. The course of this conversion was monitored, principally by the method of Hottenroth, which determined the volume of 10% NH_4Cl that would precipitate 20 g of viscose in 30 mL of water. The lower the requirement of NH_4Cl, the riper is the viscose. Usually, the ripening was carried out in temperature-controlled chambers for 2 to 5 days.

The ripened viscose was conveyed to spinning tanks, vacuumed to remove any air bubbles, filtered again, and fed to the spinning machines. The spinning apparatus, adopted mainly from the apparatus used for other manufactured fiber processes of the time, consisted of eight parts, as shown in Figure 1. These were: 1) viscose supply line; 2) spinning pumps with a common shaft; 3) filters; 4) glass viscose delivery tubes; 5) spinning nozzles or spinnerets; 6) coagulation trough or bath; 7) the formed filaments; and 8) a take up mechanism.

The viscose supply line fed a number of spinning lines each of which has a spinning pump, tubes, and spinnerets. Both piston and gear pumps were used originally, but the gear pumps soon became more widely used because of their more even action. A common shaft was used for all of the pumps to maintain the same output from each. The filter most often used was the candle filter developed by Topham [42], which consisted of a cylindrical case containing a filter of cotton batting and fabric. The filter was located in the spinning line between the pump and the spinneret and served to remove impurities and aggregates. For corrosion resistance glass tubes were used to carry the viscose from the filter into the acid bath and to the spinnerets.

The spinnerets usually consisted of a metal nozzle with at least 12 to 40 perforations through which the filtered viscose solution was extruded into the acid spinning bath. The perforations were on the order of 100 μm in diameter and were arranged in concentric circles. The nickel spinning nozzles used for the other rayon processes could not be used for viscose rayon production because of the corrosive nature of the sulfuric acid coagulation bath. Therefore, the viscose spinnerets were made of platinum or gold alloyed with platinum or palladium. Some producers even experimented with spinnerets of molybdenum or porcelain. The porcelain spinnerets were made with up to 1,000 spinning holes but were susceptible to damage if carelessly handled or dropped [34, p. 72]. The coagulation bath in which the spinnerets were immersed, was often in the form of a trough and was common to all of the spinning lines on a spinning machine.

During the development of these early spinning processes, some producers experimented with revolving spinnerets to introduce twist to the yarns. This innovation was not practical, however, because the revolving nozzles produced agitation in the coagulation bath which in turn caused breakage of some of the recently formed filaments. Further the twisted fibers, often not fully coagulated, would stick together. Since the centrifugal spinning box imparted sufficient twist, the further development of revolving spinnerets was abandoned.

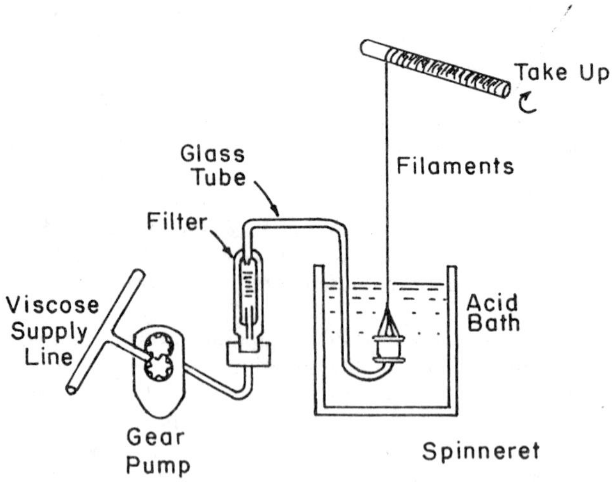

Figure 1. Schematic of spinning apparatus.

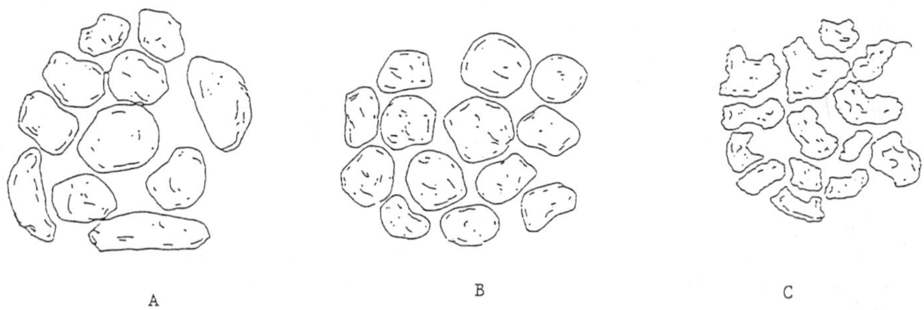

Figure 2. Cross sections of early rayons.
(A) Cuprammonium (B) Viscose (C) Viscose spun in acid solution with excess of salt.

Several types of collection devices were used for production of viscose fibers: reeling, bobbin spinning, and centrifugal (i.e. box or pot) spinning machines. The reeling machines were particularly appropriate for making staple fibers since the take up reels were designed to be collapsible for fiber removal in hanks. Because the viscose fibers tend to shorten in length upon drying, the collapsing reels permitted easy removal of fiber hanks. Bobbin spinning machines used multiple, simultaneous take up devices called bobbins that, because of their small diameter, could rotate at high speeds without being especially sensitive to slight irregularities. These machines were used primarily for simultaneous production of higher denier fibers. The centrifugal spinning machines, attributed to Topham, had the advantage as mentioned earlier of imparting twist simultaneously with take up. Since the yarns were taken up, washed and dried on the inside of the box or pot, they developed less tension during drying. Conversely, the yarns tightened on the reeling or bobbin machines because they were wound on the outside of the take up. Because of the weight of the centrifugal boxes or glass bobbins, men were usually employed in production with these spinning machines [34, p. 84]. However, women workers could handle lighter weight aluminum or pasteboard bobbins and reels.

An important part of the viscose process was the finishing treatment applied to the viscose yarns. This finishing treatment consisted of an initial washing, a desulfurizing step, a bleaching step, and a final washing. Desulfurizing was done with sodium sulfide (Na_2S), to remove any sulfur or sulfur containing compounds in the fibers. The yarn was then bleached to a pure white, usually with sodium hypochlorite ($NaOCl$). The final washing was with a dilute acid solution to neutralize any residual alkaline solution remaining from the bleaching step. These purification steps were usually done on the yarn cakes produced by the centrifugal spinning machines or on the bobbins or hanks produced by the other spinning apparatus.

COMMERCIAL DEVELOPMENTS

The first rayon fibers were truly a wonder to early 20th century consumers. Their silk-like appearance and hand assured their success, yet sometimes obscured their disadvantages. These fibers, being filaments, were lustrous and smooth. However, they also exhibited extensive swelling in water, resulting in high shrinkage of rayon fabrics. The shrinkage and swelling were higher than that of cotton fibers because the rayon fibers were less oriented and crystalline than cotton, allowing higher moisture absorption. Another significant disadvantage of rayon was its low strength, generally less than half that of silk. There was very little stretch applied to the filaments during production, and therefore, the fibers were relatively unoriented. Hottenroth [21, p. 356] speaks of the strength of a rayon yarn as being about 30% greater than the denier of the yarn. For example, a 100 denier yarn would normally have an absolute breaking strength of 130 g. Rayon fibers had even lower strength when wet. Finally, viscose and other early rayon fibers had low elastic recovery and bending recovery. This latter property, however, was probably not seen as a serious disadvantage, because other commonly used fibers at that time, cotton, flax, and silk, also had low wrinkle recovery.

The cross sections of these first viscose rayon fibers were slightly irregular and similar to the cross sections of cuprammonium and of silk. When the fibers were spun into an acid bath, which also included an excess of neutral salt, smaller irregularities developed producing cross sections which were serrated and had a greater surface area as shown in Figure 2.

Because of viscose rayon's high luster and limited suitability for weaving purposes when it first appeared on the market at the turn-of-the-century, early uses included lustrous tassels and pom-poms, knitted ties, scarves, and ribbons. Many consumers enjoyed the fiber despite the fact that it required special care techniques, and it was widely accepted by those who could not afford silk. However, the greatest early success was the use of viscose rayon in the production of women's hosiery. The replacement of black cotton stockings with rayon hosiery is partially credited with changes in fashion and the rising hemlines of the first decade of this century. These developments were "... important because it brought 'silk' stockings within the reach of the mass markets for everyday wear, pointing out vast commercial possibilities for rayon" [2, p. 47]. American knitting mills were already using imported viscose rayon in the production of women's hosiery in 1908 and by 1912 were beginning to use domestically produced rayon. These mills significantly increased their consumption of domestically produced rayon during World War I as difficulties increased in importing rayon. The hosiery market continued to act as a stimulus for the growth of rayon after World War I.

In 1916, the market for women's knitted rayon outerwear and underwear opened, creating competition for rival natural fibers. The all-rayon knit fabric called tricolette gained immediate market share in women's dresses even though its construction was simple compared with today's knit garments, and due to the low elastic recovery of the fiber, it tended to sag and stretch in finished garments. By 1920, tricolette was being widely used in underwear as well.

The American Viscose Company, reorganized in 1915 by Courtaulds and given the new name of the Viscose Company, prospered during World War I and in the years following the war. Demand for viscose rayon was so great that Courtaulds opened its second United States plant in 1917 at Roanoke, VA, selling viscose yarn for $4.00/pound. In 1918, annual viscose rayon production in the United States was at 8 million pounds, and the fiber was selling for $5.50/pound [27; 11].

The United States economy flourished following the war, and this economic strength combined with the expiration of the basic patents for the viscose process and rapidly changing fashions of the Roaring '20s encouraged the development of new markets for viscose rayon and expansion of existing markets. In 1920, E.I. du Pont de Nemours Inc., of Wilmington, DE, established its first viscose rayon production plant in Buffalo, NY, and the National Artificial Silk Company of Cleveland, OH, later known as the Industrial Rayon Company was organized. Courtaulds' United States subsidiary, the Viscose Company, continued to grow, and additional United States facilities were constructed in Pennsylvania and West Virginia. Courtaulds also expanded its production facilities in the United Kingdom. The first loom to weave rayon fabrics was built in the United States by the Draper Corporation in 1921.

NAMING OF RAYON

Until 1924, all manmade fibers including viscose rayon were known as artificial silk and were promoted and sold under brand names that dominated the market, leading to widespread confusion of the general public. Since these fibers had properties and uses far different from, and in some instances superior to, silk, and were different from each other, clarification of terms used to describe the fibers became imperative.

Despite public knowledge of viscose rayon fiber, producers feared the qualitative effect of "artificial" lent their product negativity, which could only dissuade consumer approval and confidence. Many promoters felt that by attaching the prestige of worm silk to the fiber its reputation would be enhanced. Thus, "fiber silk" was their choice for the new name. Others suggested "art silk", short for artificial silk, and wrote it stylishly as "art SILK". Some shortened the "art" to a simple A, producing the term "A SILK". Still other titles were considered: GLOSS, GLISTRA, FILATEX, FILACEL, FLOS, GLOWTEX, and even KLIS, or silk written backwards. Identifying rayon fibers grew increasingly complicated for producers and consumers [10].

A letter to the Editor of the *New York Daily News Record*, written by Herman Blum, calling for a single name for rayon was the impetus needed for a name search. Mr. Blum, who ran an upholstery fabric mill, denounced the term "Art Silk" as misleading and insisted on a new term. The National Retail Dry Goods Association offered him the chance of renaming the fiber. He organized a committee, which included the presidents and owners of many retail and fiber-producing companies, including Macy's, duPont, Industrial Fiber Company, Galey & Lord, Inc., and was chaired by Samuel Salvage, president of the Viscose Company. They gathered with the intention of naming the fiber, and after a few meetings, they decided to create a nationwide contest with a $1000 prize for the winner. The committee made the final decision from the 10,000 suggestions received within the month, including one offered by Thomas Edison. The decision on a name was followed closely by newspapers and trade associations and attracted much public attention [10].

The name agreed upon was "rayon", a French term meaning "a ray of light," which had been submitted by Kenneth Lord of Galey & Lord, Inc. [10]. Lord suggested that the term "conveyed the meaning of the radiance of bright sunshine tempered with the soft glimmer of rippling waters in the moonlight" [20, p. 11]. Some say it derived from "...'ray,' due to its sheen, and 'on,' to suggest a fiber such as 'on' in cotton" [26, p. 5]. Producers and consumers adopted the name, and in 1926, the American Society for Testing Materials (ASTM) approved the first official definition of rayon:

> Rayon (formerly known as 'artificial silk')--A generic term for filaments made from various solutions of modified cellulose by pressing or drawing the cellulose solution through an orifice and solidifying it in the form of a filament, or filaments, by means of some precipitating medium [36, p. 194].

A rayon institute was formed at this time, which later became the Man-Made Fiber Producers Association, and with the Silk Association began policing the world industry to control use of the word, rayon. The first industry-compiled data concerning rayon were also made available to the industry in 1926.

INDUSTRIAL GROWTH

Rayon experienced a significant growth period after 1925, and many new plants were established in the United Kingdom, Europe, Japan, and the United States. The rapid production expansion resulted in a flooding of the market with rayon of all qualities. A brief depression in rayon sales resulted from consumer response to the inferior quality rayons that had been introduced. This caused general quality improvements to be made quickly in order to regain market momentum. The North American Rayon Corporation and ITT Rayonier were founded in 1926. Courtaulds opened another plant at Parkersburg, WV, and the Franklin Rayon Corporation was founded in Rhode Island in 1926. In 1928, American Enka Corporation was established. The name was derived from the pronunciation of the initial letters of the parent Dutch company, Nederlandsche Kunstsijdefabriek. N.K. is pronounced En-Ka in Dutch. The plant was located in Asheville, N.C. and began operation in 1929. Chatillon also constructed a plant in Rome, GA. Even with increasing competition from new companies entering the rayon market, the Viscose Company was regarded as the world's leader in the production of rayon in 1928 with 54 million pounds [11].

Improvements in the production of viscose rayon continued as consumer demands increased and fiber prices fell, and during the 1920s, 1930s, and 1940s several important developments were made. These included unlimited variations in fiber luster; staple length fiber production, which enabled rayon to be used on cotton processing equipment; high twisted yarns for use in crepe; economical continuous spinning processes; and stretch spinning to produce high tenacity rayon.

In 1926, the Viscose Company first created dull viscose yarn by adding a delustering agent such as TiO_2. The yarn was sold under the tradename, Delusco, and the name, rayon, was restricted to shiny fabrics because of the high luster of their yarns. Creation of a dull filament yarn allowed the production of sheer, flesh-colored women's stockings. Soon, a wide assortment of fiber lusters was developed, which permitted the production of fabrics from dull finished broadcloths to shimmering satins.

Experimentation with the use of staple length rayon had begun during World War I, when the Germans began cutting rayon into short staple lengths as a substitute for wool. Fibro, a rayon staple fiber first manufactured by Courtaulds, was introduced in 1918. In 1934, the Viscose Company began the first widescale production of viscose rayon staple in the United States based on the process developed earlier by Courtaulds. By 1936, the Viscose Company was producing 270 million pounds of Fibro annually [3]. The staple fiber was produced using the same technology as the filament fiber, except that the filament fibers were cut to desired end-use lengths as they left the spinnerets. Rayon staple fibers permitted the spinning of 100% rayon yarns as well as multiple blending possibilities with other staple fibers. Staple fiber production soon expanded to include a wide range of fiber lusters, filament sizes, and staple lengths. The production of rayon staple fibers was especially strong in Germany, Italy, and Japan due to the respective governments' support. Wear testing of garments produced from rayon staple fiber began in the United States in 1936. The testing project was a joint effort of duPont, Sears, and Montgomery Ward and involved the distribution of 3,000 garments to consumers for evaluation. In 1937 the first United States facility was established for the sole purpose of production of rayon staple at Nitro, WV.

In 1928 the Viscose Company introduced a method for crimping viscose yarns into high-twist crepe fabrics. Because of its high moisture absorption, rayon was ideal for production of crepe fabrics with latent crimp. Highly twisted rayon yarns were woven into fabric under tension, and when moisture was applied, the fabric developed extensive crimp yielding a crepe texture and good drape. These fabrics appealed to designers and consumers of women's apparel.

Development of a continuous spinning/finishing process contributed to cost effective production. Since the first commercial facilities were based on batch processes, regardless of which take-up mechanism was used, there was an economic incentive for development of a continuous process. The problem was that putting the extensive post-spinning treatments in-line would require the spinning line to be longer than a mile [28,

p. 184]. To overcome this, one of the first continuous processes eliminated the desulfurizing step. It was maintained that this could be done since the residual sulfur in the fibers was very low and could be removed by routine finishing of the final fabric. Bleaching was also eliminated as it too could be done at the fabric stage. The finishing time was decreased by washing the fibers in yarn form rather than as cakes. Finally, the length of the continuous line was shortened, yet the residence time in various stages maintained, by passing the yarns over rollers a number of times. Yarns were wound around two rollers placed skew to each other, and as they advanced down the rollers, they were washed.

Another continuous process retained the desulfurizing and bleaching steps, by multiple passes of the yarn ends around small reels. The reels were stacked in decks, with the different finishing stages occurring as the fibers progressed through the sets of reels in each deck. These continuous processes, since they used single yarns, necessitated the use of bobbins for spinning, which replaced the centrifugal spinning boxes at many production facilities.

In an effort to increase the strength of rayon yarns, stretch spinning processes were developed. These processes involved stretching the not fully coagulated filaments between two rollers, with the second roller revolving at a higher speed than the first. The molecules within the fibers were oriented by the stretch occurring between the rollers, imparting a higher strength to filaments. The stretch spinning processes were used frequently for cuprammonium rayon, but were not at first considered feasible for viscose rayon production because the precipitation of the filaments in the acid bath was so rapid that more than a slight stretch would break the fibers.

Much of the credit for adapting stretch spinning to viscose production goes to Lilienfeld [24], whose process was later used by Courtaulds and other European producers [34, p. 91; 28, p. 263]. He suggested the use of spinning solutions with a higher viscosity, but a lower concentration of cellulose, which could be stretched before complete regeneration in the bath. The higher viscosity solutions were achieved by less ageing of the alkali cellulose in the first step of viscose production, by reaction with larger amounts of CS_2, and/or by decreasing the ripening time. The most significant feature of Lilienfeld's process was the use of much stronger acid baths, which formed a sulfate ester with the xanthate [38; 32, p. 416]. This compound could be stretched to a considerable extent before the cellulose was regenerated, allowing more orientation within the fibers and resulting in very fine filaments, which could not be attained with regular viscose processes. A further advantage was that, since the filaments were being significantly attenuated during the stretching process, spinnerets with larger holes could be used to produce fine fibers. The larger holes were less likely to clog during spinning, reducing fiber breakage or defects. Courtaulds' Durafil fiber was based on this process.

It was soon determined that the structure and resulting properties of viscose rayon could be altered by changing the viscose preparation or the spinning bath. This realization led to development of a number of modified or high performance rayon fibers beginning in the 1940s [35]. The Truman Commission named rayon as superior to cotton for fiber reinforcement in tires in 1944. This provided impetus for the development of viscose rayon fibers with increased tenacity and toughness.

An old technology that had been presented in the early part of the century [30], addition of zinc sulfate ($ZnSO_4$) to the spinning bath, was revived. When present in an acid regeneration bath for viscose, $ZnSO_4$ formed a more stable compound with the cellulose xanthate, retarding its regeneration to cellulose. The zinc xanthate compound could be stretched, orienting the molecules before conversion to cellulose. In the early processes, such as those used to produce Tenasco and Cordura, formation of zinc xanthate was usually confined to the outside of the filament, and so the oriented structure formed in this outer layer resulted in a "skin". The inner portion, or "core", of the filament was regenerated to cellulose, with subsequent removal of water. As this happened, the core decreased in volume, and the skin collapsed around it resulting in a crinkled or serrated cross section for the fibers. The skin had higher orientation, but smaller crystalline areas, and exhibited higher tenacity and elongation and lower swelling than the core.

The advantages of this skin structure, especially for reinforcing fibers in rubber tires, were soon recognized, and producers began searching for techniques to increase the ratio of skin to core in viscose rayon fibers. The most successful methods utilized polyethylene glycol and amine modifiers added to the viscose solution before spinning. These modifiers greatly retarded the regeneration process by slowing the diffusion of acid and allowing the zinc to penetrate further into the filament. The spinning bath was heated and a high degree of stretch applied to orient the polymers before complete regeneration. The result was fibers, which were all skin with rounder cross sections. They exhibited good strength and toughness.

Although the all skin fibers produced by the above process had high orientation and strength, their modulus or stiffness was lower. This was due to a structure with small crystalline areas, but high orientation. Core structures, on the other hand, had larger crystalline areas, but lower orientation. Modifications of the bath and spinning conditions could produce fibers with different amounts of skin and core, giving varying modulus values. These processes generally used the viscose modifiers, but less zinc, slower spinning speeds, and lower bath temperatures [39]. Under those conditions, thick, highly oriented skins were formed, while the core structure had larger crystallites to provide more stiffness. These fibers were termed high wet modulus (HWM) rayon.

Daul [16] distinguished between several types of modified rayon fibers. The all skin fibers were high tenacity, low wet modulus. Fibers with varying amounts of skin and core were designated as either intermediate wet modulus (IWM) or HWM. The higher modulus fibers of these two were made with viscose solutions with a higher degree of polymerization (DP). Therefore, the HWM fibers had better strength as well as higher modulus. However, most fibers, which exhibit a higher modulus are, now called HWM.

A final type of modified rayons were the polynosic fibers, which are often classified as HWM. The first of these fibers was Toramomen, introduced in Japan by Tachikawa, and later patented in the United States [41]. The process was similar to some of those described above in that viscose with a higher DP was used. No modifiers were added to the viscose, which was spun into a weak acid bath with a low concentration of sulfate and a lower temperature. All these factors retarded the regeneration rate, allowing a high degree of stretch. The filaments were then passed through a hot acid bath, where the cellulose was completely regenerated. The polynosic rayons are more crystalline than the other HWM types, and although the crystallites are larger, there is a high degree of orientation within the fibers [35]. Resulting properties are high strength as well as modulus.

The viscose solutions used for most of these high performance fibers were not aged, and therefore, the cellulose maintained a higher DP. There was more CS_2 added during xanthation than was usual for regular viscose production to assure dissolution of the larger size molecules. In addition, the ripening step was omitted, giving a viscose that was not as readily coagulated in the spinning bath.

Fabrics of HWM and polynosic rayon fibers were said to be more similar in properties to cotton, unlike the earlier rayons. Because of the higher modulus, they were crisper and not as limp. They also exhibited less swelling in water, due to their higher molecular orientation. The high tenacity rayon fibers were used for tire cords, where their higher strength and toughness were important.

MARKET POSITIONING

In 1937, the Viscose Company changed its name to the American Viscose Corporation, and the Franklin Rayon Corporation was renamed the Atlantic Rayon Corporation. The United States continued to dominate world production of viscose rayon prior to World War II followed by Japan, Germany, the United Kingdom, Italy, and France. Avtex's Front Royal, VA, plant opened in 1939.

In 1941, Courtaulds was forced to sell its holdings in the American Viscose Corporation. Because of Britain's extensive armaments purchases on credit up to 1940, the United States Secretary of the Treasury suggested to the British Treasury representative in Washington that Britain begin selling some of its wholly owned industrial assets in the United States. The American Viscose Corporation was Britain's largest holding, and Courtaulds had steadfastly refused the sale of shares of company stock to American interests. Pressure from the American government increased, and in 1941, the American Viscose Corporation was acquired by an American banking consortium. Shares in the corporation were later sold to the public [12].

During World War II, rayon took the place of cotton as the filament in tire cord, although steel and polyester replaced it soon after the war. Rayon was used in many military applications during the war in addition to tire cord; including self-sealing tanks and hoses, flares, fragmentation bomb and cargo delivery parachutes, navy flotation suits, linings for pilot precision gloves, and blood plasma filters. Rayon's use in women's hosiery continued during the war but was quickly replaced by nylon after the war because of the latter's higher elastic recovery. Germany and Italy also used rayon staple extensively as a substitute for cotton and wool due to extreme shortages of natural fibers during the war. Rayon's misuse as a general substitute for all natural fibers adversely affected its performance qualities and led to serious marketing problems for the fiber for several decades after World War II.

Despite increasing competition from the newly emerging synthetic fibers and consumer dissatisfaction with the fiber because of its misuse during the war, United States rayon production continued to expand after the end of World War II. In 1947, rayon production exceeded 975 million pounds representing about 48% of world production [8]. The continued growth and expansion of the rayon market was due, in part, to the continued technical refinements, which resulted in high performance fibers and to intense advertising campaigns by major rayon producers and users promoting the positive aspects of the fiber. Rayon fiber was described as being thinner and much stronger than ever before, increasing its wearability and appeal to consumers. Burlington Mills advertised,

"Because rayon's lustre can be planned and controlled, rayon goes into 2/3 of all women's dresses!...Practically every other characteristic of rayon can be controlled, too. For rayon has no fixed form or 'natural' state nor is its quality dependant on unpredictable plants or animals.... 25 years ago Burlington was already weaving rayon into new and different fabrics. Today, Burlington research and experiments go right on. Progress is amazing. For almost every day rayon shows new aptitudes." [4, p. 67].

By 1948, demands were beginning to be made for increased clarification of the use of the term "rayon" and the labeling of products made from rayon. Tennessee Eastman Corporation claimed that viscose rayon and acetate rayon were very different, and hence, no longer could the term "rayon" be used for the two fibers. An advertisement in 1949 appealed to the public to demand the term "rayon", which at that time referred to all cellulosic manufactured fibers, be limited. The ad went on to note that many shoppers were confused by the use of the same term for two fibers that exhibited such different strength, thermal, and wrinkling properties [5]. The manufacturers of wool fabrics and garments demanded the mandatory labeling of rayon suitings to assist consumers who were supposedly experiencing confusion from the advertising of rayon suitings, which used the term "worsted types". However, manufacturers of rayon suitings countered the claims of the wool industry by noting that the manufacturers of better rayon clothing labeled their garments as 100% rayon and emphasized the word rayon in advertising [8].

The beginning of the 1950s brought continued growth and success for rayon producers. Companies maintained promotional campaigns stressing the dependability and quality of rayon fabrics. Consumers valued its reasonable price, the positive and informational advertising, and the availability and adaptability of rayon. United States rayon production for 1950 was slightly over 1 1/4 billion pounds, an increase of 27% over 1949, and world production was 3 1/2 billion pounds [6; 25; 8]. Many United States producers were reinvesting part of their profits in plant expansions since they anticipated continued growth in the market for viscose rayon. In January, 1952, approximately 70% of the 1 billion pound annual United States production of rayon was located in the South, in the Carolinas, Georgia, Tennessee, and Virginia [9].

To end the continuing confusion concerning the terminology used to describe rayon, the Federal Trade Commission (FTC) in 1951 officially designated new names for acetate rayon and viscose rayon, which were acetate and rayon, respectively. In addition, producers decided that rayon would no longer be called a synthetic due to the nature of its production. One alternative was to create a new word, or simply "to make no reference to the origin of rayon...but to emphasize ...(its) virtues" [7, p. 110b]. In January of 1952, the Rayon Yarn Producers Group adopted the new name of rayon in accordance with the revised FTC rules. The FTC announced the adoption of the following generic definition for rayon in 1959: "A manufactured fiber composed of regenerated cellulose, in which the substituents have replaced not more than 15% of the hydrogens of the hydroxyl groups" [17, p. 83]. In defining the fiber in such a way, no distinction was made between the different kinds of viscose rayon, which had been invented, like modified viscose fibers with increased strength and reduced shrinkage. Due to this fact, companies such as Courtaulds, American Viscose, and Enka decided to actively promote their specialized rayon products, Corval, Avril, and Fiber 700, respectively.

Courtaulds reentered the United States market with the establishment of a rayon staple plant near Mobile, AL, which began operations in 1953. The facility was the largest rayon production site in the world when constructed. Since the relinquishment of its ownership of American Viscose Corporation, Courtaulds had been shipping its British-made rayon to the United States.

Rayon came under such heavy criticism by consumers during the mid- to late-1950s that the industry suffered a severe slump. Many consumers perceived the low price of rayon as its only purchasing advantage. A cycle was created: rayon was purchased because of its low price, but the performance was very poor, so its price would drop even further. When nylon, polyester, and acrylic were introduced, many rayon producers abandoned the fiber, among them, duPont, which completely phased out its rayon production. One producer even attempted to introduce a modified rayon with a different, unassociated name. The decline in popularity was attributed to the lack of further technological improvements in the fiber to keep it competitive with the newly introduced synthetic fibers.

Rayon's popularity slump continued into the 1960s. ITT Rayonier, American Viscose Corporation, and other major producers worked to stimulate interest in rayon by actively promoting the results of their research and development campaigns that had resulted in high performance viscose fiber. Many rayon fabrics were stabilized by compressive shrinkage, which overcame the progressive shrinkage tendencies of older rayons. New marketing efforts promoted these new fibers and dispelled prejudices attached to the older, poor performance rayon. In response to consumer demand and anticipation, new end uses were devised. The industry also began to search for non-viscose methods of production because of the capital intensity of traditional production methods.

Blends of rayon with polyester, triacetate, and acrylic fibers were increasingly popular in the late 1960s and presented advantages over "regular rayon". The polyester/HWM rayon blend was very popular with the dyeing and finishing industry at this time, but was not able to penetrate the volume yardage market. This blend was used to make fabrics that were traditionally made of all cotton or polyester/cotton blends such as batiste, corduroy, denim, and duck. At the end of the 1960s, an estimated 15-20% of all viscose fibers were of the HWM type.

The disposable market of the 1960s opened up new marketing possibilities for rayon in fashion apparel. American consumers were demanding freedom as well as fashion in their wardrobes. Disposable clothing was viewed as a viable alternative to buying clothes that eventually sat in the closet after being worn once. Wedding and graduation gowns were seen as a good market for disposable garments since these are worn only once. Disposable rayon was described as wearable and not unattractive, but not so luxurious that the consumer should be bothered when throwing it away.

Further expansion of rayon technology was slowed in the early 1970s by the many production problems encountered concerning pollution and waste disposal. However, several advances were made during this period in the preparation and spinning of rayon, all which were less costly. Some rayon fiber users believed that the rather limited definition of rayon was to blame for the lack of development and research. In the 1970s, when HWM fibers such as Avril and Fiber 700 were marketed, "extensive advertising campaigns and promotions of these fibers were necessary to overcome the handicap of the name rayon" [17, p. 83]. A new generic term, "modal", was created by the International Standards Organization (ISO) to distinguish between regular rayon and the newer, improved fibers that included HWM. Modal referred to "...regenerated cellulose fiber with wet tenacity of at least 2.5 grams/denier at 15% maximum elongation..." [17,

p. 84]. Since modal fiber had properties similar to cotton, it was often used in cotton or polyester blends.

Even with the introduction and marketing of HWM, rayon continued to lose ground to other fibers in certain markets. Rayon's use in disposables continued to grow because of its advantages over cotton. In this use rayon has better moisture absorption, does not need bleaching, and is more uniform than cotton.

By the mid-1970s many rayon producers were forced to reduce production due to changing market needs and a serious over supply problem, and rayon production dropped 46%. Public interest shifted from the lustrous look of rayon to a more natural hand and, in turn, cotton and cotton blends were the most popular fashion. American Viscose Corporation ceased production of textile-grade rayon filament yarn and closed its facility at Parkersburg, WV. The FMC Corporation had acquired the American Viscose Corporation in 1963. In 1976, Avtex was formed from a merger of the FMC Corporation's Fiber Division with Avtex Fibers, Inc., a privately owned company. The merger gave Avtex 50% of the total United States rayon capacity and made it the second largest producer in the world. The number of United States rayon producers continued to decline in the 1970s, but facilities were being expanded in other locations throughout the world: Spain, Portugal, France, Taiwan, South Korea, The Philippines, Egypt, Malaysia, Japan, and the Eastern bloc countries.

RESURGENCE OF RAYON

A noticeable resurgence in rayon use occurred during 1976-1977. The first markets to realize its new capabilities were junior dresses, blouses, and sportswear. Since these markets catered to the young, wrinkle resistance was not a prime concern. More importantly, this was a "new" idea for most consumers, and because of its good drape rayon became very fashionable, especially spun rayon challis. The new attention sparked other markets' interests, both domestic and international, and the exportation of United States-made rayon expanded. Demand increased in the "West" for drapable, soft, disco-style clothes, and in the "East," viscose/cotton blends were popular. Major uses of rayon in apparel in 1976-1977 were: athletic uniforms, blouses, career apparel, children's wear, coats, dresses, fake furs, jackets, jeans, lingerie, linings, rainwear, robes, shirts, skirts, slacks, sportswear, suits, ties, and work garments. Major uses in home furnishings were in bedspreads, curtains, draperies, sheets, slipcovers, tablecloths, and upholstery. Major uses in other items included: air hoses, filters, medical products, military uses, non-wovens, stuffed toys, teabags, tents, tire cords, and v-belts.

Research and marketing efforts increased significantly from 1978-1981 in an attempt to attain several goals, most notably to expand the market for rayon by educating consumers, improving its efficiency in production, and by supplementing cotton production with better rayons. To facilitate consumer education, the Rayon-Acetate Council was created to disseminate information to educators. Continued modifications of regular rayon were made to increase its moisture absorbency, and improvements in the HWM rayons produced new fiber variants. America Enka's Absorbit and Avtex's P.A. were two new super absorbent regular rayon fibers considered ideal for underwear, socks, jogging suits, and non-wovens. Three of the new cotton-like HWM variants included Viloft, Prima, and Avril III. Courtaulds' Viloft was less dense than ordinary rayon and was designed with a hollow center, much like cotton, that remained hollow when wet. Prima, marketed by ITT Rayonier, had a spiraled, bilobal cross section, low-amplitude crimp, high strength, high wet modulus, and resistance to caustic soda. Avtex's Avril III had a modified multilobal cross section, a dulling agent designed to create opaqueness in fabrics, and a more cotton-like, crisp hand. When these fibers were blended with polyester a wide variety of goods could be produced: challis, linen, knits for underwear, t-shirts, terry cloth and toweling, and imitation wool [17].

The production of improved rayons was boosted with predictions that cotton production would soon be limited, due to increased land use for food production to feed an expanding population worldwide. In India, where cotton is a major industry, land was already being taken from cotton producers and used for food production. Modal fibers were of particular interest to Indian producers at this time because they could replace cotton. In Egypt, the Aswan Dam changed the quality of the land, reducing the acreage of cotton. Fiber producers and users there, too, looked to modal fibers as a substitute for cotton [17].

Rayon ended the decade on a positive note with increasing production and consumption and all major producers predicting continued growth of market share. The industry felt that it had effectively changed rayon's reputation by stressing its value and versatility as a fiber and discrediting its "cheapness". Rayon began to be sold at a "fair" price. It was also seen as being safer for the environment than petroleum-based synthetic fibers. The nature of cellulose fiber makes it biodegradable, with full decomposition in nature, and thus, it was promoted as being "ecologically desirable." American and British producers tended to be more confident in rayon's future, whereas Europeans and Japanese were not as optimistic and were more concerned about possible wood pulp price increases. Adding to this optimism was rayon's relatively steady price, unlike polyester's, which was dependent entirely on the erratic price levels of petroleum derivatives.

Market response continued to dictate rayon's success in the 1980s. United States production of filament rayon continued to decline, and Avtex and North American Rayon were the only United States facilities producing quantities of filament rayon at the beginning of the decade. Courtaulds' importance in the world production of rayon continued, and the company was referred to as the mother of rayon with manufacturing sites on 7 continents, in 25 countries, and more than 400 associated companies [37].

Mergers and acquisitions affected rayon production in the 1980s. The Badische Corporation acquired American Enka in 1986 and renamed it BASF, which experienced immediate success. By 1988 total United States viscose rayon production was at 615 million pounds with 38% of production credited to Courtaulds, 36% to Avtex, and 26% to BASF. In 1988, Avtex closed. Its closing was attributed to problems with state and federal pollution laws (fines at $30 million) and the high cost of Canadian and United States wood pulp and caustic soda. BASF and Courtaulds were less susceptible to such cost increases because they purchased much of their raw material from South Africa. Mill and converter executives who were in need of a new source to fill their orders expressed serious concerns over the closing. Prices rose, and a fiber and fabric shortage was created. Avtex had been the largest United States producer of HWM and regular staple rayon. Producers suggested that perhaps the shortage would lead to higher standards for rayon, "...maybe it will end up making rayon exclusive again..." [22, p. 30].

Space exploration became a new market for rayon because carbonized viscose fibers were considered ideal as reinforcement for rocket motor nozzles and reentry heat shields. Avtex was the original supplier of the carbonized viscose rayon fiber. Since its closing, the newly reorganized North American Rayon Corporation has become the official NASA supplier. Rayon's use in the home furnishings market expanded in upholstery, draperies, and window coverings because of its drapability, blendability, and durability. In the non-woven market, rayon's absorbent, non-static, and easy processing characteristics contributed to increases in its use. Emphasis has been on its use in surgical and sanitary products, especially those which require purity and high absorption, like tampons or feminine pads and adult incontinence products, which can be made from 100% cellulose. Other non-woven uses include felts, as diaper coverstock, wipes, fabric softener sheets, envelopes, and industrial garments. Rayon also began to be used in blends with cotton in an effort to lessen the problems with cotton dust levels. When blended 50-50, the resulting product had the hand of expensive combed cotton fabric without the associated dust levels.

During the 1980s consumers began to view rayon as an elegant, natural fiber. In 1988, Courtaulds reemphasized its commitment to the apparel market by staging a fashion show entitled "Romance of Rayon". The company invited spinners, manufacturers, converters, buyers, and anyone connected to rayon in any way to the show to create a more positive impression of rayon among fashion industry personnel [18].

Rayon's popularity by the end of the 1980s was soaring. Contributing to this popularity was the high affinity of rayon for dyes, giving rayon fabrics easy dyeability and good color yield. Rayon's revival included intimate apparel as well as outer wear due to the soft, fluid line, which could be achieved, adding a new twist to the garments typically made of cotton. Rayon is versatile in that it may resemble silk when used alone, or wool challis when blended with polyester. "In either form, rayon gives a soft, fluid silhouette, drapes well, is washable, easy to print on and holds vibrant color" [44, p. 8].

Rayon fabric's higher price than cotton or polyester was seen initially as a major drawback. However, industry strategists worked to reposition rayon as a fabric for use in upscale merchandise including designer apparel. Since consumers of this merchandise have little or no resistance to dry cleaning, the care problems associated with rayon were no longer a concern. Peter Keyloun, president of Keyloun, Inc. claims that, "It's a beautiful fabric, but there is not as wide a market for rayon as a polyester and cotton blend because of the price factor and consumer understanding" [44, p. 8]. One designer stated, "Rayon is a very contemporary fabric, and it addresses all ages. It has a year-round life because of its weight" [44, p. 8]. Many other designers agreed with this sentiment and considered and advertised it as a natural fiber.

FUTURE TRENDS

The popularity of rayon has continued to grow in the 1990s. Rayon suppliers actively apply marketing principles to their production plans and intensely study consumer response and change. Manufactured cellulosics meet the new demands required of a modern fiber. With forests as the primary source of rayon fiber, there is no competition with fertile land, which would otherwise be needed for food production, and it does not have oil as its base. Even though rayon has these environmental advantages, it also possesses environmental disadvantages. Waste management problems exist with the viscose process. The wastewater from the finishing steps contains many sulfur compounds, which must be disposed of in accordance with Environmental Protection Agency and state regulations.

Although viscose has been the mainstay of the rayon industry for most of the 20th century, new developments significantly modify the viscose process or depart from it entirely. A search for new cellulose solvent spinning systems has resulted in a new class of rayon fibers based on dissolution of cellulosic raw material in N-methyl morpholine oxide. The fibers spun from this solution have even higher strength than the high performance viscose fibers, and the production process for them is essentially non-polluting.

Experimental composite fibers have been produced with a synthetic core and a viscose rayon skin [33]. The synthetic core dominates the bulk properties, such as tenacity and bending, while the rayon skin contributes moisture absorption. This last development indicates the adaptability of the viscose process in spite of intense competition from natural and other manufactured fibers. Both the composite fibers and solvent spun rayon may eventually be considered as new classes of fibers.

Even with production difficulties, viscose rayon is still an important fiber in today's market. Viscose rayon was created as an inexpensive substitute for silk, but has now achieved virtually equal status to silk as an apparel fabric. Rayon garments today sell for as much or more than silk garments. Yet, rayon is also an integral part of industrial and utility items allowing their production and marketing at cost-competitive prices.

ACKNOWLEDGMENT

The authors wish to thank Ms. Lisa Finley, graduate student in the LSU School of Human Ecology, for her invaluable research assistance in preparation of this manuscript.

REFERENCES

1. Adams, F. Rayon and Melliand Text. Monthly 16, 9, 30-33 (1935).
2. Am. Fabrics, 3, 45-53 (Summer 1947).
3. Am. Fabrics, 4, 121-144 (Fall 1947).
4. Am. Fabrics, 8, 67 (4th quarter 1948).
5. Am. Fabrics, 12, 9 (Winter 1949-50).
6. Am. Fabrics, 13, 24 (Spring 1950).
7. Am. Fabrics, 17, 110b (Spring 1951).
8. Am. Fabrics, 20, 123-128 (Winter 1951-1952).
9. Am. Fabrics, 24, 59-73 (Winter 1952-53).
10. Am. Fabrics, 84, 51-78 (Fall 1969).
11. Am. Fabrics, 107, 46-58 (Summer 1976).
12. Coleman, D.C. "The Origin and Development of Courtaulds Between 1700-1986"; summary report of Courtaulds' history.
13. Cross, C.F.; Bevan, E.J.; Beadle, C. Eng. Patent 8700, 1892.
14. Cross, C.F.; Bevan, E.J.; Beadle, C.; J. Soc. Chem. Ind., 12, 498-501 (1893).
15. Cross, C.F.; Bevan, E.J.; Beadle, C. J. Chem. Soc., 63, 837-845 (1893).
16. Daul, G.C. Am. Dyestuff Rep., 54(22), 48-57 (1965).
17. Daul, G. CHEMTECH, 11, 83-87 (1981).
18. Davidson, W.A.B. Am. Text., 14, 9, 100-102, (1988).
19. Hegan, H.J. In Talks on Rayon and Synthetic Fibres; Cotton and Rayon Merchants Association: Manchester, 1944, pp.1-24.
20. Hertz, E.; Godehn, D. Knitting Times Ap. 9, 3-5, 10-12 (1979).
21. Hottenroth, V. Artificial Silk, Fyleman, E., Trans.; Sir Isaac Pitman: London, 1928.
22. Klapper, M. WWD, 158, 85, 30 (1988).
23. Kline, E., In Cellulose and Cellulose Derivatives, 2nd ed.; Ott, E.; Spurlin, H.M.; Grafflin, M.W., Eds.; Interscience Publishers: New York, 1954, pp. 959-1018.

24. Lilienfeld, L. U.S. Patent 1 683 199, 1928.
25. Linton, G.; Carlyle, C. Am. Fabrics, 14, 88-89 (Summer 1950).
26. Man-Made Fibers Fact Book; Man-Made Fiber Producers Association, Inc.: Washington D.C., 1978; Chapter 1.
27. McNeirny, F. Mod. Text., 14-19 (Nov. 1976).
28. Moncrieff, R.W. Man-Made Fibres, John Wiley: New York, 1975.
29. Muller, M. U.S. Patent 836 452, 1906.
30. Napper, S.S. U.S. Patent 1 045 731, 1912.
31. Organic Chemistry, 2nd ed.; Gilman, H., Ed.; Wiley: New York, 1943.
32. Peters, R.H. Textile Chemistry, Vol.1; Elsevier: Amsterdam, 1963.
33. Rabe, R.L.; Collier, B.J.; Collier, J.R. Text. Res. J., 58, 735-742 (1988).
34. Reinthaler, F. Artificial Silk, Rowe, F.M., Trans.; Chapman & Hall: London, 1928.
35. Schappel, J.W.; Bockno, G.C. In Cellulose and Cellulose Derivatives, 2nd ed.; Bikales, N.M.; Segal, L., Eds.; Wiley-Interscience: New York, 1971; Vol . 5, Part 5, pp. 1115-1149.
36. Scroggie, A.G. Text. Res. J., 20, 194-198 (1950).
37. Seidel, L. Text. Ind., 87-88 (1981).
38. Sisson, W.A. Text. Res. J., 30, 153-170 (1960).
39. Smith, D.K. Text. Res. J., 29, 32-40 (1959).
40. Stearn, C.H. Eng. Patent 1020, 1898.
41. Tachikawa, S. U.S. Patents 2 592 355, 1952; 2 732 279, 1956.
42. Topham, C.F. Eng. Patent 23 157, 1900.
43. Topham, C.F. U.S. Patent 702 382, 1902.
44. WWD, 158, 8 (1988).

HISTORY OF CELLULOSE GRAFT COPOLYMERS

DAVID N.-S. HON
Wood Chemistry Laboratory, Department of Forest Resources, Clemson University Clemson, South Carolina 29634-1003

To improve chemical and physical properties of cellulose for specific applications, research in chemical modification of this renewable polymer began in 1833. Work on graft copolymerization of cellulose with selective monomers did not start until 1943. This field has always been intrigued by scientists who wished to develop novel cellulose derivatives. Despite thousands of literature have been published, and many patents are available, very little industrial applications have been achieved. Early work on graft copolymerization of cellulose, methods of preparation, and characterization of grafts are reviewed. Problems that hamper successful commercialization of this process are discussed.

Introduction

Cellulose is the most abundant of all naturally occurring organic materials on earth. More significantly, it is renewable by fixation of solar energy by plants. Since the first oil embargo in 1973, the prices of petroleum products have gone up and down, being controlled either by political or economic reasons. Because of the instability of oil supply and oil prices, cellulosic raw materials have gained a special role in product development since they are cheap and are plentiful. Hence, attention to the development of new polymer products based on cellulose will undoubtedly be growing and the scope of their manufacture will be expanding[1,2].

Cellulose forms the major portion of the cell walls of higher plants, and is a basic constituent of plants. Its chief sources are seed hairs, such as cotton; bast fibers, occurring in the inner bark of certain fibrous plants such as flax, jute, and ramie; leaf fibers, such as hemp; stalks and straws; and wood. It can be extracted from these sources by chemical pulping and bleaching. In the case of the cotton fiber, purification is achieved by mild treatment with dilute alkali, followed by washing and gentle bleaching. Cellulose is a syndiotactic polymer possessing one primary and two secondary hydroxyl groups. These alcoholic groups undergo alcoholic reactions such as esterification, etherification, oxidation and crosslinking[3]. It can also undergo graft copolymerization reactions[4]. The principal objective of this paper is to review the historical development of cellulose graft copolymerization and methods of preparation. Characterization, properties and applications of cellulose-graft copolymers are also briefly discussed. Grafting of cellulose derivatives is not considered.

Early History of Cellulose Chemistry

Historically speaking, flax has been cultivated for its fiber for at least 10,000 years[5]. It is frequently mentioned in the Bible and was cultivated in prehistoric times by the Swiss Lake Dwellers and by the Egyptians. The use of cellulosic material for the fabrication of garment was known as early as 4000 B.C.[6]. The great writing material of the ancient world was papyrus, in use by 3500 B.C. Moreover, Hindu religious scripts reported the cultivation of cotton as early as 1500 B.C. Nonetheless, chemical modification of cellulose is of only recent origin. Nitration of cotton cellulose was first discussed in 1833[7]. "Wood cellulose" was not known until 1839, when Anselme Payen, a French botanist, chemically isolated the white fibrous material from woody plants and named it cellulose[8]. Since then, the rate of cellulose research had grown significantly. Cross et al.[9] discovered the xanthation process for making regenerated cellulose (viscose rayon) in 1892, and its limited commercial use was pioneered by Cross and Bevan in 1894[10]. The acetylation of cellulose was first carried out in 1865 by Schutzenberger[11] by heating cotton and acetic anhydride to about 180°C in a sealed tube until the cotton dissolved. Despite the opportunities in research and development for cellulose in the late nineteenth century, the development of cellulose grafted copolymer was not reported until 1943[12]. Hence, modification of cellulose by grafting monomers onto its backbone is a relatively new reaction. Nonetheless, grafting of cellulose has in fact become a popular method of modifying its properties[13]. Several excellent monographs and review articles are available[13-21].

In 1943, Ushakov[12] reported the copolymerization of unsaturated cellulose derivatives. He synthesized vinyl and allyl esters of cellulose in order to use them for subsequent copolymerization experiments with maleic ester. It is assumed that the insoluble reaction products he obtained were most probably the first graft copolymers ever prepared. In 1951, Landells and Whewell[22] used thermal and free-radical polymerization techniques to deposit vinyl monomers, such as methyl methacrylate, acrylonitrile, styrene, methacrylamide and methyl vinyl ketone, inside cellulosic fibers. In 1953, Waltcher et al.[23] prepared the orthochlorobenzyl and bromo undecanoic esters of cellulose for grafting with vinyl monomers by chain transfer. Since then, the possibility of chemically combining synthetic polymers with cellulose has attracted much attention.

Definition and Nomenclature

By suitable polymerization four different types of copolymers may be prepared, namely, alternating copolymer, random copolymer, block, and graft copolymers[24]. Structural illustrations of these copolymers are shown in Fig. 1. Stannett[25] defined that "a graft copolymer consists of a polymeric background with lateral covalently linked side chains. Both the backbone and the side chain polymers can be homopolymers or copolymers." In addition,

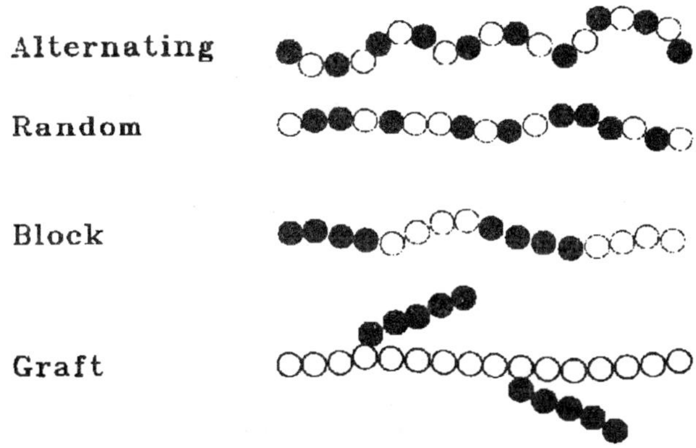

Figure 1. Classification of copolymers.

McGrath[26] defined that "all graft copolymers are comprised of two general structure features: a backbone of polymer "A" to which number of B sequences are grafted."

According to the Commission on Macromolecular Nomenclature[27], the name of the backbone polymer should be given first, and the name of the graft second, the word graft indicating the side chain. Hence, cellulose-graft polystyrene means that a cellulose backbone carries polystyrene grafts.

Methods of Preparation

The preparation of a graft copolymer requires the formation of a reactive center on a polymer molecule in the presence of a polymerizable monomer. The principal techniques frequently used are: (1) grafting by free radical polymerization, (2) grafting initiated by ionic polymerization; (3) grafting by condensation and ring opening polymerization.

Free Radical Polymerization Method

In this method, the initiation of polymerization is due to free radical formation in either an initiator or cellulose substrate, and its subsequent reaction with monomer. Free radicals can be formed by chain transfer reactions (initiators), physical (high and low energy radiations, and mechanical stress) and chemical (redox system or initiators) activations.

Chain transfer reaction was first developed by Waltcher et al.[23] who enhanced the reaction by introducing halogen-substituted ester groups into the cellulose molecule. Chaudhuri and Hermans[28] reacted cellulose and its derivatives with ethylene sulfide to introduce mercaptal groups. These groups chain-transfer very readily, and excellent yields of graft copolymer were obtained. In this reaction, monomer (M) is polymerized with the help of an initiator (I), and a growing chain radical ($P_n\cdot$) is formed:

$$I \rightarrow 2I\cdot$$

$$I\cdot + M \rightarrow IM\cdot$$

$$IM\cdot + nM \rightarrow P_n\cdot$$

In the presence of cellulose, the growing radical may terminate by abstraction of a proton from cellulose (RH) to give rise to a free radical formation on the cellulose backbone (R•). Addition of a monomer to this new macroradical produces a graft copolymer:

$$P_n\cdot + RH \rightarrow P_nH + R\cdot$$

$$R\cdot + M \rightarrow RM\cdot \overset{nM}{\rightarrow} \textit{graft copolymer}$$

Free radicals can also be generated by the radiation of cellulose with high energy sources, such as x-rays, γ-rays, electrons, and neutrons. Likewise, free radicals can also be generated by ultraviolet light irradiation with or without the use of photoinitiators. Mechanoradicals have also been proven to be effective cellulose grafting centers. They can be generated by milling or cutting.

Redox system is one of the most popular chemical means of generation of free radicals for cellulose grafting reaction. Redox systems generally comprise the transition metal ions (e.g., Cr^{6+}, V^{5+}, Ce^{4+}, Co^{3+}, Mn^{3+}, Fe^{3+}, etc.) which occur in various oxidation states. Such oxidized metal ions may directly oxidize cellulose to initiate the formation of free radicals for grafting. It should be borne in mind that the redox potential of the metal ions used would constitute an important parameter in determining the grafting efficiency. If the oxidation potential of the metal ion is high, it may react as well with the monomer which will lead to an increase in homopolymer formation. Therefore, in order to circumvent this situation, metal ions with low oxidation potential are preferred. The best known example of the redox reactions is the use of the ceric ions, originally developed by Mino and Kaizerman in 1958[29], and their applications to cellulose grafting was directed by Schwab et al. in 1962[30]. It is a rather complex reaction, but the general reaction can be condensed into the following ways:

$$RH + Ce^{4+} \rightarrow [Complex] \rightarrow R\cdot + Ce^{3+} + H^+$$

$$R\cdot + nM \rightarrow \textit{graft copolymer}$$

where RH and M denote cellulose and monomer respectively

Oxidation of cellulose by Ce^{4+} follows different mechanisms depending on the type of acid media used. In sulfuric acid, oxidation of cellulose takes place via a free radical mechanism[31]. However, in perchloric acid a preliminary complex between cerium and the glycol is first formed, followed by disproportionation of the complex as the rate-determining step[32,33]· Evidence of complex formation between Ce^{4+} and cellulose has been studied by several investigators[34-36]. Formation of ceric ion-cellulose complex exhibited a characteristic absorption maximum at 320 mμ was observed by Kurlyankina and his coworkers[34,35]. Cellulose possesses primary and secondary alcohols which are the possible sites for oxidation by ceric salts[37-39]. The result of oxidation is the cleavage of C_2-C_3 bonds[38,40-42]. Other investigators have also noted the participation of the hemiacetal units in the ceric oxidation of cellulose[43-46].

In the chain-transfer system, the redox couple would initially produce transient species which subsequently would react with the backbone polymer to produce free radicals onto the cellulose. For example, hydrogen peroxide in the presence of ferrous ion constitutes a frequently used redox chain transfer initiator. In this system hydrogen peroxide reacts with the ferrous ion to produce hydroxyl radicals which abstract hydrogen atoms from the cellulose backbone to form free radicals in cellulose[47-48]:

o Decomposition of hydrogen peroxide by ferrous ions:

$$H_2O_2 + Fe^{2+} \rightarrow \bullet OH + OH^- + Fe^{3+}$$
$$RH + \bullet OH \rightarrow R\bullet + H_2O$$

where RH denotes cellulose

Similar chain transfer reactions may also occur when organic hydroperoxides, persulfates, Fe^{3+}, Cu^{2+}, V^{5+}, etc. are used in place of hydrogen peroxide with a reducing agent such as sodium bisulfite, thiourea, and thiosulfate[49-53]. An example is illustrated below:

o Decomposition of peroxodiphosphate in 0.15 M H_2SO_4:

$$H_2P_2O_8^{2-} + H^+ \rightleftharpoons H_3P_2O_8^-$$
$$H_3P_2O_8^- \rightarrow H_2PO_4\bullet + HPO_4^{\bar{\bullet}}$$
$$H_2PO_4\bullet + H_2O \rightleftharpoons H_3PO_4 + \bullet OH$$
$$RH + \bullet OH \rightarrow R\bullet + H_2O$$

where RH denotes cellulose

The success of graft copolymerization requires the understanding of the kinetics for defining optimum conditions for grafting. The overall kinetics of the process as well as the length of the grafting chains formed are determined by the rates of the individual processes of initiation, propagation, and termination. Since redox system is the most popular initiation system for cellulose grafting reaction, a typical kinetics of such a system is illustrated here[18]. It should be noted that this kinetic can apply as well to any other initiation system discussed above.

o Radical Generation:

$$RH + Z^{n+} \overset{K_c}{\rightleftharpoons} [Complex] \overset{k_d}{\rightarrow} R\bullet + Z^{(n-1)^+} + H^+$$

o Initiation:

$$R\bullet + M \overset{k_i}{\rightarrow} R\text{-}M\bullet$$

o Propagation:

$$R\text{-}M\bullet + M \xrightarrow{k_p} R\text{-}M_n\bullet$$

o Termination

$$R\text{-}M_n\bullet + Z^{n+} \overset{k_t}{\rightleftarrows} R\text{-}M_n + Z^{(n-1)^+} + H^+$$

o Oxidation:

$$R\bullet + Z^{n+} \overset{k_o}{\rightleftarrows} oxidation\ products + Z^{(n-1)^+} + H^+$$

$$R_p = \frac{k_c k_p k_i k_d [RH][M^2]}{k_o k_t [Z^{n+}]}$$

M: monomer, Z^{n+}: transition metal ion,
Kc: equilibrium constant, k: rate constant
RH: cellulose, R_p: rate of polymerization

Initiation of free radicals can also be generated by high energy radiation, such as γ-radiation and electron beams, and ultraviolet light irradiation. The major difference between radiation chemistry and photochemistry lies in the energy of the radiation that initiates the reaction, the energy of the particles and photons concerned in radiation chemistry being much greater than the energy of the photons causing photochemical reactions. The high-energy photons and particles are not selective and may react with any molecule lying in their path, raising it to any one of its possible ionized or excited states. Subsequently the different energy-rich species react to give a complex mixture of products, in contrast to the relatively small number of products from a photochemical reaction. Hence, in many cases when energy absorption is not sufficient to initiate free radical formation, photoinitiators are used to help absorbing specific energy[54].

Free radicals of cellulose have been produced with both $\gamma(Co^{60})$ radiation and ultraviolet light irradiation, and two techniques have been used to initiate graft copolymerization[13,55]. In the simultaneous irradiation technique, mixtures of cellulose and monomer are irradiated together, whereas in the preirradiation method, cellulose is irradiated first and the activated cellulose is then followed to react with monomer. The preirradiation technique usually produces less homopolymer than simultaneous irradiation, since monomer is absent during the irradiation step. With simultaneous irradiation, however, there is a better opportunity for reaction of short-lived free radicals with monomer.
The general schemes of the grafting reactions initiated by high energy radiation and ultraviolet light irradiation are illustrated below:

$$RH \xrightarrow{\gamma} R\bullet + H\bullet \quad or$$

$$RH \xrightarrow{h\nu} R\bullet + H\bullet$$
$$R\bullet + nM \rightarrow graft\ copolymer$$
$$M\bullet + nM \rightarrow homopolymer$$

where RH and M denote cellulose and monomer, respectively.

If absorption of light does not lead to the formation of free radicals in cellulose, a photosensitizer may be used to promote free radical formation or can transfer its absorbed energy to other molecules in the system to promote copolymerization reactions[56,57]. The most effective types of sensitizers used are benzophenone, benzoin ethyl ether and many metallic ions[54,58].

For ultraviolet light with a photosensitizer, the follow scheme is illustrated:

$$S \xrightarrow{h\nu} S^*$$
$$S^* + RH \rightarrow [RH]^* + S$$
$$[RH]^* \rightarrow R\bullet + H\bullet$$
$$R\bullet + nM \rightarrow graft\ copolymer$$

where RH, M, and S denote cellulose, monomer and photoinitiator, respectively.

Many studies have been devoted to identify free radical species formed in cellulose during high energy radiation[59] and ultraviolet light irradiation with or without the presence of photoinitiators[60]. The nature of cellulose as well as the irradiation environments are all influencing the fate of free radical generation. Owing to the difference in energy between the two irradiation systems, high energy radiation produces free radicals within the bulk and the ultraviolet irradiation produces free radicals predominately at the surface of the polymer[61,62].

When cellulose is subjected to an applied stress or shear, mechanoradicals are formed[63]. These macroradicals have been proven to be very active and are capable of initiating graft copolymerization[64]. Mechanoradicals in cotton and wood cellulose which contribute to the grafting reaction have been identified by electron spin resonance spectroscopy[65,66] In comparison of the efficiency of grafting between mechanoradicals and photo-induced radicals, it has been found that the former radicals are more efficient than the latter. Hon [66] observed a lower degree of homopolymer formation resulting from the mechano-induced grafting of monomer such as methyl methacrylate on pulp in comparison with photo-induced grafting.

Ionic Copolymerization

The active site in grafting chain-growth polymerization can be an ion instead of a free radical. Ionic reactions are much more sensitive than free radical processes to the effect of solvent, temperature, and adventitious impurities. Theoretically, ionic grafting can be initiated either by radiation or chemical means. In the course of the interaction of ionizing radiation with cellulose, ions, radicals and electronically excited molecules can be generated. It is reasonable to believe that the positive and negative ions formed in the ionizing radiation may initiate graft copolymerization, and accordingly one can distinguish, respectively, cationic and anionic grafting. Although this ionic graft copolymerization method has been used for various synthetic polymers[67,68], no report has yet been found which uses this method for grafting suitable monomers onto the cellulose backbone. The use of chemical means to initiate ionic graft copolymerization has, however, been studied in some systems. The basic requirement in such processes is the selection of a suitable ionic polymerization initiator. Such initiators may be alkali metal suspensions in liquids that are Lewis bases, organometallic compounds and sodium naphthalenide. A typical example of anionic grafting involves the formation of an alkoxide of cellulose by the interaction of the polymer with anionic catalyst such as sodium-ammonia or methoxide of alkali metals and, subsequently, this alkoxide derivative of cellulose is treated with vinyl monomer. Feit et al.[69] use this system to react alkoxides of cellulose with acrylonitrile, methacrylonitrile, and methacrylic esters. Similar studies were performed by Schwenker and Pacsu[70], and Avny and Rebenfied[71]. The reaction scheme for acrylonitrile as an example of ionic polymerization is presented as follows:

o Initiation:

$$R\text{-}O^-Na^+ \;+\; CH_2{=}CHCN \;\rightarrow\; R\text{-}OCH_2\text{-}\bar{C}HCNNa^+$$

o Propagation:

$$R\text{-}OCH_2\text{-}\bar{C}CHCN \;+\; nCH_2{=}CHCN \;\rightarrow\; R\text{-}O(CH_2\text{-}\underset{\underset{CN}{|}}{C}H)_nCH_2\text{-}\bar{C}HCN$$

o Termination (by chain transfer to monomer):

$$R\text{-}O(CH_2\text{-}\underset{\underset{CN}{|}}{C}H)_nCH_2\text{-}\bar{C}HCN \;+\; CH_2{=}CHCN$$

$$\rightarrow\; R\text{-}O(CH_2\text{-}\underset{\underset{CN}{|}}{C}H)_n\text{-}CH_2CH_2CN \;+\; CH_2{=}\bar{C}\text{-}CN$$

o Termination (by chain transfer to free OH of cellulose):

$$R\text{-}O(CH_2\text{-}\underset{\displaystyle CN}{\underset{|}{C}}H)_nCH_2\text{-}\bar{C}HCN + R\text{-}OH$$

$$\rightleftarrows\ R\text{-}O(CH_2\text{-}\underset{\displaystyle CN}{\underset{|}{C}}H)_nCH_2CH_2CN + R\text{-}O^-$$

Condensation and Ring-Opening Polymerization

Methods of this type have been little studied. Schurz and Rebek[72] used the diisocyanate and a diol such as hemamethylene diisocynate and 1,2 diethyl alcohol or 1,4 dibytyl alcohol for the polyaddition reaction. The highest grafting yield was obtained when monomer was included in the cellulose fibers. Polycondensation between cellulose and dimethylchlorosilane in acetone was also conducted by Schurz and Rebek[72].

Rogovin[73-75] carried out direct polycondensation grafting of aminoenanthic acid to cellulose and its derivatives. The acid chloride of aminoenanthic acid hydrochloride was reacted with cellulose in the presence of pyridine or triethylamine, using solvent-exchange techniques. A grafted cellulose with low degree of substitution and low DP of side chains was obtained.

Grafting of cyclic compounds, such as ethylene oxide[76], ethylenimine[77], and β-propiolactone[78] have been attempted. The reaction of the former two chemicals are shown here:

o Grafting cellulose with ethylene oxide:

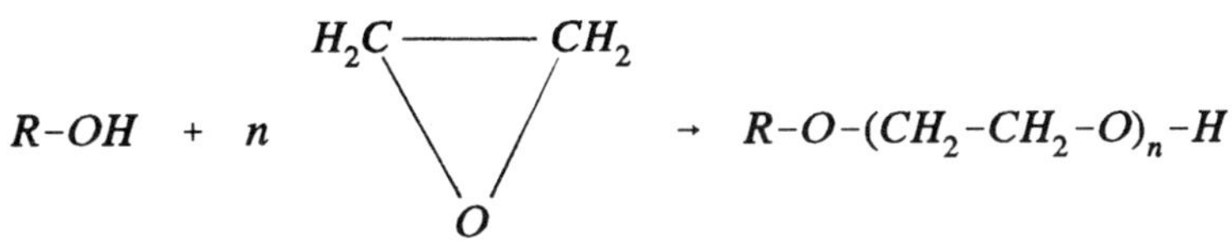

o Grafting cellulose with ethylemine:

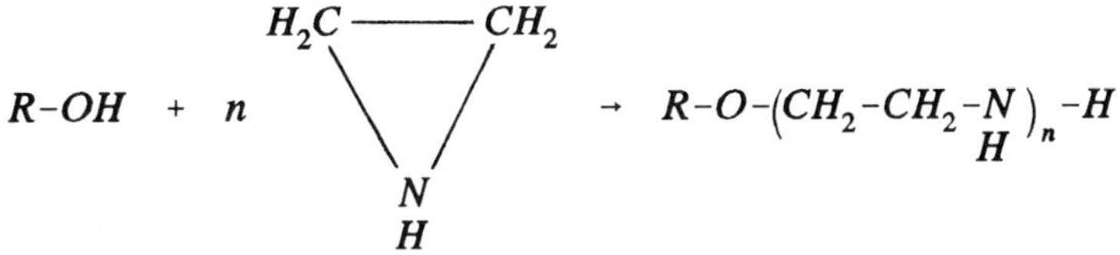

Although this type of reaction has not received much attention, a cellulose-ethylene oxide graft copolymer has achieved commercial success[79].

Characterization of the Grafts

The characterization of grafted cellulose is important in that the changes in chemical structures will result in changes in the morphology, structure, and physical and mechanical properties of cellulose due to grafting. In order to characterize a graft copolymer, the first step is to assess whether the monomer is added to the base polymer and this is usually done by measuring the weight increase after homopolymers have been removed by extensive solvent extraction. Hence, the grafting percentage and grafting efficiency can be calculated. The following described parameters are frequently used:

$$\textit{Percent Grafting} =$$

$$\left(\frac{\textit{Final cellulose weight} - \textit{Initial cellulose weight}}{\textit{Initial cellulose weight}}\right) x\ 100$$

$$\textit{Grafting Efficiency} =$$

$$\left(\frac{\textit{Final cellulose weight} - \textit{Initial cellulose weight}}{\textit{Total polymer weight after reaction} - \textit{Initial cellulose weight}}\right) x\ 100$$

Although the purification and characterization of the graft copolymer are cumbersome, they must be made since they are needed in basic research as well as for the industrial applications of graft copolymers. The presence of pure grafts can be confirmed by physical methods if the graft chains contain specific elements or functional groups. For example, the presence of polyacrylonitrile or polyacrylamide in the cellulose grafts can be characterized by using the Kjeldahl method to determine the amount of nitrogen in the pure cellulose grafts[80]. The percent grafting can then be estimated. The presence of carbonyl or ketone, ester, amide, amine, nitrile, isocyanate, and imine groups, and aromatic rings can be determined by means of infrared spectroscopy[81]. Typical absorption bands of these functional groups can be used to assess the presence of grafts in cellulose. The IR spectra of grafted cellulose has been examined by several investigators[43, 82-84]. For example, acrylonitrile has a characteristic absorption band at 2250 cm^{-1}, hence, observation of such an absorption band in the IR region would definitely prove that the graft copolymer consists of such a monomer[82]. Likewise, ultraviolet spectroscopy may be used if the graft chains absorb particular wavelengths in the ultraviolet region[85].

The presence of pure grafts can also be verified by observing the changes in the insolubility of the grafted products in cellulose solvents such as cuene and cupriethylene diamine[86-89]. This change in solubility was found to be dependent upon the degree of grafting.

Characterization of graft products is normally conducted after the crude products have been extracted with selective solvents to remove homopolymers. However, despite this effort homopolymers may not be completely removed from the graft copolymer. In many cases homopolymers are so intimately bound in the cellulose substrate that it cannot be separated by normal extraction or fractionation. This has been shown by Guthrie et al.[90] in their study of the dilute solution properties of cellulose-graft polystyrene using light scattering techniques. Studies of the heterogeneity of the copolymers suggested that they contained occluded homopolymeric species which careful solution and fractionation procedures had failed to remove. This problem has also been pointed out by Ende and Stannett[91] during their work to analyze the completeness of separation of the products by means of a density-gradient ultracentrifuge technique. Nonetheless, Kobayashi[92] made one of the best separations with cellulose-graft polystyrene. It was found that after acetylation the homopolystyrene could be removed by extraction with benzene, the ratio of homopolystyrene to grafted polystyrene being 1.25. Taga and Inagaki[93] have successfully separated the true graft from the homopolystyrene by means of a TLC technique. They first hydrolyzed the graft copolymer by acid hydrolysis and the residue was then subjected to TLC. It was found that the homopolystyrene that was associated with the real graft chain migrated up to the solvent front when tetrahydrofuran was used as the developing solvent, whereas the real graft component remained immobile. With this separation technique they could estimate the extent of real graft in the copolymer.

Molecular Weight of the Graft Chains

Since molecular weight plays an important role in copolymer properties, it is imperative to characterize the molecular weight of grafted side chains so that cellulose graft copolymers can be use for specific purposes[94,95]. A popular method is to hydrolyze the cellulose by acid and isolate the graft chain. Then the isolated side chain can be dissolved in a suitable solvent and its intrinsic viscosity can be determined. By using the Mark-Houwink equation[96], molecular weight or degree of polymerization of the graft chain can be measured. The solution can also be injected into a column of a gel permeation chromatograph to determine the molecular weight of the graft chain[97-101]. In addition, more importantly, distribution of molecular weight and polydispersity index of the graft chain can also be determined[97,102].

The average number of grafted branches in cellulose-graft copolymer N_g and the number-average molar mass of the backbone M_{bg} can be determined from the following equations if the number-average molar mass of the graft copolymer M_g and of the branch M_b are known together with the weight fraction of the branch portion in the graft copolymer w_b:

$$N_g = \frac{M_g w_b}{M_b}$$

$$M_{bg} = M_g(1-w_b)$$

Properties and Applications

The properties of cellulose can be dramatically changed by graft copolymerization. The improvements in properties depend upon the choice of monomer and grafting condition. Grafting of hydrophilic monomers has been found to lead to an increase in wettability, adhesion, dyeing and rate of release of oil stains by detergent solution. On the other hand, if the monomer is hydrophobic the result will be decreased wetting by all liquids. Depending on the specific applications, graft copolymer can be tailor-made by selecting a specific monomer. Hence, a wide variety of properties of cellulose can be improved. The improvement included tensile strength of fiber[103-107], resistance to microbial degradation[30,86,87,108], thermal stability[109,110], photostability[111,112], abrasion, and acid[108,113,114], dye receptivity[110,113-118], wet strength of paper; adhesion, absorbency[103,119-122], water repellency[123-125], oil repellency[126], fire retardation[127,128], thermoplasticity[129], and electrical properties[104]. Cellulose graft copolymers have demonstrated various ion exchange properties[115,130,131-136]. Details of preparation and processing for these applications are discussed elsewhere in this volume.

Future

Chemical modifications of cellulose to improve its chemical and mechanical properties began in 1833. It is an old chemistry[137]. Synthesis of graft copolymer of cellulose started in approximately 1943[12]. Since then, many techniques to graft monomers onto cellulose substrates have been explored, and thousands of technical papers have been published. Although cellulose graft copolymers have important applications both in an academic and industrial capacity, ironically, commercial products derived from cellulose-graft copolymers are very limited. Many problems have been recognized throughout these long years of research. In 1971, Z.A. Rogovin[94] made the following statements with regard to the requirements for a promising grafting method:

1. The synthesis of a graft copolymer without the simultaneous formation of homopolymer or with a minima of this side reaction.

2. The minimum destruction of cellulose and its esters in the grafting process.

3. The ready availability of the initiation, and the possibility of using it while carrying out the reaction in an aqueous medium.

4. A wide field of application, that is the possibility of carrying out the grafting of a large variety of monomers.

Again, in 1982, V. Stannett[138] made the following statements with regard to the challenges of developing viable commercial processes for graft copolymerization of cellulose:

1. The elimination, or at least minimization of concurrent homopolymer formation.

2. The involvement of all or most of the cellulose molecules in the grafting process.

3. Better control of the molecular weights and molecular weight distribution of the grafted side chains.

4. Better reproducibility of the grafting yields, properties and other features of the graft copolymers.

Recently, in 1988, R. Narayan[139] summarized the problems of graft copolymerization synthesis using free radical approaches:

1. It is difficult to control or change the molecular weight of the side chain grafts.

2. The molecular weight of the grafts are very high and very polydisperse.

3. Grafting of only a few high molecular weight chains occurs.

4. Considerable amounts of homopolymer are sometimes formed, some of which may get embedded in the cellulose matrix and difficult to remove.

5. No knowledge exists of the nature of the linkage between the cellulose backbone and the graft, or the ability to control it.

6. Reproducibility and control of the grafting yields, properties and other features of the graft copolymers not realized.

It is obvious that in order to bring cellulose-graft copolymers to commercial success, fundamental problems have to be solved. There is no doubt that cellulose-graft copolymers exhibit excellent properties for certain areas of application. Without improvement in the processing condition to reduce homopolymer formation to reduce the cost, industry is still very reluctant to explore its marketability.

References

1. D.N.-S. Hon, Polymernews, 13, 134 (1988).
2. D.N.-S. Hon, "New Developments in Cellulosic Derivatives and Copolymers" in Emerging Materials and Chemicals from Biomass (R. Rowell, R. Narayan and T. Schultz, eds.), ACS Symposium Series, ACS, Washington, D.C., in press.
3. N.M. Bialkes and L. Segal (eds.), Cellulose and Cellulose Derivatives, Wiley-Interscience, New York, 1971, Part V, two volumes.
4. D.N.-S. Hon, "Cellulose: Chemistry and Technology" in Concise Encyclopedia of Wood & Wood-Based Materials (A.P. Schniewind, ed.), Pergamon Press, 1989, pp.39-44.
5. Encyclopedia Britannica, Inc., The New Encyclopedia Britannica, Chicago, 1983, vol. 7, p. 271.
6. Genesis 3,7
7. H. Braconnot, Ann., 1, 242, 245 (1833).
8. A. Payen, Compt. Rend., 8,51 (1839).
9. C.F. Cross, E.J. Bevan and C. Beadle, Ber., 26, 1090 (1893).
10. C.F. Cross, E.J. Bevan and C. Beadle, U.S. Pat., 520,770 (1894).
11. P. Schutzenberger, Compt. rend., 61, 485 (1865).
12. S.N. Ushakov, Fiz.-Mat. Nauk (USSR), 1, 35 (1943).
13. A. Hebeish and J.T. Guthrie, The Chemistry and Technology of Cellulosic Copolymers, Springer-Verlag, Berlin, 1981, pp. 351.
14. H.A. Krassig and V. Stannett, Adv. Polym. Sci., 4, 111 (1965).

15. V.T. Stannett and H.B. Hopfenberg, "Graft Copolymers" in Cellulose and Cellulose Derivatives (N.M. Bikales and I. Segal, eds.), Wiley-Interscience, New York, 1971, part V., pp.907-936.
16. J.C. Arthur, Jr. (ed), Proceeding of the Symposium on Graft Polymerization onto Cellulose, Wiley-Interscience, New York, 1972, pp. 361.
17. D.N.-S. Hon (ed.), Graft Copolymerization of Lignocellulosic Fibers, ACS Symposium Series #187, Washington, D.C., 1982, pp. 381.
18. B.P. Morin, I.P. Breusova and Z.A. Rogovin, Adv. Polym. Sci., 42, 139 (1982).
19. D.J. McDowall, B.S. Gupta and V.T. Stannett, Prog. Polym. Sci., 10, 1 (1984).
20. S.N. Bhattacharyya and D. Maldas, Prog. Polym. Sci., 10, 171 (1984).
21. O. Mansour and A. Nagaty, Prog. Polym. Sci., 11, 91 (1985).
22. G. Landells, C.S. Whewell, J. Soc. Dyers and Colourists, 67, 338 (1951).
23. I. Waltcher, R. Burroughs and E.C. Jahn, IUPAC Abstracts, Stockholm, 1953.
24. J.M.G. Cowie, Polymers: Chemistry & Physics of Modern Materials, Intertext Books, Aylesbury, 1973.
25. V. Stannett, Raidat. Phys. Chem., 18, 215 (1981).
26. J.E. McGrath, Pure Appl. Chem., 55, 1573 (1983).
27. Commission on Macromolecular Nomenclature, Pure Applied Chemistry, 40, 479 (1974).
28. D.K.R. Chaudhuri and J.J. Herman, J. Polym. Sci., 51, 373 (1961).
29. G. Mino and S. Kaizerman, J. Polym. Sci., 31, 242 (1958).
30. E. Schwab, V. Stannett, D.H. Rakowitz and J.K. Magrane, Tappi, 45, 390 (1962).
31. G. Mino, S. Kaizerman and N.E. Rasmussen, J. Polym. Sci., C, 403 (1963).
32. G. Mino and S. Kaizerman, J. Polym. Sci., 31, 242 (1958).
33. M. Ardon, J. Chem. Soc., P2, 1811 (1957).
34. V.I. Kurlyankina, V.A. Molotkov, O.P. Koz'mina, A.K. Khripunov and I.N. Shtennikova, Eur. Polym. J., Suppl., 5, 441 (1969).
35. V.I. Kurlyankina, O.P. Koz'mina, A.K. Khripunov, V.A. Molotkov and T.D. Novoselova, Dokl. Akad. Nauk., SSSR 172, 341 (1967).
36. A.Y. Kulkarni and P.C. Mehta, J. Polym. Sci., Part B 5, 209 (1967).
37. F.R. Duke, J. Am. Chem. Soc., 69, 2885 (1947).
38. G. Mino, S. Kaizerman and N.E. Rasmussen, J. AM. Chem. Soc., 81, 1494 (1959).
39. R.J.E. Cumberbirch and J.R. Holker, J. Soc. Dyers Colour., 82, 59 (1966).
40. C.R. Pottenger and D.C. Johnson, J. Polym. Sci., Part A-1, 8, 301 (1970).
41. J.C. Arthur, P.J. Baugh and O. Hinojosa, J. Appl. Polym. Sci., 10, 1591 (1966).
42. G. Mino and S. Kaizerman, J. Polym. Sci., 31, 242 (1969).
43. A.Y. Kulkarni and P.C. Mehta, J. Appl. Polym. Sci., 12, 1321 (1968).
44. I. Terasaki and M. Matsuki, Sen-i Gakkaishi, 18, 147 (1962).
45. L. Neimo and H. Sihtola, Pap. Puu, 47, 369 (1965).
46. A. Takahashi and S. Takahashi, Kobunshi Kagaku, 30, 239 (1973).
47. V.A. Kargin, J. Polym. Sci., C4, 1601 (1964).
48. M.S. Brains, J. Polym. Sci., C37, 125 (1972).
49. A. Hebeish, A. Waly, S.A. Bari and S. Bedewi, Cellulose Chem. Technol., 15, 441 (1981).
50. O.Y. Mansour and A. Nagaty, J. Polym. Sci., Polym. Chem. Ed., 13, 2785 (1975).
51. S. Mukhopadhyay, J. Prasad and S.R. Chatterjee, Makromol. Chem., 176, 1 (1975).
52. A.K. Mukherjee and S. Sanyal, Cellulose Chem. Technol., 13, 13 (1979).
53. S.K. Kar, P.L. Nayak and G. Sahu, J. Polym. Sci., Polym. Chem. Ed., 19, 1581 (1981).
54. B. Ranby and J.F. Rabek, Photodegradation, Photo-oxidation and Photostabilization of Polymers, John Wiley & Sons, London, 1975, pp. 289-332.
55. J.T. Guthrie, Polymer, 16, 134 (1975).
56. J.L. Garnett, Radiat. Phys. Chem., 14, 79 (1979).
57. C.H. Ang, J.L. Garnett, R. Levot, M.A. Long and K.J. Kenneth, J. Macromolec. Sci. Chem., A17, 87 (1982).
58. N.-S. Hon, J. Polym. Chem., Polym. Chem. Ed., 13, 1933 (1975).
59. J.C. Arthur, Jr. "Reactions Induced by High-Energy Radiation" in Cellulose and Cellulose Derivatives (N.M. Bikales and I. Segal, eds.), Wiley-Interscience, New York, 1971, part V., pp.937-975.
60. D.N.-S. Hon, "Photochemistry of Wood" in Wood and Cellulosic Chemistry (D.N.-S. Hon and N. Shiraishi, eds.), Mercel Dekker, New York, 1991, pp. 525-555.
61. N.P. David, J.L. Garnett and R.G. Urguhart, J. Polym. Sci., Symp., 55, 286 (1976).
62. Gh. Rozmarin, R. Lazro, J.P. Quiles and U.F. Le Nest, Bul. Inst. Politechic Iasi, 16, 363 (1970).

63. D.N.-S. Hon, "Mechanochemically Initiated Copolymerization Reactions in Cotton Cellulose" in Initiation of Polymerization (F.E. Bailey, Jr., ed.), ACS Symposium Series 212, ACS, Washington, D.C., 1983.
64. D.N.-S. Hon, J. Appl. Polym. Sci., 23, 1487 (1979).
65. D.N.-S. Hon, J. Polym. Chem., Polym. Chem. Ed., 18, 1957 (1980).
67. V. Ya. Kabanov and R.E. Aliev, J. Polym. Sci., Polym. Chem. Ed., 17, 1255 (1979).
68. V. Ya. Kabanov and R.E. Aliev, J. Polym. Sci., Polym. Chem. Ed., 18, 871 (1979).
69. B.A. Feit, A. Bar-Nun, M. Lahav and A.J. Zikha, J. Appl. Polym. Sci., 8, 1869 (1964).
70. R.F. Schwenker and E. Pacsu, Tappi, 46, 665 (1963).
71. Y. Avny and L. Rebenfeld, Textile Res. J., 38, 684 (1968).
72. J. Schurz and M. Rebek, Papier, 20, 664 (1966).
73. Z.A. Rogovin, J. Polym. Sci., 48, 443 (1960).
74. Z.A. Rogovin, Periodica Polytech., 5, 65 (1961).
75. T. Sun and Z.A. Rogovin, Vysokomol. Soedin., 4, 714 (1962).
76. S.G. Cohen and H.C. Haas, J. Amer. Chem. Soc., 72, 3954 (1950).
77. W. Cooper and R.K. Smith, Makromol. Chem., 40, 148 (1960).
78. G.C. Daul, r.M. Reinhardt and J.D. Reid, Textile Res. J., 24, 738 (1954).
79. Chem. Eng. News, 16, 39 (1961).
80. R.B. Bradstreet, The Kjeldahl Method for Organic Nitrogen, Academic Press, New York, 1965, pp. 239.
81. D.L. Pavia, G.M. Lampman and G.S. Kriz, Jr., Introduction to Spectroscopy, Saunders College, Philadelphia, 1979, chapter 2.
82. I.M. Trivedi and P.C. Mehta, Cellulose Chem. Technol., 7, 401 (1973).
83. L. Meimo and H. Sihtola, Pap. Puu 47, 369 (1965).
84. J.S. Chawla and G. Shankar, Indian Pulp. Pap. Tech. Ass., 10, 39 (1968).
85. P. Manson and L. Westfelt, J. Polym. Sci., Polym. Chem. Ed., 19, 1509 (1981).
86. S. Kaizerman, G. Mino and L.F. Meinhold, Text. Res. J. 32, 136 (1962).
87. S.R. Rao and S.L. Kapur, J. Appl. Polym. Sci., 13, 2649 (1969).
88. F. Ide and Y. Takayama, Kogyo Kagaku Zasshi, 64, 213 (1961).
89. V.N. Sharma and E.H. Daruwalla, Indian J. Test. Res., 1, 120 (1976).
90. J.T. Gruthrie, M.B. Huglin and G.O.Phillips, Polymer, 18, 521 (1977).
91. H.A. Ende and V. Stannett, J. Polym. Sci., Part A, 2, 4047 (1964).
92. Y. Kobayashi, J. Polymer Sci., 51, 359 (1961).
93. T. Taga and H. Inagaki, Angew. Makromolek. Chem., 33, 129 (1973).
94. Z.A. Rogovin, J. Polym. Sci., C, 37, 221 (1972).
95. V. Stannett, H.B. Hopfenberg and J.L. Williams, Polymer Science and Technology (R.W. Lenz and R.S. Stein, eds.), Plenum Press, New York, 1973, vol. 1, p. 321.
96. R. Houwink, J. Prakt. Chem., 157, 15 (1940).
97. T.I. Min and H. Inagaki, Polymer, 21, 309 (1980).
98. C.I. Simionescu, F. Denes, M. Macoveanu, G. Cazacu, M. Totolin, S. Percec and D. Bahaur, Cellulose Chem. Technol., 14, 869 (1980).
99. A. Moze and S. Lapanji, Makromolek. Chem., 180, 1599 (1979).
100. M. Tsuzuki, I. Hagiwara, N. Shiraishi and T. Yokota, J. Appl. Polym. Sci., 25, 2721, 2909 (1980).
101. K.S.V. Srinivasan and D. Sudhakar, Polym. Bull., 2, 321 (1980).
102. N.I. El-Awady, N.A. Ghanem. W.B. Pedersen and K. Singer, Eur. Polym. J., 15, 1017 (1979).
103. M. Wahimi, I. Sakamoto and S. Tani, Kami to Purasuchikku, 6, 10 (1980).
104. O.I. Smolyakov, B.P. Morin, Z.A. Rogovin, M.A. Linyaeva and A.P. Perel'dik, Khim. Volokna, 4, 47 (1975).
105. B.V. Kokta, J.L. Valande and C. Daneult, Transactions, 7, TR5 (1981).
106. J.T. Guthrie and J.A. Perceival, Polymer, 18, 531 (1977).
107. V.N. Sharma and J.H. Daruwalla, Cellulose Chem. Technol., 13, 433 (1979).
108. J.H. Daniel, Jr., S.T. Moore and N.R. Segro, Tappi, 45, 53 (1962).
109. D.S. Varma and V. Narashimhan, J. Appl. Polym. Sci., 16, 3325 (1972).
110. S.B. Lee and M. Fujii, Kami-pa Gikyoshi, 18, 413 (1964).
111. H. Tonami, H. Watamoto and S. Chidamberam, Kogyo Kagaku Zasshi, 74, 1462 (1971).
112. N.-S. Hon, J. Appl. Polym. Sci., 23, 3591 (1979).
113. A. Nagaty, S.E. Shakra, S.T. Ibrahim and O.Y. Mansour, Cellulose Chem. Technol., 14, 177 (1980).
114. P. Cremonesi, Ric. Doc. Tess., 5, 142 (1968).
115. Y. Iwakura, T. Kurosaki, K. Uno and Y. Imai, J. Polym. Sci., Part C, 4, 673 (1963).
116. I.M. Trivedi and P.C. Mehta, Cellulose Chem. Technol., 7, 401 (1973).

117. Y. Nakahara, Sen-i Gakkaishi, 20, 319 (1964).
118. M.H. El-Rafie, M. El-Kashouti, F. El-Sisi and A. Hebeish, Cellulose Chem. Technol., 15, 427 (1981).
119. V.T. Stannett, W.M. Doane and G. Fanta, Polymer Grafted Cellulose and Starch in Absorbency (P.K. Chatterjee, ed.), Elsevier Scientific Pub. Co., New York, 1985.
120. V. Stannett, Raidat. Phys. Chem., 18, 215 (1981).
121. A.H. Zahran, J.L. Williams and V.T. Stannett, J. Appl. Polym. Sci., 25, 535 (1980).
122. S.H. Hui and P. Lepoutre, J. Appl. Polym. Sci., 19, 1771 (1975).
123. R.H. Yusupatiev and N.M. Idriskhodzhaeva, Uzbek. khim. Zh., 5, 23 (1981).
124. S.E. Artemenko, L.G. Glukhova and M.P. Bereza, USSR Pat., 834027 (1981).
125. S. Dumitriu and R. Butnaru, Ind. Usoara: Text., Tricotaje, Confectii Text, 31, 355 (1980).
126. Z.A. Rogovin, L.S. Sletkina, F.A. Baibikov and A.V. Kashkim, Cellulose Chem. Technol., 14, 623 (1980).
127. P. Wakayama, Jap. Kokai Tokkyo Koho, 8,153,278 (1981).
128. N. Inagaki and K. Katsura, J. Polym. Sci., Polym. Chem. Ed., 18, 441 (1980).
129. J. Erdelyi, E. Szekely and A. Haman, Papiripar, 23, 212 (1979).
130. R. Chatelin, Inf. Chim, 213, 181 (1981).
131. G.N. Richards and E.F.T. White, J. Polym. Sci., Part C, 4, 1251 (1963).
132. A. Gangneux, D. Wattiez and L.E. Marechal, Eur. Polym. J., 12, 535 (1976).
133. E. Riande, A.M. Mateos and G.M. Guzman, Eur. Polym. J., 6, 437 (1970).
134. E. Riande, A.M. Mateos and G.M. Guzman, Eur. Polym. J., 6, 1247 (1970).
135. D. Wattiez, P. Fabre, H. Lapie and A. Gangneux, Ger. Offen., 2,335,213 (1974).
136. W.J. Brickman, Tappi, 60, 99 (1977).
137. D.N.-S. Hon, "Chemical Modification of Lignocellulosic Materials: Old Chemistry, New Approaches," Polymernews, in press.
138. V. Stannett, "Some Challenges in Grafting to Cellulose and Cellulose Derivatives" in Graft Copolymerization of Lignocellulosic Fibers (D.N-S. Hon, ed.), ACS Symposium Series #187, Washington, D.C., 1982, pp. 3-20.
139. R. Narayan, "Synthesis of Controlled Cellulose - Synthetic Polymer Graft Copolymer Structures" in Cellulose and Wood: Chemistry and Technology (C. Schuerch, ed.), John Wiley, 1989, pp. 945-961.

DEVELOPMENT OF APPLICATIONS AND PROPERTIES OF CELLULOSE GRAFT COPOLYMERS

JEANETTE M. CARDAMONE
Department of Clothing and Textiles, College of Human Resources, Virginia Polytechnic Institute and State University, Blacksburg, VA 24061-8396

Introduction

Biobased materials are currently under development. Cellulose is the most abundant biomaterial. It is the chemical composition of cotton, rayon, and wood. With sustainable harvesting, this renewable resource can be used as starting material in synthesis for higher-value-added products having material characteristics similar to metals, plastics, and resins with the specific properties of high strength, low weight, corrosion resistance, biodegradability, and biological resistance. The modification of cellulose for improved properties has an extensive history. Since 1926, crosslinking applications with oligomers for improved appearance retention and dimensional stability in cotton fabrics have been reported [1,2,3]. These treatments generally result in reduced strength and abrasion resistance [4,5,6]. Other properties such as water resistance [7], flame retardency [8,9], and dyeability [10,11] were improved. Oligomer crosslinking reactions with paper improved dimensional stability and wet strength [12,13,14]. Early crosslinking reactions with rayon improved resilience and strength [15,16]. Early attempts to improve the properties of wood involved reinforcement with polymers to stabilize the cell wall and increase mechanical properties [17,18]. Graft copolymerization is another route for significant alterations in the chemical, physical, and mechanical properties of cellulose polymers. For 30-40 years, the traditional applications for cellulose in fibers and film have been in intense competition with synthetic markets. This led to the onset of early studies in the 50's and 60's to link synthetic polymers chemically to cellulose by graft copolymerization. Over the years various syntheses-related technical problems have been overcome so that grafting has become a significant commercial application. There has been a proliferation of reports on the effects of graft polymerization on morphology [19-22], surface and organo-chemical properties [23-26], structure and physico-mechanical properties [27-29], and tailoring for specific end use [27-30]. The existing methods of grafting onto cellulose have been described in several reviews [31-34]. Processes and mechanisms for graft copolymerization reactions on cellulose have been the topics of hundreds of reports. This has led to a better understanding of the structures of grafted cellulosics and the proper evaluation of their applications and properties.

Background

Cellulose. Cellulose is a naturally occurring linear polymer, a polysaccharide, formed from glucose, the source of anhydroglucose units linked by 1,4-glycosidic bonds. Cellulose is chemically, poly (1,4-ß-D anhydrogluco pyranose). The glycosidic linkages are hydrolized easily by acid but are stable to alkali. The polymer has been characterized as having a viscosity - average molecular weight of 800,000 and a degree of polymerization of 5,000 [35]. Structural elucidation studies show two polymorphic lattices, I and II, of which I is most prevalent and contains 70-80% crystallinity and 20-30% amorphous content. Type II is typified by mercerized cotton with 60% crystallinity. The fine structure of cellulose may be described as a lateral array of polymer chains arranged both in the ordered crystalline regions and in the less ordered amorphous regions where these regions may be interconnected or embedded. This lateral order of associated polymer chains is imposed through inter- and intramolecular hydrogen bonding though cellulose's hydroxy group functionality. The hydroxyl groups on cellulose on carbons 1,2, and 6 impart hydrophilicity through hydrogen bonding to water. Water is tightly bound in the accessible amorphous regions. Moisture absorption in these regions at standard conditions is 6.38% and in highly amorphous cellulose is 18.17% [36]. The strong intermolecular bonding forces in cellulose result in useful and important properties such as high thermal stability, good mechanical properties, and good moisture absorption. Major weaknesses include rapid burning behavior and microbiological susceptibility, poor solvent solubility, poor dimensional stability and lack of thermoplasticity.

Cellulose reacts chemically as a trihydric alcohol having C-1 and C-2 secondary hydroxyl groups and one primary hydroxy group on C-6. Chemical reactivity is somewhat limited to the accessible amorphous regions and because of the heterogeneous nature of cellulose, by accessibility, cellulose exhibits different extents of reactivities to regents. The products formed are thus heterogeneous. To achieve uniformity of reactivity, cellulose is generally reacted in the swollen state. The broad range of accessibilities for cellulose as measured by hydroxy-group hydrogen deuterium exchange showed the order of accessibility to moisture as: rayon > finely ground cotton > mercerized cotton > jute > bleached cotton [37]. The hydrolysis-oxidation method for quantitative determination of accessibility showed viscose rayons - 25 to 35%; wood cellulose, mercerized cotton, and high-crystallibity rayons - 10 to 15%, and cotton celluloses - 5 to 10% [38,39]. The property-behavior of cotton, rayon, jute and wood are influenced by the extent of order and amount of accessibility. For example, from X-ray diffraction measurements, the crystallinity of cotton was 70%; viscose rayons, 40%; and more highly crystalline rayons (Fortisan), 50%; mercerized cotton 51%; wood pulps, 60% [40,41]. The amount of cellulose accessible to water by hyrolytic studies in cotton fibers was determined as 30-40%, in wood pulps: 40-50%; in Fortisan: 44-65%, and in viscose and high tenacity rayons: 65-75% [42]. Order/disorder studies consistently show the inverse relationship between crystallinity and accessibility. Whereas cotton is the purest form of cellulose at approximately 95% content, the cellulose content of wood is 40-55% with associated 15-35% lignin and 25-40% hemicellulose [43]. Jute contains 65-75% cellulose; 20% lignin, and about 11-12% xylan [44].

The hydroxy groups on the 2, 3, and 6 positions of the anhydroglucose units are the functional reactive sites for modification to tailor cellulose for specific properties. As a polyhydroxy compound, these reactions are typical of alcohol groups. The extent of reactivity and cellulose heterogeneous accessible/inaccessible regions are limitations. Common reactions include derivitization to form cellulose ethers and esters. Cotton cellulose reacts topochemically from the surface to permeate the inside of the accessible regions. Wetting or swelling pretreatments which disrupt intermolecular-hydrogen bonding contribute to greater accessibility and more uniform reactivity.

Chemical modification of cellulose causes changes in fine structure and fiber morphology which influences fiber properties and the properties of manufactured yarn and fabric structures. Cotton fibers have a complex morphology consisting of primary and secondary walls made up of cellulose fibrils which spiral 70° and 20-45° respectively across the fiber axis with intermittent reversal of direction. The fiber has frequent convolutions which result in part from a dehydrated, collapsed central canal or lumen. The fiber appears to be a convoluted cylinder of tightly packed concentric spiralling fibril sheets which can be separated somewhat by swelling. These gross morphology features are visible by electron microscopy at 10,000X magnification. Chemical modification can affect this structure through surface or interior modifications as determined by accessibility and pores which have been described as "discrete spaces bounded by microfibrillar surfaces" [45].

Jute. Jute is a cellulosic fiber found in the stalk of the plant which grows 2-4 mm tall and 10-20 mm in diameter and is highly lignified. Lignin provides in jute and wood a stress-transferring polymeric matrix and protection against microorganisms in rotting and mildew action. In use, jute can be left as a lignocellulosic fiber. It differs from cotton cellulose in angle of orientation which is 7° to 11° and thus has fibril planes highly oriented along the fiber axis. Unlike other bast cellulosic fibers (flax, hemp, ramie), jute is weaker when wet and has the lowest elongation at break, 1.8% and exhibits low resistance to bending. The moisture regain of jute is high at 13.25%. It is degraded by light; and has poor aging properties, and high resistance to microorganisms. The elementary fibers of jute, contained within the lignin matrix are so short that they cannot support stress, thus requires that jute remain in the lignified state during processing and use. Removal of lignin reduces the strength of jute by 83%. [46]

Wood. The properties of wood are highly dependent on morphology. In wood, the cellulose microfibrils are embedded in a hemicellulose and lignin matrix. The matrix is an intercellar substance which binds mature wood cells with layers of cell walls. These cell walls contain heterogeneous arrangements of fibers where some are oriented parallel to the fiber axis and some are in a loosely organized network. Lignin is found in the middle lamella region between and within the cell wall. Microfibrils are aggregates of cellulose chains and their highly oriented arrangements provide rigidity for supporting stress. Wood may be thought of as a natural composite material or a reinforced structure since these microfibrils are embedded in the hemicellulose, amorphous lignin matrix. Electron microscopy is used to follow change in the parallel-oriented microfibrils and their arrangements with treatments and biological change. The microfibrils contain both crystalline and amorphous regions. Intercrystalline swelling is believed to involve only the accessible regions of the cellulose microfibrics. Intracrystalline swelling involves both amorphous and crystalline regions. A 15-20% sodium hydroxide solution is used to swell cotton in the process of mercerization which physically changes the supramolecular cellulose structure so that it becomes stronger and accepts dye more readily. It is used in wood to remove hemicelluloses in the formation of viscose rayon. Solvents for dissolving wood and cotton cellulosics are metal complexes of bases which can be used in the absence of oxygen to prevent degradation. These include Cuoxam$[Cu(NH_3)_4](OH)_2$, Cuen$[Cu(en)_2](OH)_2$, and Cadoxen$[Cd(en)_3](OH)_2$.

Rayon. Rayon is regenerated cellulosic fiber based on the modification of natural cellulose to make soluble derivatives which can be extruded to give textile fibers, films and industrial products. In the process of rayon production, cellulose from wood pulp or cotton linters in sheet form are exposed to concentrated sodium hydroxide solution which causes oxidation and aging. This results in depolymerization and molecular weight loss and the formation of carboxyl groups. After aging, cellulose is derivatized to the xanthate form by treatment in carbon disulfide. The xanthate is soluble in alkali and the viscose solution is formed and forced through a spinerette into an acid bath where the xanthate groups are decomposed and regenerated cellulose is obtained as a filament or sheet. This process causes pronounced changes in cellulose morphology and properties

The original viscose rayon proved limited in end use by strength loss and lack of dimensional stability. The viscose process resulted in a lack of fibril structure for rayon and low degree of polymerization. Modification of the fiber production processes led other more serviceable rayons such as Polynosic, High-Wet-Modulus (HWM), High-Performance (HP) including Fostisan and High-Tenacity tire cord. Improved rayons with a higher range of tenacities are more competitive with cotton. Morphology features of the rayons vary. For example HWM is well-ordered with smaller micropores and less internal surface area. [47] The irregular fiber cross section after extrusion though the spinnerette results from skin-core development where the skin content, representing the strongest area of the fiber, can be regulated to the extreme of 100% skin-like texture found in round cross-sections typical of HP tire core rayon. In polynosic and HWM rayons, a fibrillar cellulose structure is developed in fiber formation. Extruded rayon fibers of various degrees of orientation of polymer chains are altered by various amounts of stretching to induce a wide range of mechanical properties for specific end uses. [48]

Paper. The important features affecting the properties of paper include fiber strength, bond strength, and bonded area. Cellulosic fibers within a wet web are aligned with various degrees of straightness and are under various tensions when dried. Estimation of failure under strain requires measuring shear (torsional and tensional) and bending stresses of bonded areas within the web. The strength considerations are influenced by the relationships among fiber and bond properties, sheet geometry, and load-elongation behavior. [49]

Cellulose Graft Copolymerization. The chemical modifications of cellulose by etherification and esterification, oligomer crosslinking and graft copolymerization have become traditional. Crosslinking was investigated to improve dimensional stability and crease resistance. The crosslinking resins although small enough in molecular size to permeate the fibers, locked the cellulose polymer chains into a rigid three-dimensional network which led to brittleness and severely limited durability. Grafting and crosslinking treatments on cotton fabrics have been ongoing concurrently since the 50's. Grafting or creating branches on the cellulose polymer backbone was perceived as a way to improve the properties of cellulose especially in cotton cloth without destroying its desirable properties. The applications of cellulose graft copolymerizations extend also to rayon, jute, paper, and wood. Through graft copolymerization the properties of the grafted synthetic polymer are superimposed upon those of cellulose. Cellulose then take on different properties and becomes suitable for a wider range of end uses. Grafting is not precluded by the incompatibility of properties of branching polymers vs. those of the backbone polymer. [50] Property modification by grafting onto cellulose can depend upon whether the grafted polymer is located on the surface or in the bulk of cellulose. [51]

Since natural cellulosics have limited solubility, they are said to undergo solid-phase or heterogeneous graft copolymerization reactions. Thus, the physical structure state of aggregation and crystallinity are important. In order to maintain the natural state of cellulose, homogeneous or solution-phase grafting is seldom carried out. Factors which influence the heterogeneous grafting process include the contact area of cellulose and grafter monomer, the structure of cellulose, the rate of monomer diffusion, reaction mechanism and initiation into the cellulose backbone. The kinetics of the graft copolymerization reactions are determined by the overall structural and chemical changes in cellulose during reaction. Certain unsaturated monomers have been identified as useful for conferring desirable properties to cellulose: vinylchloride for flameproofness; butadiene for water-repellency and crosslink; styrene for water-repellency, and acrylonitrile for water-repellency and rigidity. In an early report, successful grafting was cited as synthesizing a graft copolymer where the monomer is added to cellulose in 15-30% by weight so that a cellulose backbone has 2 to 5 grafted chains each with a degree of polymerization of 20 to 50 and where no more than 5-10% homopolymer forms. [52]

Since the 1950s graft copolymerization of vinyl monomers onto cellulosic fibers has been the subject of intense reporting directed toward simulating the synthetic fibers, nylon and polyester [53,54,55] and ion exchange materials. [56] Problems with grafting included lack of improvement in wrinkle resistance of cotton cloth, tenacity loss, concomitant oxidation and cleavage of the cellulose backbone and concurrent homopolymer formation. These problems led to extensive reporting on various graft copolymerization reaction mechanisms, conditions, solvent systems, monomer type and concentration, control of extent of grafting, property studies, and tailoring for end use application.

Since the 1960's feasibility studies have led to industrially useful syntheses procedures for the formation of cellulose graft copolymers. The classically efficient method utilizing free radical polymerization initiated redox systems led to the industrial production of Mtilon-V (cellulose polyacrylonitrile graft copolymer) by the Textile Institute of Moscow. This fiber was produced by grafting viscose rayon from solutions of the monomer using oxidation-reduction initiation by Fe^{2+} - H_2O_2. Mtilon-V fiber was designed for carpeting with 50 to 80% Mtilon by weight. At 50 5% polyacrylonitrile grafting onto cellulose, maximum physico-mechanical properties for the end use were achieved. Although the graft copolymer showed 10 to 15% fiber strength loss, abrasion resistance increased 13 to 20 times and cyclic bending strength increased 16 to 18 times. There was an improvement in color fastness. [57]

Grafting methods have involved free radical polymerization of monomers with the radical site being initially formed on the cellulose. Grafted chains form from ionic polymerization of monomers and from polycondensation reactions or addition polymerization from further polymerization or crosslinking of a preformed polymer which was initially of relatively low molecular weight. The free radical method is most common and usually vinyl monomers are polymerized in the presence of cellulose. The properties of cellulose graft copolymerized vinyl derivatives often depend on the length of the side chains. To begin the typical reaction sequence involving free radical initiation, propagation, and termination; initiator radicals are formed by dehydrogenation, depolymerization or oxidation from typically, a Cobalt-60 source of gamma radiation. Other sources of free radicals are high energy and ultraviolet radiations, chemical redox systems with, for example, cerium ions, decomposition of peroxy compounds or azo compounds, chemical modification, use of carona or arc discharge, electrochemical, mechanical, ultrasonic, or thermal methods. These radicals combine with monomers to form free radical monomers which propagate and which can abstract a hydrogen atom from cellulose, thereby activating it and causing chain transfer of monomers to cellulose for cellulose graft copolymerization formation. Because of the poor chain transfer ability of cellulosic hydroxyl groups, thiol modification using ethylene sulfide has been used as the source of thiol groups.

Ideal conditions for graft copolymers include adequate accessibility of grafted monomer for cellulose substrate and adequate diffusion, little homopolymer formation, synthesis without organic solvents or dissolution of cellulose, no appreciable cellulose degradation, strict control of chain length of the grafts, and the ability to modify properties depending upon the length of these grafted chains.

The role of lignin in graft copolymerization of the lignocellulosics: wood and jute has been examined. [58-61] Some studies show that by oxidation with ceric ion free radical initiation, lignin may contribute to grafting up to a certain percentage content but may also contribute to grafts with lower molecular weight [61]. Lignin was found to limit grafting onto jute by causing more oxidation in the reaction with ceric ion. [59] It is possible that lignin as a potential source of free radicals may interfere with the initiation or polymerization reaction by termination or chain transfer in free radical reactions from phenolic groups. Conflicting reports indicate that high lignin content may enhance graft copolymerization on pulps by higher grafting efficiency with lower homopolymer formation. [60]

The common monomers employed in graft copolymerization may be classified by functional group. Vinylic acid monomers include acrylic, methacrylic, styrene sulfonates, and vinyl phosphate. Amine and amide monomers include acrylamide, methacrylamide, amino acrylates, N-methylolacrylamide, and vinyl pyridines. Aromatic monomers include styrene, alphamethylstyrene, and vinyl toluenes. Divinyl crosslinking monomers include butadienes, isoprene, divinylbenzene, and divinyl esters. Ester monomers include acrylates, methacrylates, maleates, and vinyl acetate. Nitrile monomers include acrylonitrile and methacrylonitrile. Tailoring for specific end use is exemplified by grafting polyacrylonitrile homopolymer onto cellulose to provide protection from mildew and durable press; polymethacrylonitrile for increased strength; polyacrylic - and polymethacrylic acids for cation - exchange reaction; polymethyvinylpyridine for anion-exchange properties; polystyrene and polyfluoro compounds for water repellency and acid resistance, polybutylacrylate for thermoplasticity, and polyvinylidene chloride for flame resistance. Biodegradability is an important issue and studies to achieve this end by graft copolymerization have involved grafting monomers with hydrolyzable and/or oxidizable functional groups onto cellulose to pattern biological systems which degrade large molecules by hydrolysis and oxidation.

When modified celluloics undergo grafting; the extent reaction will depend on many factors. One factor is the variability of fiber structure in relationship to its openness and availability. Openness and availability will affect accessibility of hydroxyl groups to reaction and especially to reaction with the newly introduced functional groups of the monomer. Accessibility will determine the location of these new groups on the anhydroglucose units. It will also determine the reaction specificity of cellulose with a free-radical initiator and indirectly, the reaction of the monomer with the initiator. Many variables must be recognized, understood fully and controlled.

When tailoring for specific applications and properties, various reaction conditions have proved successful for localizing the grafting homopolymer chain onto the substrate. These conditions become important for specific applications such as surface grafting for water and oil resistance and bulk grafting for resistance to microbiol degradation or for flame resistance. [62] Thus grafting will cause changes in fiber morphology and structure. Ultimately these changes derive from the specific features of these grafted homopolymer chains, for example, homopolymers by their chemical composition and macrostructure can be hard, rigid, and brittle or soft, rubbery, and thermoplastic; mono- or difunctional with the same or with different functional groups; with polar or nonpolar groups; with or without hydrogen-bonding capability; with high or low vapor pressures, etc. [63]

The following tables which include cotton, jute, rayon, wood, and paper cellulosics point out these features as the myriad of diverse choices available for tailoring cellulose graft copolymers for specific applications and properties.

COTTON:

Substrate/Monomer	Initiation/ Treatment	Application/ Property
Cotton yarn • Acrylonitrile	• gamma radiation from cobalt 60 • monomer 20% to 40% in 75% aqueous $ZnCl_2$ • 0.11% to 25% homopolymer add-on	elongation strength
Remarks: Slight descrease in breaking strength; significant increase in elongation at break [64]		
Cotton Fabric • 1-N-methylolacrylamide • acrylamide, hexamethylene-diamine, HCL-salt, softeners formalin	• $K_2S_2O_8$, $(NH_4)_2S_2O_8$ or H_2O_2 • acid condensation catalysts • organic solvents • pad, dry, cure	crease recovery abrasion resistance tear strength
Remarks: Grafted polymer deposited on outside without permeation inside fiber improved crease recovery. Aftertreatment with hexamethylene diamine followed with formalin is most effective in improving crease and abrasion resistance; no visible change in crease resistance after laundering; crease recovery and abrasion resistance improved; tear strength decreased [65]		
Cotton Fabric • N-Methylolacrylamide (MAM)	• Crosslink partially acrylamido methylated cotton either by addition of cellulose hydroxyls to double bonds or by anionic or free radical MAM homopolymerization • catalysis treatments (alkaline and free radical)	Crease recovery moisture regain strength elongation modulus
Remarks: MAM reacts to eitherify C6-hydroxyl groups on cellulose which is 20% accessible; homopolymerization of MAM for a cross- linking polymer to cellulose gives maximum crease recovery with 0.2mmole/gram crosslink concentration; highest crease recoveries required one crosslink per 4 to 5 accessible anhydroglucose units; wet and dry crease recovery improved with swelling of substrate; wet crosslinking increased moisture regain; wet and dry tensile strength losses; loss in breaking elongation; no change in modulus of cotton [66]		

Substrate/Monomer	Initiation/ Treatment	Application/ Property
Cotton fabrics • Acrylamide	• Ceric ammonium nitrate catalyst • cuprammonium solution • acid-cure of methylolated polyacrylamide grafted fabric alone and with adipamide	Crease resistance abrasion resistance tear strength tensile strength elongation
Remarks: Methylolated PAM-grafted fabrics (I) and methylolated-cured-PAM grafted fabrics (II) had excellent crease resistance whem compared to PAM-grafted fabric. Both I and II had comparable abrasion resistance to untreated fabric, lower tear strength but comparable tensile strength and slightly greater elongation at break. [67]		
Cotton Fabric Viscose Rayon • Acrylonitrile • Acrylamide • Methylacrylate	• Ceric ion Redox System • Solution and vapor saturation techniques	• rot and mildew resistance • tear strength • abrasion resistance • shrinkage control
Remarks: At 21.7% polymer concentration polyacrylonitrile grafted cotton maintained 100% tensile strength after 6 weeks burial; tear strength and abrasion resistance decreased; increasing polymer concentration of polyacrylonitrile from 13.4% to 29.2% imparted shrinkage control to rayon but decreased tensile and tear strengths. [69]		
Cotton Fabric cellophane • Acrylonitrite (AN)	• sodium thiosulfate - potassium persulfate redox system • saponification of PAN grafted cellulose fabric to form $-CONH_2$ for subsequent formolyation with methylol groups ($-CH_2OH$) by alkaline formaldehyde treatment at 60°-70°C followed by acid cure produced fabrics containing modified cellulose graft copolymer	microbial resistance wrinkle recovery
Remarks: PAN graft copolymerized cotton showed resistance to microbial degradation; formaldehyde treated PAN graft copolymer improved wrinkle recovery in cotton fabric, 62-67% tear strength retention and 65-70% tensile strength retention. [70]		

Substrate/Monomer	Initiation/ Treatment	Application/ Property
Cotton Fibers Cellophane Styrene	• cobalt 60 mutual and pre-irradiation in vacuo	moisture regain
Remarks: moisture regain increased with increase in % grafting; grafting increases accessibility of cotton by disruption of crystalline regions; lower crystalline cellophane exhibited greater swell- ing and thus greater accessibility over cotton fibers. [71]		
Cotton Fabric Viscose Rayon Fabric Styrene Methyl Acrylate (MA)	H_2O_2 in situ	crease resistance
Remarks: Dry and wet crease resistance of styrene graft copolymers was improved over untreated cotton; crease recovery of PMA grafted cotton in wet state increased with increasing graft level but did not improve at the same graft level on dry cotton; hydrophobic styrene grafting caused stiffening of the wet fabric and reduced water imbibition; effects of grafting PMA and styrene are similar in cotton and rayon, qualitatively. [72]		
Cotton Fabric, Phosphoryated • Vinyl acetate • Methyl acrylate • Styrene • N-methylolacrylamide • N-glycidyl acrylate	• ferrous ion- hydrogen peroxide initiators	abrasion resistance wrinkle recovery strength
Remarks: Above 30% polymer add-on with methyl acrylate, abrasion resistance increased; methyl aceylate and N-methylol acrylamide improved wrinkle recovery and abrasion resistance but lowered strength; phosphorylated cotton and acrylic graft vinyl copoly-mers were resistant to alkaline hydrolysis. [73]		
Cotton fabric • Acrylonitrile • Methyl methacrylate • Butyl methacrylate • Lauryl methacrylate	cobalt 60 irradiation • copolymerization with subsequent crosslinking to cotton cloth employing dimethylol dihydroxyetheleneurea (DMDHEU) by pad, dry, cure	Durable-Press wash-wear ratings wrinkle recovery strength
Remarks: Wash-wear ratings, abrasion resistance, tear strengths, wrinkle recovery were improved over DMDHEU crosslinked fabrics by graft copolymerization. [74]		

Substrate/Monomer	Initiation/ Treatment	Application/ Property
Cotton yarn • Acrylonitrile	ceric salt initiator 0.25 to 1.0N nitric acid	Resistance to microorganisms • test resistance to pencillium SP, Aspergillus Fumigatus, Aspergillus niger, Chaetomium globosum Memnoiella chinata, Rhizopus sp.
Remarks: Resistance to microorganisms was obtained with 20% polymer add-on; grafting increased tensile strength with conditions below 0.6N acid and less than 4 hours exposure. [75]		
Cotton fabric • Acrolein	cobalt 60, gamma irradiation • solvents: water, ethyl alcohol, acetone, diethyl ether, benzene • grafting in vapor phase	crease resistance shrink-proof, strength, elongation rot resistance • test resistance Asper gillus terrus, Penicillium asperulum, Penicillium cherisogenium, Penicillium notatum, Penicillium soppi
Remarks: Cellulose crosslinks acrolein at the double bond and at the aldehyde groups; graft copolymerization in solvent is inefficient as homopolymer forms - no homopolymer forms in vapor phase; all concentrations (0.6 to 20%) monomer in all solutions increased crease recovery; weight gain was followed by some loss in strength and elongation; acrolein graft copolymerized cellulose is fungus and rot resistant at 4% and 10% acrolein graft-on after burial 30 days at 30°C with no weight loss. [76]		
Cotton fabric DMDHEU Methacrylic acid	post irradiation, 1 Mrad solvents: methanol, water • after grafting, pad with DMDHEU, dry, cure	crease resistance anti-soil soil-release strength
Remarks: Methyacrylic acid graft when applied with DMDHEU exhibited anti-soil and soil-release properties greater than fabrics treated with DMDHEU alone; cellulose graft copolymer without DMDHEU maintained strength, elongation at break above untreated and DMDHEU treated fabrics; tear strength for graft copolymer is less than untreated fabric but much greater than DMDHEU-treated fabric; wrinkle recovery is improved the most by DMDHEU alone. [77, 78]		

Substrate/Monomer	Initiation/ Treatment	Application/ Property
Cotton fabric, fibers Fluoroalkyl (meth) acrylates	• chain transfer in pure monomers • aqueous monomer emulsions • emusifyers	non-wettability oil-repellency
Remarks: Uniform surface distribution of nonomer was achieved by reaction of cellulose in pure monomer and in emulsion with emulsifier; type of emulsifier influences molecular weight of graft chains and surface properties of the graft copolymers. These results show that non-wettability was achieved by grafting 4.5 to 6.0 of polyfluoroalkyl (meth) acrylate onto cellulose. [79]		
Cotton, 50/50 polyester blend Methyl trimethoxy silane	• in situ emulsion polymerization by silanol condensation with cellulose • pad, dry, cure • organo metallic catalyst	appearance retention abrasion resistance water repellency strength shrinkage fabric hand
Remarks: Application of the graft copolymer from the same finish bath with DMDHEU and compared with DMDHEU-treated fabric; appearance retention was comparable, abrasion resistance increased, water repellency was imparted, strength retention was comparable, shrinkage was improved over the untreated fabric but was less than DMDHEU-treated fabric; fabric hand was altered by the composition of the pad bath but was improved over the untreated fabric but comparable to DMDHEU-treated fabric. [80]		
Cotton fabric • Methyacrylamide	• photoinitiation • pad fabric with monomer in solvent and photalyze to form fabric copolymer	moisture regain elongation strength wrinkle recovery
Remarks: Appearance, durability, comfort aspects improved; improvement in moisture regain, increase in elongation at break; stength retention was 90% of breaking strength and 70% of tearing strength; wrinkle recovery was highest at the highest graft-on level, 14%; moisture regain increased with increase in polymer content. [81]		
cotton fabric, fibers • Butyl methacrylate	• cobalt 60 radiation • graft copolymerization and crosslinking with DMDHEU • two step copolymerization; irradiate cotton and immerse in solutions of monomer	durability strength elongation abrasion resistance wrinkle recovery shrinkage

Substrate/Monomer	Initiation/ Treatment	Application/ Property
Remarks: Grafted copolymer added strength to DMDHEU crosslinked cottons, improved elongation at break but less than untreated fabrics; tear resistance and abrasion resistance were increased slightly but not at the levels of the control; wrinkle resistance improved over the control but not as much as DMDHEU-treated fabric. [82]		
cotton yarn • Methyl acrylate • Ethyl acrylate • n-Butyl acylate • Methyl methacrylate	ceric ion initiation	thermal behavior
Remarks: Thermal stability increased with various acrylate monomers to various percentages of grafting; Order of thermal stability for grafted copolymers was ethyl > methyl > methacrylate > n-butyl acrylate; thermal stability of cellulose was increased up to 350°C by graft copolymerization; decomposition for cellulose alone was 210°C. Increasing graft-on % of poly(methyl methacrylate) raised the decomposition temperature (Td) of cellulose as follows: at 15%, Td = 282°C; at 62%, Td = 298°C; at 74%, Td = 296 °C. [83]		
Cellulose Samples Acrylic acid Methacrylic acid	ceric salt initiator	hydrophilicity water absorption thermal behavior
Remarks: Glass transition temperature (Tg) of grafted copolymers was water-dependent above 50% graft-on; shifts in Tg resulted from monomer type, changes in moisture regain, and bound water in the graft copolymer; at 93% relative humidity, maximum regains were 8% for acrylic acid graft copolymer and 18% for methacrylic acid copolymer; at 66% relative humidity moisture regain of copolymers was independent of graft-on percent. [84]		
cotton yarn, fabric Methyl vinyl ketone	cobalt 60-gamma irradiation solvents - water, methanol, N, N-dimethylformamide and solvent combinations	strength elongation toughness stiffness
Remarks: Mechanical properties of graft copolymers were less than untreated samples and independent of add-on; stiffness decreased with add-on, small losses in breaking strength and tearing strength and fex abrasion but slight increase in elongation over untreated cotton was achieved at 9-16% add-on of poly(methyl vinylketone). [85]		

<table>
<tr><th>Substrate/Monomer</th><th>Initiation/
Treatment</th><th>Application/
Property</th></tr>
<tr><td>cotton fabrics
Ethyl acrylate
Butyl acrylate
N-methylolacrylamide
Methyl methacrylate
Styrene

modified cottons
• acrylamidomethyl cotton (ACMC)
• cotton maleate (CM)
• allyl cotton (AC)</td><td>ceric ion initiation amine hydrochloride catalyst, hydroquinone inhibitor
• copolymerize modified cottons with ethylenic monomers</td><td>resiliency
strength</td></tr>
<tr><td colspan="3">Remarks: Methyl methacrylate is less effective than ethyl acrylate and butyl acrylate in improving wrinkle recovery, tensile strength and tear strength in ACMC cottons. In contrast, ethyl acrylate and butyl acrylate crosslinked to ACMC showed increased wrinkle recovery and tear strength retention. Ethyl acrylate was superior; DMDHEU crosslinked cottons had greater durable-press ratings and wrinkle recovery but less strength when compared to ACMC ethyl acrylate fabrics; crosslinking CM and AC fabrics did not improve properties; introduction of polymers with double bonds into cotton provided sites for crosslinking and increased resiliency; physical property differences were attributed Tg differences; lower Tg homopolymer branches contributed. elastomeric properties. [86]</td></tr>
<tr><td>cotton yarn
Cyanoethylated cotton
Formaldehyde cotton
HCHO
Methyl acrylate MA
Ethyl acrylate EA
n-Butyl acrylate BA
Methyl methacrylate MMA</td><td>ceric ion initiation
ceric ammonium nitrate</td><td>moisture regain
strength
modulus
elongation</td></tr>
<tr><td colspan="3">Remarks: MA-, EA-, BA-, MAA grafted cottons and HCHO crosslinked grafted cottons exhibited from 3.3% - 4.7% moisture regain compared to 6.2% for untreated cotton; HCHO crosslinked cottons lost little moisture regain (5.5-5.7%); grafting cyanoethylated cottons lowered moisture regain; grafting with acrylate monomers decreases modulus and tensile strength with increased graft-on; HCHO alone decreased breaking strength and increased modulus above untreated cotton; both HCHO and crosslinked acrylate grafted cottons showed decreasing elongation with increase in bound HCHO; MMA grafted samples showed lowest elongation and this was associated with the homopolymer rigidity where its Tg is greater than 20°C; the acrylates with Tg below 20°C impart rubbery characteristics resulting in greater elongation for cellulose. [87]</td></tr>
</table>

Substrate/Monomer	Initiation/ Treatment	Application/ Property
cotton fabric binary and ternary mixtures of vinyl monomers containing Acrylamide Acrylonitrite Methacrylate Methylmethacrylate Ethylacrylate	• ceric ammonium nitrate initiation • in situ methylolation crosslinking of grafted samples • impregnate acrylamide - ethylacrylate mixture containing ceric salt onto cotton by padding • emulsify monomers in padding liquor Acrylamide-acrylonitrile Acrylamide-methacrylate Acrylamide-ethylacrylate Acrylamide-methylolated Acrylamide-methylolated, crosslinked Acrylamide-acrylonitrile, methylolated Acrylamide-acrylontrile, methylolated, crosslinked	wrinkle recovery strength abrasion resistance
Remarks: In crosslinked fabrics, wrinkle recovery increased with simultaneous reduction in tear and breaking strength and abrasion resistance; grafting and subsequent crosslinking showed good wrinkle recovery with moderate strength losses and good abrasion resistance except for the acrylamide-acrylonitrile treatments. [88]		
cotton fabric rayon fabric rayon/polyester blend (50/50) Triallyl phosphate (I)(TAP) Bis (B-chloroethyl) vinyl phosphonate (II) N-methylolacrylamide (MAM)	electron beam initiation • N-methylol acrylamide grafted cellulosics • phosphorous-containing monomer copolymerized with MAM groups onto the cellulose backbone	flame retardance crease recovery
Remarks: Fabric blend and weight influenced the copolymerization reaction and rate; multifunctional vinyl phosphonate oligomers with high phosphorous content imparted good flame retardancy and permanent press; grafted copolymers exhibited high crease recovery with 50% strength retention. [89]		

Substrate/Monomer	Initiation/ Treatment	Application/ Property
cotton fabrics N-methylolacrylamide (NMA) Vinyl phosphonate oligomer (VP)	photoinitiated ultraviolet source VP:NMA mole ratios 1:0.60 to 1:1.20	flame resistance flame retardancy strength
Remarks: Breaking strength was greatest at 91% retention at the highest VP:NMA ratio; tear strength retention was 44% and VP:NMA treated fabrics were 6 times stiffer; only 21% of the original abrasion resistance was retained; wrinkle recovery was improved over the untreated control; char length after 50 wash-dry cycles for mole ratios 1:1.20 was 107 mm passing FF5-74 flammability test. [90]		
cotton yarn rayon filament yarn Acrylic acids Methacrylic acids	Co-60 gamma radiation • follow graft copolymer-copolymerization by decrystallization of the grafted cotton samples with 70% zinc chloride or hot sodium hydroxide	absorbancy supersorbency
Remarks: in cotton, acrylic acid leads to greater sorbency than methacrylic acid; under conditions of decrystallization the grafted samples have high absorbency. [91]		
cotton linter Methylacrylate Ethylacrylate Acrylonitrile	ceric ammonium sulfate disperse, vat, direct basic, acid, reactive and mordant dyes	light fastness dye receptivity
Remarks: Cellulose was not dyed with disperse dyes but poly(ethylacrylate) grafted cellulose was dyed; cellulose was stained by mordant dye but poly(methylacrylate) grafted cellulse was dyed; reactive and basic dyes dyed untreated cellulose and all poly vinyl grafted celluloses. Lightfastness increased to 5 with 98.2% graft yield for dyed poly(acrylonitrile) and increased to 7-8 for 147.4% graft yield for dyed poly(ethylacrylate) grafted cellulose. [92]		
cotton fabric Diethylphosphatoethyl-methacrylate (PEPM)	cobalt-60 gamma radiation and electron beam radiation	fire retardancy
Remarks: The 3% phosphorous content necessary for adequate retardancy was achieved by a grafting level of 25-35% by radiation initiation. [93]		
cotton yarn Styrene	ceric ion initiator	strength moisture regain

Substrate/Monomer	Initiation/ Treatment	Application/ Property
Remarks: Moisture regain of graft copolymerized cotton was less than untreated cotton and decreased with increasing levels of grafting; at more than 4% grafting, there was improved water resistance of graft copolymers; ungrafted cotton showed an endothermic peak at 345°C while graft copolymerized cotton at 9% graft level showed the endothermic peak at 355°C; hydrophobic polystyrene homopolymer formed grafts with the accessible cellulose hydroxyl groups; and these grafts were themselves hydrophobic. [94]		
Cotton fabric acrylonitrile (AN)	• Co-60 gamma radiation • Fe^{2+}-thioureadioxide - H_2O_2 redox system • mutual irradiation • soil composition: aqueous and nonaqueous oily dispersions of carbon black • DMDHEU crosslinker • nonionic softener	soil resistance soil release
Remarks: Soil resistance depends on the grafting method and the percent add-on of grafted homopolymer; degree of soiling of copolymers from irradiation are higher than those of redox initiation; grafting impairs ease of soil removal; in DMDHEU crosslinked grafted fabrics, aqueous and oily soil removal was enhanced but was still less than the crosslinked control; DMDHEU crosslinking on grafted copolymers imparted higher aqueous soil resistance than the untreated fabric; nonionic softeners in the crosslinking bath decreased soiling in copolymers and increased soil removal; PAN grafted cellulose has decreased resistance to aqueous soil and decreased soil removal; PAN has high affinity for aqueous soil and may open up the cellulose structure for susceptibility; PAN decreased the affinity for nonaqueous oily soil, the amount of soiling decreased as % add-on increased; percentage soil removal is half that of the untreated fabric; DMDHEU crosslinking without softener on PAN cotton grafted by the redox system, increased degree of soiling; for the Fe^{2+} redox system when compared to untreated fabric, oily soil affinity is lowered after crosslinking in the absence of softener at add-on's up to 9%; copolymers by irradiation have higher oily soil resistance than those by the redox system after crosslinking with or without softener unlike aqueous soiling behavior; crosslinking PAN graft copolymers decreased moisture regain and water of imbibition while copolymers increased water or imbibition and reached high absorbency levels when compared to the crosslinked control; after crosslinking moisture regain of the copolymers and the control were comparable; tensile strength and elongation at break decreased considerably after crosslinking for the copolymers but durable press ratings increased. [95]		

Substrate/Monomer	Initiation/ Treatment	Application/ Property
cotton fabrics Acrylamide (AA)	Fe^{2+}-thioureadioxide-H_2O_2 redox system mutual irradiation • soil: aqueous dispersion of carbon black	resistance to aqueous and oily soils
Remarks: Soiling and soil release were examined at different graft-on percentages of homopolymer for aqueous and oily soils and stains for graft copolymers from redox and irradiation initiation; comparisons were made to untreated fabrics and to graft copolymerized fabrics crosslinked with DMDHEU with and without softener in the application bath; redox system graft copolymer up to 20% graft-on resisted aqueous soiling and at 5% graft-on, soil resistance resistance was maximum; at 9% graft-on, irradiation induced PAA resisted aqueous soiling; copolymers at 25% graft-on have much higher soiling; crosslinking treatments with DMDHEU with and without softener decreased soiling when compared to untreated fabric and decreased soiling when compared to the copolymerized cellulose; PAA copolymers have reduced soiling by crosslinking with DMDHEU. [96]		
cellulose material • vinylacetate (VA) and maleic anhydride (MA) copolymer (copolymer I) • vinylpyrrolidone (VP) and MA copolymer (copolymer II) • pentachlorophenyl ester of acrylic acid (PPAA) (copolymer III) • pentachlorophenyl ester of methacrylic acid (PPMA) (copolymer IV) • trichlorophenyl ester of methacrylic acid (TPMA) (copolymer V)	• graft polymerization reactions of cellulose by the addition of the antimicrobic substances: • phenol(HDP) pentachlorophenol (PP) • 2,4,5-trichlorophenol (TP) • levomycetin (LV) • cellulose graft copolymers were modified by reactivity with the antimicrobic additives through the -OH group functionalities of these additives	antimicrobiotic
Remarks: Characterization of the hydrolysis of the ester bonds was made to determine antimicrobic activity by the cleavage of the bond between the antimicrobic compound and the polymer; rate constants for the hydrolysis of ester bonds in cellulose derivatives synthesized from graft copolymers I and II: I-LV, I-TP, I-PP, II-PP and for initial stages of hydrolysis of ester bonds in cellulose derivatives synthesized by grafting PPAA, PPMA, TPMA: III, IV, V were reported. [97]		

Substrate/Monomer	Initiation/ Treatment	Application/ Property
cotton cellulose Acrylamide (AM)	• acidic peroxo-salt potassium monopersulfate initiator • catalyzed by Co II) • Conditions for maximum yield (30.8% graft-on) are reported. • graft from water-soluble organic solvents	rot resistance water retention
Remarks: Graft yield was examined by reaction times, temperatures, monomer concentrations, catalyst and initiator; water retention increased with increasing percent grafting; no improvement in rot resistance; acrylamide graft copolymers showed affinity for alkalis, and moderate resistance to acids at room temperature. [98]		
cotton linters cellulose thiocarbamate Ethylacrylate Acrylonitrile	• ceric ammonium sulfate initiator • microbial strains: • • gram-negative bacteria Escherichia Coli • • gram-positive bacteria Bacillus cereus • • Bacillos subtilis and Sarcina lutea • • Yeasts-Candida albicans and Saccharomyces cervisiae • • fungi-Fusarium oxyporium V. Vasifectum, Delodia ohryzae, Aspergillus flavus, Botrylis alii, and Helminthosporium turcicum	antifungal activity
Remarks: Ethyl acrylate and acrylonitrile graft copolymers of cellulose thiocarbamate exhibit highly specific antifungal activity; grafted cellulose thiocarbamates were highly effective against the different fungi strains but did not affect gram-negative, gram-positive bacterial or yeast; grafting for 2 hours provided maximum inhibition; acrylonitrile graft copolymer was more effective against fungi than ethyl acrylate copolymer of cellulose-thiocarbamate. [99]		

Substrate/Monomer	Initiation/ Treatment	Application/ Property
Cellulose, aminoethyl (AE-Cell) Cellulose, aminopropyl (AP-cell) γ-benzyl-L-glutamate (PBLG) N-2-hydroxyethyl-L-glutamine (PHEG)	AE-cell and AP-cell were used as macroinitiator off γ-BLG,NCA	antithrombogenic blood compatibility biomedical materials • cell-g-PBLG polymer-coated sutures tor implantation into biological systems • the graft copolymers exhibited excellent blood compatibility • cellulose graft copolymers with polypepides are absorbable
Remarks: Peptides were introduced into cellulose by polymerization; N-Carboxy-γ-benzyl-L-glutamate (γ-BLGNCA) by cellulose derivatives having primary amino groups at the chain end or as the side chain; the preparation and characterization of Cell-g-PBLG graft copolymers and their blood compatibility are described. [100]		
cotton fabric Acrylonitrile (AN)	ferrous cellulose thio-carbonate-hydrogen peroxide redox system	resistance to chlorine bleach (oxidation by sodium hypochlorite)
Remarks: PAN graft chains introduced onto the cellulose substrate promoted susceptibility of cellulose to oxidation with hypochlorite, the extent of which depended on the level of grafting; carboxylic groups form during hypochlorite oxidation by oxidation of functional end groups from cellulose hydroxyls and oxidation of -CN groups on the PAN-grafted fabric; tensile strength after hypochlorite treatment is lower than ungrafted cotton at all graft levels of homopolymer. [101]		
cotton fabrics, sulfonated acrylonitrile	• ammonium cerium (IV) nitrate • absorpiton of basic dyes	dye receptivity color fastness
Remarks: Dyeing properties and color fastness were compared over sulfonated and grafted cotton fabrics with two extents of sulfonation and three levels of percent graft-on; an increase in sulfonation resulted in increased dye uptake; unsulfonated grafted fabrics showed little affinity for dyeing; the increase in graft-on results in decreased sulfonate groups; extent of grafting and amount of dye uptake vary inversely; the introduction of PAN onto cotton significantly improves color fastness to xenon lamp exposure; the highest graft-on percent fabrics had the best color fastness. [102]		
cotton fibers Vinyl acetate (VA) Methyl acrylate (MA)	ceric ion	thermal stability

Substrate/Monomer	Initiation/ Treatment	Application/ Property
Remarks: Theromogravimetric analysis and differential thermal analysis showed that grafting vinyl acetate onto cellulose imparts a decrease in thermal stability of cellulose. Temperature values for PVA cellulose graft copolymer are less than those for ungrafted cotton; the stability of the graft copolymer increased with increased grafting. With grafting higher than 67% the stability of the graft copolymer is higher than that of PAN; the thermal stability of cotton grafted with methyl acrylate is higher than that of cotton; the thermal stability of methylacrylate-grafted cellulose is higher than the ungrafted cellulose if the grafting yield of the copolymer is 430%. Grafted cotton samples having 33, 83, and 174% grafting yield begin initial decomposition at lower temperatures than vinylacetate grafted cellulose but decomposition beyond 360°C occurs at higher temperature than the original cotton. [103]		
cotton fabric Acrylic acid	potassium persulfate free-radical initiator basic dyestuffs	• behavior of grafted cotton samples toward printing with basic dyes • tensile strength • elongation • crease recovery
Remarks: Color strength increased with carboxy content as sites for dye molecule interaction; a maximum carboxyl content of 126.34% was found for dye uptake; ungrafted dyed cotton fabric exhibited wash fastness of 2-3 where 5 is maximum whereas grafted cottons reached 4 where the graft yield did not exceel 126.34 m. eq. -COOH/100g cellulose; fastness to perspiration and light was higher for graft copolymerized cottons than untreated cottons; tensile strength and elongation of the printed graft copolymerized cottons were lower than the untreated cotton; crease recovery angle of printed grafted cottons increased with increased graft yield. [104]		
cotton fabric Lignin polyurethane (LPU)	Condensation polymerization • Synthesis of monocapped aromat, phenol-tolylene diisocyanate blocked derivative (I) • Reaction of I with ethoxylated hydroxypropyl lignin to form LPU for application to cotton fabric and crosslinking by pad/dry/cure	• Durable-press • strength • abrasion resistance • dimensional stability

Substrate/Monomer	Initiation/ Treatment	Application/ Property
Remarks: Durable-press properties were improved over untreated fabric (2.5 to 3.0) but did not reach the level of conventional DMDHEU resins at 4.0-4.5. Tensile strength was maintained and improved over DMDHEU. Dimensional stability was greater than untreated fabric but not as improved as DMDHEU. Abrasion resistance for LPU-treated fabrics was improved over DMDHEU fabrics and was slightly higher than the untreated control. [105]		
cotton print cloth degraded by high-voltage electrons to strength levels of two Peruvian archaeological textiles: 1200 A.D. and 1000 B.C. • Acrylamide • N-methylolacrylamide • Carbowax acrylate • 50% acrylamide/ 50% Carbowax acrylate	• electron voltage irradiation • doses: 50 and 100 Mrad • Consolidating materials: • • acrylic resins • • monomers (in situ polymerization by pad, dry, cure)	• consolidation of degraded textiles • conservation • strength • flexibility
Remarks: After degradation by 50-Mrad and 100-Mrad doses, control fabric retained 26.6% and 16.2% of original strength, respec. Tensile strength of these fabrics retained after grafting was 22% (50-Mrad) and 13.9% (100-Mrad) whereas resins increased strength to 41.0% (50-Mrad) and 16.4% (100-Mrad). Acrylic resins also increased tear strength more than grafting with monomers. Flexural rigidity of degraded fabrics and grafted fabrics was low. An acrylic resin with Tg 33°C showed highest levels of tensile strength and extremely high flexural rigidity with an average increase of 1300%. [106]		
cotton print cloth degraded by high voltage electrons to strength levels of two Peruvian archaeological textiles: 1200 A.D. and 1000 B.C. Acrylonitrile (AN) Dimethyldichlorosilane	• electron voltage irradiation • doses: 50-, 100-, 150-Mrad or degradation by 5N hydrochloric acid for 5.5 hours at 20°C • vapor phase reaction • impregnation with free-radical-forming catalyst: 2% potassium persulfate • exposure of degraded fabrics to monomer vapor to add on of 2-5%	• consolidation of degraded textiles • conservation • strength
Remarks: No significant improvements were found with PAN graft copolymerized degraded fabrics. Graft copolymerization reactions with silicone polymer formed acidic HCl gas from the reaction of methylchlorosilanes with the hydroxyl groups and this caused degradation of cellulose. Strength losses were 33-80% lower than untreated fabrics. [107]		

Substrate/Monomer	Initiation/ Treatment	Application/ Property
RAYON		
cellulose pulp fiber composites Allyl acrylate Aallyl methacrylate	• commerical composite of natural rubber grafted cellulose fibers • cellulosic reinforcement of rubbers	• wettability • adhesion to matrix in composites
Remarks: Fiber content influences stress/strain behavior; grafted fibers had better adhesion for the rubber matrix; the stiffness of the allylacrylate and allylmethacrylate composites indicated reactions between the matrix, the cure system, and the double bonds at the fiber surface; grafting increased wettability. [108]		
Rayon, Fortisan, High-Tenacity tire yarn, High Wet Modu!us filament Styrene	cobalt 60-gamma-rays	fine structure and morphology
Remarks: After solvent extraction of rayon graft copolymers, polystyrene was present on the surface in a pattern following longitudinal striations aligned along the fiber axis; crosssections by electron micrography revealed three distinct components - cellulose-styrene graft, polystyrene homopolymer, and ungrafted cellulose; for Fortisan and tire yarn, this homopolymer was located in the filament fiber voids; in Fortisan, extractable homopolymer occupied 64% of the cross-sectional area and 43% in the case of High Wet Modulus rayon; at the macroscopic level, styrene graft copolymers of rayon are homogeneous. [109]		
Rayon, filament yarns 120 denier, 50 filaments Ethyl acrylate	high-energy electron beam irradiation	elastomers tensile strength elongation recovery modulus
Remarks: Rayon graft copolymers exhibited elastomeric behavior with poly(ethylacrylate) uniformly distributed among the fibrils; tensile strength and modulus decreased with extent of grafting; elastic recovery and breaking elongation increased which could be due to depolymerization of cellulose and homogeneous grafting within the microstructure; stiffness of the graft copolymer decreased sharply at -20°C. The transition temperatures of grafted fibers were about -20°C; the Tg for poly(ethyacrylate) was - 23°C to - 29°C. [110]		
Rayon yarn Ethyl acrylate	• $Ce(NH_4)_6(NO_3)_6 \cdot 6H_2O$ pre-irradiation • chemical initiation	elastomers mechanical properties

Substrate/Monomer	Initiation/ Treatment	Application/ Property
Remarks: At 1000% graft-on, elastomeric properties developed with accompanying strength loss; bifunctional monomers when added to the grafting solution prevented loss in extension; elastomeric properties in graft copolymerized rayon were developed with 100% graft-on followed by decrystallization treatment; disruption of crystalline regions in the substrate and uniform distribution of grafting throughout the micro structure are essential for the development of elastomeric properties; grafting decreased breaking strength and modulus. [111]		
Rayon, viscose staple fibers Acrylonitrile	redox system, Fe^{2+}-H_2O_2	Mtilon-V photo- and thermo-stability and resistance to weathering and the action of microbes mechanical behavior, moisture regain
Remarks: Optimum graft-on of PAN was 50 5% on the weight of rayon to develop satisfactory physico-mechanical properties; tenacity decreased slight by 10-15%; elongation at break, abrasion resistance, and cyclic bending strength increased appreciably; dye fastness was improved. These properties are important for the carpet end use. [112]		
Rayon, viscose cord fibers Isoprene (rubber) Acrylonitrile	diazonium salts, redox system	strength abrasion resistance modulus adhesion to rubber
Remarks: Grafting rayon with polyisoprene did not change the breaking strength but increased breaking elongation and modulus with increasing graft-on percent; at 15% polyisoprene graft-on adhesion of rayon cord to rubber increased 1.6 times over the untreated rayon cord; tenacity and abrasion resistance in polyacrylontrile graft copolymers of rayon increased 4.0 times and in polyisoprene copolymers, 2.5 times; at over 40% graft-on, abrasion resistance declined. This was attributed to changes in the supramolecular structure of rayon by grafting; degree of crystallinity by X-ray diffraction showed 44.5% for the ungrafted fiber, 39.1% for polyisoprene graft rayon copolymer and 29.5% for the polyacrylonitrile copolymer. [113]		
Rayon, fibers, fabrics phosphorous-containing monomers • triphenyl phosphite • 2-phenoxy-1,3,2-dioxaphospholone • triethyl phosphite • tris (dichloropropyl) phosphate	RF cold plasma • grafting by aromatic cyclic opening mechanisms of the "monomers" and/or through dehydrogenation and dehydrochlorination	flame retardancy

Substrate/Monomer	Initiation/ Treatment	Application/ Property
Remarks: Graft copolymerization with all phosphorous-containing monomers and nitrogen plasma treatment alone lowered the Activiation Energy (Ea) for the thermo-oxidation destruction of rayon. Self-extinguishing burning behavior was exhibited for triphenyl phosphite, triphenyl phosphate, 2-phenoxy-1,3,2-diox aphospholane, and triethyl phosphite grafted rayons, at 0.13 to 0.66% graft-on levels; grafting was localized on the surfaces of the fibers. [114]		
Rayon, viscous staple fiber cellulosic anion-exchanger TsM-Al 2-methyl-5-Vinylpryridine (MVP)	H_2O_2	sorption filter materials
Remarks: for optimal sorption properties, 45-55% grafting of PMVP onto TsM-Al was prescribed to ensure effective swelling and accessibility of the active groups and high mechanical strength. [115]		
Jute, fibers, lignified Methyl methacrylate (MMA) Ethyl acrylate (EA) Acrylonitrile	• $FeCl_3$- H_2O_2redox system for MMA and EA • ceric ammonium sulfate for Acrylonitrile • lignified jute and bleached fibers (50% lignin removed) • grafting at 50°C with 2% monomer concentration	tenacity elongation at break
Remarks: PMMA grafting increased breaking tenacity significantly and elongation slightly. PEA grafting increased elongation to a greater extent (more than 50%) and the increase in tenacity was not as great as with PMMA; these differences were attributed to PEA segments being "soft" and PMMA segments being "hard"; with these differences the ratio of the moduli of these two graft copolymers was unchanged. Graft copolymerization confers homogeneity to jute microstructure. Grafting with polyacrylonitrile increased elongation at break by about 70%. [116]		
Jute, fibers, lignified Acrylamide	acidic peroxosalt "potassium monopersulfate" initiator under catalysis with Fe (II) in nitrogen	water absorption acid, base resistance tenacity stiffness modulus
Remarks: Absorption of water and water vapor increased; acid and base resistance increased; tenacity, elongation at break, and tensile modulus increased with increasing grafting percent in jute-polyacrylamide graft copolymers. [117]		
Jute, fibers, lignified Methyl methacrylate	potassium permangamate initiator	thermal behavior tensile strength

<table>
<tr><th>Substrate/Monomer</th><th>Initiation/
Treatment</th><th>Application/
Property</th></tr>
<tr><td colspan="3">Remarks: In untreated jute, by thermogravimetry, weight loss began with a 10% loss at 100°C and showed maximum weight loss at decomposition temperature (T_D) = 300°C. Any percent graft-on decreased tenacity. This was attributed to reaction conditions including the choice of initiator. [118]</td></tr>
<tr><td colspan="3">WOOD</td></tr>
<tr><td>Kraft pulp bleached handsheets

Vinyl chloride</td><td>ceric ion redox system</td><td>tensile strength
elongation at break
work of rupture
moisture regain</td></tr>
<tr><td colspan="3">Remarks: Grafting resulted in reduced tensile strength, elongation and work of rupture. This was interpreted as a reduction in hydrogen-bonding potential by the grafted poly vinyl chloride; it was suggested that acrylonitrile and acrylamide vinyl monomers, having hydrogen bonding capability should be examined. Moisture regain as a function of relative humidity decreased. This was attributed to the hydrophobic nature of polyvinyl chloride. [119]</td></tr>
<tr><td>wood, wafers of Ponderosa pine
sapwood

Styrene
Styrene-Ethyl acrylate mix</td><td>Co-60 irradiation source
• Aqueous-organic solvent systems
• • dioxane-styrene
dioxane-styrene
ethanol-styrene
acetone-styrene</td><td>strength
wood stabilization
abrasion resistance</td></tr>
<tr><td colspan="3">Remarks: Graft copolymerization occurred within the cell lumen only as the wood was not swollen; stabilization by graft copolymerization was considered an internal coating. Styrene graft copolymers reached maximum swelling in 10-12 hours whereas untreated samples took 21 minutes. Copolymerization in aqueous dioxane-styrene produced ASE (antishrink efficiency) up to 70%. This was attributed to homopolymerization in the cell lumen and changes in microstructure leading to more accessible regions for swelling. Overall changes in tensile strength and abrasion resistance were small and water sorption was slow. The styrene-ethyl acrylate system did not differ from the styrene system. [120]</td></tr>
<tr><td>wood pulp, high-yield Kraft, groundwood and blended pulps

Acrylamide (AM)</td><td>• peroxide initiation
• pre- and mutual irradiation</td><td>strength</td></tr>
<tr><td colspan="3">Remarks: Lignin containing pulps failed to graft. Bleached Kraft pulp was grafted but could not form sheets. When blended with ungrafted pulps, there was improved strength in samples containing 25% graft-on and 12.6% lignin. Other grafted pulps were copolymerized with PAM with shortened side chains, and when blended with ungrafted pulps, strength developed. [121]</td></tr>
</table>

Substrate/Monomer	Initiation/ Treatment	Application/ Property
wood, pulp and hand sheets Styrene Styrene-acrylonitrile copolymer	Co-60 gamma irradiation	moisture regain bursting strength breaking length folding endurance
Remarks: Styrene graft copolymers resulted in handsheets which were brittle and frialite with low folding endurance and poor bursting and tensile strength. Poly (styrene-acrylonitrile) copolymers grafted to handsheets improved these properties up to 30% graft-on with an absorption of 23% more water than polystyrene alone grafted to wood. [122]		
wood pulp Styrene • Dimethylaminoethyl methacrylate • Acrylamide-Diethylamino-ethyl methacrylate copolymer (DMAEMA) • Acrylamide and N-dimethylamino-ethyl methacrylate double graft	peroxide preirradiation mutual grafting • graft blends with pulps	tensile strength bursting strength folding endurance
Remarks: Acrylamide was used to improve strength because of its high hydrogen-bonding capacity. It was ineffective with high-yield pulps and ground wood. Thus, N-dimethyl-aminoethyl methacrylate being cationic was used for partial ionic bonding with the lignosulfonates and other lignin moieties in wood pulp. Cationic-grafted high-yield (20.8% lignin content) pulp showed improved strength. There was not much improvement with poly(acrylamide) grafting. Double grafts were not superior to blended strong bleached Kraft long fiber pulp. In the case of ground wood pulp, grafted pulps improved paper strength but did not give higher strength than the same amount of blended bleached Kraft pulp. Cationic grafting improved strength over double grafting and over grafting with poly acrylamide. Styrene grafting was recommended for paper directly as the hydrophobic nature of styrene prevented defibrillation. [123]		
wood pulp fibers Acrylamide (AA)	ceric ammonium initiator	Viscoelasticity
Remarks: Grafted poly(acrylamide) chains were long enough so the graft copolymer exhibited viscoelastic behavior. Twist of the fiber structure under a load during heating and wetting was attributed to the softening of the grafted chains. [124]		

Substrate/Monomer	Initiation/ Treatment	Application/ Property
pulp, Kraft, bleached Acrylonitrile Acrylic acid Ethylacrylate	Xanthate - H_2O_2 • Graft with Acrylonitrile with subsequent hydrolysis to yield amide and carboxylate copolymer polyelectrolytes: • • sodium poly-acrylate-poly-acrylamide copolymer (anionic) • • Dimethylethylamine ethyl methacrylate hydrochloride (cationic)	ion-exchange absorption
Remarks: Hydrolyzed poly(acrylonitrile) grafted fibers showed high swelling in water and high water retention. Water and saline absorption increased with % polymer graft level. Water retention capability varied with pH and was attributed to formation of carboxylate groups which increased with graft-on and with pH. Functional group type contents were regulated by pH of the swelling medium. At pH 7 to 9, the carboxylate salt form content was high and the amine form, low. These ionized carboxylate forms contribute to high water retention and ion-exchange properties. [125]		
wood • methylmethacrylate (MMA) • Butylmethlacrylate (BuMA)	• Co-60 gamma radiation • peroxide • in situ polymerization	• consolidation • conservation • hygroscopicity • water absorption • swelling • compression strength
Remarks: Resistance to moisture absorption increased with increasing polymer content in wood where MMA had the greatest effect at 60% graft-on and 45% relative hygroscopic resistance. BuMA graft copolymerized wood had the least moisture resistance. Hygroscopicity of consolidated wood was reflected by swelling which decreased in PMMA grafted wood. Compression strength of the consolidated wood increased with the amount of polymer, PBuMA having the least effect. The lower moisture resistance of PBuMA compared to PMMA was explained by differences in Tg, 20 °C vs. 105 °C respec. The Tg value of 20 °C and ease of deformation was used to explain the lower compressive strength of PBuMA when compared to PMMA. Copolymerization of PMMA and PBuMA did not contribute to improving the mechanical properties of graft polymerized wood. [126]		

Substrate/Monomer	Initiation/ Treatment	Application/ Property
Wood, Southern pine • ethyl α-hydroxymeth-acrylate (EHMA) • EHMA-2-vinyl,4-dimethyl-2-oxazolin-5-one (vinyl azlactone) comonomer, EHMA-VAL • EHMA-methyl methacrylate comonomer EHMA-MMA • EHMA-Styrene comonomer EHMA-Sty	2-(carbamoyl-isobutyronitrile (V-30) radical initiator • impregnation of wood by PEHMA for hydrogen bonding to cellulose and inter- and intratransesterfication by reaction of hydroxyl groups on the polymer and cellulose with monomer ester groups	dimensional stability (antishrink efficiency, ASE) compression modulus
Remarks: ASE values for EHMA at 68.8% were highest followed by 40.9% for impregnation with 90:10 EHMA-VAL comonomer. As the proportion of VAL increased in the comonomer, ASE decreased. At 25:75 EHMA-VAL, a negative value was obtained for ASE indicating possible opening up of the substrate and exposure of carboxylate groups from the grafted chains as sites for water absorption. Compression modulus for EHMA grafted wood was low at 23.9 GPa and highest for Styrene (30.6) and 50:50 EHMA-Sty (29.8) and 75:25 EHMA-VAL (28.7). Styrene treated wood has relatively low dimensional stability. SEM micrographs confirmed impregnation with EHMA homopolymer filling wood cell lumens and good adhesion of polymer to the cell wall. [127]		
• paper pulp sheets (high hemicellulose content) • acetate grade pulp (low hemicellulose content) Acrylonitrile (AN) Methyl- and Ethylacrylate Acrylic acid Butylacrylate 2-Ethylhexylacrylate Acrylamide	$Ce(NH_4)_2(NO_3)_6$ initiator • graft in the form of suspension • graft in the form of laboratory sheets	• breaking length • tear, burst, double fold strengths
Remarks: Weight increase of grafted PAN was slightly higher for sheets higher in hemicellulose content and higher for paper pulp than acetate grade pulp. The presence of hemicellulose provided more grafting capability. Wet strength of sheet from graft copolymers increased with weight increase during grafting and wet strength for graft copolymers formed in suspension form were considerable lower and at low values with increasing weight %. Grafting resulted in no significant reduction in paper tensile, tear, and burst strength, and led to increased wet strength. Double fold strength decreased significantly and the paper became brittle. Grafting from suspension gave higher graft yeilds than from sheet form but resulted in significant strength loss, thus grafting in suspension form destroyed sheet forming properties. [128]		

Substrate/Monomer	Initiation/ Treatment	Application/ Property
bleached sulfate pulp papers from softwood and hardwood Methyl methacrylate (MMA) Styrene (SE) Acrylonitrile (AN) Vinyl acetate (VAc)	ceric salt initiator ceric ammonium nitrate • beaten and unbeaten pulp	dimensional stability strength water retention moisture regain
Remarks: Water retention values (WRV) were not different for soft- and hard woods. WRV increased with degree of pulp beating and decreased with increasing graft-on. However, water retention expressed as % of weight of cellulose contained in the grafted sample (WRV*) increased slightly, indicating the hydrophobic character conferred to the graft copolymer by vinyl polymers. Decrease in WRV of grafted pulps showed the order: ST < MMA < AN < VAc and moisture regain showed the same trend. Lowered water retention yielded higher dimensional stability of the paper. Grafted papaers were weak. [129]		
commercial printing papers Methyl methacrylate (MMA) Acrylonitrile (AN)	ceric salt initiator	dimensional stability moisture regain breaking length stiffness opacity brightness
Remarks: AT 60% weight increase of papers by grafting, the dimensional stability increased two fold. Moisture regain decreased and hydrophobic charcter developed. Breaking length decreased with grafting as did density and brightness. Stiffness and opacity increased. Water retention valves and moisture regain for graft copolymers follwed the monomer order: Styrene < MMA < vinyl acetate. For the same graft-on level, PAN graft copolymers exhibited lower moisture regain and slightly higher dimensional stability than those from PMMA monomers. Although grafting improved the dimensional stability of paper, it caused a loss in strength. This was attributed to the permeation of hydrophobic homopolymers into the fibers with the loss of hydrogen bonds. PAN caused a slightly lower decrease than PMMA for breaking length. During grafting, paper thickness increased and this correlated with decrease in density and with weight increase. Grafted samples were more bulky. [130]		
filter paper Whatman No. 41 Acrylamide	preirradiation in air electron accelerator, 200-keV dose rate: 5 X 10^4Mrad/hr total dose: (3.1 0.3) X 10^{20}eV/g	moisture regain handle

Substrate/Monomer	Initiation/ Treatment	Application/ Property
Remarks: Above the 20% level of grafting, the paper was hard and brittle and thicker than ungrafted paper. Moisture regain increased with extent of grafting, reaching 9% for the 100% grafting level. [131]		
filter paper Whatman No. 41 Styrene Methyl methacrylate (MMA) Ethyl acrylate (EA)	UV radiation gamma ray radiation • simultaneous irradiation of monomer and substrate • organic acid additives for increasing UV grafting yields of acrylates to cellulose using styrene comonomers • simultaneous grafting of styrene and acrylate monomers	conservation
Remarks: Formic acid additive to the graft copolymerization reaction solution enhanced copolymerization only slightly whereas with styrene, grafting yield increased. Addition of styrene in a comonomer mixture during graft copolymerization resulted in grafted cellulose having properties of both monomers. Recommendation was made to adapt the copolymerization to grafting without solvent, use sensitized (possibly uranyl nitrate) UV or electron beam rapid-oxidation methodology for quick polymerization. Choice of radiation initiator may be determined by the reversibility aspects of the polymerized monomer. [132]		
wood pulp: dissolving pulp softwood and hardwood bleached Kraft pulps cellulose powder carboxymethylcellulose Acrylamide (AM) Acrylic Acid (AA) Dimethylaminolethyl-methacrylate (DMAEM) Dimethylaminopropyl-methacrylamide (DMAPMA) 2-Acrylamido-2-methylpropyl-sulfonic acid (AMPSA)	cellulose Xanthate Fe^{2+}-H_2O_2 Redox	Flocculation activity removal of iron and zinc

Substrate/Monomer	Initiation/ Treatment	Application/ Property
Remarks: All cellulosic products improved removal of iron. Dissolving pulp also removed zinc. Cellulose graft copolymers compared favorably to commercial Flocculants in the removal of iron after 30 minutes, but PAN copolymerized dissolving pulp was less effective. There was no improvement in flocculation activity from grafts of DMAEM, DMAPMA, AMPSA, AM, AA, and their mixtures onto dissolving pulp [133]		
Whatmans filter paper Ethyl acrylate (EA) Methyl methacrylate (MMA) EA-MMA mixture	gamma ray irradiation	conservation folding endurance
Remarks: Folding endurance increased with increasing % weight of EA:MMA(5:1) in graft copolymerization of filter paper and mechanical paper. A one-stage deacidifying strengthening process was developed by incorporating a basic amino monomer to neutralize the acid present in aged paper, and a multifunctional monomer with more than one double bond to form more crosslinks to the mixture of EA:MMA monomers in the graft copolymerization reaction with paper. [134]		
Wood, grand fir (Ablies grandis) Combinations of: • tri-n-butyltin meth-acrylate, maleic anhydride (TBTMA-MAnh) • tri-n-butyltin meth-acrylate-glycidyl methacrylate (TBTMA-GMA)	• benzoyl peroxide of azobisisobutyronitrile free radical initiation • copolymerization of organotin monomers carrying the bioactive tributyltin group • in situ polymerization of vinyl monomers in wood Resistance to: • Psudomonas nigrifaciens (marine bacterium) • Sarcina lutea (soil fungus) • Giomeralla cingulata (soil fungus)	• Bioactive wood-polymer composites • decay resistance • dimensional stability (ASE) • moisture absorption • mechanical behavior • fexural and impact strengths • white-and brown-rot, fungi resistance • resistance against marine organisms
Remarks: Comonomers containing anhydride and epoxy functionality were grafted to wood through esterification and etherification of wood hydroxyl groups. Decay resistance is based on the controlled release of the toxic TBT moiety chemically linked to graft copolymerized wood. Impregnation of graft copolymers increased antishrink efficiency (ASE) and decreased water absorption in the order: TBTMA-MAnh > TBTMA-GMA >GMA. Mechanical properties in the transverse direction were improved but only moderately improved in the longitudinal direction. Organotin polymers increased microbiol decay resistance. This was confirmed by field tests where resistance after 2 years expsure surpassed the resistance of creosote-treated samples. [135]		

Substrate/Monomer	Initiation/ Treatment	Application/ Property
Waterlogged wood beech, oak monomers (water soluble) N-Vinylpyrrolidone (nVP) Methylacrylamide (MAID) resins (non water soluble) Styrene-unsaturated polyesters • Norsodyne Resin-Isophthalic type • Ludopal-Butyl Acrylate formulation butyl acrylate-styrene-polyester 20:27:53	• Gamma rays from Co-60 source • Impregnation followed by gamma irradiation for in situ polymerization with monomers and resins	preservation of waterlogged wood
Remarks: Poor results with water-soluble monomers led to cracking and splitting of graft copolymerized woods. Resins gave best results without cracking but in some cases radial cracks developed and some surfaces were tacky. The best resin is one which produces a hard, non-tacky surface but is flexible to prevent cracking. [136]		

References

1. Porkerson, F. S.; Reeves, W. A.; Tripp, V. W. Textile Res. J.. 944 (1960).
2. Andrews, B. A. K.; Kimnel, L. B.; Bertoniere, N. R.; Herbert, J. J. Textile Res. J.. 59, 675 (1989).
3. Frick, J. G.; Harper, R. J. J. Appl. Polym. Sci. 27, 983 (1982).
4. Welch, C. M.; Danna, G. F. Textile Res. J.. 52, 149 (1982).
5. Welch, C. M.; Andrews, B. A. K. Text. Chem. Col. 21, 13 (1989).
6. Cousins, E. R.; Bullock, A. L.; Mack, C. H.; Rowland, S. P. Textile Res. J.. 34, 953 (1964).
7. Rooks, R. J. Text. Chem. Col.. 4, 14 (1972).
8. Jarvis, C. J. Coat. Fabrics. 6, 182 (1977).
9. Godfrey, L. E. A. Textile Res. J.. 40, 116 (1970).
10. Harper, R. J.; Lambert, A. H. Amer. Dyestuff Reptr. 78, 15 (1989).
11. Blanchard, E. J.; Reinhardt, R. M. Text. Chem. Col.. 21, 19 (1989).
12. Caulfield, D. F.; Weatherwax, R. C. TAPPI J. 59, 114 (1976).
13. Stamm, A. J. TAPPI J. 42, 44 (1959).
14. Kennedy, R. J. TAPPI J. 45, 738 (1962).
15. Cooke, T. F.; Dusenburg, J. H.; Kienle, R. H.; Lineken, E. E. Textile Res. J.. 24, 1015 (1954).
16. Nuessle, A. C.; Fineman, M. N.; Heiges, E. O. J. Textile Res. J.. 25, 24 (1955).
17. Werner, A. E. A. "Consolidation of Deteriorated Wooden Artifacts." In Proceedings of International Symposium on the Conservation and Restoration of Cultural Property Conservation of Wood. Tokyo National Research Institute of Cultural Properties, 17 (1977).
18. Grattan, D. W. "Consolidants for Degraded and Damaged Wood. In Proceedings of the Furniture and Wooden Objects Symposium. Ottawa, Canadian Conservation Institute. 27 (1980).
19. Kaizerman, S.; Mino, G.; Meinhold, L. F. Textile Res. J.. 32, 136 (1962).
20. Rollins, M. L.; Cannizzaro, A. M.; Blouin, F. A.; Arthur, J. C. J. Appl. Polym. Sci. 12, 71 (1968).
21. Dimov, K. and Pavlon, P. Cellul. Chem. Technol. 7, 451 (1973).
22. Focher, B.; D'Angiuro, L.; Cremonesi, P. Cellul. Chem. Technol. 6, 277 (1972).
23. Guthrie, J. T.; Percival, J. A. Polymer. 18, 531 (1977).

24. Ogiwara, Y.; Kubota, H.; Murayama, H.; Sakamoto, A. TAPPI J. 54, 1279 (1971).
25. Hoffman, A. S. Isotopes and Radiation Technology. 8, 84 (1970).
26. Fernandez, M. J.; Fernandez, M. D.; Casinos, I.; Guzman, G. M. J. Appl. Polym. Sci. 39, 2219 (1990).
27. Zahran, A. H.; Stannett, V.; Liepins, R.; Morosoff, N. Radiat. Phys. Chem. 16, 265 (1980).
28. Ibrahim, N. A.; Haggag, K.; Abo-Shosha, M. H. Amer. Dyestuff Reptr. 77, 34, (1988).
29. Kamogawa, H.; Murase, R.; Sekiya, T. Textile Res. J.. 774 (1960).
30. El-Zairy, M. R. Amer. Dyestuff Reptr. 79, 48 (1990).
31. Kassig, H. A.; Stannett, V. Advan. Polymer. Sci. 4, lll (1965).
32. Battaerd, H. A. J.; Tregear, G. W. "Graft Copolymers", Interscience, NJ (1967).
33. Hebeish, A.; Guthrie, J. T. "The Chemistry and Technology of Cellulosic Copolymers". Springer-Verlag, NY (1981).
34. Samal, R. K.; Sahoo, P. K.; Samantoray, H. S. J. Macromol. Sci. Chem.. Rev. Macromol: Chem. Phys. C. 26(1), 81 (1986).
35. Arthur, J. C., Jr. Appl. Polym. Symp. 36, 201 (1981).
36. Zeronian, S. H.; Coole, M. L.; Alger, K. W.; Chandler, J. M. J. Appl. Polym. Sci. Appl. Polym. Symp. 37, 1053 (1983).
37. Skachkov, V. M.; Sharkov, V. I. Izv. Vyssh. Ucheb. Zaved, Les. Zh. 10, 127 (1967). (Chem. Abs. 68 96906C).
38. Nickerson, R. F. Ind. Eng. Chem. 33, 1022 (1941); 34, 85, 1480 (1942); 37, 1115 (1945); 38, 299 (1946).
39. Nickerson, R. F.; Habrle, J. A. Ind. Eng. Chem. 39, 1507 (1947).
40. Herman, P. H.; Weidinger, A. J. Polymer Sci. 4, 135 (1949); 5, 656 (1950); 6, 533 (1951).
41. Sjostrom, E. "Wood Chemistry Fundamentals and Applications." Academic Press, Florida, 170 (1981).
42. Howsmon, J. A.; Sisson, W. A. In "Cellulose and Cellulose Derivatives," Part I. Ott, E.; Spurlin, H. M.; Grafflin, M. W. Interscience, NY, 277 (1954).
43. Nevell, T. P.; Zeronian, S. H. In "Cellulose Chemistry and Its Applications." Wiley, NY, 15 (1987).
44. Ward, K., Jr. In "Cellulose and Cellulose Derivatives," Ott, E., Spurlin, H. M.; Grafflin, M. (eds.) Interscience, NY, 18 (1954).
45. Rowland, S. P.; Bertoniere, N. R. In "Cellulose Chemistry and Its Applications." Nevell, T. P.; Zeronian, S. H. (eds.). Wiley, NY, 122 (1987).
46. Zylinski, T. "Fiber Science." Scientific Publications Foreign Cooperation Center of the Central Institute for Scientific, Technical and Economic Information. Poland, 414 (1964).
47. Carter, Mary E. "Essential Fiber Chemistry." Marcel Dekker, Inc. NY, 43 (1971).
48. Morton, W. E.; Hearle, J. W. S. "Physical Properties of Textile Fibers." London: The Textile Institute, 49, (1975).
49. Jayne, B. A. "Theory and Design of Wood and Fiber Composite Materials." NY: Syracuse University Press, NY, 179 (1972).
50. Hopfenberg, H. B.; Stannett, V.; Kimura-Yeh, F.; Rigney, P. T. Appl. Polym. Sym. 13, 139 (1970).
51. Stannett, V. ACS Symp. Ser., 187, American Chemical Society, Washington, D.C. 1(1982).
52. Atlas, S. M.; Mark, H. Cellulose Chem. & Tech., 3, 325 (1969).
53. Walsh, W. K.; Jin, C. R.; Armstrong, A. A. Textile Res. J.. 35, 648 (1966).
54. Stannett, V.; Hoffman, A. S. Am. Dyestuff Rep. 57, 91, (1968).
55. Sakurada, I.; Okada, T.; Ikada, Y. Cellul. Chem. Technol. 6, 35 (1972).
56. Mansour, O. Y.; Schurz, J. Sven. Paper. Stidn. 76, 258 (1973).
57. Cherepanov, V. N.; et al. Khimicheskie Volokna, 5, 68 (1973).
58. Hornof, V.; Kokta, B. V.; Valade, J. C. Cellul. Chem. Technol. 12, 593 (1978).
59. Trivedi, I. M.; Mehta, P. C. Cellul. Chem. Technol. 1, 401 (1973).
60. Hornof, V.; Kokta, B. V.; Valade, J. C. J. Appl. Polym. Sci. 20, 1543 (1976).
61. Kubota, H.; Ogiwara, Y. J. Appl. Polym. Sci. 13, 1569 (1969).
62. Rogovin, Z. A. In Proceedings of the Symposium on Graft Polymerization onto Cellulose. J. Polym. Sci.Pt. C. Polym. Symp. No. 37, 233 (1971).
63. Bhattacharyya, S. N.; Maldas, D. "Graft Copolymerization onto Cellulsics." In Progress in Polymer Science. 10, 254, Pergammon Press, N. Y. (1984).
64. Arthur, J. C., Jr.; Demint, R. J.; McSherry, W. F.; Jurgens, J. F. Textile Res. J. 29, 759 (1959).
65. Kamogawa, H.; Murase, R.; Serkiya T. Textile Res. J.. 30, 774 (1960).
66. Gardon, J. L. J. Appl. Polym. Sci., 18, 734 (1961).
67. Kamogawa, H.; Sekiya, T. Textile Res. J.. 31, 585 (1961).
68. Blouin, F. A.; Arthur, J. C., Jr.; Orr, R. S.; Ott, V. J. Textile Res. J.. 31, 597 (1961).
69. Kalzerman, S.; Mino, G.; Meinhold, L. F. Textile Res. J.. 32, 136 (1962).
70. Kulkarni, A. Y.; Chitale, A. G.; Validya, B. K. J. Appl. Polym. Sci. 1, 1581 (1963).
71. Staudinger, P. H.; Kern, P. W. Die Makromolekulare Chemie. 65, 248 (1963).
72. Richards, G. N.; White, E. F. T. Textile Res. J.. 34 810 (1964).
73. Gallagher, D. M. Textile Res. J.. 40, 621 (1970).

74. Harris, J. A.; Mares, T.; Arthur, J. C., Jr.; Hoffman, M. J. Text. Ind. 133, 117, 1969.
75. Rao, S. R.; Kapor, S. L. J. Appl. Polym. Sci. 13, 2649 (1969).
76. Ishanov, M. M.; Azizov, V. A.; Nigmankhodzhayeva, M. S.; Usmanor, Kh. V. J. Polym. Sci. Pt. A-1, 9, 1013 (1971).
77. Byrne, G. A.; Arthur, J. C., Jr. Textile Res. J.. 41, 788 (1971).
78. Byrne, G. A.; Arthur, J. C., Jr. Textile Res. J.. 41, 271 (1971).
79. Aleksandrova, T. M.; Rogovin, Z. A.; Slerkina, L. S.; Cheburkov, Yu. A. Vysokomol.soyed, A13, No. 11, 2595 (1971).
80. Rooks, R. J. Text. Chem. Col.. 4, 47 (1972).
81. Reine, A. H.; Arthur, J. C., Jr. Textile Res. J.. 42, 155 (1972).
82. Harris, J. A.; Carra, J. H.; DeGruy, I. V.; Arthur, J. C., Jr. Textile Res. J.. 42, 15 (1972).
83. Varma, D. S.; Narasimhan, V. J. Appl. Polym. Sci. 16, 3325 (1972).
84. Ogiwara, Y.; Kubota, H.; Hayashi, S.; Sekine, K. J. Appl. Polym. Sci. 16, No. 9, 2197 (1972).
85. Reine, A. H.; Arthur, J. C., Jr. J. Appl. Polym. Sci. 16, 1893 (1972).
86. Temin, S. C.; Lifland, L. Ind. Eng. Chem. Prod. Res. Develop. 11, No. 3, 325 (1972).
87. Varma, D. S.; Narasimhan, V. J. Appl. Polym. Sci. 18, 3745 (1974).
88. Sharma, V. N.; Daruwalla, E. H. Textile Res. J.. 46, 398 (1976).
89. Walsh, W. K.; Bittencount, E.; Miles, L. B.; Rutherford, H. A. J. Macromol, Sci-Chem. A10(4), 695 (1976).
90. Harris, J. A.; Arthur, J. C., Jr.; Goynes, W. R. J. Appl. Polym. Sci. 24, 201 (1979).
91. Zahran, A. H.; Williams, J. L.; Stannett, V. T. J. Appl. Polym. Sci. 25, 535 (1980).
92. Nagaty, A.; Shakra, S. E.; Ibrahim, S. T.; Mansour, O. Y. Cellul. Chem. Technol. 14, 177 (1980).
93. Shiraishi, N.; Williams, J. L.; Stannett, V. Radiat. Phys. Chem 19, No. 1, 79 (1982).
94. Thejappa, N.; Pandey, S. N. J. Appl. Polym. Sci. 27, 2307 (1982).
95. Hebeish, A.; Waly, A.; Abou-Zeid, N. Y.; El-Alry, E. Am. Dyestuff Reptr. 15 (1983).
96. Hebeish, A.; Elkharadly, E.; Abdel Fattal, S. H.; Nasr, H. Amer. Dyestuff Reptr. 72, 48 (1983).
97. Virnik, A. D.; Gorcharova, V. A.; Budzan, B. I.; Plotkina, N. S. Vysokomol.soyed. A26, No. 6, 1264 (1984).
98. Sahoo, P. K.; Samantaray, H. S.; Samal, R. J. Appl. Polym. Sci. 32, 5693 (1986).
99. Nagieb, Z. A.; El-Gammal, A. A. J. Appl. Polym. Sci. 31, 179 (1986).
100. Miyamoto, T.; Takahashi, S.; Tsuji, S.; Ito, H.; Inagaki, H. J. Appl. Polym. Sci. 31, 2303 (1986).
101. Hebeish, A.; El-Rafie, M. H.; Khalil, E. M.; El-Hafiz, S. A.; ABD. Cellul. Chem. Technol, 20, 523 (1986).
102. Sabaa, M. W.; Sakamoto, H. W. M.; Furuhata, K. J. Appl. Polym. Sci. 31, 1131 (1986).
103. Fernandez, M. J.; Fernandez, M. D.; Casinos, I.; Guzman, G. M. J. Appl. Sci. 39, 1101 (1990).
104. El-Zairy, M. R. Amer. Dyestuff Report V, 48 (1990).
105. Cardamone, J. M. Presentation, "Crosslinking Cotton Cellulose with Lignin-Based Polyurethane." 1990 American Chemical Society SE/SW Combined Regional Meeting, New Orleans, LA, December 5, 1990.
106. Berry, G. M.; Hersh, S. P.; Tucker, P. A.; Walsh, W. K. Adv. Chem. Ser. 164, 249 (1977).
107. Hutchins, J. K.; Hersh, S. P.; Tucker, P. A.; McElwain, D. M.; Kerr, N. Adv. Chem. Ser. 193, 315 (1981).
108. Flink, P.; Stenberg, B. British Polymer Journal 22, 147 (1990).
109. Kaeppner, W. M.; Huang, R. Y-M. Text. Research Journal 35, 505 (1965).
110. Nakamura, Y.; Arthur, Jr., J. C.; J. Appl. Polym. Sci. 14, 929 (1970).
111. Williams, J. L.; Stannett, V.; Roldan, L. G.; Sello, S. B.; Stevens, C. V. Internat. J. Appl. Radiation and Isotopes 26, 169 (1975).
112. Cherepanov, V. N.; Volodko, E. F.; Lobanova, N. N.; Shur R. D.; Bershova, N. V.; Semenov, V. P.; Stanchenko, G. I.; Morin, B. P.; Garf, E. V. Khimicheskie Volokna 5, 68 (1973).
113. Bu, K. J.; Dumitriu, S.; Simionesco, C. I. Cellul. Chem. Technol. 11, 655 (1977).
114. Simionescu, C. I.; Denes, F.; Macoveanu, M. M.; Cazacu, G.; Totolin, M.; Popescu, F.; Perceus, S. Cellul. Chem. Technol. 13, 463 (1979).

115. Asaulova, T. A.; Lutsyk, R. V.; Morin, B. P.; Ennan, A. A. Zhurnal Prikladnoi Khimii 59, No. 6, 1410 (1986).
116. Mukhopadhyay, A. K.; Sarkar, B. K.; Dutt, A. S. J. Appl. Polym. Sci. 27, 4525 (1982).
117. Samal, R. K.; Samantaray, H. S.; Samal, R. N. Polymer Journal 18, No. 6, 471 (1986).
118. Moharana, S.; Mishra, S. B.; Tripathy, S. S. J. Appl. Polym. Sci. 40, 345 (1990).
119. Cronell, R. H. TAPPI J. 45, No. 7, 145A (1962).
120. Kenaga, D. L.; Fennessy, J. P.; Stannett, V. T. Forest Prod. J. 12, No. 4, 161 (1962).
121. Kobayashi, A.; Phillips, R. B.; Brown, W.; Stannett, V. T. TAPPI J. 54, No. 2, 215 (1971).
122. Oraby, W.; Hopfenbery, H. B.; Stannett, V. J. Appl. Polym. Sci. 16, 2985 (1972).
123. Radi, F. S.; Hopfenbery, H. B.; Stannett, V. J. Appl. Polym. Sci. 16, 2685 (1972).

124. de Ruvo, A.; Brende, L. Cellul. Chem. Technol. 7, 191 (1973).
125, Lepoutre,, P.; Hui, S. H.; Robertson, A. A. J. Macromol. Sci. Chem. A10(4), 681 (1976).
126. Simunkova, E.; Smejkalova, Z.; Zelinger, J. Studies in Conservation 28, 133 (1983).
127. Mathias, L. J.; Wright, J. R.; Warren, S. C. ACS Polym. Preprints 31 No. 1, 646 (1990).
128. Bergem, N.; Ellefsen, O.; Kringstad, K.; Kroken, P. Norsk Skogindustri 17, No. 9, 347 (1963).
129. Ogiwara, Y; Kubota, H.; Murayama, H.; Sakamoto, A. TAPPI J. 53, No. 9, 1687 (1970).
130. Ogiwara, Y.; Kubota, H.; Murayama, H.; Sakamoto, A. TAPPI J. 54 No. 8, 1279 (1971).
131. Lawrence, K. D. N.; Verdin, D. J. Appl. Polym. Sci. 17, 2653 (1973).
132. Davis, N. P.; Garnett, J. L.; Long, M. A.; Major, G.; Nicol-Csiro, K. J. Adv. Chem. Ser. 197 (1981).
133. Graczyk, T.; Hornof, V. J. Appl. Polym. Sci. 29, 1903 (1984).
134. Butler, C. E.; Millington, C. A.; Clements, D. W. G. In "Historic Textile and Paper Materials II Conservation and Characterization" ACS Symposium Series 410, 34 (1989).
135. Subramanian, R. V. Adv. Chem. Ser. 207, 291 (1983).
136. Tran, Q-K.; Ramiere, R.; G.-G. A. Adv. Chem. Ser 225, 217 (1988).

HISTORY OF CELLULOSE ACETATE FIBERS

MASTURA RAHEEL, PH.D.
University of Illinois at Urbana-Champaign, Division of Consumer Sciences, School of Human Resources and Family Studies, 237 Bevier Hall, 905 South Goodwin Avenue, Urbana, IL 61801-3852

Introduction

Cellulose acetate, an ester of cellulose and acetic acid, was first produced in 1865 by Paul Schuetzenberger. However, 35 years passed before a practical, safe, and a relatively inexpensive technique for manufacturing the product was developed by Henri and Camille Dreyfus. At that time the primary use of cellulose acetate was in the form of lacquers, plastic film, and "dope" for use on early airplane fabrics. The first acetate fibers were produced in 1921 by the Dreyfus firm. In 1924, an allied organization began to produce acetate in the United States. Today, acetate is produced in the United States by Hoechst (Celanese Corporation), Avtex Fibers, Inc., and Eastman Chemical Products, Inc..

Triacetate fibers were developed along with regular or secondary acetate. However, commercial manufacture of triacetate was delayed until 1952, at which time a satisfactory and relatively safe solvent became available in sufficient quantity to make production economically profitable. Another problem that had plagued the production of triacetate had been dyes that would work on the fiber; this, too, was solved by the 1950s. The only triacetate fiber in the United States was Arnel^R, produced by Celanese Corporation, which introduced it in 1952 but production was discontinued in 1986.

The nomenclature used for rayon, acetate, and triacetate fibers has been a problem. All through the first half of the twentieth century, both cellulose acetate fibers and rayon (regenerated cellulose fiber) were identified first as "artificial silk" during the 1920s and then as "rayon." As the fibers are quite different in behavior, these names were misleading and frequently led to consumer dissatisfaction. With the passage of Textile Fiber Products Identification Act, in 1960 [1] the problem of names has been alleviated to a great degree.

Acetate is defined by the TFPIA as "a manufactured fiber in which the fiber-forming substance is cellulose acetate. Where not less than 92 percent of the hydroxyl groups are acetylated, the term triacetate may be used as a generic description to the fiber". Acetate and triacetate fibers are similar in chemical structure with acetate having 83% of the hydroxyl groups of cellulose acetylated (2.4 DS.), also known as secondary or regular acetate. Triacetate is also known as primary acetate.

The wide use of acetate fibers in textiles is the result of attractive economics, uniform high quality, color, styling versatility, drape, hand and other favorable aesthetic properties.

History and Developments: An Overview

The first reference in the technical literature regarding the acetylation of cellulose appeared in 1865 and is credited to Schuetzenberger [2] who heated cellulose in a closed tube with acetic anhydride. Worden, [3] in an excellent review and discussion of the art of preparing cellulose acetate up to 1919, divides the advances in the preparation of the ester into three periods:

Figure 1. Paul Schuetzenberger
Source: Worden, E. C. Technology of Cellulose Esters. Vol. VIII. D. Van Nostrand Company: New York. 1916, p. i.

1. The academic period, covering from the work of Schutzenberger in 1865 to that of Cross and Bevan in 1894 [4].

2. The period from 1894 to 1911, covering important advances indicated by the Miles patents [5].

3. The period 1911 to 1919, covering important advances indicated by the Dreyfus patents.

Worden [6] has given a detailed account of the first period and methods of preparing cellulose acetate. An important development after Schuetzenberger's original work appears to be that of Franchimont [7] who discovered the so-called catalytic action of sulfuric acid or zinc chloride in the acetylating reaction. That this discovery marked an important step is evident from the fact that even today most of the industrial methods of acetylating cellulose utilize sulfuric acid, in one form or another, as the catalytic agent. The next important development in the field of cellulose acetate appears to be the 1894 patent of Cross and Bevan [4] which marks the beginning of Worden's second period in the development of acetate. In this patent the industrial value and technical applications of cellulose acetate are mentioned for the first time.

Undoubtedly the most important development during the second period is covered by the patents of G. W. Miles, particularly his patents of 1905. Miles's early patents covered the direct acetylation of fibrous cellulose, for example, the United States Patent Application No. 24,575, dated June 23, 1900, which was allowed December 18, 1990, abandoned and then included in his application of May 27, 1901. His United States Patent No. 733,729 (1903) and Canadian Patent No. 90,848 (1905), while important, covered the production of cellulose acetates soluble only in expensive, and toxic solvents, such as chloroform and tetrachloroethane. However, his later discovery-covered by British Patent No. 19,330 (1905), United States Patent No. 838,350 (1906), reissued 12,637 (1907), French Patent No. 358,079, Belgian Patent No. 187,308 (1905) Hungarian Patent No. 35, 866 (1905), Austrian Patent No. 41,461 (1909), Canadian Patent No. 103,045, and German Patent No. 252,706-covers a hydrating or Hydrolyzing step in the preparation of cellulose acetate, wherein the chloroform soluble (primary) acetates are converted into acetone soluble (secondary) products. The German Patent covering this process was issued only after years of litigation and sale to the Bayer Company, who held German Application No. E-20,963, which was added to Miles's Application M-28,289, and covered by the one patent, No. 252,706. The value of this process lies in the solubility of the hydrolyzed product in comparatively less expensive solvent, from which it gives filaments and films of good strength and flexibility. Many modifications of the original hydrating process have been patented, and described in the literature, e.g., British Patent No. 24,067 (1906); French Patents No. 371,447 (1906), and No. 428,554 (1910); etc.

Worden's third period covers the first Dreyfus patents of July 5, 1911, British Patents No. 20,975, No. 20,976, and No. 20,977, etc. These patents cover the preparation of highly viscous cellulose acetates. They cover the entire range of cellulose acetate manufacture, including the solvents used, etc. The first of these patents points out that high viscosity, high tensile strength, and elasticity are complementary. The product is soluble in alcohol plus chloroform, or in acetone, but insoluble in chloroform alone.

The original patents for the manufacture of acetate fiber appear to be those issued to A.D. Little, W. H. Walker and H. S. Mork, (the latter became vice-president to the Lustron Company of Boston). These were United States Patents No. 712,200 of 1902, and No. 792,149, June 13, 1905, which covered the use of cellulose acetate prepared according to United States Patents No. 709,922, etc.

Figure 2. George Wellington Miles
Source: Worden, E. C. Technology of Cellulose Esters. Vol. VIII. D. Van Nostrand Company: New York. 1916, p. 2531.

The scientific, technical, and patent literature from the time of Schuetzenberger up to the world war I, is replete with references to research on cellulose acetate, acetate fiber, dyes for, and dyeing methods, and related subjects.

Possibly Grandmonyeu was the first to prepare acetate fiber in 1898, but the Lustron Company of Boston appear to have been the pioneers in the manufacture of acetate fiber on a commercial scale, with Bayer and Company of Germany. The Societe Chimique des Usines du Rhone of Lyons, France, founders of the "Rhodiaseta" Company, followed closely. Henckle von Donnersmarck Artificial silk and Acetate Works also appear to have investigated the field to some extent and probably manufactured some acetate silk as it was called at that time.

Acetate fiber was produced in a small way by Dr. Camille and Dr. Henri Dreyfus in Basle, Switzerland, in 1910 to 1913. During the World War I these investigators were called to England to handle the production of cellulose acetate for use in the preparation of airplane dopes. At the close of the war their attention was again given to the production of acetate fiber, which resulted in the formation of British Celanese, Ltd., and associated companies in various countries. Although the plants of this company in England had been the world's largest producers of cellulose acetate during the war, they did not start the production of the fiber Celanese^{R} until about 1918. In the early part of 1921, they were producing about 1000 pounds of Celanese^{R} daily but doubled their capacity during the year.

Celanese^{R} was produced in the U.S. on a small scale in 1919, and large-scale production was started in 1922, with a constant increase in capacity. Within a few years Celanese^{R} was also produced in both France and Belgium. Later in 1927, the Canadian Celanese, Ltd, was built with Dr. Camille Dreyfus as president.

Acetate fiber had been manufactured for some years in France by the Societe pour la Fabrication de la Soie "Rhodiaseta." This company was founded by the Comptoir des Textiles Artificiels and the Societe Chimiques des Usines du Rhone, the latter company having had some experience in the production of high grade, stable, cellulose acetate since 1917. While Rhodiaseta yarn did not appear to be very well known in the U.S., it enjoyed an excellent reputation and sale in France and neighboring countries during 1920s. Acetate was also manufactured in Switzerland.

By the early 1930's advances in acetate filament production for non-industrial uses began to assume significant proportions. Progress from 1930 onwards was most rapid in the U.S.A. where chemicals required could be produced under especially favorable conditions.

In February 1929, the first of cellulose acetate fiber spun by the Eastman Co was wound on a bobbin in Kingsport, Tenn. Perley S. Wilcox, then president of Tennessee Eastman Co, took a nail and scratched the time and date on a roofing paper wall covering: "4:40 p.m., Feb. 8, 1929." Textile yarn was an additional outlet for Eastman's cellulose acetate, which was originally produced as a raw material for parent company Eastman Kodak's film operations. In 1937, Eastman became the first commercial producer of acetate staple fiber in the U.S. and in 1948 a new trade mark Estron^{R} was given to Eastman acetate fibers. In 1951 Eastman introduced chromspun^{R}, the first U.S. solution dyed acetate fiber. Today, Eastman Chemical Products, Inc., markets over 150 variations of acetate and polyester fibers.

In Japan prior to 1914 production was negligible, but from 1919 when Teikoku was formed in association with Snia Viscosa (Italy), production increased rapidly. Asahi was formed in 1925 in association with V. G. F., and from then onwards production was taken up by a number of established textile firms.

Figure 3. Camille Dreyfus
Source: Worden, E. C. Technology of Cellulose Esters. Vol. VI. Part 4. D. Van Nostrand Company: New York. 1921, p. 2889.

Figure 4. Henri Dreyfus
Source: Worden, E. C. Technology of Cellulose Esters. Vol. VI. Part 4. D. Van Nostrand Company: New York. 1921, p. 2890.

Figure 5. Perley S. Wilcox, then president of TEC, took a nail and scratched the time and date of the first acetate yarn spun at TEC, "4:40 pm, February 8, 1929." Courtesy of Eastman Chemical Products, Inc.

By 1939 the worldwide production of acetate fiber was 184 million lb. (178 filament, 6 staple) and increased to 461 million lb. (371 filament, 90 staple) by 1956. Cellulose triacetate textile fiber was commercialized later in the 1950s. The only triacetate fiber produced in the U.S. was Celanese Corporations Arnel^{R} , which was introduced in 1952. However, due to problems with OSHA requirements production was discontinued in 1986.

The combined cellulose acetate/triacetate textile fiber production continued to grow and in 1971 peaked at 938 million lb world-wide. Then, there was a decline during the 1970s, resulting in 756 million lb in the year 1979 [8]. The impact of synthetic fibers, namely, polyester, nylon, and acrylics on the world market during the 1960s and 1970s was evident and is believed to be the major cause for the decline in cellulose acetate/triacetate textile production during the 1970s. However, the production of cellulose acetate cigarette tow increased steadily during the 1970s, and this tended to counterbalance the decreasing trend of cellulose acetate/triacetate for textile application during that period. The production of about 680 million lb [9] in 1979 world-wide for cigarette tow came close to the 756 million lb for world textile acetate. At present the world wide production of man-made cellulosic fibers (including rayon) is 7.044 billion lb.

Acetylation of Cellulose

About the year 1860 Paul Schuetzenberger commenced his investigations on the process of acetylation. He published memoirs on the acetylation of cyanogen and Iodine in 1862 [10]. The action of acetic acid and acetic anhydride on various carbohydrates, including fibrous cellulose [11] was published in 1865. He heated cotton in a closed tube with acetic anhydride to temperatures of 130° - 140°C, until the cotton had entirely passed into solution. When water was added to this solution the acetylated product precipitated out, which was washed, neutralized and dried. This white, amorphous powder was found to be soluble in alcohol and conc. acetic acid, and readily hydrolyzed with KOH to dextrin and an alkaline acetate.

Schuetzenberger and L. Nadin [12] in 1869, and Schuetzenberger in 1870 [13] continued their investigations on acetylation of cellulose to understand the principles governing the yield and the chemical properties of the acetylated cellulose. In their preferred method, cellulose (cotton or filter paper of Berzelius) was heated in a closed vessel to 180°C with 6-8 times its weight of acetic anhydride until a thick syrup resulted. When poured into a large volume of water, this syrup precipitated as large white flakes of cellulose triacetate. Although it was evident that the cellulose triacetate obtained by this manner was profoundly modified and degraded, yet these experiments aroused the interest and laid the foundation for further investigations.

Liebermann and Hoermann [14] in 1878 devised an improvement upon the Schuetzenberger acetylation process by heating cellulose with sodium acetate and acetic anhydride for a more complete esterification results. They also concluded that there were less bye-products, and acetylation could be conducted at lower temperatures with a considerably higher quantitative yield.

A. Franchimont, in 1879, in repeating the work of Liebermann, was unable to obtain a reaction between sodium acetate, acetic anhydride and cellulose (Swedish filter paper), at least the cellulose did not dissolve [15]. He regarded sodium acetate as of value as a probable dehydrating agent only, and from this reasoned that a more energetic dehydrating agent, such as sulfuric acid, would be correspondingly more efficient. It was found that when a little sulfuric acid was added to a mixture of one part of cellulose and four parts of acetic anhydride, brisk action immediately commenced, even with-out the application of external heat, and the cellulose passed into solution, forming a viscous and colored liquid. The acetylated cellulose could be recovered from this liquid by pouring into water, and washing free from uncombined acid. The product after purification from hot ethyl alcohol, gave upon elementary analysis figures corresponding to an eleven times acetylated triglucose. When fused zinc chloride was substituted the reaction was less energetic, the final solution being colorless, from which a white amorphous powder was precipitated which gave analytical figures corresponding to cellulose triacetate. With sulfuric acid as the condensing agent, Franchimont in 1881 was able to obtain at least two dissimilar compounds as indicated by their solubility in hot and cold ethyl alcohol. Whereas when the sulfuric acid was substituted by zinc chloride, only the derivative soluble in hot alcohol was formed, while the colorless acetylating solution showed that the cellulose decomposition had been less severe [16]. C. Guignet [17] in his investigations of artificial parchment and the action of sulfuric acid on cellulose by impregnation, accumulated considerable data, which was extended by subsequent workers such as E. Erwigs and W. Koenigs. These investigators studied the various acetylation methods which had been brought forward up to that time and especially that proposed by Franchinmont in 1879, in which zinc chloride was added to acetic anhydride in place of sulfuric acid or sodium acetate.

The subject was next advanced by C. Cross and E. Bevan [18], who following the method of Franchimont heated acetic anhydride to the boiling point, dropped in a fragment of zinc chloride, after which the specially purified fibrous cotton was added in successive small portions, the digestion being continued from two to three days. Preliminary observations on the crude ester of the reaction, separated by treatment of the filtered solution with water, showed that the acid found by saponification was considerably in excess of the quantity theoretically obtainable from cellulose triacetate.

The product obtained was purified by solution in glacial acetic acid, but the solution was not viscous as stated by Franchimont in the case of the cellulose triacetate which he was able to prepare.

Cross and Bevan embodied the results of their experiments in a patent in 1894, [19] in which cellulose hydrate prepared from precipitation of solutions of cellulose in zinc chloride, or cuproammonium or from viscose, was heated with crystallized zinc acetate and acetyl chloride. The product being soluble in chloroform, from which transparent coherent pellicles were obtained from evaporation, and which suggested to the inventors the use of acelytaled cellulose for the replacement of collodion in pharmacy and surgery. C. Weber, [20] working along similar lines to Cross and Bevan, produced an acetylated cellulose by means of magnesium acetate and acetyl chloride which was described as insoluble in methyl or ethyl alcohols, ethyl acetate, amyl acetate, acetone and ether, but soluble in ethyl benzoate, chloroform, acetic anhydride and acetic acid, and in ethyl benzoate, nitrobenzene and epichlorhydrin. On account of the non-flammability, resistance to moisture and high temperature, and their high insulating capacity, Weber predicted the acetylcelluloses would soon become powerful rivals of celluloid.

In 1895 Cross and Bevan [21] in a contribution to the general chemistry of cellulose, summarized their work in this field up to that date. Also, Cross and Weber combined their activities in the patent field, assigned their interests to I. Frankenburg, and patents were taken in nearly all the countries which at that time had protective patent laws [22].

In the year 1900 appeared the Lederer patents [23] which consisted of using hydrocellulose with four times the quantity of acetic anhydride but no acetic acid. The reaction took place in a short time with the evolution of heat. The cellulose acetate thus obtained was soluble in chloroform and in nitrobenzene.

In May, 1901, the first Miles [6] patent was filed in which he disclosed that in the acetylation process, the addition of minute quantities of sulphuric acid or a monobasic fatty acid not only enabled the formation of cellulose ester directly from cellulose (Cotton), but that the reaction took place at a temperature not exceeding 70°C.

The most important contribution in the field of cellulose acetylation in 1905 undoubtedly was the partial hydrolysis disclosure of G. W. Miles [24]. This partial hydrolysis of acetylated cellulose by the successive withdrawal of acetyl from the ester opened up new and much broader applications, in that it permitted the use of solvents of less cost, lower boiling point, greater speed of evaporation, and solvents which were readily obtainable in unlimited quantities, such as acetone.

The credit for the discovery of the commercially valuable cellulose acetates, and those of maximum strength, pliability and elasticity as suitable for film formation, had been erroneously ascribed to A. Eichengruen and the Farbenfabriken vorm. F. Bayer & Co., to whom many of Miles' patents had been assigned. Miles was the first to recognize and establish priority of conception by patent application for the process of partial hydration whereby less expensive solvents could be used and the maximum of strength for film formation still be obtained. His German Application, embodying the partial hydration of acetylated cellulose, made Sept. 30, 1905, as Application No. M-28289, was, after seven years of continuous litigation, passed by the German Patent Office for issuance, and was then purchased by the litigants and issued by them as D. R. P. 252706 to F. Bayer & Co. This firm had an application covering substantially the same idea, which was applied for on November 29, 1905, as F-20963, and elected by them as an Addition to Miles Application M-28289, which was later issued as D. R. P. 252706. The Bayer Co. did not purchase the Miles F. P. 358079, 1905, covering partial hydration in France. It is true that Bayer & Co. was granted a patent in England (E. P. 21628, 1901; 24067, 1906) for the partial hydration of acetylated cellulose by precipitating the acetylated cellulose, washing to neutrality, drying, and then hydrating. But anyone who has attempted to hydrate dry, normal cellulose acetate will at once appreciate the futility and commercial impossibility of such a method. According to Miles patent, 100 parts of dry, scoured cellulose (preferably bleached cotton) was treated at a temperature below 40°C. with a mixture of 270 to 310 g of acetic anhydride, 390 to 410 g of glacial acetic acid, and 3 to 9 mL of concentrated sulfuric acid, until the cellulose disintegrated to form a white, nontransparent, pasty mass, the temperature being allowed to rise to 50°C. As the reaction is exothermic, the temperature was controlled by refrigeration and held at this temperature for 36 to 40 h or until the mass became more or less transparent, less viscous, and the cellulose apparently passed in solution, forming a heavy, nearly transparent, light brown mass. A test portion was removed, precipitated in a large bulk of water, washed neutral, and dried. It was completely soluble in alcohol-free chloroform, but practically insoluble in acetone. At this point the acetate was ready for the hydrolyzing or ripening process.

A mixture of 60 or 65 mL of water and 60 mL of glacial acetic acid, per 100 g of original cellulose was added slowly to the cellulose acetate solution while stirring vigorously to avoid precipitation. This mixture was held at 40 to 50°C. for 12 to 16 h or until precipitated. The washed and dried sample was only plastic in warm chloroform but entirely soluble in pure acetone. The entire mass was then precipitated by pouring it into cold water, then washed until neutral, and dried at 35 to 40°C. The yield from 100 g of cellulose was 130 to 145 g of partially hydrolyzed cellulose acetate. The resulting white, fibrous mass of secondary cellulose acetate was soluble in formic or acetic acids with some difficulty; readily soluble in tetrachlorethane, acetone, chloroform containing alcohol; less soluble in ethyl acetate, nitrobenzene and pyridine. It was insoluble in water, ethyl or amyl alcohol, amyl acetate, carbon tetrachloride, benzene, toluene or xylene. It was plastic in hot chloroform;

soluble in hot 70 per cent aqueous ethyl alcohol from which it gelatinized on cooling; plastic in glycerol at 125°C.; and soluble in a mixture of hot ethyl alcohol and benzene from which it precipitated on cooling.

In the year 1906, there appeared an important paper on the state of the art up to that time by H. Ost [25]. Ost had independently investigated the processes of cellulose acetylation of Cross and Bevan by the action of acetyl chloride and zinc acetate; L. Lederer's process by the action of acetic anhydride and sulfuric acid on hydrocellulose; and the Bayer/Miles process by the action of the same reagents on normal cellulose, all yielding the same product, a cellulose triacetate.

Although his results were somewhat uncertain in view of the inadequacy of the methods, certain fundamentals were formulated which were of general acceptance at that time. Ost maintained that the production of triacetates was preceded by the formation of diacetates which could be isolated by stopping the reaction while still incomplete.

Knoll & Co. in 1911 patented the use of sodium acetate, [26] and nitrates such as ammonium nitrate [27] to temper the destructive action of sulfuric acid and other contact substances in the acetylating bath, potassium acetate having been patented for this purpose.

The year 1914 saw a determined effort to minimize the drastic action of sulfuric acid by the use of other catalysts, a further search for suitable solvents and high boiling combinations, and an increase in endeavors to replace the cellulose nitrates by cellulose acetate, in those industrial fields where diminished flammability was of prime importance.

The period between 1915-1919 saw applications of acetate in the war efforts. C. Dreyfus [28] showed that by impregnating textile fabrics in the stretched condition with cellulose acetate, an exceedingly taut and impervious surface was produced when the lacquer dried. Thus acetate dope was used for aeroplane wing fabrics during the war. A series of suggested formulas for cellulose acetate aircraft dopes was published by C. Dreyfus in 1915 [29].

The first acetylated cellulose manufacturing process using pre-cooling of the components in order to ensure uniformity of esterification was developed by H. Dreyfus [30] and issued in 1915. It was one of the stellar contributions in this field. He found that cellulose acetates insoluble in chloroform and in alcohol-chloroform, but soluble in acetone could be prepared by acetylating cellulose, using sulfuric acid as catalyst. The acetylating mixture being cooled prior to the introduction of the cellulose to a temperature usually 10° or lower. In fact, incipient crystallization of glacial acetic acid at these temperatures fulfilled the function of disintegrating the cellulose by attrition while acetylation was proceeding, and thus facilitated the production of an unusually uniform ester. In order to obtain esters of relatively high viscosity, the acetylating bath was subjected to prolonged cooling prior to the introduction of cellulose, and acetylation stopped when the product showed insolubility in chloroform and solubility in alcohol-chloroform. Instead of using sulfuric acid as catalyst, it was suggested that a bisulfate such as sodium bisulfate may be employed. The bisulfate being preferably formed in the acetylating mixture by the addition of sodium carbonate or acetate [30].

After the war, during the 1920s intense activity occurred in the field of dyes and dyeing of cellulose fibers, and fiber modifications, such as flame resistant fibers by incorporating borates or chloro and bromo substituted acids; and highly pliable fibers by incorporative oils. During the 1930s, much academic efforts focused on catalysts other than sulphuric acid, such as chlorides of zinc, antimony, calcium or sulfuryl chloride. Studies on acetylating with perchloric acid and perchlorates in the presence of acids led Kruger [31] to establish a hypothesis on the effect of various catalysts in the acetylating mixture. Nevertheless, the process of cellulose acetylation was well established by the 1920s and 1930s, and the manufacture of cellulose acetate filaments was controlled mainly by Dreyfus patents.

To bring this discussion to a close, the basic principles of acetylation as it is understood at present are described here briefly.

Since cellulose is highly crystalline and its polymer chains are held tightly together in an ordered manner through extensive hydrogen bonding, it is insoluble in the reaction medium until practically complete acetylation is achieved. Thus a full acetylation is needed, using the known commercial processes. Although for acetate fiber some acetate groups are stipped off, which goes back to when the Dreyfus brothers commercialized cellulose acetate.

The accessibility of the cellulose is very important because most of the acetylation is a heterogenous reaction. The greater the accessibility, the easier it is for the reactants to diffuse into the interior of the cellulose. Therefore, adequate shredding of the cellulose sheet (woodpulp) and adequate swelling of the cellulose through a pretreatment step are necessary operations prior to the acetylation. The basic principle behind pretreatment is that the penetrating medium, which always contains some water, disrupts the hydrogen bonding in the bundles of cellulose chains. The cellulose becomes swollen and, therefore, the diffusion of reactants into the swollen cellulose is enhanced. It is well known that monomeric alcohols can be acetylated with acetic acid using a mineral acid catalyst to prepare an acetate. However, this does not apply to cellulose, because there is a competing reaction during acetylation of cellulose, namely, degradation or depolymerization, in which cellulose chains are cleaved. If one were to attempt to acetylate cellulose with acetic acid and sulfuric acid catalyst, very low-molecular-weight acetylated products would be obtained [32]. These would not have sufficient average chain length to be useful as fibers, films, or plastics. Acetic anhydride is needed to supply driving force for speeding up the acetylation process so that target viscosity can also be achieved. Although some chain depolymerization occurs, but it can be controlled so that the end products have adequate average chain length.

Acetylation of cellulose is a highly exothermic reaction using acetic anhydride and acid catalyst. Therefore, it is necessary to control the temperature during acetylation, as the rate of depolymerization becomes excessive resulting in a loss of target viscosity. However, the extent of cooling needed depends on the particular process. For example, some processes operate at low catalyst level and comparatively high temperatures where the exotherm is actually used purposely to raise the reaction temperature. However, high or moderate catalyst processes require both cooling of the reactants and jacket cooling to control the temperature.

A strong mineral acid catalyst is needed for the acetylation. Sulfuric acid is generally used for commercial processes. Sulfuric acid catalyst actually combines with the cellulose forming sulfate linkages; however, most of these are removed during acetylation via an exchange with acetyl groups. It is important that the final cellulose triacetate or cellulose acetate contains only a very small residual amount of sulfate groups. The presence of sulfate groups is undesirable because the residual sulfate groups adversely affect the properties of the fibers, films, or plastics, especially the color.

Since sulfuric acid catalyst is used commercially, a full cellulose triacetate with a DS of 3 is never realized for the manufacture of cellulose triacetate, as there always are some residual sulfate groups present after the acetylation. Most of these are removed in a special desulfation step. First excess acetic anhydride is hydrolized (killed) by addition of an aqueous acetic acid solution, sometimes called stop acid. This causes hydrolysis of sulphate groups and generation of some hydroxyl groups. Thus, the cellulose triacetate with DS in the range of 2.91-2.96 results, depending on the catalyst level used during acelytation.

Perchloric acid is an excellent catalyst for the acetylation of cellulose. It does not combine with the cellulose; therefore, a full cellulose triacetate (DS of 3) can be made with this catalyst. Sulfuric acid is preferred over perchloric acid because it is lower in cost and also because it does not present a potential hazard. There is concern that for commercial processes perchloric acid could represent an explosion hazard should there be a build-up of perchlorates in the acetic acid recovery system.

The production of secondary cellulose acetate is done in a separate operation, wherein the cellulose triacetate is hydrolyzed under carefully controlled conditions to a DS of about 2.4. Hydrolysis time depends upon the specific catalyst level and temperature range. The higher the temperature, the lower the catalyst level in order to preserve viscosity. Patents have been issued to Campbell et. al. of Celanese Corp. in 1973 [33] and Yabune et. al. of Daicel Chemical industries in 1984 [34] on high temperature hydrolysis. In high temperature hydrolysis (120 - 160°C), the catalyst is completely hydrolyzed, otherwise, severe chain degradation would occur. In the case of cellulose triacetate sulphate groups are hydrolyzed with minimum amount of water, followed by neutralization with a suitable base. The acetate is then precipitated with dilute acetic acid as uniform flakes, washed, dried and blended. Acetic acid recovery system is vital to the economics for manufacturing cellulose acetate.

Mechanisms of Cellulose Acetylation

The overall reaction in the conversion of cellulose to cellulose triacetate is:

$$\text{Cell}\begin{matrix}\diagup \text{OH}\\ -\text{OH}\\ \diagdown \text{OH}\end{matrix} + 3\text{Ac}_2\text{O} \xrightarrow[\text{Catalyst}]{\text{Acid}} \text{Cell}\begin{matrix}\diagup \text{OAc}\\ -\text{OAc}\\ \diagdown \text{OAc}\end{matrix} + 3\text{HOAc} \tag{1}$$

were Cell is the anhydroglucose ring (without -OH groups) and Ac is an abbreviation for acetyl, $COCH_3$. Equation 1 shows that 3 mol of acetic anhydride react with 1 mol of cellulose to give 1 mole of cellulose triacetate and 3 moles of acetic acid. During hydrolysis some of the acetate groups are hydrolyzed. This is represented approximately by the following equation:

$$\text{Cell}\begin{matrix}\diagup \text{OAc}\\ -\text{OAc}\\ \diagdown \text{OAc}\end{matrix} + \text{H}_2\text{O} \xrightarrow[\text{Catalyst}]{\text{Acid}} \text{Cell}\begin{matrix}\diagup \text{OH}\\ -\text{OAc}\\ \diagdown \text{OAc}\end{matrix} + \text{HOAc} \tag{2}$$

Equation 2 shows the formation of a diacetate. Actually, the DS is higher at 2.4, which means that less than one acetyl group per anhydroglucose unit is hydrolyzed on an average. In other words, an average of three acetyl groups are hydrolyzed for every five anhydroglucose units.

In 1946 Malm et al. [35] showed that the acetylation of cellulose with sulfuric acid catalyst follows the same pattern as when methanol is sulfated in the presence of sulfuric acid and a mixture of acetic acid and acetic anhydride [36]. The sulfuric acid catalyst combines quantitavely with the cellulose forming cellulose sulfate:

$$\text{Cell-OH} + \text{H}_2\text{SO}_4 \rightarrow \text{Cell-O-SO}_2\text{OH} + \text{H}_2\text{O} \tag{3}$$

Apparently, the acetic anhydride removes the water just as soon as it is formed, which drives Equation 3 to the right.

The work of Malm et al. [35] showed that during the intermediate stage of the acetylation the catalyst was combined quantitatively as shown in equation 3. However, as acetylation approached completion and was extended beyond completion, the sulfur content continually dropped, there being a replacement of the sulfate groups with acetyl groups.

The replacement of - SO_2OH group by $-COCH_3$ group was proposed by Laamanen and Sihtola [37] in 1946, as follows:

$$\text{Cell-O-SO}_2\text{OH} = \text{Cell-O}^- + {}^+\text{SO}_2\text{OH} \quad (4)$$

$$\text{Cell-O}^- + {}^+\text{SO}_2\text{OH} + \text{Ac}_2\text{O} \rightarrow \text{Cell-OAc} + \text{AcO-SO}_2\text{OH} \quad (5)$$

Equation 4 implies that the sulfate linkage is not 100% covalent, but that there is also present some polarization resulting in some ion separation. Equation 5 implies that the $^+SO_2OH$ ion cleaves a molecule of acetic anhydride:

$$\text{Ac}_2\text{O} + {}^+\text{SO}_2\text{OH} \rightarrow \text{Ac}^+ + \text{AcO-SO}_2\text{OH} \quad (6)$$

The strong acetylium ion Ac^+ reacts readily with the negatively charged alkoxide ion:

$$\text{Cell-O}^- + \text{Ac}^+ \rightarrow \text{Cell-OAc} \quad (7)$$

In order that the acetylation continue according to the above mechanism, it is necessary that the sulfuric acid catalyst be regenerated. Akim in 1967 [38] stated that there are actually two mechanisms for acetylating cellulose, one through the sulfate group such as described above and the other through the acetyl-sulfuric acid intermediate. Acetylsulfuric acid can form in the reaction mixture:

$$\text{Ac}_2\text{O} + \text{HO-SO}_2\text{OH} \rightarrow \text{AcO-SO}_2\text{OH} + \text{HOAc} \quad (8)$$

And if Laamanen and Sihtola's proposed mechanisms expressed by Equations 4 and 5 are correct, acetylsulfuric acid is also formed via replacement of sulfate groups with acetyl. According to Akim [38] the acetylsulfuric acid also acetylates cellulose:

$$\text{Cell-OH} + \text{AcO-SO}_2\text{OH} \rightarrow \text{Cell-OAc} + \text{H}_2\text{SO}_4 \quad (9)$$

In fact, he states that the degree of acetylation through Equation 9 and the degree of acetylation through cellulose sulfate depends to a great extent on the reactivity of the cellulose. When the reactivity is good, acetylation through sulfate formation predominates. When the reactivity of the cellulose is not as good, the reaction mixture will contain a higher percentage of acetylsulfuric acid and then acetylation via Equation 9 will predominate.

It seems most probable that the sequence of events and mechanism for the acetylation of cellulose using sulfuric acid catalyst is as follows:

1. The sulfuric acid combines covalently with cellulose in the presence of acetic anhydride-acetic acid forming cellulose sulfate.

2. The sulfate group in the cellulose is replaced by an acetyl group (acetylation) via a mechanism such as shown by Laamanen and Sihtola in Equation 4 and 5.

3. The acetylsulfuric acid by-product from Equation 5 then acetylates cellulose regenerating sulfuric acid catalyst, as shown in Equation 9.

4. The regenerated sulfuric acid combines with cellulose to start the process over again. The driving force in the early stages of the acetylation is the fast removal of acetylsulfuric acid as in Equation 5, the rapid regeneration of sulfuric acid according to Equation 9, and the rapid sulfation reaction according to Equation 3.

5. As acetylation becomes complete the available cellulose hydroxyl groups are depleted more and more. However, there is excess of acetic anhydride. Therefore, there is a continual build-up of acetylsulfuric acid because the sulfuric acid cannot find cellulose hydroxyl with which to combine so it reacts with acetic anhydride.

6. As the acetylsulfuric acid builds up with the simultaneous depletion of cellulose hydroxyl groups, it is more and more difficult for the sulfate groups to be replaced by acetyl groups. Therefore, there will be residual sulfate groups present after termination of the acetylation reaction. These will have to be removed after acetylation. Some of them are hydrolyzed by slow addition of stop acid when killing excess anhydride, as pointed out previously. Others are removed during hydrolysis to the secondary acetate or by a special desulfation step for cellulose triacetate manufacture.

The probable mechanism for the degradation of cellulose during acetylation with sulfuric acid catalyst is that first the proton attacks the glycosidic linkage thereby cleaving the chain as follows:

$$\text{—Cell—O—Cell—O—} + H_2SO_4 \rightarrow \text{—Cell—O—}SO_2OH + \text{HO—Cell—O—} \qquad (10)$$

The mechanisms for acetylating the chain ends of these fragments are similar to those described for acetylating the cellulosic hydroxyl groups.

Dope Preparation

The basic principles are the same for the preparation of the cellulose acetate dope compared with the cellulose triacetate dope. Acetone is the preferred solvent for acetate and methylene chloride/methanol (9:1) is generally used for triacetate. The flakes, solvent mixture, and a filtering aid are added to a heavy duty mixer. The solution is prepared in a fully enclosed system to minimize solvent losses and also to meet strict exposure levels regarding the workers. In the case of cellulose acetate the main solvent is acetone, which is highly flammable. Therefore, the vapor/air ratio must be maintained at a level that meets safety regulations. Strict fire codes are maintained in the dope preparation department as well as in the fiber-spinning department.

Serad and Sanders [39] reported in 1979 that the typical solvent composition for cellulose acetate is about 95% acetone and 5% water and that the typical solids concentration in the extrusion solution is about 20-30%, depending on the polymer molecular weight. The viscosity of the solution at room temperature is reported to be about 100-300 Pa.s (1000- 3000 poises). Concentration, temperature, and mixing uniformity are closely controlled through several mixing stages.

Moelter and Steele [40] reported that this method for cellulose acetate dope preparation is a continuous operation. Since acetone presents a fire hazard at the relatively low concentration of about 2.5% in air, the operations from dope preparation through to extrusion are continuous and closed. Precise control of concentration is required to extrude fibers with uniform weight per unit length. Also, the acetone/water ratio requires close control because of its influence on fiber properties, such as tensile strength and elongation. Moncriff [41] reported that fibrous or flake cellulose triacetate is made into a 20% solution in methylene/methanol (9:1).

Both bright and dull yarns for cellulose acetate and cellulose triacetate are produced. In the case of bright yarn, the spinning dope contains cellulose acetate or triacetate and the solvent mixture. For dull yarn about 1-2% finely ground titanium dioxide pigment is added to the dope. It is necessary that the pigment be milled to a very small particle size, 1-2 um, or even smaller, so that it can pass through the filtration system as well as the small jet holes of the spinnerette.

There are two methods that can be used to prepare the spinning dope containing the delusterant (titanium dioxide). One is to add the TiO_2 in the initial mixing stage, a procedure that has the advantage of giving a very uniform dispersion of the pigment in the dope. However, a disadvantage is that completely separate dope preparation systems from the very initial stages of dope preparation are necessary. Greater flexibility to the dope preparation is achieved by injecting a pigment slurry after filtration [40]. However, it is necessary that thorough mixing be obtained by the latter method as well as strict control of the pigment slurry.

Besides dull yarn, black fiber is also produced. It is produced in a similar manner to that of dull yarn by adding 2% [on the weight of cellulose acetate) of very finely dispersed carbon black to the extrusion dope with an average particle size of 1-2 um [40].

Sprague and Horner [42] in 1965 reported that acetate and, to a limited extent, triacetate fibers were being produced in solution-dyed colors. This method of coloration is accomplished by adding suitable coloring agents, usually organic or inorganic pigment, to the spinning dope prior to extrusion, which is called "solution" or "dope dyeing." This method was used to a limited extent as far back as 1911, when the first patent was issued to Bayer and Co. in which cellulose was dyed before acetylation with dyestuffs capable of withstanding acid treatment and then acetylated in the usual manner.

Extensive filtration of the dope is needed to remove mechanical impurities, unacetylated particles, gels, and dirt. The filtration must be thorough enough so that the dope passes through the fine holes in the spinnerette without causing jet blockage. The fine jet holes range in size from 30 to 80 um in diameter [39].

Spinning of Fibers

Dry-Spinning. Most of the Cellulose acetate and triacetate fibers are produced by dry extrusion. The preheated dope is extruded through the fine holes of the spinnerette into a heated column or cabinet. Each of the emerging fine streams encounters a flow of hot air which brings about an instantaneous evaporation of the solvent from the surface of each fine dope stream resulting in immediate filament formation with a solid skin forming over the still plastic interior of the filament. As the filaments pass down the column, more and more solvent is removed from the interior of the filaments, thereby continually solidifying and strengthening them.

Near the bottom of the column the individual filaments converge to form a yarn that really is a bundle of parallel filaments. The yarn passes over a lubrication wheel and is pulled by the feed roll and wrapped around it once. It then passes through a balloon guide and is taken up on a ring and traveler and wound on a bobbin.

The production of cellulose acetate or triacetate staple follows the same basic principles as those for continuous filament yarn. Dry-spinning by extruding the dope through a spinnerette into a hot air column is still done, there usually being more holes per spinnerette for staple compared with continuous filament. Also, instead of collecting the yarn from a single spinnerette, the yarn filament ends from a number of spinnerettes are gathered together into a strand or tow. Tow is a continuous band composed of several thousand filaments held loosely together by crimp, a wave configuration set into the band during manufacture. A tow is formed by combining the output of a large number of spinnerettes and crimping the collection of filaments to create an integrated band of continuous fibers. Staple is produced by cutting the tow into short lengths (about 4-5 cm). Both acetate and triacetate staple fibers are produced. Cigarette tow is produced from acetate fibers.

Wet-Spinning. Although wet-spinning of cellulose acetate or triacetate is feasible, it is not practiced to any large extent because the economics are not as favorable compared with dry-spinning. In wet-spinning the spinnerette is submerged in a liquid spinning bath. The coagulating bath has an optimum composition and temperature to obtain proper coagulation of the fine dope streams extruding from the spinnerette and also proper stretch of the resulting yarn filaments. Wet-spinning can produce cellulose acetate and triacetate fibers having higher tenacities compared with those from dry-spinning. However, the wet-spinning process is limited to a staple or tow and even then the economics are not favorable compared to staple or tow from dry-spinning.

Melt-Spinning. When cellulose acetate or triacetate is heated to the melting point, neither of these polymers has good stability in the melt. Severe discoloration and considerable decomposition of the melt occurs, especially if held long as a melt. Therefore, melt-spinning of cellulose acetate or triacetate is not considered to be an attractive method for producing fibers. However, a relatively small amount of triacetate yarn has been made by melt-spinning. As described by Sparague in 1965 [42], special techniques are necessary to prevent degradation at high temperatures at which triacetate melts (>300°C). Therefore, a primary requirement is to hold the polymer in the molten state for only a short time.

Cellulose acetate are produced from 36 to >900 denier or 4 to >100 tex. Individual filament deniers are usually 2-4 dpf. Common continuous filament yarns have total deniers of 55, 60, 75, and 150 [39].

Besides total denier and dpf, the cross section of the filament is another factor that relates to the geometry of the spinnerette hole (round, triangular, etc.). Therefore, the finished fabric will have a variety of esthetics such as hand, luster, and cover which are influenced by the fiber cross section, total denier, and dpf.

Fiber Properties: An Overview

Cellulose acetate and triacetate fibers are not particularly strong like polyester and nylon, although they have ample strength for their intended end uses. The key words for cellulose acetate and triacetate fibers and fabrics are silk like softness, pleasing appearance, and comfort. They are characterized in the trade as having superb, soft hand and good draping quality. The overall characteristic for both fibers is simply an excellent balance of properties.

Cellulose acetate is not too hydrophobic nor is it hydrophilic. This results in moisture regain properties that relate to the excellent comfort factor. It also results in easy dyeability of the fibers and the resistance of these fibers to microorganisms and insects.

Cellulose triacetate is more hydrophobic than cellulose acetate but not as hydrophobic as polyester. It has enough hydrophobicity that the heat-treated fabrics possess "wash-and-wear," "ease-of-care" characteristics, but it is not too hydrophobic so as to interfere with its aesthetic properties like softness and comfort.

Fine Structure of Fibers

Cellulose Acetate. Although cellulose from either cotton linters or wood pulp is highly crystalline, dry-spun cellulose acetate fiber has very little crystallinity and fiber orientation. It is apparent that the acetylation process followed by partial hydrolysis of the triacetate to a secondary acetate destroys the fine structure of the original cellulose.

Work [43] studied the X-ray diffractions of cellulose acetate fibers in 1945, and pointed out conclusively that the structured order of cellulose disappears during the normal process of esterification. Even annealing the fibers or dry-stretching the yarn just short of the breaking point did not increase the crystallinity of cellulose acetate. Only the orientation of the already existing small amount of crystallinity was increased by these treatments. Even swelling the fiber in solvents such as dioxane-water followed by stretching did not increase the crystallinity, although this treatment gave the most orientation.

Work pointed out that the basic reason for the lack of crystallinity and order in cellulose acetate is the heterogeneity in the molecular segments along the chain molecules. Because of the heterogeneity, the acetyl groups and the hydroxyl groups are unable to pack together with any great amount of regularity using heat treatment. He concluded that this is an example of stearic disorder due to heterogenous side groups-the acetyls and hydroxyls randomly occurring along the chain molecule.

Cellulose Triacetate. Work [43] showed that cellulose triacetate dry-spun fiber before any heat treatment had very low crystallinity. However, annealing the fiber at elevated temperatures markedly increased the crystallinity. Stoll, [44] in 1955 showed conclusively that cellulose triacetate fiber (Arnel) can be made highly crystalline by treating the fiber at 280°C under tension, as evidenced by the very sharp X-ray diffraction patterns for the heat-treated fiber. The formation and growth of the crystallites, resulting in large, perfected crystals in the fiber, were believed to take place during the heating and cooling cycle.

The development of high crystallinity in the fiber has beneficial effects on the fiber and fabric properties. Heat-treated cellulose triacetate yarns possess a higher softening point, thereby increasing their resistance to heat. It also lowers the moisture regain, thereby improving its dimensional stability. The practical result of heat treatment is to impart ease-of-care characteristics to fabrics and garments made from cellulose triacetate. An important point is that although the moisture regain is lower, it is not too low. This results in outstanding comfort for fabrics made from cellulose triacetate, which is much superior to that for polyester. High crystallinity and orientation are related to strength and abrasion resistance of fibers and fabrics. Acetate fibers are not particularly strong which accounts in part for their low abrasion resistance.

Dyeing Characteristics

The extensive commercial use of acetate fibers was somewhat delayed, owing to the difficulties in dyeing these fibers and attaining fast colors. Failures in this respect were simply the consequence of attempts made to apply the same dyes, by the same methods as those used for dyeing cotton or rayon. Whereas the first acetate fibers consisting of primary cellulose acetate (triacetate), were of such a nature that it was not possible to dye them with acid or basic dyes. After World War I (1914-15), intense research activity was directed toward the area of dyes and dyeing of cellulose acetate fibers as well as fiber modifications. In 1920, J. Briggs [45] published an article on "dyeing of acetate silk", and in 1921 appeared the dyeing process of J. Briggs and C. Palmer [46], in which cellulose acetate was pretreated with a thiocyanate solution to augment its dyeing properties, so that acid dyes, basic dyes, and vat dyes could be used. Also dyestuffs of aryl-or substituted arylbenzo - or naphtho-quinone - monoimide series were used for dyeing acetate [47]. A series of papers were published in 1925 [48] giving a list of precautions and recipes for dyeing of acetate and a description of the "soluble vat" dyestuffs for dyeing acetate and union fiber fabrics. However, before the discovery of the ionamines in 1922, there was no special class of dyestuffs for acetate fibers which would give reasonably fast shades in all colors [49].

The early difficulties in dyeing acetate fibers not only led to research on new dyeing methods and special dyes, but also resulted in several attempts to manufacture solution or dope colored fibers as discussed under dope preparation. Thus the first solution dyed Chromspun^R fibers were introduced in the U.S. in the year 1951, by Eastman Co.

For the most part the classes of soluble dyes are not applicable to the more hydrophobic cellulose acetate and triacetate fibers and fabrics. There is not enough swelling of these fibers to absorb the soluble dye molecules into their interior. However, Egerton in 1949 [50], reported that the water-soluble Solaced dyestuffs by I.C.I. were applicable to cellulose acetate. But Moncrieff [41], in 1975 commented that although they gave good penetration of heavy fabrics their wet-fastness was inferior to that for dispersed dyes. Another point is that the alkaline conditions under which certain water-soluble dyes like the sulfur and vat dyes are applied can possibly saponify some of the acetate groups.

The preferred dyes for cellulose acetate and triacetate are the disperse dyes. These consist mainly of azo, anthraquinonoid, and nitrodiphenylamine dyes that are insoluble or at best slightly soluble in hot aqueous dye bath systems. These crystalline dyes must be ground to a very small particle size such as 2 um. They are then made into a paste with water and a wetting agent and added to the dye bath through a fine mesh screen. A dispersing agent and a carrier may be present in the dye bath to enhance diffusion of the dye molecules into the fibers. For cellulose acetate dyeing a temperature range of 70 - 80°C is used [50]. Higher temperatures such as 85°C are avoided because they could cause delustering of the yarn [41]. For cellulose triacetate (Arnel), a higher temperature in the range 96-100°C is used [51].

The fundamental aspect of acetate fiber dyeing is diffusion of the dye molecules into the interior of the fibers. Diffusion is the rate-controlling factor in dyeing with disperse dyes. Diffusion rate depends on both the dye and the substrate. The more hydrophobic the substrate, the more difficult it is for the dye molecules to penetrate the fibers. Cellulose triacetate is more difficult than cellulose acetate regarding the ease of diffusion of the dye molecules into the fibers. Therefore, dispersing agents and carriers or accelerators are used to increase the rate of diffusion of the dye molecules into the fibers.

Dispersing agents such as Turkey red oil, synthetic detergents of alkyl sulfates or alkylaryl sulfonates, and fatty alcohol-ethylene oxide condensates have been used [52]. They tend to solubilize the water-insoluble dye. Hartley [53] proposed that the oil-soluble dye dissolves in the spherical micelles of the paraffin-chain salts. This not only aids in the diffusion of the dye into the fiber, but importantly, prevents a deposition of the dye particles on the surfaces of the fibers, which is undesirable.

Carriers or accelerants are not necessarily used for cellulose acetate dyeing. But they are used for more hydrophobic fibers like cellulose triacetate and polyester. Typical carriers are tripropyl phosphate, butyl benzoate, methyl salicylate, diallyl phthalate, and diethyl phthalate. According to Fortess [51] tripropyl phosphate acts as an accelerant for dyeing Arnel^{R} triacetate with dispersed dyes by (a) solubilization of the crystalline dyes in the aqueous phase, (b) adsorption onto the fiber surface providing a dye solvent at the fiber-liquor surface, and (c) absorption into the fiber, swelling the fiber and weakening molecular bonds, thus permitting easier access for dye diffusion into the fiber.

Fortess and Salvin [54] in 1958, reported that the rate of dyeing increases markedly with increasing temperature. Cellulose triacetate may also be dyed with disperse dyes at temperatures above 100°C where suitable pressure equipment is available. According to Boulton [55] the advantages to be gained are improved overall fastness, particularly to dry cleaning with trichloroethylene, owing to the better penetration of dyes into individual fibers, and the possibility of using very slow diffusing dyes, which may be necessary to obtain certain colors and fastness and which do not build up sufficiently well at the boil. Temperatures up to 130°C may safely be used in such cases.

The requirements for the dyed fibers and fabrics are very demanding. Most important features are: attractive colors and shades, good wash fastness, good lightfastness, resistance to gas fading (oxides of nitrogen), resistance to ozone fading, and fastness to sublimation when fabrics undergo thermal finishing operations such as heat setting and pleating. These features depend upon dye selection, dyeing conditions, and after treatment of the dyed fibers or fabrics.

As for dye selection Fortess and Salvin [54] in 1958 reported that lowering the basicity of anthraquinone blue dyes greatly improved the lightfastness and gave excellent resistance to atmospheric fading due to oxides of nitrogen. Specifically, Structure I of 1,5-dihydroxy-8-nitro-4-anilidoanthraquinone has excellent resistance to gas fading

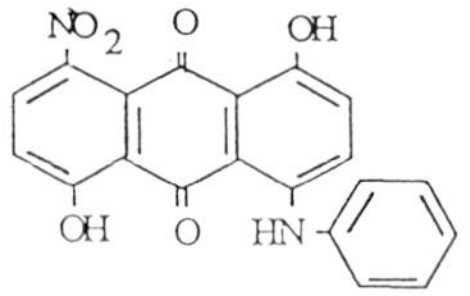

I - Good

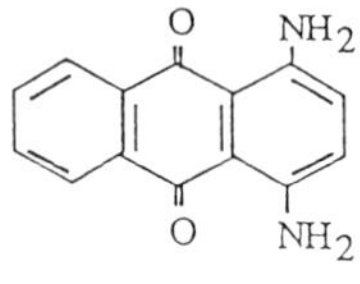

II - Poor

because the low basicity of the phenylamine nitrogen does not form nitroso compounds under the conditions of atmospheric fading. On the other hand, Structure II of 1,4-diaminoanthraquinone represents a typical gas fading blue because it has high basicity and forms nitroso compounds readily with oxides of nitrogen. Fourness [56] in 1956 showed that attaching a hydroxyl group to an alkyl radical improves the fastness to sublimation. Thus, when the two ethyl groups in Structure III of 4-diethylamino-4-nitroazobenzene are replaced by two hydroxyethylene groups giving Structure IV of

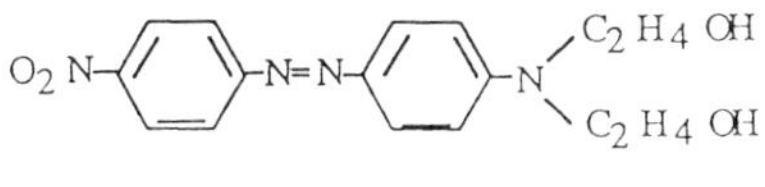

III - Good

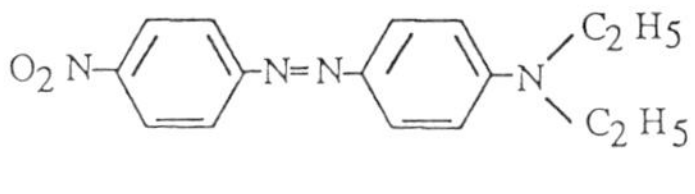

IV - Poor

4-bis(hydroxyethylene)amino-4-nitroazobenzene, the latter has improved fastness to sublimation. Fastness to sublimation is important in blends of fibers making up fabrics; the dye of one fiber may mark-off to the other fiber during thermal finishing operations without good fastness to sublimation.

An example of an aftertreatment on a dyed fiber to improve the dye fastness and other characteristics is the heat treatment of dyed cellulose triacetate. Tesi [57] showed that heat-treating dyed Arnel^R fiber at 425°F for 10 sec caused the dye to penetrate the interior of the fiber completely. In other words, there was further migration of the dye molecules into the interior of the fiber during the heat treatment. Such a heat treatment imparts extremely high washability performance for the dyed fabrics. According to Fortess [51] there is also improvement in gas-fading resistance, Ozone fading resistance, and cracking resistance.

For cellulose acetate/viscose or cotton fabric blends, cross-dyeing can be done because cellulose acetate is unstained by many direct dyestuffs and because disperse dyes used for acetate do not dye viscose or cotton. Two-color and resist effects can be obtained on such blends from one dye bath.

Cellulose acetate and triacetate can be solution or dope dyed. The main advantage of dissolving or dispersing dyes in the spinning dopes is that there results maximum intimate contact between the dye molecules and the cellulose acetate or triacetate molecular chains. Therefore, there is no concern for dye diffusion and penetration as in the dye bath method. Dope-dyed acetate and triacetate yarns are extremely colorfast to washing, dry cleaning sunlight, perspiration, sea water, and cracking and are very resistant to gas or fume fading [39]. However, dope-dyed fibers are more expensive than fibers dyed by the dye-bath method.

Fiber Modifications

Spin Dope-Modified Cellulose Acetate

Since cellulose acetate has been a major commercial textile fiber for almost 80 years, considerable research efforts have been made to enhance service properties. Major areas are improvements in strength, abrasion resistance, shrinkage resistance, sunlight resistance, and flammability. Although dyeability and static resistance characteristics are generally good, there is still room for improvement with these.

Historically modifications of acetate fibers go back to 1929 when Clarke and Malm [58] in an endeavor to reduce flammability of acetate fibers while maintaining stability, described mixed esters of acetate with chloro and bromo substituted propionates and butyrdes etc. However, other approaches have been used and are described briefly.

Polymer additives. One approach for modifying cellulose acetate is to add polymer additives to the spin dope. The critical parameter using this approach is compatibility. A broad range of polymer additives have been used such as polyorganosiloxanes to enhance abrasion resistance, thermal stability, and reduced static propensity; acrylic copolymers to enhance dyeability, abrasion resistance, thermal and chemical stability; polyurethane and epoxides to enhance abrasion resistance and fatigue; polyvinylchloride to reduce flammability; polyethylene glycol and polyethylene oxide to enhance abrasion resistance, tenacity, fatigue, and to reduce static propensity [32].

However, in no case has complete compatibility between the polymer additive and cellulose acetate phases been achieved.

Graft Copolymer Additives. One method of improving the compatibility between polymers in solution is by introducing graft copolymers of each polymer. Thus, a graft copolymer of a monomer with cellulose acetate has better compatibility with cellulose acetate than the homopolymer. Furthermore, the graft copolymer increases the compatibility of the respective homopolymer with cellulose acetate.

Various grafted polymers onto cellulose acetate have been added to the cellulose acetate spin dope. For example polyvinyl fluoride has been reported by Yulchibaeva and coworkers [32], to enhance thermal stability; polyacrylonitrile [59-61] to enhance tenacity and thermal and chemical stability; acrylics [32] to enhance abrasion resistance, dyeability, hydrophilicity (hence reduced static problems), as well as thermal stability; and polystyrene [62] for abrasion resistance and chemical stability. However, not all grafted copolymers showed good compatibility. For example, cellulose acetate polystryrene graft copolymers existed in separate domains in the cellulose ester fiber.

Liquid Crystals

It was pointed out previously that dry-spun cellulose acetate has very little crystallinity. Commercial spinning dopes are isotropic and have no molecular chain order. However, it is possible to prepare concentrated solution dopes that are anisotropic and thus have chain order resulting in liquid crystals. When such solutions are dry-jet wet-spun or wet-spun, the result is highly crystalline fibers in which the crystals are oriented along the fiber axis. Such fibers have high birefringence.

A patent, issued to DuPont in 1977 [63], describes the use of anisotropic cellulose derivative solutions for the production of high-tenacity fibers. Trifluoroacetic acid solutions of cellulose acetate were dry-jet wet-spun into alcohols. Dry jet wet-spinning means that the spinning jet is positioned above the liquid spin-bath so that the extruded filaments first encounter air before entering the bath. In wet-spinning the jet is immersed in the spinbath. Fibers were obtained that had superior tenacity at about 5g/den, but very low elongation at about 5%. A key point is that the highly polar trifluoracetic acid solvent produced the anisotropic solutions at reasonable concentrations and viscosities.

Bheda et al. [64] investigated the wet-spinning of cellulose triacetate from anisotropic solutions in trifluoroacetic acid. The highest birefringence and mechanical properties were obtained in fibers when the highest draw-down ratio was used. Spinning from a dope concentration of 0.30 g/ml and using a draw ratio greater than 4, they obtained cellulose triacetate fibers having a Young's modulus of 61.1 g/den, tensile strength of 4.1 g/den and elongation to break of 10%. These fibers possessed very high birefringence. Also, the wide-angle X-ray spectroscopy patterns for these fibers appeared to correspond to Cellulose triacetate I crystal structure reported by Sprague et al. [42].

Auerbach et al. [65] in 1980 reported their investigations of cellulose and cellulose-derivative solutions regarding characterization of the solutions and spinning studies. Realizing the disadvantages of trifluoroacetic acid (cost, toxicity, and corrosiveness), these authors studied solutions of cellulose acetate in acetone. However, in the acetone system the critical concentration for liquid crystals was much higher (50% cellulose acetate) than the trifluoroacetic acid. This caused very high visocity of the dope and problems in spinning. Also fiber properties were poor due to slow relaxation time of these very viscous streams resulting in imperfections in the final fiber. At present, trifluoracetic acid seems to be the most effective solvent for producing liquid crystals and retaining these in the dry-wet jet-spun or wet-spun fiber solid state. From a practical standpoint a more attractive solvent system is needed.

Segmented Block Copolymers

There is another approach for modifying cellulose acetate and triacetate, particularly triacetate, namely, to cleave the main polymer chains giving low-molecular-weight blocks, having preferred difunctionality, and then rebuilding the blocks with other difunctional low-molecular-weight polymers that result in segmented block copolymers. This approach was taken by Steinmann [66], who in 1968, prepared low-molecular-weight blocks of cellulose triacetate by the controlled depolymerization of a full cellulose triacetate, giving substantially difunctional blocks with hydroxyl end-groups, and then reacting the blocks with low-molecular-weight polyesters (aliphatic) or polyethers capped with isocyanate end-groups. Santangelo and Steinmann [67] dry-spun the segmented block polymers from methylene chloride solution giving elastomeric fibers of the Spandex type. Fibers from 37% cellulose triacetate blocks, 55% polyester blocks, and 7.5% methylene bis (4-phenyl isocyanate) had a tenacity of 0.53 g/den and elongation of 510%, with permanent set of 10% after stretch. An interesting point is that although low-molecular-weight blocks of cellulose triacetate were used, the resulting fibers had the brilliant luster and soft hand typical of cellulose triacetate. Also, the films cast from methylene chloride solution had sparkling clarity typical of cellulose acetate or triacetate.

The problem with the cellulose triacetate blocks was that they were not strictly difunctional. Therefore, solution dopes had relatively poor shelf life (tended to gel) requiring that spinning be done soon after dope preparation. Strictly difunctional blocks of cellulose triacetate were never achieved.

Textile Applications

A unique combination of desirable qualities and low cost accounts for the demand for acetate in textiles. In the U.S., acetate and triacetate fibers are used in tricot-knitting and woven constructions, with each accounting for approximately half the total volume. This distribution changes slightly according to market trends. The main markets are women's apparel, eg, dresses, blouses, lingerie, robes, housecoats, ribbons. Decorative homefurnishings applications are draperies, bedspreads, and ensembles. A variety of fabrics having pleasing aesthetic properties are produced from acetate and triacetate fibers. Examples are dress velvets, terry-like structures, fine-gauge single jersey, taffetas, and silk-like to bulky wool-like knitwear.

Cellulose triacetate offers the unique combination of "ease-of-care" and aesthetic properties. A particularly important application of triacetate is in velour and suede-like fabrics for robes and dresses. These fabrics offer superb aesthetic qualities at reasonable cost. Triacetate is also desirable for print fabrics as it produces bright, sharp colors [39].

Another textile end use is linings. Fabrics are plain-woven taffetas, but satins and twills are also used. Linings are used in apparel, such as for men's suits, and in nonapparel applications, such as for curtains and caskets.

Acetate and triacetate fibers have lower strength and abrasion resistance than most other synthetic fibers and are frequently combined with nylon or polyester in yarns or specific fabric constructions. This permits them to be used in applications not suitable for 100% acetate fabrics, eg, men's shirts. Combination yarns can be prepared by twisting or air-entanglement and bulking. Yarns prepared by air-entanglement and bulking have unique characteristics and aesthetic properties that allow their use in casement and upholstery fabric markets. With chemical additives, both acetate and triacetate fibers can pass U.S. Government flame-retardant fabric regulations, eg, DOC FF 3-71.

Non-Textile Applications

Cigarette Tow

Acetate fiber used in the production of cigarette filters is supplied in the form of tow [68]. Tow is a continuous band composed of several thousand filaments held loosely together by crimp, a wave configuration set into the band during manufacture. A tow is formed by combining the output of a large number of spinnerets. The wide range of available acetate filter tow products permits control of properties in the finished cigarette-filter rod.

A tow product is characterized by cross section, tex (denier), and crimp. The shape of the cross section is related to the shape of the minute orifices in the spinneret used to form the filament. The presence of crimp in the tow is necessary to ensure that the tow can be packaged, processed, and handled easily and to impart bulk to the finished filter. To achieve the latter in the production of the filters, it is necessary to open the tow band to the desired bulk so that the fibers completely fill the paper wrap without voids and soft spots.

A linear relationship, constant for any given tow item, exists between the weight of tow in a cigarette-filter rod and certain of its properties, such as pressure drop and smoke-removal efficiency. By preparing filter rods over a range of rod weights and testing the rods for pressure drop, a linear curve characteristic of the specific tow item is obtained. This characterization provides a means of designing the required cigarette-filter characteristics.

Cellulose acetate is unique for cigarette filters. Although it is a hydrophobic polymer it has a fairly high moisture regain resulting in an excellent balance regarding absorption of substances in smoke and maintaining uniform pressure drop across the filter rod. Another important point is that cellulose acetate per se is nontoxic.

Hollow Fibers

Semipermeable membranes are produced from cellulose acetate films. Again, the excellent balance of the hydrophobic-hydrophilic characteristic for cellulose acetate makes this polymer ideal for semipermeable membranes used for reverse osmosis.

Research has been done on hollow cellulose acetate fibers that are semipermeable. Leonard [69] dry-jet wet-spun cellulose acetate from 25% solution in dioxane-formamide into water. The extrudate passed into a solvent-enriched evaporation zone before coagulation in water. The fibers were stretched to an external diameter of 350 um and inside diameter of 120 um by spinning at 36 m/min. Membranes made from these fibers were charged with water containing 200 ppm $MgSO_4$ at a flow rate of 435.98 l/m^2 membrane surface per day (reverse osmosis), whereby 98.1% of the salt was removed.

Although hollow cellulose acetate fibers are not yet commercialized, they represent an excellent potential end use for the future such as in the desalination of water.

REFERENCES

1. Rules and Regulations Under the Textile Fiber Products Identification Act, effective March 3, 1960, U.S. Federal Trade Commission, Washington, D.C., 1960, p. 4.
2. Schuetzenberger, P. Compt. Rend 61, 485-6 (1865).
3. Worden, E.C. J. Soc. Chem. Ind 38, 370-4T (1919).
4. Cross, C., Bevan, E., British Pat. 9,676.
5. Miles, G.W. U.S. Pat. 835350, Dec 11, 1906, Chem Abstr, 1, 653 (1907).
6. Worden, E.C. Technology of Cellulose Esters, Vol VIII, D. Van Nostrand: New York, 1916.
7. Franchimont, A. Compt. Rend, 89, 711 (1879).
8. Textile Economics Bureau Textile Organon, March 1981, p. 41.
9. Textile Economics Bureau Textile Organon, June 1981, p. 86.
10. Schuetzenberger, P. Comp. Rend., 54, 154-156, 1026-1029 (1862).
11. Schuetzenberger, P. Comp. Rend., 61, 485-6 (1865); abst. Bull. Soc. Chim., 5 (2), 291 (1866).
12. Schuetzenberger, P. Nadin, L. Comp. Rend. 68, 814 (1869); abst. Chem. Centr. 11, 681 (1869).
13. Schuetzenberger, P. Ann. Chim. Phys. 21 (4), 235-64 (1870); abst. Chem. Centr. 568 (1871).
14. Liebermann, C; and Hoermann, O. Ber. 11, 1618 (1878); abst. Bull. Soc. Chim. 32, 338 (1879).
15. Franchimont, A. Rec. Trav. Chim. 18 (2), 465,472 (1899); abst. J.C.S. 78, 141 (1900).
16. Franchimont, A. Compt.Rend. 92, 1053 (1881); abst. Chem. Centr. 424 (1881).
17. Guignet, C. Compt. Rend. 108, 1258 (1889); abst. J.C.S. 51, 847 (1889).
18. Cross, C; and Bevan, E. Chem. News, 60, 163, 254 (1889).
19. Cross, C; and Bevan, E. E. Pat. 9676, 1894, abst. J.S.C.I. 14, 987 (1895).
20. Weber, C. Chem. News, 80, 247 (1899).
21. Cross, C; Bevan, E; and Beadle, C. Proc. Chem. Soc, 31 (1895).
22. Cross, C; and Webber, C. Dan. Pat. 2558, 2636, 1899, Abst. Chem. Ztg. 23, 1001 (1899).
23. Lederer, L. U.S. Pat. 654988; Abst. Mon. Sci. 57, 98 (1901).
24. Miles, G.W. Ger. Pat. M-28289, Sept 30, 1905.
25. Ost, H. Zts. ang. Chem. 19, 993 (1906); Abst. J.S.C.I. 25, 606 (1906).
26. Knoll and Co. U.S. Pat 981574, 1911; Abst. C.A. 5, 1192 (1911).
27. Knoll and Co. D.R. Pat. 196730, 1906. Abst. C.A. 2, 2301 (1908).
28. Dreyfus, C. E. Pat. 127678; Abst. C.A. 13, 2444 (1919).
29. Dreyfus, H. E. Pat. 3321, 1915; abst J.S.CI, 36, 28 (1917).
30. Dreyfus, C. E. Pat. 100009; abst C.A., 10, 1269 (1916).
31. Kruger and Roman, Angew Chem, 47, 58 (1934).
32. Steinmann, H.W. In Handbook of Fiber Sci and Tech; Lewin, M. and Pearce, E. M. Eds., International Fiber Science and Technology Series, Marcel Dekker: New York, 1985, Vol. 4, Chapter 12.
33. Campbell, K. C., Davis, J. M., and Woods, R. E., Jr., U.S. Pat. 3,767,642, (to Celanese Corportation) 1973.
34. Yabune, H., Ikemoto, Y., Kato, Y., and Uchida, U.S. Pat. 4,439,605, 1984 (to Daicel Chemical Industries, Ltd.). 1989.
35. Malm, C. J., Tanghe, L. J., and Laird, B. C. Ind. Eng. Chem. 38, 77 (1946).
36. Malm, C. J., Tanghe, L. J., and Schmitt, J. T., Ind. Eng. Chem. 53 363 (1961).
37. Laamanen, L., and Sihtola, H., Paper and Timber 46, 159 (1946).
38. Akim, E. L., Pure Appl. Chem. 14, 475 (1967).
39. Serad, G. A., Sanders, J. R., Kirk-Othmer: Encyclopedia of Chemical Technology, John Wiley: New York, 1979, Vol. 5, 3rd ed. p. 89.
40. Moelter, G. M., and Steele, R., In Encylcopedia of Chemical Processing and Design. J. McKetta, Ed. Marcel Dekker: New York, 1976, Vol. 2, p. 171.
41. Moncrieff, R. W., Man-Made Fibers, 6th ed. Interscience: New York, 1975. p. 232.
42.. Sprague, B. S., and Horner, L. T., Encyclopedia of Polymer Science and Technology, John Wiley: New York, 1965 Vol. 3, p. 419.
43. Work, R. W. Text. Res. J, 19, 381 (1949).
44. Stoll, R. G. Text. Res.J, 25, 650 (1955).
45. Briggs, J. J.S.C.I. 39, 286, R (1920); abst. Faser. 3, 54 (1921).
46. Briggs, J., and Palmer, C., U.S. Pat. 1398357; abst. C.A., 16, 836 (1922).

47. British Cellulose and Chemical Mfrg. Co. Ltd., and Ellis, G., E. Pat 220,505; abst. J.S.C.J., 1924, 43, 865B.
48. Mullin, C., Amer. Dyestuff Rept 14, 173, 214, 237, 243 (1925).
49. Mullin, C., Acetate Silk and Its Dyes. D. Van Nostrand: New York, 1927, p. 9.
50. Egerton, G. S., Am. Dyest. Rep. 38, 608 (1949).
51. Fortess, F., Am. Dyest. Rep. 44, 524 (1955).
52. Bird, C. L., J. Soc. Dyers. Colour. 70, 68 (1954).
53. Hartley, C. S. J. Chem. Soc., p. 1972 (1938).
54. Fortess, F. and Salvin, V.S., Text. Res. J. 28, 1009 (1958).
55. Boulton, J. J. Soc. Dyers. Colour. 71, 45 (1955).
56. Fourness, R. K., J. Soc. Dyers Colour. 72, 513 (1956).
57. Tesi, A. F. Am. Dyest. Rep. 45, 512 (1956).
58. Clark, H., and Malm, C., U.S. Pat. 1,690,620, 1929; Abst. C.A., 23, 512 (1929).
59. Siahkolah, M., Walsh, W. K., Text. Res.J. 44, 895 (1974).
60. Hagemeyer H., and co-workers, U.S. Pat. 3,298,979, 1967 (to Eastman Kodak).
61. Cates, D., and White, H., J. Polym. Sci. 20, 155 (1956).
62. Wellons, J. D., Williams, J. L., Stannett, V. J. Polymer Sci. 31, 1341 (1967).
63. Panar, M., and Wilcox, O. D., Ger. Pat. 2,705,382, 1977, (to E. I. duPont).
64. Bheda, J., Fellers, J. F., and White, J. L., J. Appl. Polym. Sci. 26, 3955 (1981).
65. Auerbach, A. B., Dyer, J., and Virgin, R. T., Anisotropic Cellose and Cellulose Derivative Solutions, Presentation at the 179th ACS Meeting, Houston, Texas, 1980.
66. Steinmann, H. W., U.S. Pat. 3,386,932, 1968 (to Celanese Corporation).
67. Santangelo, J. G., and Steinmann, H. W., U.S. Pat. 3,386,930, 1968 (to Celanese Corporation).
68. Acetate Tow Production and Characterization, World Smoking Products Technical bulletin WSP 2.1, Celanese Fibers Marketing Co., Charlotte, N.C., 1981.
69. Leonard, R. L., Ger. Pat. 2,388,786, 1974 (to Monsanto Co.)

HISTORY OF DEVELOPMENT OF TEXTURING YARNS FOR CHEMICAL FIBERS

WANG SHANYUAN
Department of Textile Technology, China Textile University, 1882 West Yan-An Rd., Shanghai, China

The new era of production of chemical fibres began at the end of 1930s. Since then, there have been, in the past seven decades, more than 100 varieties of chemical fibres produced in different ways, made from various raw materials and thus being of particular performance characteristics respectively. But only a small number of them have come into industrial production, including man-made fibres such as rayon and acetate, and synthetic fibres such as polyamide, polyester, acrylic and polyolefin.

Owing to the rapid development of production of chemical fibres since 1930, especially that of synthetic fibres, their proportion in the total world output of textile raw materials has kept steep rising, resulting in the significant change of the constitution of textile raw materials in various countries, while the proportion of natural fibres has correspondingly been declining.

Table 1 shows the output of main textile fibres over years.

Table 1. The world output of textile fibers (Unit: 10^4t).

Yarn	Chemical Fibres			Natural Fibres			Total
	Synthetic Fibres	Man-made Fibres	Total	Cotton	Wool	Silk	
1950	6.9	161.2	168.1	664.7	105.7	1.9	940.4
1960	70.2	260.8	331.0	1011.3	146.3	3.1	1491.7
1970	470.0	343.6	813.6	1178.2	160.2	4.1	2156.1
1980	1049.2	324.2	1373.4	1426.6	158.2	5.6	2963.7
1982	1014.5	294.5	1309.0	1473.8	162.0	5.5	2950.8
1983	1184.9	309.4	1494.0	1552.8	167.2	5.5	3219.8
1984	1256.0	303.0.	1559.0	1661.9	167.6	5.5	3393.9
1985	1311.3	298.3	1609.6	1718.1	168.4	3.6	3581.8

Since 1960, the increase of total world output of chemical fibres has been chiefly the result of the increase of the output of synthetic fibres; the proportional distribution of man-made fibres and that of synthetic fibres in chemical fibres have changed thereby. The reason why synthetic fibres have won rapid development is that they possess high strength so that the textiles made are strong, durable and elastic, being suitable for various purposes. Synthetic fibres have become cotton fibre's comparative. They are almost as important as cotton fibre. In addition, the perniciousness brought about by the production of synthetic fibres is not so serious as that by the production o man-made fibres, especially rayon, which will cause air pollution and water pollution to some extent. On the other hand, synthetic fibres have their disadvantages. They do not have desirable moisture pick-up, hygiene performance and heat resistance, and they generate strong static electricity making textile process difficult. However, there deficiencies can be made up by means of modification, texturing, and/or other processes. Synthetic fibres hold promise of being more competitive.

At the early stage of chemical fibres production, the representative product was filament known as rayon, whose raw materials was natural high molecular one. During World War I, those countries in which cotton and wool were not available because of blockade and embargo had to seek a way out that they cut filaments into short staple fibres of various lengths as "man-made cotton fibre" or "man-made wool fibre". Short staple chemical fibres are of low cost of production and of many uses. Fabrics made from them possess good handing, covering power and comfort as well. Staple chemical fibres can be not only processed into yarn, but also blended with natural fibres. their output has become greater than that of filaments over a period of time. Among varieties of synthetic fibres, the proportional distribution between staple fibre and filament varies greatly: Polyamide is mainly of filament production whereas polyester and acrylic are mainly of staple fibre production. Of all the chemical fibres, polyester has the greatest output. The features pertaining to the development of polyester filament are as follows: production scale is expanding with increasing determination, with fully drawn yarn and textured yarn developing as twin lotcus flowers on one stalk and produced as either thick tex filament or fine tex filament, the former being the major; linear density is getting finer and finer; wider and wider applications are being found in the fields of clothing articles, decoration and ornamentation, and industrial production. The rapid development is due to the fact that polyester filament production can save money and labour power in that there is no spinning process as the filament can be directly used to weave fabrics. The development of high-efficiency warp knitting machine, weft knitting machine, shuttleless loom and new type finishing machine has greatly improved the productivity of post-processing. The mighty advance in production technology also paves the way for the rapid development. The spinning speed has risen from about 1000 m/min to about 3500 m/min and is marching forward to super high one. Spinning--texturing technology has also quickened its pace by first simplifying the "low speed spinning--drawing and twisting--texturing" three-step process into the "high speed spinning--drawing and texturing" two-step process, and then, on the basis of wide application of two-step process,sublimating to the "spinning--drawing and texturing" one-step process. Both investment and cost of production are considerably lowered; both productivity and quality are substantially heightened. What should be particularly mentioned is that polyester filament, as it has made up its deficiencies by means of various texturing methods, now possesses those particularities staple fibre processes so that it is finding wider and wider application.

Table 2 and Table 3 show the total world output of staple fibre and filament of various kinds of synthetic fibres and man-made fibres.

As it is known to all of us, the fundamental constituent unit of any kind of yarns and textiles is fibre. It is the gravity of the fibre (esp. the geometric properties of the fibre) that determines the bulkiness of textiles. From the traditional point of view, most of textiles are asked to possess the essential properties of full handle and warmth retention. Generally speaking, fibre and yarn with good bulkiness consequently give good bulkiness to textiles. Bulkiness is the essential and fundamental requirement of textile manufacturers. It is just because many textile fibres such as cotton fibre and wool fibre possess, in view of their geometric properties, bulkiness that their textile products possess the basic functional properties of full handle and warmth retention. However, a number of textile fibres lack the necessary bulkiness. Efforts have long been made so as to impart bulkiness to such fibres by modifying their geometric properties. As early as 1930s, Smith & Son Carpet Co., USA used stuffer box technique (U.S. Patent 2,311,174) to improve the crimp in poorer South American wools. This crimping texturing of natural fibre is, in essence, to improve the bulkiness of the fibre (Fig. 1).

Table 2. The output of chemical filament and staple fibre in various countries (districts) in 1988 (Unit: 10^4t.).

	PET		Nylon		Acrylic	Cellulose Fibre	
	Filament	Staple Fibre	Filament	Staple Fibre	Staple Fibre	Filament	Staple Fibre
South Korea	41.2	33.8	16.8	6.0	19.1	2.1	---
USA	55.7	71.2	78.4	42.7	26.7	9.7	18.1
West Europe	48.7	51.5	54.4	14.5	86.5	18.8	35.2
Japan	32.9	29.3	26.7	15.0	39.5	10.0	19.1
Taiwan of China	54.5	55.9	19.1	1.0	13.9	0.2	12.3
China	26.3	52.5	10.0	---	8.5	5.0	11.0
Other ctries. (Dist.)	83.5	117.7	96.6	12.2	52.5	47.7	99.3
The total world output	343.0	451.9	302.0	71.6	246.6	93.5	195.0

Table 3. The output of chemical filament and staple fibre in various countries (Districts) in 1989 (Unit: 10^4t).

	PET		Nylon	Acrylic
	Filament	Staple Fiber	Filament	Staple Fiber
USA	54.8	108.2	79.8	24.6
Taiwan of China	59.5	59.4	19.0	14.3
South Korea	45.5	36.6	17.3	18.9
Japan	36.7	30.8	26.6	36.3
China	30.0	60.0	10.4	10.0
West Germany	24.0	8.9	20.6	25.0
Italy	9.7	5.0	11.8	30.9
G.B.	4.7	---	4.4	15.4
The total world output (including other countries/districts)	263.2	479.9	304.4	234.8

Rayon was known to be short of bulkiness, so that its application was limited. The activities with the view to improving its bulkiness took place before World War II. In 1932, Heberlein AG, Wattwill Co., Swiss developed a texturing method for filament yarns and got the patent right DP 618050 of the crimping texturing of rayon (the application was made in March, 1932). The process was as follows: first, winding the highly twisted filament yarn onto the bobbin, then steaming it under high temperature and drying it, and finally untwisting it. By this process, crimped rayon with improved bulkiness was obtained. The product was marketed under the trade name Nigrila.

Meanwhile, the Toyo Rayon Co., Japan, succeeded in developing the three-step process similar to Heberlein's. What was different was that the former used formalin to set the twisted filament yarn (JP 130429--Special Permit 13--5330; application was made in December, 1937).

The above mentioned processes did not find popular acceptance, however, owing to the fact that such cellulose fibres as rayon, cuprammonium cuprene and acetate filament yarns failed to get the satisfactorily stable crimping effect.

The rapid development of texturing process is closely related to the industrial production of thermoplastic synthetic fibres. As synthetic fibres are thermoplastic, their filament yarn is textured generally by mechanical means, and then undergoes heat setting under textured configuration. A sufficiently stable crimping effect is obtained through the change of the yarn's chemical structure to control the texturing effect. That is how texturized synthetic fibres have become an ideal material for manufacturing textiles with high bulkiness.

It can be said that the texturing process goes forward side by side with the rapid development of synthetic fibres and is of increasing importance. A glaring example is the texturing process of polyamide and polyester.

After World War II, the Swiss Heberlin Co., the USA Marlonett Co., and the Japan Toyo Rayon Co. adopted the "twisting--heat setting--untwisting" intermittence technique to process polyamide filament yarn. The method was later called the classic process. The process enable polyamide filament yarn to possess high bulkiness, stretch and stability, and thus to become stretch yarn. From 1940s through 1950s, this classic process found much acceptance as it could be performed on any kind of twisting equipment provided with a steamer. However, it had the following disadvantages: too many working procedures, high production costs and low productivity (Fig. 1).

Celanese Co., G. B., advanced two-step texturing, in which the two procedures twisting and setting were merged into one, and further developed a new texturing process in which the three procedures twisting--setting--untwisting were merged into a single whole. The latter got patent right BP 424880 (application was made in June 1936).

The patented process is known as False Twist Texturing, short procedure and low production cost being its advantages.

Polyamide came into industrial production next. In 1950, the Swiss Heberlein Co. began to use the above mentioned one-step process--false twist texturing to process polyamide since the yarn was of thermoplastic property (thermoplasticity). The Japan Toyo Reyon Co. followed in 1951, and produced "Woolic Nylon". However, it was not until 1955-1956 that the industrial application of false twist texturing process really began.

Figure 1. The crimping pattern of stretch yarn produced by the classic process

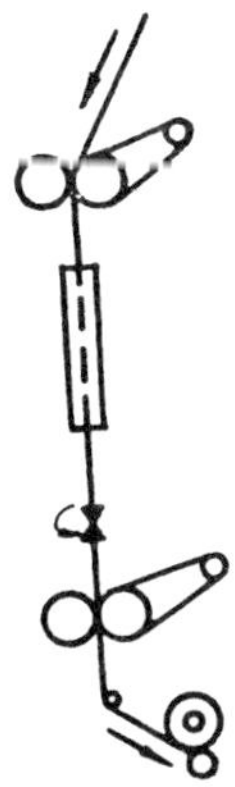

Figure 2. False-twist texturing process

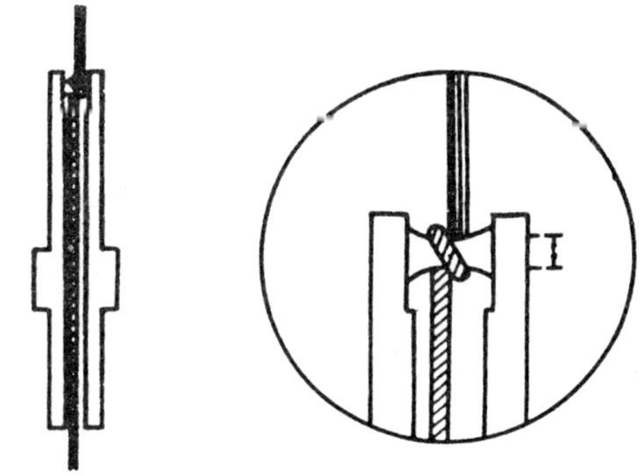

Figure 3. Pin-type spindle

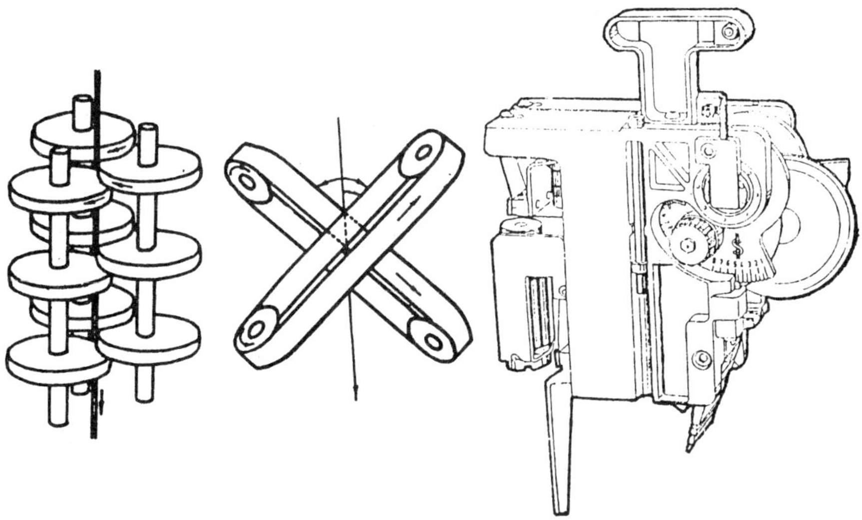

Figure 4. Friction twisting devices

Step by step, the process has been entering the stage characterized by high speed and high performance and taking the place of the classic process (Figs. 2 and 3).

In 1954, the Swise Co., Heberlein advanced friction false twist texturing (Patent DP-797051). In 1958, British Nylon Spinner Co., developed the first friction type false twist spinning frame (Fig. 4).

During the late 1950s, on the basis of one-step false twist texturing process, second heat setting procedure was adopted in various countries (the second heat box or setting heater was equipped). The textured filament yarn which had been of high bulkiness and high stretch now became one which was of high bulkiness and low stretch and low stretch rate. The G. B. Imperical Chemical Industry began to produce this kind of false twist textured polyester filament yarn in 1959 (Fig. 5).

In the mean time, there were other texturing processes coming into being to keep abreast of high productivity times and meet the requirements of various textured products. Air jet texturing was the fruit of the USA Du Pont Co. in 1952. It took the advantage of the effect of turbulent flow to cause the superfed filament yarn to form irregular entangled loop shape with the staple--fibre--like appearance besides being bulked. Du Pont CO., first set up an experimental workshop to trial--produce air jet textured yarn--ATY or Taslan, and then went into operation in 1952. Four or five years later, air jet textured yarn was spread across Europe and various countries in other districts. The air jet textured yarn then was either of medium tex or of coarse tex, suitable for manufacturing decorative fabrics and carpeting. The air jet texturing technique did not have dramatic development during the next 25 years or so because of the following reasons: the texturing was of low speed, the consumption of compressed air was great, and the production cost was high (Fig. 6).

In the mid 1930s, the USA, Jos. Bancrost Co., adopted stuffer box (U.S. Patent 2,311,174), which had been used to improve wool's crimp, to crimple the filament bundle of chemical fibres. This process was later applied to thermoplastic polyamide filament, with the effect that the filament became indented crimplized. It was known as Ban lon and met with favour because the production speed was high and the filament yarn produced was with wool-like crimp. This textured yarn was used as clothing material in knitting and weaving.

Later, several other texturing processes were developed one after the other. In 1943, the G. B. ICI patented the edge crimping texturing process, in which the heated filament brushed past the edge of a knife with very small curvature radius (Fig. 7).

The texturing on the inner side in contact and that on the outer side were of different effects: the orientation of high molecules on the inner side part in contact with the knife edge was decreased since the inner side was under pressure, while that on the outer side was increased since the outer side underwent drawing. Therefore, the filament at the knife edge was of acute curve in shape and remained to be crimped after cooling. In industrial production, 15 D textured yarn was first used to knit women's stretch hose. Over 1000 D textured yarn was later trial--produced for heavy denier fabrics such as carpets and for knitted outwear and men's hosiery.

Fig. 8 shows the gear--shaping texturing process in which filament passes through a pair of closely engaged heating gears with the result that the filament is gear--shaped and permanently rippled (corrugated) through heat setting at the same time. The textured yarn can be produced is various patterns of ripple (corrugation) and various

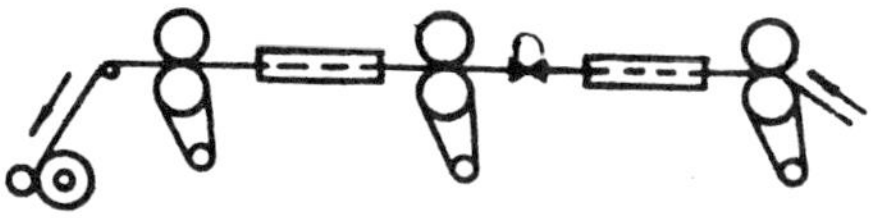

Figure 5. Two-heaters false-twist texturing process

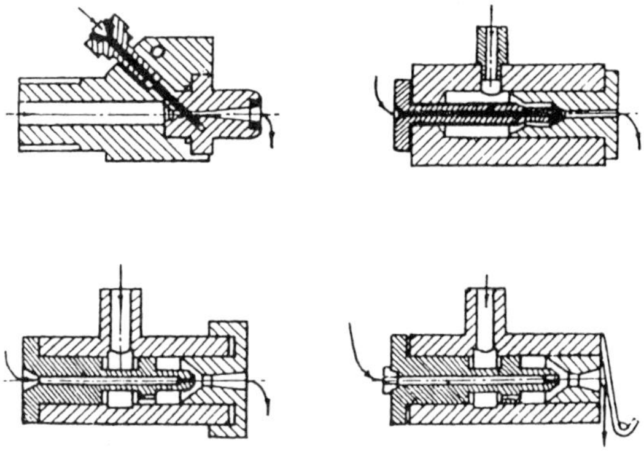

Figure 6. DuPont air-jet

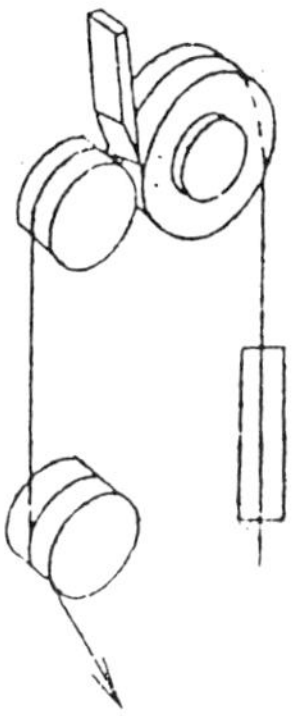

Figure 7. Edge crimping process

Figure 8. Gear crimping process

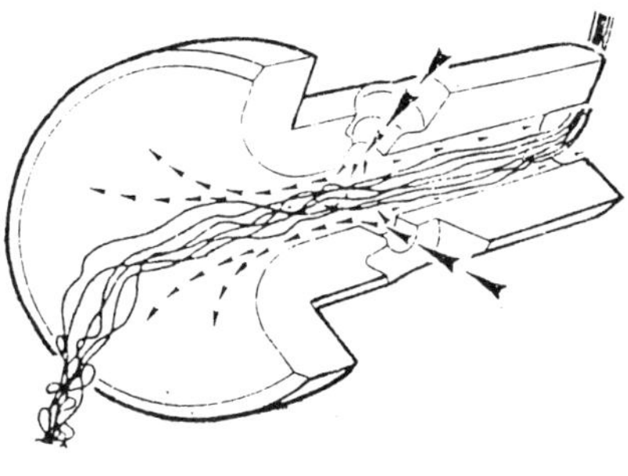

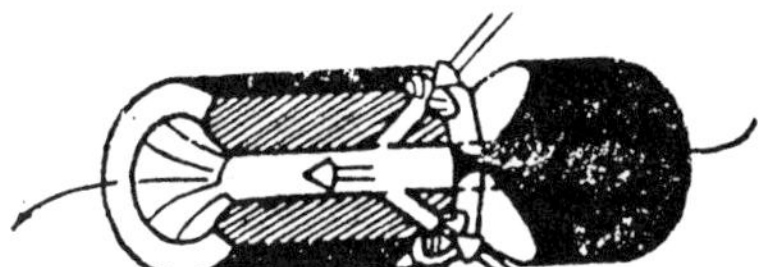

Figure 9. Hema Air-jet

crimping frequencies by means of gears with different specifications. The production is generally used for women's and children's knitted outwear, sportswear, and women's dust coat or blouse.

In 1956, Munden introduced in his article knit-de-knit method. The crimping textured filament yarn was obtained by first knitting filament on the knitting machine into circular knit and then de-knit the circular knit after heat setting treatment. The crimping was a two-dimensional one with large and regular ripples, so that the textured yarn had the concave--corvex crimp effect on its fabrics as the stitch lengths on the knitted fabric were all the same and the knit loop coincided with the cycling of the crimping length. The knit-de-knit textured yarn could be of various crimping shape and crimping frequencies by varying the up=rising distance of the knitting machine and the structure of the fabric. The yarn was used to knit various kinds of knitted fabrics suitable for women's undershirt, blouse, dress and summer skirt.

Textured yarn production entered a new stage in 1970s. With the steady increasing of the output of textured yarn, texturing process was impressively developing. The outstanding achievement in the field of synthetic fibres industry was the advent of "high speed spinning--drawing and false twisting" texturing. It was indeed, a constituent part of vital importance to "yarn revolution". The process shortened the procedure, heightened the output, and lowered the cost, and, what was more, it produced high quality false twist textured yarn. Consequently, it was rapidly spread across various countries.

The industrial production of POY-DTY, when spread, won its rapid development to such an extent that in the mid 1970s, the market became a saturated one, with the signal that the production, especially in USA, West Europe and Japan, stagnated. How to give variety to textured yarn and its shape and enable it to possess desirable properties in view of extending its uses was the key problem people made efforts to solve. Air jet texturing process was further studied and bettered.

Du Pont's improved air jet and the air jet (Fig. 9) advanced by Heberlein at the end of 1970s raised the speed of air jet texturing process and lowered the air consumption, and the yarn produced was of fine tex suitable for clothing materials. Correspondingly, there was a sharp increase in the output of air jet textured yarn. In recent years, the multiple texturing process which combines air jet texturing process and other texturing processes on one hand, and the blended filament yarn made from different raw materials on the other hand have attracted people's attention and stimulated successful development of such new fancy varieties of textured yarn as loop yarn, pile yarn, corn yarn, multiwall yarn, bamboo-like yarn.

Synthetic fibres have also found application in the field of carpet manufacturing. Woolen carpets are expensive. They are used in limited range. There has been a great demand for textured synthetic yarns to meet the needs of the development of new type pile carpets. Big companies including the U. S. Du Pont Co., the G. B. Imperical Chemical Indurly Co., the France Roan Pulanh Co., and the Federal Germany Nenmag Co. and Barmag Co. have all used "spinning--drawing--texturing" one step process (Fig. 10) to produce bulked continuous filament--BCF as carpet yarn.

The continuous hot fluid jet texturing (Fig. 11) now finds wide application since it has the advantages of having high texturing speed, being a good fit for producing coarse tex yarn, and eliminating the mechanical damage to the filament. And it is still finding wider and wider applications.

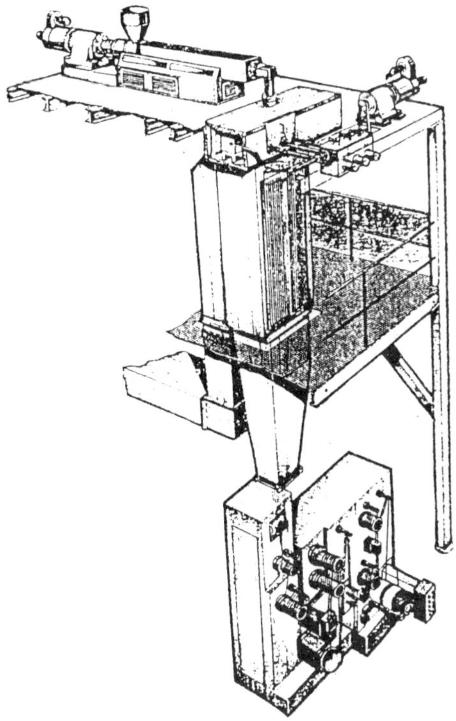

Figure 10. BCF technology process

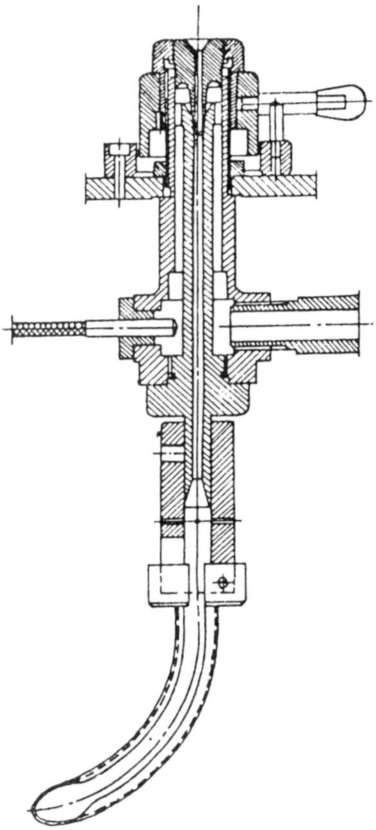

Figure 11. BCF Hot-fluid texturing jet

The production of textured yarn has been developing with the development of texturing technique. Table 4 shows the proportional change of the output of several main textured yarns from 1974 through 1984. Table 5 shows the annual output of several main textured yarns.

Table 4. The proportional Change of the output of several main textured yarns from 1974 through 1984.

Textured Yarn	The output in percentage	
	1974	1984
False twist textured yarn	85	72
Hot fluid textured yarn	8	19
Air Jet textured yarn	2	4
Stuffer box textured yarn	5	5

Table 5. The annual world output of several main textured yarns from 1983 through 1984.

Textured Yarn	Annual output (ten-thousand)
False twist polyester fiber	154.0
False twist nylon	67.5
BCF nylon and polypropylene fibre	54.0
Air jet textured yarn	11.0
Other textured yarns	14.3

REFERENCES

1. B. Piller, Bulked Yarns Production, Processing and Application, 1973.
2. B. C. Goswami, J. G. Martindale and F. L. Scardino, Textile Yarns Technology, Structure, and Application, 1977.
3. M. P. Nosov, production of textured yarn.
4. D. K. Wilson, Text. prog., 9(3), (1977).
5. D. K. Wilson, Text. prog. 10(3), (1978).
6. D. K. Wilson, Text. prog., 16(3), (1987).

CHEMICALLY-MODIFIED NATURAL FIBERS

CHARLES H. FISHER
Chemistry Department, Roanoke College, Salem, VA 24153

Who was the first to synthesize new man-made fibers? And when did this happen? Christian Friedrich Schonbein, a professor of chemistry at the University of Basel, Switzerland, made the new fibers; the year was 1846, almost 100 years before nylon.

In 1846, Schonbein (1799-1868) was experimenting with a mixture of nitric and sulfuric acids in his kitchen. He had been forbidden to experiment there, but his wife was absent. By accident, he broke a flask containing nitric and sulfuric acids; the corrosive liquid spilled over the kitchen floor. Unable to find a mop, Schonbein seized the first thing at hand, his wife's cotton apron, and sopped up the mixture. He then washed the apron and hung it in front of a hot stove to dry. This caused the treated apron to ignite and burn [1-3].

In subsequent planned work, Schonbein found that the cotton cellulose had been converted into cellulose nitrate, a highly flammable material. He also explored fully-nitrated cellulose (trinitrate or "gun cotton") as an explosive. In 1847, a factory making gun cotton exploded, killing twenty-one persons [3].

Chemical research to develop new or improved textile products from natural fibers may be divided into two major types:

1. The cellulose or protein fibers are altered chemically without loss of the fibrous form.

2. The cellulosic fibers, usually wood pulp or cotton linters, are brought into solution. The solutions are extruded to obtain new fibers. Proteins also have been dissolved and the resulting solutions extruded to give new fibers.

Both of these approaches have been used successfully. As already stated, Schonbein used the first approach. The second approach was used extensively, beginning in the 19th century, to produce soluble cellulose derivatives (e.g., nitrates, acetates, and xanthates) that were used to produce films, plastics, and fibers (the last by extrusion through small orifices) [1,4].

The 20th century witnessed a renewal of interest in modifying natural fibers chemically without loss of the fibrous form; most of this work was done with cotton fibers, yarns, or fabrics. Chemically-modified cottons were made by esterification, etherification, graft polymerization [5], or replacement of the toluenesulfonate group with ammonia or amines [6].

The desire to have dye resist yarns apparently dominated the early work. At the beginning of the 20th century, Cross and Schneider prepared dye resist yarns by acetylating and nitrating cotton yarns [6].

Three dye resist products from cotton were made commercially in the 1920's: the benzoate (reserve cotton), toluenesulfonate (immunized cotton), and acetate (passive cotton).

Cotton textile products having amine groups and increased receptivity to some dyes were made by causing the fibrous toluenesulfonate of cellulose to react with ammonia or amines.

Acetylated cotton (Cotopa 30), based on patents of A. Rheiner of Sandoz A. G., was manufactured in England beginning in 1928.

In 1928 patents issued to Tootal Broadhurst Lee in England showed that some degree of wrinkle resistance and dimensional stability could be imparted to rayon and cotton by the application of ureaformaldehyde resins.

Soluble cotton gauze, made by oxidation with nitrogen dioxide, has been used as a surgical dressing. Nitrated cotton yarns have been manufactured and used because of their property of becoming soluble under some conditions in alkaline solutions [6].

Hydroxyethylated cotton was made and studied intensively in about 1940. This product has a crisp hand, linen to parchmentized appearance, and enhanced reactivity and dyeability.

Water repellent cottons, made by treating cotton with stearamidomethylpyridinium chloride, octadecylethyleneurea, ketene dimers, alkyl pyridinium salts, or certain betaines, methylolamides, and methylolureas have been available.

Aminized (AM) cotton, made by treating cotton with 2-amino-ethylsulfuric acid in the presence of alkali, has been studied extensively. Because it has a primary amine group (Cell-$OCH_2CH_2NH_2$), AM cotton has ion-exchange properties and increased chemical reactivity and affinity for some dyes.

Following the dissemination of information on cyanoethylated cotton by the Institute of Textile Technology, several organizations made this product in relatively large quantities. It has been extensively evaluated for the many end uses indicated by its improved resistance to rot and heat and enhanced affinity for some dyes [6].

Extremely durable flame resistant cotton products have been made by treatment with a mixture of equimolar portions of APO (triethylene phosphoramide) and THPC [tetrakis(hydroxymethyl) phosphonium chloride].

Both the Navy and NASA have used flame-retardant cottons in preparing some of their clothing [7].

Chemically-modified cottons esterified or etherified to suitably high degrees of substitution (more than about 25% acetyl for acetylated cotton) have thermoplastic properties [6].

The conversion of cellulose fibers into new fibers by graft polymerization has been reviewed at length by Arthur [5].

The chemically-modified cottons mentioned above, including fully acetylated cotton (Cotopa 60), carboxymethyl cotton, aminized cotton, cyanoethylated cotton, flame-resistant cottons, and weather-resistant cotton shade cloth, have been reviewed [6,8,9]. Treatments for wool and mohair have been reviewed by Lundgren [10].

A recent and important chemical treatment gives clothing that will help keep the wearer cool under warm conditions or warm under cool conditions. Tyrone Vigo and Joe Bruno of USDA's Southern Regional Research Center in New Orleans developed the new process and products [11-13]. The key chemical used in the process is polyethylene glycol (PEG-1000), a phase-change substance.

Vigo and Bruno prepared fabrics with durable, enhanced thermal properties by insolubilizing phase-change materials on the fabrics. Using polyfunctional crosslinking agents, they bound PEG-1000 to cotton, wool, polyester, cotton-polyester blends, polyamide, polypropylene, and even glass fibers. Presumably there is some grafting of glycols to cotton and wool, in addition to insolubilization of the polymer by crosslinking as occurs with other fiber types.

NeutraTherm Corp., in Des Moines, Iowa, sells thermal underwear and socks made by the Vigo-Bruno process; Intertex Inc., a subsidiary of Mitsui & Co. in Tokyo, is developing ski jackets, pants, gloves, and shoe linings.

Whether made from cotton, synthetics, or blends, the growth and success of "wash-wear" products, introduced in 1952, has been remarkable. Cotton-based and other cellulosic "wash-wear" or "durable press" products are made by the intermolecular crosslinking of the cellulose molecules in the fibers. Commonly-used crosslinking chemicals are formaldehyde or formaldehyde adducts of urea, melamine, and carbamates.

In 1906 Eschalier reported that cellulose fibers treated with formaldehyde had improved wet strength. Subsequently, numerous publications and patents have described investigations of the crosslinking of cellulose by many crosslinking chemicals, usually formaldehyde or N-methylol derivatives. According to recent reports [14-16], a non-formaldehyde chemical, namely, butanetetracarboxylic acid (BTCA) is a superior crosslinking chemical for making durable press products.

Arthur, Vail, Drake, Andrews and DeGruy have published excellent reviews of chemical treatments for cotton and cellulose [5,9,17-21].

The new fibers made from cotton by chemical modification have helped cotton tremendously in its competition with other fibers. The production of U.S. cotton in the 1990-91 crop year is estimated to be approximately 15.5 million bales (nearly 8 billion pounds), the largest since 1981. It is anticipated about half of the 15.5 million bales will be used domestically and that about half will be exported [21].

REFERENCES

1. Seymour, R. B., and C. E. Carraher, Giant Molecules, John Wiley and Sons, Inc., New York, 1990.
2. Mark, H. F., American Scientist, 72, 156-162 (1984).
3. Fisher, C. H., in Organic Coatings: Their Origin and Development, R. B. Seymour and H. F. Mark, Editors, Elsevier Science Publishing Co., New York, 1990.
4. Seymour, R. B., and C. E. Carraher, Jr., Polymer Chemistry, Second Edition, Marcel Dekker, Inc., New York, 1988.
5. Arthur, J. C., Jr., in Encyclopedia of Polymer Science and Engineering, Editors, H. F. Mark et. al., Editors, Second Edition, Vol. 3, John Wiley & Sons, Inc., 1985.
6. Fisher, C. H., and F. S. Perkerson, Textile Res. J., 28, 769-778 (1958).
7. Cotton Gin and Oil Mill Press, April 28, 1984, page 18; and Nov. 26, 1983, page 16.

8. Jones, M., Annotated Bibliography of Cotton Research at the Southern Utilization and Development Division, (Misc. Publication 893), U.S. Department of Agriculture, 1962; Supplement 1, 1972.
9. Vail, S. L., Textile Finishing, in Kirk Othmer Encyclopedia of Chemical Technology, Third Edition, Vol. 22, John Wiley & Sons, Inc., New York, 1983.
10. Lundgren, H. P., Yearbook of Agriculture, U.S. Department of Agriculture, Washington, D.C., 1965, pages 361-362.
11. Vigo, T. L., and J. Bruno, Popular Science, May 1991, page 85.
12. Seltzer, R., Chemical & Engineering News, Oct. 20, 1986, pages 15-16.
13. Vigo, T. L., and A. F. Turbak, High-Tech Fibrous Materials, American Chemical Society, Washington, D.C., 1991.
14. Rowland, S. P., C. M. Welch, and M. A. F. Brannan, U.S. Patent 3,526,048, 1970.
15. Welch, C. M., and B. A. K. Andrews, U.S. Patent 4,820,307, 1989.
16. Welch, C. M., and B. A. K. Andrews, Textile Chemist and Colorist, 21, 13, (1989).
17. Arthur, J. C., Jr., in Comprehensive Polymer Science, G. Allen et. al., Editors, Vol. 6, Pergamon Press, 1989.
18. Kottes-Andrews, B.A., pages 99-117, in Encyclopedia of Textiles, Fibers, and Non-Woven Fabrics, John Wiley and Sons, New York, 1984.
19. Drake, G. L., Jr., pages 188-212 in Encyclopedia of Textiles, Fibers, and Non-Woven Fabrics (See Ref. 18).
20. Vail, S. L., pages 466-498 in Encyclopedia of Textiles, Fibers, and Non-Woven Fabrics (See Ref. 18).
21. Kottes-Andrews, B.A. and I. V. DeGruy, 326-328 in Concise Encyclopedia of Chemical Technology, Editor, M. Grayson, John Wiley and Sons, Inc., New York, 1985.
22. Lange, M., Cotton Gin and Oil Mill Press, June 29, 1991, page 10.

HISTORY OF ACRYLONITRILE FIBERS

MASTURA RAHEEL, PH.D.
University of Illinois at Urbana-Champaign, Division of Consumer Sciences, School of Human Resources and Family Studies, 237 Bevier Hall, 905 South Goodwin Avenue, Urbana, IL 61801-3852

Introduction

The history of acrylonitrile fibers goes back to 1893 when the monomer was first synthesized by Moureu, and attempts were made to polymerize it. The polymer came out of obscurity during the World War II when it became a component of synthetic rubber. Intensive research in Germany and the U.S.A. made acrylic and modacrylic fibers a commercial reality in the 1940s.

The term acrylic fiber specifically refers to a fiber containing at least 85% by weight acrylonitrile comonomer [1]. The first commercial acrylic fiber "orlon" was introduced by DuPont in 1944. A similar class of fibers, the modacrylics, were included in the acrylic category until 1960, however in 1960, the Federal Trade Commission established a separate category called modacrylic and defined it as fibers containing at least 35% but less than 85% by weight acrylonitrile comonomer. The first modacrylic fiber was developed by Union Carbide in 1948 under the trade name Vinyon N. In Europe, modacrylic fibers were originally defined as those containing 50-85% by weight acrylonitrile units, but in 1977 the International Organization for Standardization [2] adopted the broader 35-85% acrylonitrile units definition also. The nature of the other components is not specified for either fiber. Commercially, however, the only modacrylics marketed have 25-60% of monomers, such as vinyl chloride or vinylidene chloride, and consequently possess a high degree of flame resistance.

Much of the growth in acrylic fibers usage has come from the replacement of wool, as they have many of the wool's desirable properties and cost much less. Acrylics are valued for their warmth, soft hand, generous bulk and loft, reasonable abrasion resistance and the elastic recovery qualities. In addition, compared to wool, acrylics are more resistant to attack from chemical and biological agents, are more stable toward degradation from light and heat, and are non-felting. Thus, the sustained commercial success of acrylic fibers over the past four decades can be attributed to their desirable balance of properties. Nevertheless, some deficiencies in the physical properties of the fibers such as poor hot-wet strength, only modest tensile strength and abrasion resistance have prevented penetration of acrylics into some markets.

The major acrylic markets are in the apparel and home furnishings sectors. Also, a relatively small amount goes into industrial uses. However, promising applications in non-textile sectors continue to emerge and may account for substantial acrylic fiber growth in the future.

History and Developments

The first reported synthesis of acrylonitrile and polyacrylonitrile (PAN) was by the French chemist, C. Moureu [3]. In 1893 Moureu, who was working in Germany, reported two methods for synthesizing acrylonitrile and a year later reported the polymerization of acrylonitrile. From that time, up to almost the outbreak of World War II, acrylonitrile remained a laboratory curiosity. The major reasons being that no solvents were known for the polymer at that time which would allow dry spinning of the fibers like acetates. Also, the polymer decomposes before reaching its melting point thus precluding melt spinning like nylons. The first break-through in developing solvents for PAN occurred in 1938, in Germany, when H. Rein of I.G. Farbenindustrie described fibers made from the polymer dissolved in aqueous solutions of quaternary ammonium compounds, such as benzylpyridinium chloride or of metal salts, such as lithium bromide, sodium thiocyanate, and aluminum perchlorate [4].

During the late 1930s, acrylonitrile acquired a new status by becoming a constituent of one of the more important types of synthetic rubber under development in Germany and the U.S.A. In 1937, I.G. Farbenindustrie introduced the first acrylonitrile-butadiene rubber. Later synthetic rubber compounds based on acrylonitrile were developed in the United States during the early 1940s in response to wartime needs. American Cyanamid remained the sole producer of acrylonitrile in the U.S.A. In addition to acrylonitrile-butadiene rubber, polyblends of acrylonitrile-butadiene with acrylonitrile-styrene copolymers were developed by the United States Rubber Co. By the end of the war, acrylonitrile was a relatively inexpensive industrial chemical, and was available in large quantities. Thus polymers of acrylonitrile came under intensive review in the rubber, plastics and synthetic fiber industries.

In 1943, DuPont [5] and I.G. Farbenindustrie [6] simultaneously reported solvents suitable for spinning acrylonitrile fibers. Based on this solvent breakthrough, DuPont developed a commercial process and produced the first acrylic fiber, initially called Fiber A and subsequently named Orlon. In 1944, a "semiworks" was built under the auspices of the Research Section of DuPont's Acetate Division. Until August, 1945, the development of Orlon acrylic fiber was concerned primarily with its possible aid in the war effort, but was not brought to fruition before World War II ended. By early 1946, Orlon looked promising enough for commercialization. Process development continued and in October, 1948, DuPont built a plant near Camden, South Carolina. Production of Orlon acrylic fiber started in May, 1950.

Following the lead of DuPont, the Chemstrand Corp. (now known as Monsanto Co.) developed another acrylic fiber, which they named Acrilan. A plant was built in Decatur, Alabama; and operations were started late in 1952.

For many years, the American Cyanamid Co. manufactured acrylonitrile for use by other companies in the manufacture of acrylic fibers. The company eventually decided to experiment in the production of its own acrylic fiber. After much experimentation, it produced, in 1953, a fiber tentatively designated as X-51 at its pilot plant in Stamford, Connecticut, and began commercial production in 1958 of its own acrylic fiber, Creslan, in a new plant near Pensacola, Florida.

The development of Zefran acrylic fiber was begun in 1949. Research in the laboratories of the Dow Chemical Co. was concerned with polymers and particularly with polyacrylonitriles. One of the problems associated with polyacrylonitriles had been their low affinity for dyes and the limited fastness of the colors. After several years of intensive research, a chemical alloy with acrylonitrile was produced having properties similar to those of acrylic fibers plus superior dyeing properties. This fiber could be dyed with a wide variety of dyes including those used for natural fibers as well as those used for manmade fibers. This not only meant greater affinity for dyes but also greater possibilities for blending this new fiber with a wide array of other fibers without being hampered by dyeing problems frequently encountered with blending. Satisfied that they had a desirable product which was commercially feasible, the Dow Chemical Co. established the trademark of that fiber as Zefran. Dow introduced its new fiber as a nitrile alloy in 1958, thereby differentiating its composition as well as its properties from the other acrylonitrile fibers. Upon establishment of the Federal Trade Commission's ruling in 1960, Zefran was classified as an acrylic fiber. It is now manufactured by the Badische Corp. in Williamsburg, Virginia. Subsequent to the development of Zefran another acrylic fiber under the trademark of Zefkrome was introduced. However the trademark was later dropped in favor of the identification Zefran for all its acrylic variants [7]. In Europe, Suddeotsche Chemiefaser (Hoechst) introduced Dolan; and Bayer introduced Dralon.

The first modacrylic fiber was commercially produced by Union Carbide in 1948 under the trade name Vinyon N. [8] It was a continuous filament yarn from a copolymer of 60% vinyl chloride and 40% acrylonitrile. The staple form was introduced in 1949 under the trade name Dynel [9]. Production of continuous filament Vinyon N was discontinued in 1954, and Dynel was discontinued in 1977 [10]. Another modacrylic fiber, Verel, was introduced in 1956 by Eastman Kodak [11]. It was based on acrylonitrile vinylidene chloride composition, however, its production was also discontinued in 1983. The British company Courtaulds, Ltd., introduced another modacrylic fiber in 1962, which was also based on acrylonitrile vinylidene chloride copolymer, first as a continuous filament yarn, then as a staple fiber [12]. In subsequent years several modacrylic fibers followed, such as Dralon MA by Bayer AG, Kanekalon by Kanegafuchi, Vonnel H-704 by Mitsubishi Rayon, and SEF by Monsanto Company.

Thus the acrylic fiber industry experienced a spectacular growth in the 1950s with at least 18 companies introducing acrylic fiber products during that period. By 1960, annual worldwide production had risen to over 200 million pounds, and was concentrated in Western Europe and the United States. Early products were largely filament yarns for industrial and outdoor applications, however, staple processes were soon developed and acrylic fibers became a major competitor in markets held primarily by woolen fibers. Much of the initial demand for acrylic fibers can be attributed to the development of high-bulk processes for sweater yarns and heavy- denier carpet yarns. By 1963 these two markets accounted for almost 50% of the total acrylic market [13].

During 1960 to 1970 there was rapid growth in acrylic fiber production and consumption. This was due to wool-like appearance and handle of acrylics, and their favorable economic position owing to the relatively low cost of acrylonitrile [14]. During the 1970s, however, the growth rate decreased from 23% in 1969 to only 8%. This was due primarily to the maturing of the wool replacement market in the United States. In addition, nylon became the dominant carpet fiber, reducing the acrylic carpet market share from 25% at its peak to just 6% by 1976. Easy-care fibers and blends also cut into the acrylic share of the synthetic fibers market [13]. The slow growth during the late 1970's and 1980's has been primarily a result of worldwide recession that affected all man-made fibers, a swing in fashion, as well as the rapid rise in world wide production capacity of acrylics relative to total world wide consumption. Major growth occurred mainly in Japan, Eastern Europe, and the developing countries. By 1981, an estimated overcapacity of approximately 20% had developed. This overcapacity is expected to decrease through the 1990s with the June 1990 announcement by DuPont [15] to cease production of acrylic in the U.S.A., thus leaving Monsanto, American Cyanamid and Mann Industries as major producers of acrylic fibers in the U.S.A. In western Europe, Courtaulds Fibers is consolidating its output of acrylic fibers with the shut down of a plant in Calais, France, due to acrylic unprofitability. Losses in acrylic business have also been reported by Enimont in Italy [16], and Bayer in Germany during 1990. DuPont's withdrawal is expected to balance supply/demand in the U.S.A., but even after Calais plant closure there still may remain overcapacity in Europe. However, new markets for acrylics are opening in China, Eastern Europe, USSR, and the South American countries. Also, improved acrylic textile products are now targeted for broader markets, such as cotton replacement fibers, carbon fibers for sports equipment and aircraft, and high-tech products such as hollow fibers for gas separations. These developments plus shifts in product emphasis by the many acrylic fiber producers should help restore world balance of supply and demand. Globally acrylics represent a 6.5 billion lb market as of 1990.

Monomer Manufacture

Acrylonitrile readily undergoes addition polymerization. In 1893, Moureu described two routes for the synthesis of acrylonitrile [3]. These were based on the dehydration of either acrylamide or ethylene cyanohydrin.

$$\text{(a)} \quad \underset{\text{acrylamide}}{CH_2 = CH\text{-}CONH_2} \xrightarrow{^{-}H_2O} \underset{\text{acrylonitrile}}{CH_2 = CHCN}$$

$$\text{(b)} \quad \underset{\text{ethylene cyanohydrin}}{OHCH_2CH_2CN} \xrightarrow{^{-}H_2O} CH_2 = CHCN$$

Later, Union Carbide developed the ethylene cyanohydrin process [17] for commercial use by reacting ethylene oxide with hydrogen cyanide in an aqueous alkaline solution. The process is carried out at 50-60°C for a period of up to 10 hr in the presence of a catalyst, such as diethylamine. The resulting ethylene cyanohydrin decomposes spontaneously into acrylonitrile and water when heated to 200°C in the presence of magnesium carbonate.

Another process based on the addition of hydrogen cyanide to acetylene was developed in Germany in 1939 for the synthetic rubber industries [18]. This method was used for commercial production of acrylonitrile until 1960.

$$\text{(c)} \quad CH{=}CH + HCN \rightarrow CH_2{=}CHCN$$

In this process acetylene is reacted with hydrogen cyanide in an aqueous medium in the presence of catalytic amounts of cuprous chloride. The reaction is maintained at 80-90°C at a pressure of 1-2 atmospheres. The reaction is highly exothermic, forming a gaseous reactor effluent. This crude product is water scrubbed and the pure acrylonitrile product is recovered from the resultant 1-3% aqueous solution by fractional distillation. The major drawbacks of this process are the large number of by-products formed by hydration and the loss of catalyst activity from hydrolysis reactions and the build-up of ammonium chloride and tars.

The acetaldehyde process was developed in 1958 by Knapsack-Griesheim [19]. In the first stage, acetaldehyde is reacted with hydrogen cyanide to form lactonitrile, which in turn is mixed with catalytic phosphoric acid and heated rapidly to 600-700°C. The mixture is then chilled rapidly to 50°C in a solution of 30% phosphoric acid to form the crude acrylonitrile product.

$$\text{(d)}\quad CH_3CHO + HCN \rightarrow CH_3CH\begin{matrix} OH \\ CN \end{matrix} \rightarrow CH_2 = CHCN$$

Today most cost-effective processes are those based on propylene as the starting material. There are three major variations of propylene processes, the Distillers process developed in 1959 [20-22], the Sohio process also developed in 1959 [23] and the DuPont process developed in 1957 [24,25]. All three processes are based on the ammoxidation of propylene. The Distillers process is carried out in two stages. In the first stage, propylene is oxidized in air to form acrolein and water. These intermediate products are allowed to react in the second stage with ammonia in the presence of molybdenum oxide and air to form crude acrylonitrile. The pure monomer is recovered by a series of azeotropic distillations.

$$\text{(e)}\quad CH_2{=}CH\text{-}CH_3 \xrightarrow{O_2} CH_2{=}CH\text{-}CHO \xrightarrow{NH_3} CH_2{=}CH\text{-}CH\begin{matrix} NH_2 \\ OH \end{matrix} \rightarrow CH_2{=}CHCN$$

The Sohio process is carried out in just one stage. Ammoxidation of propylene takes place in air at 2-3 atmospheres pressure and 425-510°C. With concentrated bismuth molybdate or phosphor molybdate as catalyst, the reaction takes place with 53% yield in a reaction time of only about 15 sec. In the DuPont version of this process, the ammoxidation is brought about with nitric oxide at 500°C using a silver or silica catalyst.

Polymerization

In the preparation of acrylic and modacrylic fibers, both polymerization and spinning help determine the ultimate properties of the fiber. The polymerization process determines the composition and molecular weight of the polymer which affect the properties of the fiber as well as the spinning process. Acrylonitrile and its comonomers can be polymerized by any of the free-radical methods. Bulk polymerization is the most fundamental of these, but its commercial use is limited due to its autocatalytic nature which makes it difficult to control the process. This feature, combined with the highly exothermic nature of acrylonitrile polymerization, make large-scale commercial bulk polymerization difficult. However, a commercially feasible process for bulk polymerization in a continuous stirred tank reactor has been developed and patented by Montedison Fibre in 1972 [26]. However, the major commercially practiced methods include aqueous dispersion, emulsion, or solution polymerization. Aqueous dispersion polymerization is the most common commercial method, whereas solution polymerization is used in cases where the spinning dope can be prepared directly from the polymerization reaction product. Emulsion polymerization is used primarily for modacrylic compositions where a high level of a water-insoluble monomer is used or where the monomer mixture is relatively slow reacting.

Aqueous dispersion polymerization. Redox initiation is normally used in the commercial production of polymers for acrylic fibers. This type of initiator can generate free radicals in an aqueous medium at relatively low temperatures. The most common redox system consists of ammonium or potassium persulfate (oxidizer), sodium bisulfite (reducing agent), and ferric or ferrous iron (catalyst). This redox system works at pH 2 to 4 at temperatures of 35 to 60°C [27]. Chain growth is limited in the aqueous phase, however, because the monomer concentration is normally very low and the polymer is insoluble in water. Nucleation occurs when aqueous chains aggregate or collapse after reaching a threshold molecular weight [28]. If many polymer particles are present, as is the case in commercial continuous polymerization, the aqueous radicals are likely to be captured on the particle surface by a sorption mechanism. Since the particle surface is swollen with monomer, the polymerization continues in the swollen layer and the sorption becomes irreversible as the chain end grows into the particle. Extensive work in this fundamental area has been reported by Bero, Sugimori, Thomas, Dainton, and Lenka [29-34].

Since polymer swelling is poor and the aqueous solubility of acrylonitrile is relatively high, the tendency for radical capture is very limited. Consequently, the rate of particle nucleation is high throughout the course of the polymerization and particle growth occurs predominantly by a process of agglomeration of primary particles. As the initiating radicals are incorporated into the polymer chain, they may serve as dyesites for the final fiber product.

Emulsion polymerization. The mechanism of emulsion polymerization was first developed qualitatively by Harkins in the mid 1940s [35], and later, quantitatively, by Smith and Ewart during 1948-49 [36,37], and by Gardon in 1970 [38,39]. In the 1960s it was shown that the emulsifier disperses a small portion of the monomer in aggregates, called micelles, of 50-100 molecules, approximately 50 A in diameter. The majority of the monomer stays suspended in droplet form. These droplets are typically 10,000 A in diameter, much larger than the micelles. Since a water-soluble radical initiator is used, polymerization begins in the aqueous phase. The micelle concentration is normally so high that the aqueous radicals are rapidly captured [40]. The micelle is essentially a tiny reservoir of monomer, therefore, polymerization proceeds rapidly, converting the micelle to a polymer particle nucleus. This particle nucleus continues to grow in size, being supplied with monomer by diffusion from the monomer droplets.

The ability of emulsion polymerization to segregate radicals from one another is of great importance commercially. The effect is to minimize the rate of radical recombination, allowing high rates of polymerization to be achieved along with very high molecular weight. In practice, many commercial processes employ a chain-transfer agent to control molecular weight at a reduced level. Processes of commercial importance are for the production of synthetic rubbers and latexes, and for adhesives and paints. Its use in the textile industry is limited to the manufacture of modacrylic compositions. The most notable is the 1947 Union Carbide process for Dynel [41], and the Bayer process of 1972, also for modacrylic vinyl chloride copolymers [42,43]. Commercial processes have been developed by DuPont in 1961 and 1973 [44,45], and Asahi in 1970 [46]. Emulsion processes have also been claimed by Toray Industries [50,51] in 1973 and 1974, Japan Exlan in 1977 [48,49], and Mitsubishi in 1978 [47].

A potential advantage of emulsion polymerization is that the emulsion can be heated to the wet melting temperature of the polymer and melt-spun. Modacrylics are more suitable than acrylics because they have lower melting points. Emulsion polymerization may also be coupled with emulsion spinning. In this case, the polymerization is carried out in an emulsion containing both monomer and a solvent such as dimethyl sulfoxide, as described by Toray Industries [50,51].

Solution polymerization. Solution polymerization is widely used in the acrylic fiber industry. The reaction is carried out in a homogeneous medium by using a solvent for the polymer. Suitable solvents can be highly polar organic compounds or inorganic aqueous salt solutions. Dimethylformamide (DMF) according to the patent granted to SINIA VISCOSA [52] in 1966, and dimethylsulfoxide [53] patented by Toray Industries in 1973 are the most commonly used commercial organic solvents. Examples of suitable inorganic salts are aqueous solutions of zinc chloride [54,55] and aqueous sodium thiocyanate solutions [56]. Other aqueous inorganic salt solutions that have been studied are calcium thiocyanate [57], nitric acid [58], and aqueous solutions of sodium [59] or magnesium [60] perchlorate.

The advantage of solution polymerization is that the polymer solution can be converted directly to spinnable dope by removing the unreacted monomer. There are a number of disadvantages, however. One disadvantage is that it is more difficult to achieve high molecular weight. The solvents required are often chain-transfer agents, and because chain mobility is high compared with the heterogeneous polymerization systems, chain termination is more rapid. Incorporation of nonvolatile monomers, such as the sulfonated monomers commonly used as basic dye acceptors, can also be a problem. The sulfonated monomers, in particular, have poor solubility in organic solvents and must be solubilized by converting them to a soluble form, such as the amine salt form [61].

Spinning of Fibers

Acrylonitrile copolymers used for acrylic fibers are white, easy flowing powders. The average molecular weight is commonly in the range of 100,000-150,000. The polymers decompose before melting, but indirect methods indicate a melting point of 340°C for PAN and 260-280°C for copolymers containing 90-95% acrylonitrile. Above 180°C exothermic decomposition becomes rapid. It results from intrapolymerization of nitrile groups along the chain to form cyclic structures [62]. This characteristic, undesirable for acrylic fiber because it causes discoloration, is used advantageously for the manufacture of carbon fibers [63].

Solvents and solution preparation. Of fourteen solvents given for polyacrylonitrile [64], seven are used commercially. They are: dimethylformamide, dimethylacetamide, aqueous NaSCN, aqueous HNO_3, aqueous $ZnCl^2$, dimethyl sulfoxide, and ethylene carbonate. Dimethylformamide is the principal solvent since it is used in both dry and wet spinning processes. Second is dimethylacetamide. Aqueous solutions of sodium thiocyanate and nitric acid are the most important inorganic solvents. Fiber production from sulfuric acid solution has also been reported but no commercial processes are known.

Effective solvents for polyacrylonitrile must break the intermolecular nitrile-nitrile attraction. Solubility and solution properties are affected by the polymer composition and molecular weight. The presence of 3-10% neutral comonomers increases acrylic polymer solubility and decreases solution viscosity at a given concentration. Molecular weight, composition, spinning concentration, and coagulation conditions are adjusted to give acceptable fiber formation and good tensile properties.

Numerous soluble and insoluble solution additives are used for a variety of purposes in both acrylic and modacrylic fibers. Dull or semi-dull fibers are delustered with TiO_2. Colored fibers may be produced by the incorporation of either insoluble pigments or soluble cationic dyestuffs which react with dye sites in the fiber during coagulation. Flame resistance of fibers is increased by additives containing chlorine, bromine, or phosphorus. Anitmony compounds are used for their synergistic effect on flame resistance in fibers containing a halogen [65,66].

Heat stabilizers are added to minimize discoloration of the solution prior to spinning and the fiber during production. Typical heat stabilizers were patented during 1966 - 1974, and include organic acids and sequestering agents [67], organophosphorus compounds [68,69], tin compounds [70], zinc compounds [71], and zirconium salts [72]. Other additives may be used to improve fiber light stability. In general, heat and light stabilizers are particularly useful in modacrylic fiber production.

Fiber production. Acrylic fibers are produced mostly by wet spinning (78%) and about 22% by dry spinning. Each major spinning system has been developed over 30-40 years to produce competitive and commercially useful products. Research and development by each major fiber producer has focused on optimizing fiber structure and properties, increasing production rate, and improving process stability and fiber uniformity. The combination of process conditions to achieve all these features simultaneously is strongly influenced by the desired filament size or tex.

The major difference between wet and dry spinning occurs in the initial fiber formation stage, subsequent process steps, such as washing, orientation, drying, relaxation, and finish application, are common to both wet and dry spinning.

Wet Spun Fiber Formation: In wet spinning, fine streams of viscous polymer solution are extruded through spinnerette holes into a liquid bath containing a coagulant that is miscible with the solvent, generally a dilute solution of the solvent in water. Fiber formation occurs rapidly as the polymer comes out of solution and is drawn onto the take-out roll. Water as a coagulant has advantages in cost, process simplicity, and minimum environmental concerns. Spinnerettes may have 10,000-60,000 holes (hence the number of filaments produced form each spinnerette), with hole sizes from about 0.05-0.38 mm in diameter. The holes are often divided into segments to improve coagulant diffusion across the spinnerette face.

Fiber Structure and ultimate fiber properties are largely determined by the choice of coagulation conditions, consequently, much research effort has been devoted to understanding and optimizing this critical step. Craig and co-workers [73] in 1962 reported the structure of fibers wet-spun from solutions of dimethylacetamide, aqueous zinc chloride, and aqueous sodium thiocyanate into dilute coagulation baths of the respective solvent. The fibers, prepared from the same polymer, had similar structures. The freshly coagulated fibers were described as an unoriented, fibrillar network consisting of voids and fibrils with diameters on the order of 20 nm. Superimposed on this network in some fibers were large radial macrovoids. This fibrillar network was oriented by subsequent stretching and the fibrils were also elongated. The oriented structure collapsed during drying to produce a compact fibrillar network without residual pores. Knudsen [74] in 1963 found that the best tensile properties, abrasion resistance, and fatigue life were obtained when the fiber leaving the coagulation bath had a high gel density, a homogeneous internal structure with small fibrils, and no macrovoids. These three characteristics are now generally recognized as guidelines for optimizing fiber structure.

In wet spinning two transitions may occur, gelation, and phase separation [75]. Gelation is the gradual and continuous transition of the fluid solution into a homogeneous elastic gel. To achieve the desired dense, homogeneous fiber structure, gelation should precede phase separation or solidification. Paul [76], in 1967 reported that gelation tendency increases as polymer concentration increases, and as the percent of neutral comonomer in the polymer decreases. Gelation also occurs more rapidly as temperature decreases, as was found by Arthur Gresswell in 1946 at the Stamford laboratories of The American Cyanamid Co. These observations relate directly to effects observed in fiber spinning. A large effect of neutral comonomer on gel density and average pore size was reported by Terada in 1973 [77] in the dimethyl sulfoxide process. The gel density decreased and the average pore size increased in fibers spun from copolymers containing 5 and 10% methyl acrylate. The same study found that incorporation of acidic comonomers, such as styrene sulfonate or allyl sulfonate, in the polymer increased gel density and reduced average pore size in the fibers.

Spin-bath composition and diffusion rates of the solvent and coagulant also have an important influence on fiber structure. Generally the diffusion rate increases as spin-bath temperature increases and as linear velocity of the filaments in the bath increases. It has been observed however, that as the diffusion rates increased, fiber structure became poorer as evidenced by lower gel density and larger fibrillar size. Optimum fiber properties were obtained with 55% dimethyl sulfoxide in the coagulation bath, which is in the range often reported for the other organic solvent systems. For inorganic solvent systems, aqueous spin baths containing 30-35% $ZnCL_2$ has been reported by Dow Co. scientists [78], 28-36% HNO_3 by Asahi [79], and 11% NaSCN by Courtaulds [80] during 1964-76.

Rheological properties of the spinning solution influence fiber structure and also have an important effect on spinning speeds and process stability. Cross-sections of acrylic fibers wet-spun from organic solvent systems are often bean-shaped as a result of initial skin formation around the periphery of the swollen filament. As the fiber interior deflates, the skin must fold. Fibers from the inorganic solvent systems are normally round. Shaped fibers can be produced from nonround holes by control of coagulation conditions [81].

Three modified processes also have been described for the production of acrylic fibers. The use of nonaqueous coagulants, such as glycerol, was recognized as early as 1943 from the work done at DuPont [82]. Coagulation baths containing a poly-alkylene glycol and solvent were used by Monsanto scientist in 1960s and produced fiber with very high gel densities and small fibrils, which led to superior tensile properties [83,84]. The high molecular weight glycols diffuse very slowly and a fine gel structure is formed as the solvent diffuses out of the fiber. The use of hexanetriol, low molecular weight alcohols, and cumene-paraffin mixtures have also been reported [85]. In a 1960 Badische process known as "gel state" spinning, a hot, concentrated polymer solution is extruded downward through air and the filaments allowed to gel by cooling before they pass into an aqueous bath for solvent extraction [86,87]. Consequently, homogeneous macrovioid-free fibers are claimed. This system is different from the Monsanto's dry jet-wet spinning process of 1963, in which the spinnerette is positioned only a few centimeters above the coagulation bath surface [88,89], and no gelation occurs in the air. Improved fiber properties and high spinning speeds are claimed for dry jet-wet spinning [90]. The high spinning speeds achieved with nonaqueous coagulants or by dry jet-wet spinning are particularly useful for making continuous filament acrylic products.

Dry Spun Fiber Formation: In dry spinning, fiber formation occurs as solvent is evaporated from streams of polymer solution drawn downward through heated gas. The spinning cell consists of a vertical jacketed cylinder with the spinnerette positioned in the center of the top and the fiber exits at the bottom. Depending on filament tex, exit speed, and tower length, residual solvent in the fiber may be as high as 50%.

The spinnerette may have 300-900 holes arranged in concentric circles. The number of holes is limited by the need to maintain filament separation. Turbulence in the tower is also minimized so that the filaments will not touch each other prematurely and fuse.

Fiber formation in the polymer-solvent systems used for dry spinning results only from gelation [75]. Gelation occurs gradually as the solvent evaporates and the concentration of polymer in the fiber increases. Homogeneous gelation is desirable for optimum fiber structure. The gel densities of the dry-spun fibers are much higher than those of the wet spun fibers and approach collapsed-fiber density as the residual solvent decreases. During stretching, the dry-spun fibers develop an oriented fibrillar structure similar to that of the oriented wet-spun fibers.

Monocomponent dry-spun fibers often have dogbone cross sections. This results from rapid solvent removal and skin formation around the periphery of the filament. Round cross-sections can be produced by use of a static zone with a high solvent-gas ratio in the top of the tower as in DuPont's method of 1961 [91]. Acrylic fibers with a trilobal shape are also produced.

Melt Extrusion: The polymers used for acrylic and modacrylic fibers cannot be melt spun because they degrade too rapidly at temperatures below their melting point.

Tow Processing: Wet-Spun filaments from several spinnerettes may be combined after leaving the coagulation bath and drawn directly into the tow line. The weight of water and solvent in the tow is generally 100-300% of the fiber weight. Ropes of dry-spun fibers are combined to form a tow of the desired size. The dry-spun tow may contain 10-50% solvent based on fiber weight. Tows from the different wet and dry spinning systems all undergo the essential steps of washing, orientation or stretching, finish application, and drying accompanied by collapse, relaxation, and crimping, but the sequence of these steps varies. Washing removes the solvent and minor impurities. Orientation or the stretching step is critically affected by temperature. McPeters and Paul [92] in 1976, reported a rapid increase in maximum draw ratio as the water bath temperature increased above 65°C, the fiber's approximate glass-transition temperature. Higher bath temperatures also reduced stress on the fiber at a normal draw ratio; a desirable feature for process stability.

The fibrillar network observed in wet-spun fibers leaving the coagulation bath undergoes major changes as the fibers are oriented, dried, and collapsed. Bell and Dumbleton [93] in 1971, reported that the fibrillar elements remain intact during stretching and collapse, although network junctures are destroyed and new junctures form. The unoriented fiber after washing and, to a lesser extent, the wet stretched fiber have a porous structure in which diffusion can occur rapidly. Consequently, a variety of in-line treatments have been developed either before or after stretching. In-line dyeing processes with cationic dyes have been described for DuPont's dry-spun fibers in 1976 [94-96] and wet-spun fibers from three different solvent systems used by American Cyanamid, Dow, and Monsanto respectively [97-99]. In some cases, dye pickup is increased by drawing the fibers in the dye bath. In-line treatments are also used to add flame retardants [100], apply crosslinking agents [101], incorporate dye receptor polymers [102], and add ultraviolet light stabilizers [103]. All this work was carried out during the 1960s. Finish is often applied to the fibers before drying. Standard finishes contain a lubricant to prevent filament sticking during drying, an antistatic agent and a softener to improve fiber hand.

Fiber tows are dried and collapsed by contact with a heated roll surface, by hot air flow through the tow on a perforated cylinder, or by hot air flow as the tow on a belt moves through an enclosed oven. When drying is accompanied by relaxation, special conditions of temperature and humidity may be necessary to obtain complete collapse as those devised by the American Cyanamid in 1961 [104].

Relaxation of acrylic fibers after orientation is an essential part of most processes. Relaxation increases fiber elongation, decreases the tendency to fibrillate, and increases dye diffusion rates into the fiber. These changes result primarily from a deorientation of the amouphous regions in the fiber structure.

Acrylic fibers are usually relaxed either wet or in steam since moisture reduces the temperature required. Relaxation of the collapsed fiber may be in-line or by batches in an autoclave [105]. Unrestrained relaxation is important for the full development of helical crimp in bicomponent fibers. Fiber shrinkage during relaxation may range from 10-40% depending on the amount of prior orientation, the fiber structure, and the intended use.

Dyeing

When PAN fibers first began to be marketed, it was considered impossible to dye them satisfactorily. These difficulties were attributed to a lack of specific dye sites in the fiber and to the high level of intermolecular dipolar bonding that severely limited the diffusion rate of dye molecules. In the 1950s the manufacturers learned that these barriers could be overcome through copolymerization [106]. Incorporation of vinyl monomers with bulky side-chains such as methyl acrylate and vinyl acetate was found to disrupt the intermolecular bonding (or pseudo-crystallinity) and greatly increase the level of polymer chain segmental mobility, thus permitting the dye to diffuse into the fiber.

Acrylic fibers are dyed mainly with cationic dyes. Bright, fast colors are achieved that have high aesthetic appeal. Disperse dyes can also be used to get a good range of colors, although not as bright. A small quantity of acrylic spun from special polymers containing basic groups is dyed with anoionic dyes. Dye sites for cationic dyes are acidic polymer end groups from polymerization initiators [107] and others added by copolymerization with acidic monomers. Diffusion of dye to these sites is negligible below the glass-transition temperature (about 65°C in water) but is rapid at 96-100°C, the normal dyeing temperature. No carriers are required to accelerate dyeing, instead, retarders are commonly added, to reduce dyeing rate and improve uniformity of color. Typical dyeing time is 1-2 h, but can be shortened by dyeing under pressure.

The fiber is readily dyed as staple (stock dyed), in yarn on cones (package dyed), or as skeins. As previously mentioned, some is dyed during fiber spinning (producer dyed) or as tow, using special tow dyeing equipment. Fabric or garment dyeing (piece dyeing) is used on acrylic knitted goods. To a lesser extent it is also used on woven materials, particularly where acrylics are blended with other fibers such as polyesters.

Transfer printing is a process introduced in the 1970s in which sublimable dyes are transferred to the fabric by heat from a printed design on paper. Acrylics, in general, are not well suited for this process. Dyeing in nonaqueous liquids (solvent dyeing) is also a useful process for coloring acrylics; fiber or fabric distortion is reduced, as is loss of fiber luster. Both processes require less energy than aqueous dyeing and greatly reduce water pollution.

Fiber Properties

Physical properties. The sustained commercial success of acrylic fibers over the past several decades can be attributed largely to their desirable balance of properties. Apparel goods and some home furnishings made from acrylics have appealed to the consumer because they are aesthetically pleasing, comfortable, easy to care for, and reasonably durable. The physical properties of the fibers are by no means remarkable and some obvious deficiencies such as a poor hot-wet strength and only modest tensile strength and abrasion resistance have prevented their use in many applications where nylon and polyester dominate.

Acrylic fibers are mostly sold as staple and tow, though a small amount of continuos filament is also produced. Acrylics are restricted to staple yarn processes because a good method for texturizing a continuous filament acrylic yarn has not been devised.

Staple lengths may vary from 25 to 150 mm, depending upon the spinning system that is used to convert fiber into spun yarn. The filament denier may vary from approximately 1.2 to 15. Garments intended to have a very soft hand are constructed from low-denier fibers. Most typical acrylic apparel fabrics utilize deniers near 3, and some industrial and craft knitting yarns use deniers from 5 to 6. Acrylic carpet yarns are usually made from 15 denier fiber in order to achieve the required stiffness and wear resistance.

The tenacity of acrylic fibers (breaking strength) varies from 2.2 to 3.6g/den (or 0.19 to 0.32 N/tex), and the breaking elongation varies from 33 to 64%. The only apparent pattern among the commercial fibers is the lower tenacity and higher elongation of the large-denier fibers, which suggests that the molecular orientation of the fiber may decrease with increasing denier.

The boiling water shrinkage of standard acrylic fibers is generally quite low, but may be made intentionally high for special types of fibers. High-bulk fibers that have been stretched and cooled to retain a frozen-in strain are designed to shrink up to 30% in boiling water. The bulk is usually developed during dyeing. In comparison with the other two major synthetic fibers, nylon and polyester, it has a lower breaking elongation and considerably lower work of rupture. Acrylics do not compete well with these fibers in applications where very high load-bearing properties are of paramount importance. It does, however, have a higher tenacity and work of rupture than wool. This is significant since acrylics have replaced wool in many markets.

The very significant reduction in modulus that occurs when the temperature in water exceeds T_g and the fiber becomes very rubbery points to the cause of perhaps the most glaring property deficiency of acrylics-their poor hot-wet properties.

Acrylic fibers are less moisture absorbent than cotton, wool, and rayon and somewhat more moisture absorbent than nylon and polyester. However, many methods have been developed to increase the moisture-absorbing properties of acrylics. One extremely important property of a textile fiber is the relative durability or wear resistance of a fabric made from the fiber. The acrylics as a class are much less durable than either nylon or polyester.

Chemical properties. One of the outstanding properties of acrylic fibers is their superior resistance to degradation by outdoor exposure, particularly their resistance to sunlight and the action of microorganisms [108]. This unique property has given acrylic fibers a special niche among fabrics used in outdoor applications, such as tenting and awning fabrics, or sandbags for river shore stabilization. Acrylic and modacrylic fibers are also outstanding in their resistance to heat and chemical agents. This property is thought to result from dipole effects related to the polarity of the nitrile group; these produce electrostatic repulsive forces between adjacent nitrile groups and simultaneously create strong dipole bonding between nitriles of adjacent chains. Hydrogen bonding between the electronegative nitrogen atoms and tertiary hydrogen atoms on the polymer backbone is also significant, though to a lesser degree. Strong dipole bonds must be broken before melting or solvation can occur. In addition, it is believed that the repulsive forces within chains present a barrier to bond rotation, resulting in a stiff polymer backbone. Thus, even when the hydrogen-nitrile dipole bonds are broken, the entropy gained in the melting or dissolving process is relatively small. consequently, acrylonitrile fibers are insoluble and unswollen by most chemical solvents, such as weak acids, weak alkalis, organic solvents, oxidizing agents, and dry cleaning solvents. These fibers are attacked by strong bases and highly polar organic solvents such as dimethylformamide, dimethyl sulfoxide, ethylene carbonate, and dimethylacetamide required for spinning these fibers.

Acrylic fibers discolor and decompose rather than melt when heated. These fibers, however, have very good color and heat stability at temperatures less than 230°C. Fibers used in textiles must not ignite readily when placed in contact with a flame. In carpets, sleepwear, draperies, and bedding, however, the fibers must also be self-extinguishing after removal from the ignition source. This is generally achieved in acrylonitrile-based fibers by incorporating various amounts of halogen comonomer. Vinylidene chloride is the most popular because it is inexpensive, easy to copolymerize, and has a very high chlorine content.

FIBERS WITH MODIFIED PROPERTIES

Acrylic fibers have very desirable properties when used in their standard commercial forms. They are adaptable fibers as their polymerization and spinning offer many opportunities for modifying the final product. Fiber properties that have received attention are high bulk, improved hot-wet properties, wear comfort, softness of handle, antipilling, antistatic, anti-soiling, and bacteriostatic properties.

Bicomponent fibers. Bicomponent fibers are produced when a high degree of bulk is desired in the spun yarn for applications such as sweaters and craft knitting yarns. Bicomponent acrylic fibers have become a major product type since their introduction in the early 1960s [109]. These fibers are composed of two polymers in separate areas of the cross-section. Each polymer is continuous along the fiber length and the two are permanently joined at an interface. The polymers are selected to shrink or swell differently in response to heat or moisture. As a result, when the fibers are properly treated, differential shrinkage will occur and a spiral or helical crimp forms. The extent of crimp development depends on the shrinkage differential between the components, the distribution of the components in the fiber, and translational restraints which may inhibit crimp development [110].

Two different types of bicomponent acrylic fibers are produced. One contains different amounts of water-ionizable groups in the two polymers; consequently the hydrophilic side swells more in water. Crimp develops when the fiber is dried after wet-heat treatment. This spiral crimp is water-reversible; ie, it decreases on wetting out and reforms on drying. Commercial examples are Monsanto's Paquel and DuPont's Wintuk.

In the second bicomponent type, the polymer pairs contain different amounts of nonionic comonomers. Helical crimp develops when the oriented fiber is heated to a temperature where adequate differential shrinkage between the two polymers occurs. Once formed, the crimp in this type is permanent. Example of this type of fiber is DuPont's Sayelle developed in 1959. Each bicomponent type has certain advantages in processing, esthetics, and product care. The properties of each can be modified by process changes [111-113].

A variation of the bicomponent is known as the random bicomponent. This concept is based on the fact that viscous spinning solutions can be merged without complete mixing [114]. When passed through a standard spinneret, bicomponent fibers ranging in composition from 100% component A to 100% component B are produced. This type of process yields a bicomponent fiber with good bulk development and substantially increases productivity according to the patent granted to Monsanto in 1981 [115].

Toray and Asahi have reported new types of bicomponent fibers that employ more than two functional components. Toray, for example, in 1983 announced the development of a water-repellent, antipilling, bicomponent fiber for the sport sweater market [116]. This product, which derives its water repellency from the addition of a fluorine-type resin, exhibits an exceptionally soft handle and long-lasting water repellency. Asahi developed a three-component fiber in 1980, made by spinning one polymer as the continuous phase, and intermittently injecting the remaining two polymers as dispersed phases [117]. Polymers of differing dyeability reportedly give a fiber with a unique, periodically varying shade.

Improved hot-wet properties. One of the deficiencies of acrylic fibers is their inability to retain their modulus under hot-wet conditions. Knits and woven fabrics tend to lose their bulk and shape in dyeing and, to a limited extent, in washing and drying cycles. Under high humidity conditions, the fibers may lose their resilience, and develop wrinkles that are difficult to remove. The problem has been treated in a number of fundamental studies which have demonstrated that moisture lowers the wet glass transition temperature of acrylonitrile copolymers [118], hence crimp imparted during spinning to enhance yarn bulk is lost when the yarn is exposed to conditions required for dyeing and laundering.

A number of polymer and fiber modifications were devised during 1960s and 1970s to allow acrylics to compete successfully in easy-care apparel markets. Fibers were treated with compounds such as, ammonium sulfide [119], hydrazine derivatives [120], thiosemicarbazides [121], silicone oils [122], and emulsions of polysiloxane and polyepoxide [123]. Some success was achieved during the late 1960s and early 1970s by incorporating comonomers which increase the wet glass transition temperature of the polymer or make the copolymer more water resistant [124-129]. Sheath-core fibers have been reported by Asahi in 1979, [130] in which the core polymer is stable under hot-wet conditions and the sheath polymer is used to compensate for deficiencies in dyeability.

Dope additives have also been reported. A patent issued to the American Cyanamid in 1977 [131], for example, describes a process in which a fine dispersion of a polycarbonate resin is incorporated into the spinning dope. The fibers produced in this manner contain fibrils of the polycarbonate resin that have the effect of improving the fiber's hot-wet elasticity.

Many attempts to improve hot-wet properties by cross-linking have been reported. Patents issued to Bayer AG, during 1977 and 1978 [132-134], for example, claim that excellent properties under hot-wet conditions can be achieved by incorporating N-Methylol vinyl monomers into the acrylic polymer. The fiber is spun from DMF and cross-linking is initiated by drying the fiber, tension-free, at 170°C.

The carpet market is one area in which acrylics have failed as a result of inadequate hot-wet stability. Although acrylic fibers produce a carpet of superior wool-like appearance, the pile loses its resilience and bulk during dyeing and under high humidity conditions. As a result, acrylics have lost popularity as a carpet fiber since the peak year of 1968.

Modification of handle. There are many factors that together determine the softness or handle of a fabric. For a given fabric construction the denier, bending modulus, cross-sectional configuration, degree of crimp, moisture absorption, and surface smoothness of the fibers all influence the softness of the final product. Very fine filament deniers such as those patented by Bayer AG in 1981 are effective where good draping, anti-crease properties, and softness of handle are desired [135]. Such yarns are expensive to produce and are more susceptible to abrasion than coarse filament yarns. The desired softness of handle also can be achieved through modifying the fiber cross section. Dry-spun fibers, for example, generally have flat or dog-bone-shaped cross sections. This has the desirable effect on the handle. Wet-spun fibers generally have a cross section that is only slightly deviated from round. However, nonround cross sections can be achieved in wet-spinning by modifiying the coagulation conditions and/or the spinneret orifices, developed during 1960s and early 1970s [136-140].

Softness has also been achieved by grafting long-chain esters as in the American Cyanamid company's method of 1972-73 [141-143]. Fibers with a soft, wool-like handle have been produced by copolymerizing long-chain alkyl-acrylates and alkylmethacrylates by Toray Ind., in 1967 [144] and by blending conventional fibers with fibers containing hygroscopic comonomers, such as N, N-dimethylacrylamide [145] or derivatives of N-vinyl pyrrolidone as described in patents granted to Asahi in 1969 and 1979 [146]. Toray Industries also has developed special techniques such as sheath-core spinning during 1970-80, [147,148] and Asahi has developed conjugate spinning [149,150].

Wear comfort. Although many properties contribute to wear comfort, it is believed that the ability to absorb moisture from the skin, and the extremely fine denier (both dominant properties of cotton fiber) contribute to comfort. Both properties can be achieved in acrylic fibers. However, fine-denier acrylics are achieved at the cost of reduced process productivity. Many methods have been developed to increase the moisture-absorbing properties of acrylics. Acrylonitrile polymer can be modified by incorporating hydrophilic comonomers that decrease ultimate fiber density. Improved moisture retention can also be achieved by modifying the fiber-spinning process or by using aftertreatments. An illustration of a polymer modification combined with a special spinning technique is Bayer's process patented in 1978, for making porous, moisture-absorbent fibers [151]. This process uses copolymers containing carboxyl groups in the acid form. The polymer is dry-spun from a spinning dope containing a high-boiling glycol or alcohol derivative. The glycol or alcohol derivative is retained within the fibers in the dry-spinning tower where a porous fiber morphology is established and subsequently extracted during the fiber-washing step. After spinning, the carboxyl groups are converted to the salt form, thereby enhancing their hydrophilic character. The resulting fibers have a unique porous sheath-core structure that allows water to diffuse into the fiber leaving the surface relatively dry. Fibers produced from this process can absorb large quantities of water without giving a clammy feel. Many variations of the sheath-core concept are described in the patent literature [153-156]. Dunova, an acrylic sheath-core fiber of this type marketed by Bayer, is a commercial example [157].

Reduced pilling. Acrylic fibers, like others used in staple fabrics, develop small balls of fiber or "pills" as a result of abrasive action on the fabric surface. Because of the relatively high flex life of acrylic fibers, however, pills tend to build up more on acrylic fabrics than on comparable woolens. Many efforts have been made to overcome this problem. Examples include using highly twisted yarns, minimizing the number of short staple lengths, modifying the fiber cross section, and modifying the fiber surface to minimize tangling. The most effective methods are those designed to increase the likelihood that the pills will wear off. The required fiber properties include reduced fiber strength, defects in the fiber, brittleness and lack of shear strength. All of these properties are easy to impart to the fibers, but increases in antipilling properties generally lead to corresponding decreases in the spinnability or processability of the fibers. Thus, commercial antipilling processes depend on a balance between these factors.

The polymer can be modified to give more brittle fibers by decreasing the comonomer content, adding an ionic monomer or inducing cross-linking during the spinning process. Wet-spinning processes can be modified by using low solvent concentration in the spin-bath and high spin-bath temperature to give brittle fibers with high void content. Low draw ratios result in low tensile strength; drying the fibers under tension increases fiber brittleness and annealing the fibers at reduced temperatures reduces fiber elongation to break. Skin-core fiber structures, have also been used to achieve antipilling properties.

Commercial processes have been patented by DuPont in 1971 [158], Courtaulds in 1968 [159], Mitsubishi Rayon 1970 [160], Toray Ind. during 1970-1980 [161,162], Asahi during 1978-82 [163, 164], and Hoechst in 1981 [165]. The Hoechst antipilling fiber is marketed in various forms as Dolan 40, 41, and 44.

Improved abrasion resistance and increased fiber density. The abrasion resistance of acrylic fibers is distinguished from fibrillation resistance in that fibrillation is primarily a surface-related phenomenon. Due to poor lateral cohesion, surface fibrils are torn from the fiber surface causing a whitening effect in dyed fabrics. True abrasion resistance relates more to fiber integrity and strength. It is directly related to fiber elasticity, toughness, and density. It has been found that abrasion resistance is generally improved by reducing the microvoid size, increasing the fiber density and eliminating macrovoids.

Void-free fibers have been produced by incorporating hydrophilic comonomers, such as sulfonated monomers, acrylamide derivatives, and hydroxyalkyl acrylates, described in a 1978 patent granted to Monsanto Co. [166] Polymer blends are also effective in reducing macrovoids. Examples include Bayer Company's blends of hydrophilic polymers such as polyvinyl methyl formamide [167], poly-N-vinyl pyrrolidone, and acrylonitrile-dimethylacrylamide copolymer [168]. Dense fibers can be produced by incorporating comonomers, such as vinylidene chloride, with small molar volumes relative to acrylonitrile [169].

The spinning process also has a major effect on fiber density and abraision resistance. Modified wet-spinning processes have been reported in an Austrian patent to Lenzing AG in 1981 [170]. Dry-spinning, however, is known to produce a denser fiber structure than conventional wet-spinning. Wet spinning techniques used to improve fiber density and abrasion resistance generally do so by approximating dry-spinning conditions in some way. Examples include modified spin-bath, high solvent concentration, low spin-bath temperature, and additives to the dope or spin-bath. Spin-bath additives include nondiffusing liquids, such as polyethylene glycol. Surface-active agents also have been added to the dope as well as to the spin-bath to increase fiber density. Variations in wet-spinning that approximate dry-spinning are dry-jet wet-spinning and spinning into a standard spin-bath through an inert, low-density organic liquid dispersed on the spin-bath surface. Melt-spinning of wet acrylonitrile copolymer in 120°C superheated steam is also reported to give dense, void-free fibers as described in a patent granted to Japan Exlan, in 1977 [171].

Improved whiteness and thermal stability. Fiber whiteness and thermal stability can generally be improved by producing a polymer that is linear and free of conjugated or unstable chemical structures formed by side-reactions during the polymerization process. The most straightforward approach is to maximize the probability that polymer radicals react with monomer molecules rather than with the polymer backbone itself. This is accomplished by maintaining the unreacted monomer concentration in the reactor as high as is practical. Successful application of this technique has been accomplished by operating the polymerization process at low overall conversion of monomer to polymer according to Mitsubishi's patents of 1973 & 1974 [172, 173]. Side-reactions can also be minimized by avoiding comonomers having low reactivity with acrylonitrile and those comonomers that are known to undergo side-reactions themselves. Polymerization reaction conditions are also important. High pH is known to cause hydrolysis of the nitrile groups in the polymer and high-temperature reaction conditions tend to favor undesirable side-reactions. Reaction temperatures above 50°C are normally avoided by using redox initiators, such as the persulfate-bisulfite-iron system. Additives used during polymerization include reducing agents such as organophosphorus compounds described in the 1974 patents granted to Monsanto [174], sulfinic acids and other sulfur compounds used in 1970 by Bayer AG. [175], and glycerol monoesters, and phenol-type antioxidants described in the 1970 British patent to Mitsubisi [176]. Comonomers of amides and sulfonates and other comonomers with special functional groups have also been used effectively by Hoechst AG [177].

A number of methods have been reported in the literature for improving the thermal stability of acrylic fibers. For example, Courtauld's 1969 method of impregnating wet-spun fibers in the gel state with salts of metals from Group II [178] or sodium sulfide solution. Also, dried fibers can be impregnated with sulfuric acid solutions of formamide, phosphates or borax [179], or organotin salts as in Monsanto's methods of 1960s [180].

Antistatic fibers. Acrylic fibers are naturally somewhat hydrophilic and tend to dissipate static charge more readily than other synthetic fiber types. Antistatic properties have been improved, however, by making the fibers hygroscopic with hydrolyzing and cross-linking agents developed by Toray Industries in 1967 [181], and Mitsubishi Rayon in 1979 [182].

Since static electricity is a surface phenomenon, effective antistatic surface treatments can be used as a means of dissipating static charge. Treatments that give an ionic surface character have been effective. Surface treatments empolying fatty acids, vinyl ether copolymers [183], and quaternary ammonium compounds [184] have all been reported.

Antisoiling fibers. Low dirt-absorbing fibers have been developed by incorporating paraffins into the spin bath by Toray Ind. in 1973 and Societa Italina Resine in 1967 [186]. American Cyanamid, during 1968-70 modified the polymer composition by incorporating comonomers of fluorinated acrylates [187] and copolymers of fluorinated acrylamides and other monomers [188]. Porous fibers with reduced staining tendencies have been made from acrylic copolymers containing methylacrylate and sodium allylsulfonate [189]. Antisoiling properties can also be achieved by using finishes, either by treating the fiber during spinning, or by applying finishes directly to the fabric.

Antimicrobial fibers. Acrylic fibers are very resistant to attack by micro-organisms and will not support microbial growth. However, contaminants of low molecular weight such as lubricants, and spinning oils used in the finishing of textiles may provide sufficient nutrients for mild surface growth of micro-organisms. In most cases it will not affect the strength of the fabric but may give rise to staining & discoloration. Antimicrobial finishes are not only incorporated in fiber to protect the textile itself, but to protect the wearer or user of textile material against bacteria, yeast, dermatophytic fungi, and other related micro-organisms for aesthetic, hygenic, or medical purposes.

Organometallic tin-titanium salts are incorporated into synthetic fibers to impart antimicrobial activity. Antimicrobial fabrics have been developed by American Cyanamid in 1982, Kanebo in 1983, and most recently by Courtaulds. Courtaulds [190] fiber known as Courtek M. contains encapsulate chemicals within the fiber structure, furthermore the degree of encapsulation can be controlled thereby controlling their rate of release. Courtaulds research began in 1986 and the fiber has been made commercially available in 1990. The fiber will find many applications in the fields of industrial and personal hygiene, for example in hospitals, food processing or intimate apparel.

Applications

Apparel and home furnishings. Acrylic fibers are used extensively in the apparel and home furnishings markets; a relatively small amount goes into industrial uses. Within the apparel market major uses are for sweaters, single and double knit jersey fabrics, hosiery and craft yarns, and pile and fleece goods. In home furnishings the major uses are for blankets, curtains, drapes, and to a limited extent, carpets.

A growing trend in the acrylic fibers industry however, is to develop premium products based on proprietary technology. Specialty fibers, such as ultrafine denier fibers, acid-dyeable, producer-dyed, and pigmented fibers, ion-exchange fibers, metallized and semiconducting fibers, hollow fibers for apparel, and coarse-denier fibers for wigs and simulated animal hair are either available or being developed. Another important example of this trend is the so-called "functional fabrics." One of the most recent of these is the antimicrobial fabrics discussed earlier. These fabrics are developed by American Cyanamid in 1982, Kanebo in 1983, and Coutaulds Fibers in 1990 [190]. This type of fabric is made from microporous fibers. The void structure is treated with chemicals that retard the growth of bacteria and fungi. These fibers are valuable for apparel subjected to perspiration, such as socks, sportswear, underwear, and baby wear. Other functional fabrics are made from moisture-repellent fibers, moisture-absorbant sheath-core fibers, antistatic fibers, and flame-resistant fibers.

Industrial textile applications. Acrylics and modacrylics are also useful industrial fibers. Fibers low in comonomer content, such as DOLAN 10 and DuPont's PAN Type A, have exceptional resistance to chemicals and good dimensional stability under hot-wet conditions. These fibers are useful as industrial filters, battery separators, asbestos fiber replacement, hospital cubical curtains, office room dividers, uniform fabrics, and carbon fiber precursors. The excellent resistance of acrylic fibers to sunlight also makes them highly suitable for outdoor use. Typical applications, including modacrylics, are awnings, sandbags, tents, tarpaulins, covers for boats and swimming pools, cabanas, and duck for outdoor furniture [191].

Non-textile Applications

Carbon and graphite fibers. Acrylics are used as precursors for carbon and graphite fibers. Carbon and graphite fibers are valued for their unique combination of extremely high modulus and very low specific gravity. Steel is the cheapest and most commonly used high-modulus fiber. Steel fibers, however, are very dense and, therefore, are not suitable for applications where light weight is critical. Carbon and graphite fibers, on the other hand, have Young's moduli equivalent to steel at a density of only 1.8-2.0 g/ml. Therefore, carbon and graphite fibers find wide use in applications, such as in sporting goods, aircraft, and space vehicles where high modulus and low weight are critical. Anticipated markets are automotive parts, asbestos replacement, fireproof textiles, and rapid acting heating elements. The market for carbon and graphite fibers is expected to grow rapidly as production costs fall.

Carbon fibers had been known for nearly a century when Thomas Edison first used them as filaments for incandescent lamps. These early carbon fibers were prepared by the carbonization of bamboo and rayon. The present upsurge in the development of carbon fibers stems from work carried out in the 1950s at Wright-Patterson Air Force Base where the first major effort was undertaken to make high-strength carbon fibers. Union Carbide [192] developed this process and was able to enhance the mechanical properties of rayon-based carbon fibers using stress graphitization. This process led to the production of high-modulus and high-strength carbon fibers and to Union Carbide's introduction of Thornel 25, 40, and 50 carbon fibers.

At about the same time (early 1960s), Shindo in Japan [193] and Walt, Johnson, and Phillips in the United Kingdom [194,195,196], working independently, successfully prepared carbon fibers from polyacrylonitrile. These carbon fibers had relatively high modulus but low strength. The mechanical properties of these PAN-based carbon fibers were further enhanced by carrying out certain oxidative treatments under stress. The PAN process of carbon fiber production received the immediate attention of several manufacturers because the yield of the carbon fiber was almost double that of its yield from rayon. Moreover, the process did not entail the difficulties and high cost of high-temperature stressing treatment.

Acrylic precursors are made by standard spinning conditions, except that increased stretch-orientation is required to produce precursors with higher tenacity and modulus.

Asbestos replacement fibers. Asbestos is valued as a reinforcing fiber because of its chemical and thermal inertness, fibrous structure, and high elastic modulus. However, asbestos has been found to be extremely hazardous when the microdust is inhaled and deposited on the lining of the lungs. With over 5,000,000 tons consumed annually in 3000 different products, there is great interest in finding a substitute material.

Acrylic fibers are a strong contender in this area [197]. As a reinforcing material for ambient-cured cement building products, acrylics seem to offer three key properties--high elastic modulus, good adhesion, and good alkali resistance [198]. The high modulus requires an unusually high stretch-orientation. This can be accomplished by stretching the fiber above its glass transition temperature. Normally this is done in boiling water or steam to give moduli in the 100-150 g/den range [199]. Alternatively, the stretch-orientation can be achieved by a combination of wet stretch at 100°C and plastic stretch on hot rolls or in a heat transfer fluid such as glycerol. This technique developed by Toray Ind., in 1974 is reported to give moduli as high as 200 g/den [200].

Patents for acrylic asbestos replacement fibers have been obtained by Asahi [201-203], Wuestefeld [204], REDCO [205], and Hoechst [206] during 1978-1981. The Hoeschst fiber, marketed under the trade name Dolanit (originally Dolan 10), is offered in two staple lengths and fineness. This fiber is already finding application in reinforced concrete window boxes, wall cladding, roof tiles, short corrugated sheets, and drain pipes.

Another application where acrylics have proven to be effective is in brake and clutch linings. Here thermal stability is a key property. Hoechst has developed an experimental heat-resistant acrylic fiber (VF) that can be used in combination with glass fibers [206]. The French company, Valco, has developed a friction material for brakes and clutches that is also based on the use of acrylic fibers.

Fibers for reverse osmosis and ion exchange. Tubular fibers for reverse osmosis gas separations, ion exchange, ultrafiltration, and dialysis are a significant new application of acrylic fibers and other synthetics [207]. Commercial acrylic fibers have already been developed by Nippon Zeon during 1978-80 [208-210], Asahi [211], and Rhone-Poulenc. One such fiber, developed at the California Institute of Tecnology in 1977 [212], is a hollow fiber made from copolymers of acrylonitrile and vinyl acetate or ethyl acrylate. The acrylonitrile-based hollow fiber acts as a substrat which is subsequently inbibed with liquid monomers that can be polymerized in situ. Monomers are used that form insoluble, cross-linked, ion-exchange resin particles embedded in the walls of the fiber. These fibers are useful for removing counter ions from solution. Solid fiber alkylated with epichlorohydrin, having anion-exchange properties, are reported by by Alikberova in 1982 [213] for removing toxic cyanide compounds from air. Cation-exchange fibers, containing ionogenic carboxyl groups, are described by Dorokhina and Zharkova [214].

Moisture absorbent synthetic paper. Processes for making a water-absorbent synthetic paper with dimensional stability have been developed by several companies. In a process developed in 1981 by Mitsubishi Rayon, acrylic fiber is insolubilized by hydrazine and then hydrolyzed with sodium hydroxide. The paper, formed from 100 parts fiber and 200 parts pulp, has a water absorption 38 times its own weight [215].

Other processes for making hygroscopic fibers have also been reported in the patent literature. These fibers are used in moisture-absorbing nonwovens for sanitary napkins, filters, and diapers. For example, in a process reported by Exlan in 1981 [216] the fibers are made by incorporating a hydrophilic comonomer, such as methylene-bisacrylamide, and then hydrolizing the copolymer with caustic soda. The resulting copolymer, has hydroxl groups which is then blended with a second hydrophilic acrylonitrile copolymer.

Electrically conducting fibers. A process for making electrically conducting fibers based on treatment with zinc oxide have been reported by Kanebo in 1982 [217]. Copper ions have been used by Exlan [218], Nippon Sanmo Dyeing Co. [219], and Teijin [220]. In the Exlan process, a copolymer of 86% acrylonitrile, 11% vinyl acetate, and 3% dimethylaminoethyl methacrylate is spun into fibers and then treated in a solution containing copper sulfate and $(NH_2OH)_2H_2SO_4$. The resulting fiber is then reduced in a solution containing $Zn\text{-}(HSO_2CH_2O)_2$ to give fibers with a specific resistance of only 33 ohms/cm compared to 10^{10} ohms/cm without this treatment. Electrically conducting acrylic fibers are also made by spinning fibers containing carbon black [221]. When blended at a rate of 3-4% with untreated acrylic or polyamide fibers, to make carpets and nonwovens, the resistivity of the end-products was reduced by a factor of 10,000.

Conclusions

Thus the history of acrylonitryl fibers is one of rapid growth during the first two decades of their introduction into the world market and then their maturing and leveling off during the 1980s. The wool-like properties and low cost of acrylic fibers caused a rapid growth in consumption from almost nothing in 1950 to 6.5 billion pounds in 1990. Acrylics found wide uses as a wool replacement fiber in carpets, home furnishings, and knitted apparel products. The world market share claimed by both acrylics and polyester increased sharply during 1950 - 1970 primarily at the expense of wool and nylon. For acrylics this increase was from about 14% in 1960 to approximately 23% in 1969. Thus, the period from 1950 to 1970 was one of rapid change with nylon giving way to polyester as the dominant synthetic fiber. By 1980 acrylics held approximately 20% of the world market share of fibers [222]. However, consumption of acrylics leveled off in the 1980s.

There are several reasons for the leveling off of growth in acrylic fiber consumption during the 1980s. The most important is probably the maturing of the wool replacement market. Next is the continuing dominance of nylon in the carpet sector. Due to the rapid capital investment in acrylics that occurred in the early 1970s and the plant capacity added in the developing countries, there has been a very large excess in world capacity.

Generally, the 1990s appear to be a period of reequilibration in world acrylic fibers markets. The larger, technology-oriented companies of the United States, Europe, and Japan are moving away from commodity markets where profit margins are small into special market areas such as carbon and graphite fiber precursors, fibers for reverse osmosis, and functional fibers with premium pricing, and high-volume nontextile products. Fiber producers in the developing countries, on the other hand, are likely to capture market shares in commodity products.

References

1. Rules and Regulations Under the Textile Fiber Products Identification Act, effective March 3, 1960, U.S. Federal Trade Commission, Washington, D.C., 1960, p. 4.
2. Man-made fibres-Generic names, ISO-2076, 1977(E), International Organization for Standardization, Switzerland.
3. Moureu, C. Ann. Chem. Phys. 2, 186 (1983).
4. Rein, H. U. S. Patent 2,117,210, May 10, 1938, and U. S. Patent 2,140,921, Dec. 20, 1938, (to I. G. Farbenindustrie).
5. Latham, R. C.; Houtz, R. C. U. S. Patent 2,404,714 and U. S. Patent 2,404,720, June 17, 1942, (to DuPont).
6. Rein, H. DPA 72,024 IVC/39b, April 13, 1942, (to I. G. Farbenindustrie).
7. Corbman, B. P. Textiles: Fiber to Fabric; McGraw-Hill: New York, p 383 (1983).
8. Rugeley, E. W.; Field, T. A., Jr.; Fremon, G. H. Ind. Eng. Chem. 40(9), 1724 (1948).
9. Koch, P. A. Fibres, 16, 174 (1955).
10. Tortora, P. G. Understanding Textiles, 3rd ed.; Macmillan: New York, 1987; p 169.
11. Coover, H. W. Mod. Text. 37, 68 (1956).
12. Skinner's Rec. Man-Made Fibres. 40(8), 534 (1966).
13. Frushour, B. G.; Knorr, R. S. In Handbook of Fiber Sci and Tech; Lewin, M. and Pearce, E. M. Eds., International Fiber Science and Technology Series, Marcel Dekker: New York, 1985, Vol. 4, Chapter 3.

14. Textile Organon, 54, 90 (1983).
15. Reisch, M. Chem. Eng. News 68(40), 9 (1990).
16. Greenfield, E. J. Textile Asia. 21(8), 158 (1990).
17. Hadley, D. J. Chem. Ind. Feb. 25 (1961).
18. Bayer, O.; Kuntz, P., Ger. Pat. 728,767, 1939.
19. Sennewald, K. Erdoel Kohle, 12, 364 (1959).
20. Belg. Patent 577,691, 1959, (to Distillers Co. Ltd.).
21. Bellringer, F. J.; Bewley, T.; Stanely, H. M. U.S. Patent 2,691,037, Oct. 5, 1954, (to Distillers Co. Ltd.).
22. Hadley, D. J.; Woodcock, C. A. U.S. Patent 3,009,943, Nov. 21, 1961, (to Distillers Co. Ltd.).
23. Idol, J. D. U.S. Pat. 2,904,580, Sept. 9, 1959, (to Sohio Co.).
24. England, D. C. U.S. Pat. 2,736,739, Feb. 28, 1956, (to E. I. duPont de Nemours & Co., Inc.).
25. Johnson, W. G. U.S. Pat. 2,980,726, April 18, 1961, (to E. I. duPont de Nemours & Co., Inc.).
26. Malacini, P.; Tedesco, R.; Patron, L.; Moretti, A.; Can. Patent 911,650, Oct. 3, 1972, (to Montedison Fibre).
27. Bach, H. C.; Knorr, R. S. Ency. Polym. Sci. Engn., 2nd ed. Vol 1. Wiley: New York, 1985, p. 351.
28. Fitch, R. M. Br. Polym. J. 5, 467 (1972).
29. Bero, M.; Rosner, T. Macromol. Chem. 136, 1 (1970).
30. Sugimori, T.; Nishikawa, S. Kobunshi Kakagu, 29(331), 817 (1972); Kobunshi Kagaku, Engl. Ed., 1(11), 1289 (1972).
31. Thomas, W. M.; Gleason, E. H.; Mino, G. J. Polym. Sci. XXIV, 43 (1957).
32. Dainton, F. S. et al., J. Polym. Sci. XXXIV, 209 (1959).
33. Dainton, F. S.; Seaman, P. H. J. Polym. Sci. XXXIX, 279 (1959).
34. Lenka, S.; Dhal, A. K. J. Polym. Sci.-Chem. Ed. 19, 2115 (1981).
35. Harkins, W. D. J. Am. Chem Soc. 69, 1428 (1947).
36. Smith, W. V; Ewart, R. H. J. Chem. Phys. 16, 592 (1948).
37. Smith, W. V. J. Am. Chem Soc. 70, 3695 (1948); ibid. 71, 4077 (1949).
38. Gardon, J. L. J. Polym. Sci., A-1, 6, 623 (1968); ibid. pp 643, 665, 687, 2853, and 2859.
39. Gardon, J. L. Br. Polym. J. 2, 1 (1970).
40. Fitch, R. M.; Shih, Lih-bin Progress in Col. and Polym. Sci. 56, 1 (1975).
41. Shriver, L. C.; Fremon, G. H. U.S. Pat. 2,420, 330, May 13, 1947 (to Carbide and Carbon Chemicals Corp.).
42. Koenig, J.; Sveling, C.; Boehmke, Ger. Patent 2,604,630, August 11, 1977, (to Bayer AG).
43. Ger. Pat. Appl. 2,300, 713, Jan. 8, 1973, (to Bayer AG).
44. Bechtold, M. R. U.S. Pat. 2,972,511, Feb. 21, 1961, (to E. I. du Pont de Nemours & Co., Inc.).
45. U.S. Pat. 4,028,809, Dec. 27, 1973, (to E. I. du Pont de Nemours & Co., Inc.).
46. Ger. Pat. 2,154,676, Nov. 5, 1970, (to Asahi); Brit. Pat. 1,348,838, Nov. 5, 1970, (to Asahi).
47. Iwata, H.; Otani, T.; and Kobayashi, T. Jpn. Pat. 104,689, Sept. 12, 1978, (to Mitsubishi Rayon Co., Ltd.).
48. Kohashi, N.; Shioda, H.; Umetasni, H. Jpn. Pat. 77 107,045, Sept. 8, 1977, (to Japan Exlan Co., Ltd.).
49. Kobashi, T; Masuhara K. Ger. Pat. 2,550,871, Mar. 20, 1976, (to Japan Exlan Co., Ltd.).
50. Saka, H.; Hamada, S.; Kitagawa, H.; Mukoyama, E. Jpn. Pat. 74 14,546, Apr. 8, 1974, (to Toray Industries, Inc.)
51. Saka, H.; Hamada, S.; Inove, H.; Hosaka, S. Jpn. Pat. 73 90,380, Nov. 26, 1973, (to Toray Industries, Inc.)
52. SNIA VISCOSA S.P.A., Br. Pat. 1,177,801, Feb. 7, 1966.
53. Toray Inds., Jap. Pats. 36,430 and 36,431, May 11, 1973.
54. Toyo Rayon Co., Jap. Pat. 11,248, 1961.
55. Taniyama, M; Yoshida, M. Jap. Pat. 3,892, Apr. 24, 1961, (to Toyo Rayon).
56. Courtaulds, Ltd., Br. Pat. 830,750, 1960.
57. Toyo Spinning Co., Jap. Pat. 9,597, 1959.
58. Asahi, Jap. Pat. Appl. 41,233, 1970.
59. Miller, M. L. U.S. Pat. 2,963, 457, June 12, 1960, (to American Cyanamid Co.).
60. Ulbricht, J. Z. Phys. Chem. 221, 346 (1962).
61. Grandine, L.D. Jr. U.S. Pat. 3,075,934, Jan. 29, 1963, (to E. I. du Pont de Nemours & Co., Inc.).
62. Joh, Y. and co-workers, J. Polym. Sci. Part B 3, 933 (1965); Joh, Y. and co-workers, J. Polym. Sci. Part B 4, 673 (1966); Joh, Y. and co-workers, J. Polym. Sci. Part A-1 5, 2503 (1967).
63. Hinrichsen, G.; Orth, H. Kolloid Z. 247, 844 (1971).
64. Fester, W. "Physical Constants of Polyacrylonitrile," in Brandrup, J. and Immergut, E. H. eds., Polymer Handbook, 2nd ed., Wiley, New York, 1975, p V38.
65. Chase, J. D.; Potter, R. L. U.S. Pat. 3,847,864, Nov. 12, 1974, (to American Cyanamid).
66. Yates, P. C. U.S. Pat. 3,657,179, Apr. 18, 1972, (to E. I. du Pont de Nemours & Co., Inc.).

67. Pettyjohn, B. M. U.S. Pat. 3,383,350, May 14, 1968, (to E. I. du Pont de Nemours & Co., Inc.).
68. Kirby, J. R. U.S. Pat. 3,784,511, Jan. 8, 1974, (to Monsanto).
69. Masson, J. C. U.S. Pat. 3,784,512, Jan. 8, 1974, (to Monsanto).
70. Neth. Pat. Appl. 65-1113, Feb. 28, 1966, (to Farbenfabriken Bayer).
71. Logemann, H.; Roos, E.; Suling, C. U.S. Pat. 3,436,364, Apr. 1, 1969, (to Farbenfabriken Bayer).
72. Hennes, J. H.; Pfeifer, C. R. U.S. Pat. 3,296,171, Jan. 3, 1967, (to The Dow Chemical Co.).
73. Craig, J. P.; Knudsen, J. P.; Holland, V. F. Text. Res. J. 32, 435 (1962).
74. Knudsen, J. P. Text. Res. J. 33, 13 (1963).
75. Ziabicki, A. Sen'i Gakkaishi 26, 158 (1970).
76. Paul. D. R. J. Appl. Polym. Sci. 11, 439 (1967).
77. Terada, K. Sen'i Gakkaishi 29, T-451 (1973).
78. Crozier, R. D.; Harder, R. E.; Hood, J. H.; Hurley, R. B. U.S. Pat. 3,346,685, Oct. 10, 1967, (to The Dow Chemical Co.).
79. Fujisaki, Y.; Nakayama, C.; Kobayashi, H. U.S. Pat. 3,147,322, Sept. 1, 1964, (to Asahi Kasei Kogyo).
80. Poynton, D.J. U.S. Pat. 3,944,384, Mar. 16, 1976, (to Courtaulds Limited).
81. Han, C. D.; Park, J. Y. J. Appl. Polym. Sci. 17, 187 (1973).
82. Watkins, W. W. U.S. Pat. 2,426,719, Sept. 2, 1947, (to E. I. du Pont de Nemours & Co., Inc.).
83. Knudsen, J. P. U.S. Pat. 3,124,629, Mar. 10, 1964, (to Monsanto).
84. Knudsen, J. P. U.S. Pat. 3,088,188, May 7, 1963, (to Monsanto).
85. Costa R. M.; Codignola, F. U.S. Pat. 3,449,485, June 10, 1969, (to Societa Italiana Resine).
86. Zwick, M. Appl. Polym. Sym. 6, 109 (1967).
87. Lieseberg, F. U.S. Pat. 2,957,748, Oct. 25, 1960, (to Badische-Aniline).
88. Ucci, P. A. U.S. Pat. 3,080,210, Mar. 5, 1963, (to Monsanto).
89. Vigneault, R. E. U.S. Pat. 3,523,150 Aug.4, 1970, (to Monsanto).
90. Kuratani, K.; Fukushima, K. U.S. Pat. 3,701,820, Oct. 31, 1972, (to Japan Exlan).
91. Euler, R. D. U.S. Pat. 2,975,022, Mar. 14, 1961, (to du Pont E. I de Nemours & Co., Inc.).
92. McPeters, A. L.; Paul, D. R. Appl. Polym. Symp. 25, 159 (1974).
93. Bell, J. P.; Dumbleton, J. H. Text. Res. J. 41, 196 (1971).
94. Dworjanyn, L. O. U.S. Pat. 3,932,571, Jan. 13, 1976, (to E. I. du Pont de Nemours & Co., Inc.).
95. Wilkinson, W. K. U.S. Pat. 3,944,386, Mar. 16, 1976, (to E. I. du Pont de Nemours & Co., Inc.).
96. Reinehi, U.; Nogaj, A.; Holzing, G. U.S. Published Pat. Appl. B 458,060, Apr. 4, 1974, (to Bayer).
97. Nakagawa, K.; Kuratani, K. U.S. Pat. 3,483,576, Dec. 16, 1969, (to American Cyanamid).
98. Wirth, A. R.; Murdock, S. A.; Berry, G. B. U.S. Pat. 3,111,357, Feb. 19, 1963, (to The Dow Chemical Co.).
99. Bittle, D. F; McPeters, A. L. U.S. Pat. 3,907,498, Sept. 23, 1975, (to Monsanto).
100. Zgibneva, Z. A. and co-workers, Fibre Chem. 5(6), 659 (1973).
101. Harris, S. O.; Miller, G. K. U.S. Pat. 3,544,262, Dec. 1, 1970, (to American Cyanamid).
102. Briar, H. P.; Hurley, R. P. U.S. Pat. 3,296,341, Jan. 3, 1967, (to The Dow Chemical Co.).
103. Ucci, P. A. U.S. Pat. 3,281,260, Oct. 25, 1966, (to Monsanto).
104. Robertson, T. H.; Klausner, G. K. U.S. Pat. 2,984,912, May 23, 1961, (to American Cyanamid).
105. Harder, R. E.; Ticknor, L. B. U.S. Pat. 3,895,908, July 22, 1975, (to Dow Badishe).
106. Ham, G. E. Text. Res. J. 24, 604 (1954).
107. Peters, R. H. Textile Chemistry, Vol. III, Elsevier Scientific Publishing Co. Inc., New York, 1975, pp 549-579.
108. Hobson, P. H.; McPeters, A. L.; Kirk-Othomer: Encyclopedia of Chemical Technology, 3rd ed., Vol. 1, Wiley: New York, 1978, p 362.
109. Hicks, E. M. Jr.; Ryan, J. F. Jr.; Taylor, R. B. Jr.; Tichenor, R. L. Text. Res. J. 1960, 30, 675.
110. Fitzgerald, W. E.; Knudsen, J. P. Papers of the 51st Annual Conference of the Textile Institute 1966, the Textile Institute: Manchester Eng., 1966 p 134.
111. Breen, A. L. U.S. Pat. 3,038,236, June 12, 1962, (to E. I. du Pont de Nemours & Co., Inc.).
112. Hoffman, H. A. Jr. U.S. Pat. 3,547,763, Dec. 15, 1970, (to E. I. du Pont de Nemours & Co., Inc.).
113. Sekiguchi, H.; Tsutsui, N.; Sumi, T. U.S. Pat. 3,864,447, Feb. 4, 1975, (to Japan Exlan).
114. Fitzgerald, W. E.; Knudsen, J. P. Text. Res. J. 37, 447 (1967).
115. Jpn. Pat. 134,214, Oct. 20, 1981, (to Monsanto Co.).
116. Textile Daily Apr. 8, 1983.
117. Jpn. Pats. 137,219 and 137,218, Oct. 25, 1980, (to Asahi Chem. Co., Ltd.).
118. Fuzek, J. F. A C S Symp. Ser. 127, 515 (1980).
119. Levites, L. M. and co-workers, Khim. Volokna 15(3), 21 (1973).
120. Jap. Pat. Applic. 32,432, 1971, (to Mitsubishi Rayon Co., Ltd.).
121. Zharkova, M. A.; Kudryartsev, G. I.; Khudoshev, I. F.; Romanova, T. A. Khim. Volokna 1969, 11(2), 49; Fiber Chem. USSR 1969, 1969, 191.

122. Br. Pat. 1,275,138, Dec. 19, 1969, (to Rolls-Royce, Ltd.).
123. French Pat. 1,549,347, Dec. 21, 1966, (to E. I. du Pont de Nemours & Co., Inc.).
124. U. S. Pat. 3,778,416, Jan. 7, 1971, (to Hoechst AG.).
125. U. S. Pat. 3,626,049, Sept. 2, 1967, (to Toyo Spinning Co., Ltd.).
126. Br. Pat. 1,355,701, May 2, 1970, (to Hoechst AG.).
127. U. S. Pat. 3,809,685, Sept. 26, 1970, (to Asahi Chem. Ind. Co., Ltd.).
128. Swed. Pat. 166,913, Apr. 21, 1969, (to Stockholms Superfostat Fabriks).
129. Br. Pat. 1,227,410, Apr. 7, 1971, (to Toyo Spinning Co., Ltd.).
130. Yutaka, Y.; Iwasa, T. Jap. Pat. 59, May 14, 1979, 426, (to Asahi Chem.Co., Ltd.).
131. Takeya, K.; Suzuki, H.; Yamawaki, N. U.S. Pat. 4,012,459, March 15, 1977, (to American Cyanamid).
132. Neukam, T.; Reinehr, U.; Bentz, F.; Nischk, G. U.S. Pat. 4,059,556, Nov. 22, 1977, (to Bayer AG.).
133. Neukam, T.; Reinehr, U.; Bentz, F.; Nischk, G. Ger. Offen. 2,622,921, Dec. 1, 1977, (to Bayer AG.).
134. Wolf, G. D. and co-workers, U.S. Pat. 4,100,143, July 11, 1978, (to Bayer AG.).
135. Eur. Pat. Appl. 31,078, July 1, 1981, (to Bayer AG.).
136. Bell, J. P.; Dumbleton, J. H. Tex. Res. J. 41(3), 196 (1971).
137. Br. Pat. 1,191,538, May 13, 1970, (to Toyo Rayon Co., Ltd.).
138. Br. Pat. 1,271,334, Apr. 19, 1972, (to Toray Ind., Inc.).
139. Maranci, A. U.S. Pat. 3,760,053, Sept. 18, 1973, (to American Cyanamid Co.).
140. Br. Pat. 1,230,392, Apr. 28, 1971, (to Japan Exlan Co., Ltd.).
141. U.S. Pat. 3,844,825, Nov. 30, 1972, (to American Cyanamid Co.).
142. U.S. Pat. 3,655,420, May 6, 1970, (to E. I. du Pont de Nemours & Co., Inc.).
143. Br. Pat. 1,380,422, Apr. 19, 1971, (to Ciba-Geigy Corp.).
144. Br. Pat. 1,230,174, Apr. 24, 1967, (to Toray Ind., Inc.).
145. Dutch Pat. Appl. 7,010,775, July 22, 1969, (to Asahi Chem. Co., Ltd.).
146. Jap. Pat. 33,289, Oct. 19, 1979, (to Asahi Chem. Ind. Co., Ltd.).
147. Jap. Pat. Appl. 1,651, 1970, (to Toray Ind., Inc.).
148. Jap. Pat. 84,426, June 25, 1980, (to Toray Ind., Inc.).
149. U. S. Pat. 3,701, 819, Aug. 16, 1968, (to Toho Beslon Co., Ltd.).
150. Jap. Pat. 33,290, Oct. 10, 1979, (to Asahi Chem. Ind., Co., Ltd.).
151. Radlmann, V.; Reinehr, U.; Nischk, G. U. S. Pat. 4,123,200, March 6, 1978, (to Bayer AG.).
152. Jap. Pat. 158,321, Dec. 9, 1980, (to Mitsubishi Rayon Co., Ltd.).
153. Austrian Pat. 363,168, July 10, 1981, (to Chemiefaser Lenzing AG.).
154. Reinehr, U.; Herbertz, T.; Jungverdorben, H. J. Ger. Offen. 2,706,522, Aug. 17, 1978, and Ger. Offen. 2,706,032, Aug. 17, 1978, (to Bayer AG.).
155. U. S. Pat. 3,839,520, Apr. 10, 1972, (to American Cyanamid).
156. U. S. Pat. 3,781,391, Jan. 24, 1968, (to American Cyanamid).
157. Korner, W.; Blankenstein, G.; Dorsch, P.; Reinhrand, U. Chemiefasern/Textilindustrie 29/81, June 1979, 452.
158. Ryan, J. F. Jr. U.S. Pat. 3,360,603, Feb. 2, 1971, (to E. I. du Pont de Nemours & Co., Inc.).
159. Br. Pat. 1,263,082, Feb. 7, 1968, (to Courtaulds, Ltd.).
160. Br. Pat. 1,309,051, May 27, 1970, and Jap. Pat. 71,414, May 4, 1982, (to Mitsubishi Rayon Co., Ltd.).
161. Yoshitake, M.; Autami, H.; Shimamura, M. Jap. Pat. 12,046, May 1, 1970, (to Toray Ind., Inc.).
162. Saito, K.; Kumura, S.; Sato, T. Jap. Pat. 13,958, May 14, 1971, (to Toray Ind., Inc.).
163. Honda, Y.; Yamamoto, K.; Iwasa, T. Jap. Pat. 126,321, Nov. 4, 1978, (to Asahi Chem. Ind. Co., Ltd.).
164. Jap. Pat. 25,410, Feb. 10, 1982, (to Asahi Chem. Ind. Co., Ltd.).
165. Sassenrath, B. Chemiefasern/Text. Anwendungstech. Text. Ind. 1981, 31, 808.
166. Palethorpe, G.; Poole, L. Ger. Offen. 2,825, Dec. 21, 1978, 754, (to Monsanto Co.).
167. Br. Pat. 1,329,357, Sept. 25, 1970, (to Bayer AG.).
168. Br. Pat. 1,348,044, Mar. 26, 1970, (to Bayer AG.).
169. Jonkoff, I. M. U.S. Pat. 3,618,307, Nov. 9, 1971, (to E. I. du Pont de Nemours & Co., Inc.).
170. Weinrotter, K.; Schmidt, H.; Kraessig, H.; Haslinger, G.; Verwaenger, A. Austrian Pat. 364,069, Sept. 25, 1981, (to Chemiefaser Lenzing AG.).
171. Sugiura, A.; Kohashi, N.; Ozaki, M. Jap. Pat. 99,318, Aug. 20, 1977, (to Japan Exlan Co., Ltd.).
172. Br. Pat. 1,341,868, Dec. 28, 1973, (to Mitsubishi Rayon Co., Ltd.).
173. Joh, Y.; Sugimori, T. U.S. Pat. 3,813,372, May 28, 1974, (to Mitsubishi Rayon Co., Ltd.).
174. Kirby, J. R. U.S. Pat. 3,784,511, Jan. 1, 1974, (to Monsanto Co.).
175. Br. Pat. 1,360,669, Dec. 5, 1970, (to Bayer AG.).
176. Br. Pat. 1,349,669, Aug. 1, 1972, (to Mitsubishi Rayon Co., Ltd.).
177. Br. Pat. 1,355,701, May 2, 1970, (to Hoechst AG.).
178. U.S. Pat. 3,682,004, Jan. 20, 1969, (to Courtaulds, Ltd.).
179. Bruner, A. H.; Truscott, T. B. U.S. Pat. 3,558,765, Jan. 26, 1971, (to Monsanto Co.).

180. Palethorpe, G. U.S. Pat. 3,642,628, Dec. 30, 1968, (to Monsanto Co.).
181. Br. Pat. 1,202,793, Jan. 24, 1967, (to Toray Ind., Inc.).
182. Takashi, T.; Ito, T.; Nagai, S.; Hiraoka, S.; Mizuno, H.; Kakumoto, Y. Jap. Pat. 34,500, Mar. 13, 1979, (to Mitsubishi Rayon Co., Ltd.).
183. Ichikawa, Y.; Schichijo, Y.; Sugaya, Y.; Veda, T. U.S. Pat. 3,770,494, Nov. 6, 1973, (to Asahi Chem. Ind. Co., Ltd.).
184. Br. Pat. 1,301,029, Feb. 28, 1969, (to Mitsubishi Rayon Co., Ltd.).
185. Takeda, H. U.S. Pat. 3,758,659, Sept. 11, 1973, (to Toray Ind., Inc.).
186. Lugo, L.; Remi, C. U.S. Pat. 3,608, 1967, 043, (to Societa Italiana Resine S.P.A.).
187. Br. Pat. 1,374,802, and Br. Pat. 1,373,552, Aug. 20, 1970, (to American Cyanamid).
188. Br. Pat. 1,237,628, June 12, 1968, (to American Cyanamid).
189. Jacobasch, H. J. and co-workers, Ger. Pat. (East), Oct. 31, 1979.
190. Murayama, Y.; Nakamura, M.; Yokokoshi, K. Jap. Pat. 6,926,379, Nov. 29, 1969, Chem. Abstr., 72, 112726j (1970).
191. Text. Horizon, 10(8), 38 (1990).
192. Industrial Fabric Products Review, 59(10A) (1983).
193. Shindo, A. Rpt. Osaka Inds. Res. Inst. No 317, 1961.
194. Bacon, R.; Smith, W. H. Proc. 2nd Conf. on Ind. Carbon and Graphite, Soc. Chem. Ind. London, 1965, p 203.
195. Walt, W.; Phillips, L. N.; Johnson, W. Br. Pat. 1,110,791, Apr. 24, 1964, (to National Research Development Corp.).
196. Walt, W.; Phillips, L. N.; Johnson, W. Br. Pat. 1,148,874, June 15, 1966, (to National Research Development Corp.).
197. Melliand Textilbericht, 12(2), 115 (1983).
198. High Performance Textiles, 3,(10), 3 (1983).
199. Br. Pat. 2,018,188A, Nov. 24, 1978, (to American Cyanamid).
200. Takeda, H. U. S. Pat. 3,814,739, June 4, 1974, (to Toray Ind., Inc.).
201. Asahi Chem. Ind. Co., Lts., Jap. Pat., 60,051, May 6, 1980.
202. Asahi Chem. Ind. Co., Ltd., Jap. Pat. 98,025, Feb. 13, 1976.
203. Asahi Chem. Ind. Co., Ltd., Jap. Pat. 17,966, Feb. 20, 1981.
204. Wuestfeld, A.; Wuestfeld, C. Ger. Pat. DE 3,012,998, Oct. 15, 1981.
205. REDCO, S. A. Belg. Pat. BE 889,260, Oct. 16, 1981.
206. Hoechst AG, Eur. Pat. Appl. EP 44,534, Jan. 27, 1982.
207. High Performance Textiles, 2(10), 1 (1982).
208. Yasushi, J.; Akihiko, N.; Kaneko, N.; Fukutome, A. Jap. Pat. 6,627, Jan. 21, 1978, (to Nippon ZeonCo., Ltd.).
209. Yasushi, J.; Akihiko, N.; Kaneko, N.; Fukutome, A. Jap. Pat. Dec. 19, 1979, 160, 819, (to Nippon Zeon Co., Ltd.).
210. Yasushi, J.; Akihiko, N.; Kaneko, N.; Fukutome, A. Jap. Pat. Mar.4, 1980, 30, 427, (to Nippon Zeon Co., Ltd.).
211. Uchida, Y.; Iwasa, T. Jap. Pat. 78,323, July 22, 1978, (to Asahi Chem. Inc. Co.)
212. Klein, E.; Rembaum, A.; Yen, S. P. S. U.S. Pat. 4,045,352, Aug. 30, 1977, (to California Inst. of Tech.).
213. Alikberova, L. Yi SU 000343, July 7, 1982.
214. Dorokhina, I. S.; Zharkova, M. A. Khim. Volokna 1, 1971, 69.
215. Mitsubishi Rayon Co., Jap. Pat. 144,299, Nov. 10, 1981.
216. Jap. Exlan Co., Ltd., Jap. Pat. 154,514, Nov. 30, 1982, and 118, 938, Sept. 18, 1981.
217. Kanebo Synthetic Fibers, Ltd., Jap. Pat. 39,214, Mar. 4, 1982.
218. Japan Exlan Co., Ltd. Jap. Pat. 112,909, Sept. 5, 1981.
219. Nippon Sanmo Dyeing Co., U. S. Pat. 4,336,028.
220. Tanaka, H.; Tsunawaki, K. Eur. Pat. Appl. 14,944, Sept. 3, 1980; Jap. Pat. Appl. 15,035, Feb. 14, 1979, (to Teijin, Ltd.).
221. Japan Exlan Co., Ltd., Jap. Pat. 4,713, Jan. 19, 1981.
222. Stuart, M. Textiles, 10(2), 41 (1981).

EARLY ADVANCES IN THE USE OF ACRYLIC RESINS FOR THE CONSERVATION OF ANTIQUITIES

MAUREEN A ROBSON
Consultant Conservator, Studio, 29 Park Avenue, Birmingham, B18 5ND, England.

During the post war years acrylic resins were widely used in the conservation of antiquities. The two main categories of these synthetic resins are thermosetting and thermoplastic of which the latter may offer long term reversibility, an essential prerequisite to all conservation treatments. Consolidation processes provide additional strength to fragile porous archaeological materials such as bone and terracotta, where a change in colour is undesirable. Surface applications will reattach crazed lacquer, glaze and decorative pigment, to japanned lacquerware, ceramics and polychrome wooden sculpture or consolidate friable surfaces after desalination of ceramic vessels and stone statuary. Some of the thermoplastic resins, in emulsion form with a continuous phase, will provide a suitable water soluble paint system which on drying becomes an acrylic film reversible in an organic solvent. During the mid seventies came the development of a one part acrylic system which fast cures under ultraviolet light and fulfills all the criteria to effect full restoration and conservation of antiquities.

INTRODUCTION

During the post war years, with the advent of cheap supplies of oil, the acrylic resins, poly(butyl methacrylate) (PBMA) and poly(methyl methacrylate) (PMMA) resins were widely used in the restoration and conservation of antiquities. The two main categories of these synthetic resins are thermosetting and thermoplastic of which the latter category has been proved capable of long term reversibility which is now considered an essential prerequisite to all conservation treatments. The tendency for thermosetting resins to cross link renders them other than user-friendly when restoration is required! If the tendency is proved , these resins would not be selected today for use in conservation. However, during the post first war years, both categories of acrylic resins were put to a variety of uses for the conservation of a wide range of antiquities.

CONSERVATION

ACRYLIC POLYMERS: SOLID OR IN SOLUTION

Among the early uses of the PBMA resins, such as Lucite 44, was as a varnish for oil paintings. However, in the 1930's and 1940's these conservation practices were not generally documented; so it was over twenty years before any such studies were published. (Werner 1952) The fact that PBMA showed a tendency to cross link when exposed to light negated its reversibility in the long term, although short term reversibility was achieved by its solubility in hydrocarbon solvents. It also had advantages of flexibility, colour and clarity retention both in the short and long term and was widely used as a picture varnish up until the 1950's.

By this time, long term reversibility was now regarded as an important criterion which led to the search for a more suitable resin. During the 1950's saw the emergence of Paraloid B72 used initially as a lacquer for silverware and copper alloy (Biek 1952) and subsequently put to a variety of uses in the field of conservation including the consolidation of fossils (Rixon 1955) and earth (Shorer 1964). Surface applications with various solvents and matting agents will attach crazed lacquer, glaze and decorative pigment to japanned lacquerware, ceramics and polychrome wooden sculpture. Consolidation processes provide additional strength to fragile porous archaeological materials such as bone and terracotta, where a change in colour is undesirable. During the late fifties to early sixties a new approach was introduced with the idea of complete immersion in a synthetic resin. For the first time, account had to be taken of the possible effect of the conservation process, on the external appearance of the porous material such a bronze age pottery urns or excavated bone. Any testing or sample removal must be carried out prior to treatment or authenticity would be lost. The complete pottery urn, its sherds/or bone were immersed in a suitable solution under vacuum of Bedacryl 122X [poly(butyl methacrylate)] and later Paraloid B72 (acrylic polymer) in xylene. (Robson 1988). Bedacryl 122X (PBMA) has had a chequered career. During 1955-56 it was thought well of by the British Museum's conservation scientists but fell from favour twenty years later when its long-term reversibility was in doubt. Currently, Bedacryl 122X (PBMA) is again in general use and can be purchased in two forms: either as a solid which can be dissolved to form the resin, or in a concentrated solution. Both forms may be diluted with toluene to the required concentration and when applied dries to a matt finish. These resins were effective as means of consolidation since they were used to strengthen whole vessels in both excavation and the laboratory workshop before polyvinyl alcohol emulsions were available. They were also used on individual fragments which could later be re-assembled and bonded with a suitable adhesive. However, the resin treatment did result in a certain amount of discolouration of the pot fabric. (Shorer, personal communication) A vacuum desiccator is used for this purpose and a glass vessel of small dimensions is used as the container, in order to economise in the use of the solvent. This vessel is placed inside the desiccator which is then evacuated. The object is allowed to remain under vacuum until all bubbling has ceased so that the consolidant is absorbed by the porous pot fabric. Air is then admitted and when the desiccator reaches atmospheric pressure, the urn is lifted out of the container and allowed to drain: it can then be dried without acquiring a gloss by leaving it in an atmosphere of the solvent (Plenderleith 1962) Durofix and HMG (nitro cellulose adhesives) form satisfactory and compatible adhesives for repair work. The two adhesives are easily reversible in the short term by application of acetone. Paraloid B72 dissolves more slowly in acetone than Durofix or HMG, so it is possible to dismantle joints while leaving the consolidant intact. More recently Paraloid B72 (PBMA) has also been packaged in a tube format for ease as an adhesive for both textiles (Beecher 1968) and wood (Angst 1979).

Clear paper varnish (Winsor and Newton) was another favoured acrylic polymer used in the early conservation pioneer practices within the British Museum during the 1940's. Although it has not been on the market for many years, it mixed well with levigated fine powder (Winsor and Newton) and furnished a clear colourless smooth glaze to porcelain either by brush or spray application. It required two day to harden. A camel hair brush was used to spread out the varnish across the bond area so that by the time it reached the join the varnish layer was perceptively thinner, thus eliminating the risk of a shadow being formed from any hard edge from applying the varnish by brush. (Shorer, Personal Communication)

ACRYLIC MONOMERS

The use of polymethyl methacrylate resins for the encapsulating of natural history (and other specimens) was practiced from the 1940's until the early 1960's. (Organ 1963)

During the 1950's Technovit 4004a (Germany) and its British counterpart Tensol (ICI, UK) were both widely used for gap filling perspex and thus suitable as an application for restoring glass. However, both contracted considerably and their shrinkage in the short term, after three months and after three weeks respectively, seriously limited their long term suitability in the field of restoration. (Shorer, personal communication)

Cyanacrylates too, have been suggested for glass but as they break down in the alkaline conditions on most glass surfaces, they have met with only limited success. For use in combination repairs with epoxy resins, they locate and temporary hold the glass join when applied at intervals along the break edge, before the epoxy resin cures. (Terwen, 1983)

DISPERSIONS

The use of acrylic dispersions were favoured during the 1950's; the retention of their optical properties in the long term bettered the polyvinyl acetate dispersion applied films although there was some variation from batch to batch. These dispersions have been widely used for heat-set adhesive, liquid adhesives and consolidants for paper. Texacryl 13002 and Revacryl 1A (Acrylic copolymer emulsions) have more recently been used to attach fragile woven fibres to a support fabric of suitable weight where the adhesive is applied to the support fabric in the wet state and allowed to dry. The adhesive is then reactivated by application of gently heat with a spatula iron. (Robson, 1986)

Some of the thermoplastic acrylic resins in emulsion form with a continuous phase will provide a suitable water soluble paint system which on drying become an acrylic film reversible in an organic solvent eg. Rowney Cryla Flow acrylic paints. These paint systems are light stable and widely used in the retouching of restored lacunae in a wide variety of antiquities ie ceramic, glass, stone, wallplaster etc.

During the mid seventies came the development of the one-part acrylic system which fast cures under ultraviolet light and fulfills all the criteria to effect full restoration and conservation of antiquities.

1. These products are one part systems that contain no solvents.

2. They are clear colourless liquid photopolymers which are suitable for application by syringing and will fast cure when exposed to ultraviolet light.

3. They may be applied direct to the prepared lacunae in association with the fractured enamel.

4. A short precure with ultraviolet light of sufficient duration to set the bond allows it to be readjusted, if necessary, without disturbing the alignment.

A comparative study of a range of glass adhesives was undertaken from 1983 - 1986 to establish a range of adhesives with the following criteria: retention of optical properties, non yellowing on ageing, reversibility both in the short and long term and stability under exposure to ultraviolet light and moisture. Of the ultraviolet curing (photosensitive) acrylics tested, Norland optical adhesives (urethane related prepolymers) were found to satisfy such criteria. (Unpublished research, Robson, 1986) These UV curing acrylic adhesives with a thin consistency, can be syringed directly into cracks in situ, can be realigned after a short precure of ultra violet light of sufficient duration, to set the bond, and retain absolute colour and clarity in both the short and long term. The adhesion of NOA61 and NOA63 is good initially but will become greater with ageing over a period of about two weeks. Acetone is used to remove the uncured resin, whilst after curing, any rejects can be separated in methylene chloride. The bonded area must be soaked in the solvent and normally will separate over night. The time required to break the bond depends upon the extent of cure and area of the bond.

At present a suitable dispersant is sought to enable these acrylic adhesives to pass easily though a compressed air spray attachment. Should recent trials (unpublished research, Robson, 1990) prove successful in the long term this will allow the wider application of spray techniques to be used to paint out the ground colour over much larger areas and fast cure under ultraviolet light providing the evaporation of the solvent is ensured before curing.

It is important to match the refractive index of the adhesive to that of the glass to be bonded to effect an invisible repair (Tennent, 1984) The Norland optical adhesives provide a range of refractive indices which can be successfully used to bond glass of different strengths and refractive index ranging from the structurally sound to the severely weakened (eg architectural glass). The RI of NOA61 and NOA63 under discussion is 1.56. In addition that of NOA68 is RH 1.54 and the lowest refractive index in their range of optical adhesives is 1.524 (NOA65). In this instance the cured resin is very flexible and is designed to minimise strain. Provided the glass is mechanically sound and a sample is tested for any sign of crizzling or fading of the original hand painting on exposure to ultraviolet light, this adhesive can be suggested for the lightweight bonding of extremely fragile thin sectioned glass. The latter adhesives are currently under investigation and the results will be published when a full assessment of their suitability has been completed. (Unpublished research, Robson, 1990).

The shelf life of the liquid is at least four months if kept at room temperature in the original container. This can be extended if refrigerated.

The toxological properties of NOA61 and NOA63 have not been thoroughly investigated. At present the manufacturers are not aware of any severe hazards in using this material. As with any organic chemical, care should be taken in handling it. Prolonged skin contact should be avoided and affected areas should be washed well with soap and water. If accidental eye exposure occurs, flush with copious amounts of water and seek medical attention. Use in a well ventilated area.

CONCLUSIONS

The acrylic resins, with their wide ranging and diverse applications have had in the past, a tremendous influence on solving many problems in the conservation of antiquities. For as long as they fulfill all criteria for present day conservation, they will remain in common use. The more recent development of a one part ultraviolet curing acrylic system is but one example of progress in this field of research. Lateral communication between conservators, research chemists, technologists and art historians will ensure that conservation problems and their possible solutions are successfully matched for long term consideration.

Scientific issues remain central to the conservation profession after six decades since its birth. Today as resource allocation, politics, and public awareness help shape the agenda for conservation, a review of contemporary conditions seems needed to help secure the future of the world's cultural heritage .... if progress in conservation's scientific realm is to have impact, it must be matched by increased public enthusiasm and political support. (Conservation, G.C.I. 1991)

ACKNOWLEDGEMENT

The author wishes to thank the late Distinguished Professor Raymond B. Seymour for inviting her to participate in the 201st National American Chemical Society Meeting in Atlanta, Georgia, USA; and the following for their discussion, encouragement and support: David Bailey, Leo Biek, Pat Brown, Debbie Boulton, Andrew Carson, Eileen Collins and Christopher Radbourne, Anne Hudson, Hilde Smith and Pauline Timmins; and to Wendy Firmin, Editorial critic, Bunny Warren, and Kim Sumner Secretariat Kevin and Robert Brown, Directors, Advance Office Services, UK, for practical assistance and professional advice.

DEDICATION

The author wishes to dedicate this paper to the late Lone Gedye (tutor and mentor) and an early pioneer in the establishment of the internationally recognised course in Conservation of Archaeological Materials at the Institute of Archaeology, University of London, UK, without whom early advances in conservation and material history would not have been documented for the teaching of students worldwide.

REFERENCES

Angst, W. "Problems with Objects of Historic Significance," American Institute for Conservation Preprints, 1979.

Biek, L. "Protective Coatings for Silver," Museums Journal, 1952, 60,61.

Beecher, E.R. "The Conservation of Textiles," in UNESCO 1968.

Horie, C.V. Materials for Conservation, Butterworths, 1987.

Organ, R.M. "The Consolidation of Fragile Metallic Objects," in Thomson (Ed.) 1963.

Plenderleith, H.J. The Conservation of Antiquities and Works of Art, Oxford University Press, 1956.

Rixon, A.E. "The Use of New Materials as Temporary Support in the Developments and Examination of Fossils," Museums Journal, 55, 1955, 54-58.

Robson, M.A. "The Use of Aqueous Polymer Emulsions in the Conservation of Ethnographic Material: A New Method for Supporting Fragile Woven Fibre Decoration," in Symposium 1986 The Care and Preservation of Ethnological Materials (Eds. R. Barclay, M. Gilberg, J.C. McCawley and T. Stone), Canadian Conservation Institute 1986.

Robson, M.A. "The Use of Norland Optical Adhesives in the Repair and Restoration of Fractured Glass Beads on Ethographic Materials," ICOM Ethnographic Newsletter, 1987.

Robson, M.A. "Methods of Restoration and Conservation of Bronze Age Pottery Urns in the British Museum," Early Advances in Conservation, British Museum Occasional Paper No. 65, 1988.

Robson, M.A "Interior Displays of Restored Stained and Vessel Glass: A Call for Suitable Cold Curing Paint Systems for 'Repaints' in situ," The Stained Glass Magazine, 1988.

Robson, M.A. "The Use of Norland Optical Adhesives NOA61 and N0A63 in the Repair and Restoration of Enamelware and Ormolu," CGCG (UKIC) Occasional Paper 1989.

Robson, M.A. "Historical Applications of the Use of Synthetic Resins for Passive and Interventive Conservation of Fragile Woven Vegetable Fibres and Collagenous Skins," Atlanta, USA 1990, publication forthcoming.

Robson, M.A. "One Part Acrylic U.V. Curing Systems with Fillers for Restoration of Porcelain and Enamelware," CGCG (UKIC) West Dean College, 1991.

Robson, M.A. "The Long Term Effect of Surface Treatments on the Properties of Cellulosic and Ligneous Museum Artifacts," New Orleans, USA, 1990.

Shorer, P.H.T. "Soil Section Transfers : A Method for the Transfer of an Archaeological Soil Section on to a Flexible Rubber Backing," Studies in Conservation, 9, 1964 74-77.

Tennent, N.H. and Townsend, J.H. "The Significance of the Refractive Index of Adhesives for Glass Repair," in Brommelle et al (Eds.) 1984.

Terwen, P.A. "The Mending of Stained Glass: A Technical Instruction," in Tate el al (Eds.) 1983.

Werner, A.E.A. "Plastics Aid in Conservation of Old Paintings," British Plastics. 25, 1952, 363-366.

ANTIMICROBIAL FIBERS AND POLYMERS

TYRONE L. VIGO
USDA, ARS, SRRC, 1100 Robert E. Lee Boulevard, P.O. Box 19687, New Orleans, LA 70179-0687

The historical development of antimicrobial fibers and polymers is discussed from several viewpoints and disciplines. The first topic is disinfection, sterilization, and the concept of antiseptic treatment of clothing and textiles. The protection of fibers and fabrics from microbial degradation due to fungi, algae and selected bacteria is the next topic discussed. The concept of antibacterial and germicidal finishing of textiles from the early 1940's to the present day is a third area of interest. The development of reliable test methods for evaluating disinfection, biodeterioration and antibacterial activity is also noted. The advent of synthetic fibers and how they compare to natural fibers with regard to biodeterioration and transmission of disease is the fourth topic. More current studies indicate the importance of bioactive polymers and copolymers applied to fibers to impart durable antimicrobial effects. Modern day techniques to produce antimicrobial materials have utilized the concepts of controlled release and microencapsulation. Future progress in all these areas is the final topic discussed.

INTRODUCTION

The use of antimicrobial agents for preserving fibers and preventing the transmission of disease by materials dates back to ancient times. Linen cloth containing salt as a preservative was used by Egyptians to carefully wrap mummies. Resins were used as binding agents. Ancient literature is replete with the use of silver, sulfur and other germicidal agents on clothing for health purposes. However, during the Dark and Middle Ages, all types of scientific advances fell into disuse. It was not until around the time of the American Civil War that scientific progress was again made in this area.

As with most scientific advances, modern antimicrobial materials resulted from overlapping and concurrent advances in allied and basic disciplines. The fields of bacteriology, surgery, applied microbiology, polymer science and surface science and advances in sterilization, disinfection and weathering of natural fibers contributed to the development of the status of antimicrobial fibers and polymers.

This historical review presented here lists and highlights in chronological order key concepts and references on the following topics: (a) disinfection, sterilization and development of test methods of clothing and textiles; (b) preservation of fibers from degradation and discoloration and development of test methods suitable for evaluating protection of fibers; (c) antibacterial and germicidal finishing of textiles; (d) discovery of synthetic fibers and their antimicrobial properties related to fiber preservation and disease transmission; (e) use of polymers to impart durable antimicrobial activity to fibrous substrates; and (f) development and refinement of the concepts of controlled release, microencapsulation and related methodologies that led to modern day materials having broad spectrum and specific antimicrobial properties.

DISINFECTION AND STERILIZATION OF CLOTHING AND TEXTILES

Historically, it was believed that the clothing of persons with transmissible diseases such as leprosy and bubonic plague should be burnt or destroyed to prevent further transmission of the diseases. This was remarkable since it was not until the middle of the 19th century that it was proposed that microorganisms were the causative agents of a variety of specific contagious diseases and disorders. Two important events happened in the 1860's. The first volume of Lancet, a prestigious British medical magazine, published that a man was convicted for possessing clothing previously worn by persons who died of yellow fever [1]. This occurrence set a precedent for public health ordinances to prevent the spread or transmission of disease by fomites, or inanimate objects that serve in such transmission. The second significant event, in 1867, occurred when the famous English surgeon Joseph Lister impregnated bandages with carbolic acid or phenol as an antiseptic in surgery [2].

However, it wasn't until around 1917 that intensive studies were conducted in Europe by the French and Prussian armies to introduce antiseptic or antibacterial garments [3,4]. There was a military need to have clothing that would be resistant to contamination from wounds. It was recognized that long term antiseptic action lasting at least a month was desirable, but it was also observed that garments needed to be retreated to make them continuously antimicrobial.

The introduction of standard test methods began in 1881 when Koch impregnated silk threads with spores of Bacillus anthracis, then subjected the threads to germicides for specified time periods. More elaborate and carefully controlled tests were developed throughout the first half of the 20th century, including the incorporation of germicides and antimicrobial agents into bandages and sutures to prevent or minimize infection in the body [5].

The first test specifically designed to evaluate disinfectants in a controlled manner was the Rideal-Walker test (1903). This test standardized several variables such as growth medium, pH, time, age of culture, specific species of test microorganism and a germicidal standard as a control disinfectant. Improvements in this test were made by Reddish and others in the mid- to late 1920's [5].

A concurrent development was the use of the "phenol coefficient" to determine how much more effective disinfectants were than phenol (set a standard value of unity). Only Salmonella typhosa initially was used as the test microorganism in the phenol coefficient test. However, regulatory agencies (such as EPA) later required that other microorganisms (bacterium such as Staphylococcus aureus and Pseudomonas aeruginosa and pathogenic fungi such as Trichophyton mentagrophytes) be tested against disinfectants, particularly if the disinfectants were used in hospitals and other confined environments. Although there are no specific tests for the disinfection of textiles, there are at least eight different variations of approved tests for disinfectants by the EPA [6]. Many of these tests could be adapted for evaluation of porous materials such as fabrics. The development of test methods for antiseptics paralleled that of test methods for disinfectants. However, there are no relevant standard test methods to evaluate the antiseptic activity of textiles to suppress contamination of wounds.

Residual bacteriostatic activity on fabrics may be achieved by use of sanitizers (usually as laundry additives). With the passage of FIFRA (Federal Insecticide, Fungicide and Rodenticide Act) in 1948, test methods appropriate for the determination of sanitizers developed rapidly. The method of Petrocci and Clarke [7] is recognized as a valid test for textile sanitizers by EPA. This test requires a 99.9% reduction of Staphylococcus aureus and Klebsiella pneumoniae within 10 minutes after the fabric is exposed to the sanitizer. An analogous method for evaluating carpet sanitizers has been developed and recommended by EPA [6].

TABLE I. Selected Historical and Current Definitions on Antimicrobial Action [8,9]

Term	Original Definition (Year or time)	Current Definition
antiseptic	Substance which opposes putrefaction and decay (1751)	Substance that inhibits growth of or kills microorganisms, especially for agents applied to living tissue
disinfectant	Any agent that would free anything from infection (17th century)	Products that kill the growing forms, but not necessarily the resistant spore forms of micro-organisms on inanimate objects by decontamination
sanitizer	Disinfectant without any residue harmful to subsequent users of the article or product that eliminates any esthetically objectionable contaminants (1945)	Agent commonly used with substances that are applied to inanimate objects, which reduces number of bacterial contaminants to safe levels, as judged by public health laws
sterilization	Act or process of freeing from all living microorganisms (1932)	Same as original definition

The terms "antiseptic" and "disinfectant" originated several hundred years ago (Table I) and their meaning has changed substantially over time. The first known use of the word antiseptic in English dates back to 1751 when it was defined as an agent to prevent putrefaction and decay. Its meaning was broadened when Lister used the term to describe the destruction or inhibition of microbes on living tissue by chemical agents.

The term "disinfectant" was used in the 17th century to describe the removal of infection by certain chemicals, even before it was recognized that microorganisms were the causative agents of many diseases. The current use of the term describes the removal of deleterious microorganisms from inanimate objects. Terms such as "sanitizer" and "sterilization" were introduced in the 1930's and 1940's, and their current meaning (see Table I) is essentially unchanged from their original definitions [8,9].

The concept of sterilization of materials and fabrics dates back to the turn of the of the 20th century as a variation of pasteurization of foods and other substances. The eminent French scientist Louis Pasteur used heat to prevent the spoilage of wine and beer by microorganisms. At the same time that heat was used to sterilize bandages, it was also observed that short wavelength radiation (from the recently discovered x-rays) was also capable of killing microorganisms. However, it wasn't until 1932 that an acceptable definition of sterilization was formulated.

Ultraviolet radiation [10] and gases (such as formaldehyde and ethylene oxide) were used for sterilization [11] in the 1920's, but it was not until 1949 that both types of sterilization were validated by modern scientific techniques. Currently, sterilization techniques for bandages and other fibrous materials use ionizing radiation or heat (dry or moist). However, if the fabrics (such as wool) are degraded by heat, cold sterilization chemical techniques are used.

PRESERVATION OF FIBERS AND TEXTILES

Salt as an antimicrobial agent plus a resin to improve durability was used by the ancient Egyptians to preserve mummies and deter the biodeterioration caused by fungi and bacteria. It also clearly established the need for moisture as being of primary importance for microbial growth. Similarly, cod fish had been salted by the early Vikings and Scandinavians as a similar method of preserving unrefrigerated fish for a year or longer without spoilage. Again, this illustrated the vital part that moisture plays in sustaining microbial life. It was not until 1884 that the first modern day patent was issued that described protection of cotton fabric from weathering outdoors by its mineral dyeing with lead chromate [12].
From 1900 to 1940, experiments were conducted to determine the causes (fungi, algae and certain bacteria) of degradation of cellulosic and proteinaceous fibers. The discovery and application of some effective treatments for textile preservation also occurred during this period.

In 1913, Osborn's paper [13] presented the first scientific documentation of common saprophytic molds as the causative agents for mildew on cotton fabric; the effect of moisture on promoting fungal growth in fabrics also was noted in this study. A series of papers by Thaysen and his colleagues in the mid-1920's [14-17] addressed and answered some fundamental questions about the bacterial decomposition of textile fibers; observations were made about the role of spores and enzymes in microbial degradation of cellulose and the selectivity of bacterial species in degrading cellulose. Burgess [18] assessed the status of mildew preservatives for wool (many of which were inorganic salts) and noted which preservatives were the most effective for this purpose. Veitch and Levine [19] devised a test for the mildew resistance of fabrics that were treated by the few mildewproofing processes then available.

New and more effective biocides to prevent mildew and deterioration developed concurrently with progress made in the microbiology of fiber degradation. Although several inorganic salts, such as zinc and mercuric chloride, were being used at that time, more selective approaches for chemical modification of cellulose and application of organic mildewproofing agents were patented in the 1920's. Doree' was the first to report the resistance of cellulose acetate to microorganisms (in sea water), and thus establish the blocking action of certain cellulose derivatives to prevent microbial attack [20]. Waterproofing and the addition of selected antimildew additives was also developed as an experimental approach during the 1920's. Halogenated phenols [21], salicylanilides [22] and copper naphthenate [23] were patented as effective agents for preventing mildew in fabrics.

The next period (1940-1970) may be described as the "golden age" of fiber preservation. It was during this period that most of the current test methods and mildew-resistant treatments for cellulosic fabrics were developed. During this time, synthetic fibers also were invented and commercialized; most of these fibers were noted to have excellent resistance to biodeterioration.

The impetus for rapid progress in textile biodeterioration was the numerous fundamental and tropical field studies by the U. S. Army Quartermaster Corps. This effort was conducted to preserve military clothing and supplies during World War II . The book by Siu [24] was the first important milestone in this era. It described and summarized the progress of (a) cellulose decomposition in nature, (b) morphological, physical, chemical features of cotton fibers and structural features of cotton fabrics, (c) morphological and taxonomic aspects of microorganisms that cause biodeterioration, (d) the effect of nutritional and environmental factors of cellulolytic activity of microorganisms, (e) mechanisms of fiber degradation, (e) laboratory and field methods for testing microbiological resistance of textiles, and (f) physical and chemical methods of preventing microbiological damage. Most of the test methods described in the book were developed in the early 1940's and are still standard methods today. These include the inoculation of fabrics with pure cultures of fungi (such as Aspergillus niger or Chaetomium globosum), the soil burial test, use of tropical chambers, and long term exposure in field trials. The humidity jar test is the only method in use today not specifically described by Siu that is approved for use by the EPA. The test consists of inoculating a fabric inoculated with the test fungus, followed by incubation. The fabric is then inspected visually for fungal growth every week up to a maximum of four weeks [8].

In addition to the voluminous amount of data and information produced by the Army Quartermaster Corps, the Southern Regional Research Center (USDA, ARS) conducted extensive field studies and developed several effective weather-resistant finishes for fabrics during the 1950's and 1960's. These are documented in two critical reviews that were written in 1960 [25] and 1983 [26].

By 1960, there were several methods to prevent mildew on fabrics. Many of the methods used resins or coated fabrics to improve durability of the antimicrobial finish. It was recognized by that time that outdoor degradation of fabrics was due not only to antimicrobial effects, but also to air pollution and ultraviolet degradation.

By 1970, most effective weather-resistant fabrics (cellulosic-type) contained photostabilizers, antioxidants, biocides and resins or polymers as binding agents. The concept that polymerization was essential for long term durability of all types of textile finishes was introduced in a 1983 review paper. Weather-resistant finishes were viewed as one of many categories to which this concept was applicable [26].

CONCEPTS OF ANTIBACTERIAL AND GERMICIDAL FINISHES FOR TEXTILES

In 1940 the concept and term "antibacterial finish" evolved. It was noted that test methods were needed to evaluate bacteriostatic and bactericidal activity of untreated and treated fabrics against disease-causing microorganisms. Specific finishes needed development for protection against pathogenic microbes. These finishes had somewhat different requirements than those that were effective in producing mildew-resistant fabrics [27]. Advances in antibacterial finishes for fibers and antibacterial/antimycotic polymers occurred in two stages: the period between 1940-1960 and between 1960-1980. A selected list of key discoveries and concepts for these periods is given in Table II.

TABLE II. Milestones in Antibacterial Fibers and Antibacterial/Antimycotic Polymers

Concept or developments	Year	Reference
Use of term "antibacterial fabrics" and evaluation of materials for both bacteriostatic and bactericidal activity	1941	27
Use of insoluble silver salts to impart permanent bactericidal properties to bristles, hairs and other fibrous materials	1942	28
Treatment of cotton fabrics with bis(halophenols) and claims of bacteriostatic activity after several launderings	1951	29
Study indicated that polyamide (synthetic fiber) was less effective in having bacteriostatic properties imparted than cotton when both were treated with quaternary ammonium compounds	1956	30
Use of EDTA salts to produce bandages with antibacterial properties	1958	31
First commercial application of an antibiotic (neomycin sulfate) to produce a durable antibacterial fabric	1958	32
Landmark paper by Gagliardi that described the principles and strategies for antibacterial fiber finishes	1962	33
Development of Quinn test, a quantitative method for determining antibacterial and antifungal activity directly on fiber	1962	34
Discovery of broad spectrum antimicrobial fiber Letilan using the principle of controlled release of 5-nitrofurylacrolein from the acetal of poly (vinyl alcohol)	1963	35
Production of flame retardant fabrics that also had antibacterial properties durable to 50-100 launderings	1966	36
Production of antibacterial plastics (polyolefins, acrylics, styrene-butadiene, etc.) containing hexachlorophene	1970	37
Production of inherently biostatic coated surfaces based on a silicon polymer with a pendant quaternary ammonium group	1972	38
Development of synthetic absorbable fibers for wound closure	1972	39
First use of microencapsulation technique for controlled release of bactericides from fibrous materials	1974	40
Controlled-release, inexpensive finish based on zinc peroxide polymer to produce cotton fabrics with antibacterial activity durable to 50 launderings	1978	41

In the 1970's, a series of antibacterial plastics containing hexachlorophene (now prohibited as a carcinogen) was developed [37] and a polymeric coating (organosilicon with a pendant quaternary ammonium group) was claimed to impart inherently biostatic properties onto a variety of surfaces [38]. The first synthetic and absorbable sutures (based on polylactide and polyglycolic acid structures) were patented and commercialized [39]. Microencapsulation was shown to be an effective method of releasing bactericides, fungicides and other active agents by incorporating them into a multilayer fibrous assembly [40]. The concept of using an appropriate bioactive polymer with control-release characteristics was demonstrated. The active species was a zinc peroxide polymer that provided long term antibacterial activity when it was bound to cellulosic fibers [41].

ANTIMICROBIAL PROPERTIES OF SYNTHETIC FIBERS

Regenerated celluloses (rayons) and cellulose esters (primarily acetates) were the first group of synthetic fibers to be commercialized. The Count de Chardonnet in 1880 recognized the potential of artificial silk or regenerated cellulose fibers, and ten years later, developed a commercial rayon process based on nitrocellulose chemistry. Viscose and cuprammonium processes were patented as methods for producing rayon around 1900-1905. Initial efforts also were made to produce cellulose acetate fibers on a commercial scale [42]. The poor resistance of most rayons to biodeterioration was readily noted, and documented in investigations of all types of cellulosic fibers [16]. The inherent resistance of cellulose acetate to mildew and rot was observed shortly after this fiber was made commercially available [20]. The hydrophobic nature of cellulose acetate is responsible for resistance to biodeterioration; in contrast, readily wettable rayons are easily attacked and degraded outdoors. When totally synthetic fibers were invented (Table III), causative agents and test methods for biodeterioration were well understood.

TABLE III. Commercialization of Major Synthetic Fibers

Synthetic Fiber	Time of commercialization or major production	Resistance to biodeterioration
Various rayons	1890-1905	Generally poor
Celluose acetate	1900	Superior
Aliphatic nylons	1939	Superior
Glass fibers	1937	Excellent
Polyester	1949-1953	Excellent
Acrylics and modacrylics	1949	Outstanding
Polypropylene	1961	Inert to chemical and biological agents
Aramids (aromatic polyamides)	1967-1971	Superior

During the first period (1940-1960), progress was characterized by development of treatments that claimed to have some durable antibacterial effects, including the insolubilization of silver salts on bristles [28], treatment of cotton fabrics with bis (halophenols) [29] and the first application of an antibiotic to a fabric [32]. It was also observed that synthetic fibers (such as polyamides) had no advantage over natural fibers in their ability to withstand or transmit disease-causing microorganisms [30].

In the early 1960's (second period), several important advances and developments occurred, the most significant of which was the landmark paper by Gagliardi [33] in which he defined the principles and strategies of antibacterial fiber finishes. The different types of chemical approaches required to achieve such activity were classified, and the concept of the regeneration principle was proposed. The regeneration principle states that the proper choice of a chemical species covalently bound to a fiber should provide a continual reservoir of antibacterial agent; the agent could be released by hydrolysis, laundering, ultraviolet light or other methods.

Another important advance during the second period (1960-1980) was the development of a test to measure directly the antibacterial and antifungal activity of agents on a fiber [34] rather than to rely solely on earlier, diffusion-controlled tests (such as the agar plate method). Controlled release of broad spectrum microbiocides from poly (vinyl alcohol) fibers was first demonstrated by Vol'f in the Soviet Union [35]. It was also recognized that long term antibacterial activity in fabrics was produced by application of a flame-retardant finish [36].

It was observed that synthetic fibers were superior to cotton and other cellulosic fibers in their resistance to biodeterioration. However, in the more complex process of weathering, air pollution, photodegradation and other factors are important; thus, some synthetic fibers performed better than natural fibers while other synthetics performed worse than natural fibers in outdoor environments. A comprehensive study of the comparative weathering performance (in England and Nigeria) of acrylic, nylon, polyester, cellulose acetate, cotton and wool yarns demonstrated that acrylic fibers were far superior to all other fibers because of their remarkable resistance to both ultraviolet and microbial degradation. Cotton performed as well or better than nylon in areas of low humidity and air pollution, while nylon performed better than cotton in areas of high humidity and air pollution [43]. The inert characteristics of polypropylene fibers made them suitable for geotextile uses in which the fabric is used as support for construction of roads. Fabrics made of polypropylene have been buried underground for over 20 years with no evidence of biodeterioration. The weathering resistance of polypropylene above ground is not as impressive because it is susceptible to ultraviolet degradation and requires photostabilizers as additives. Another factor in the development of plastics and synthetic fibers for outdoor uses was noted. Certain plasticizers served as nutrients for microorganisms even though the plastic or synthetic fabric was considered resistant to most microorganisms [44]. This led to a careful selection of plasticizers for incorporation into plastics and synthetic fibers exposed to outdoor environments.

The advantage of synthetic fibers over natural fibers in relation to biodeterioration was not observed for transmission or growth of disease-producing microorganisms and viruses. A comparative study of the survival rate of Streptococcus pyogenes on swabs made from different unsterilized fibers demonstrated that only cotton and calcium alginate fibers produced less than 10% survival rates of the bacteria after 4 hours. Other fibers, such as rayon, polyester, polypropylene, nylon and cellulose acetate served as excellent surfaces to promote microbial growth of these virulent bacteria [45].

There was also a series of studies conducted during this period on virus transfer with various fabrics [46-50]. Although these investigations were considered esoteric at the time, the current prevalence of other viruses (such as hepatitis and the recently characterized HIV virus) make these studies extremely valuable and useful. Important factors in virus transfer by fabrics were the type of virus, type of fabric, method of virus contact (aerosol or solution), and relative humidity. Aerosol contamination resulted in much quicker transfer for most fabrics, particularly for wool fabrics exposed to polio virus. However, all other types of fabrics were considered fairly effective in their ability to transfer viruses under a variety of end use conditions unless they were laundered and/or treated with a virucidal agent [48, 50]. Other studies demonstrated that cotton fabrics were superior to synthetic fabrics in preventing body odor [51] and growth of one of the dermatophytic fungi that causes athletes' foot [52]. Further studies are needed to validate these observations, but all types of fibrous materials should be treated with antibacterial, antimycotic and/or antiviral agents or finishes to minimize or prevent transmission of disease.

ANTIMICROBIAL POLYMERS AND APPLICATION TO FIBERS

The initial treatments of fabrics to impart resistance to biodeterioration and disease-producing microorganisms usually consisted of insolubilization of compounds or formation of stable covalent bonds with reactive groups in fibers. However, by the mid-1950's, it was becoming increasingly apparent that the best durability for an antimicrobial finish was achieved by using polymers as binding agents to retain an active biocide. Polymeric drugs and antimicrobials were discovered and commercialized at this time. By the early 1960's, polymers were used in controlled release systems to provide long term antimicrobial effects. Each of these three aspects is briefly discussed in an historical context.

Commercial, weather-resistant cotton fabrics were developed by scientists at the Southern Regional Research Center in New Orleans and at the Army Quartermaster Lab in Natick between 1955-1970 [12]. Most of the finishes contained an ultraviolet stabilizer, an antimicrobial agent, and a crosslinked resin as a binding agent to impart long term durability of the fabrics to weathering. An excellent example of this approach was the development of an acid colloid of melamine-formaldehyde resins to obtain an effective weather-resistant finish for cotton fabrics [53]. This type of approach later was used to produce flame retardant fabrics with antibacterial properties effective for 50-100 launderings [36] and nylon socks containing mercurated allyltriazines that were effective against dermatophytic fungi after 60 launderings [54].

The development of polymeric drugs and antimicrobials was concurrent with the use of polymers as binding agents to improve the durability of antimicrobial activity. Although it is impractical to discuss this vast development in detail, a few selected examples are given to illustrate how polymers have been beneficial in health and medical applications. The first commercial iodophor (controlled release of iodine as a germicide) was a complex of poly (vinylpyrrolidone) and iodine [55], and is still in widespread use today in hospital environments. Polymeric quaternary ammonium salts were found to be as effective as monomeric quaternary ammonium salts for their bactericidal and fungicidal activity [56]. The reaction of diamines with diamine salts of dicyanamide produces polymeric biguanides that had excellent antibacterial activity against representative gram-negative bacteria and other microorganisms [57]. Numerous other examples of polymeric drugs, antimicrobials, and polymers used inside the human body as replacement parts, are discussed in two excellent reviews written in the late 1960's and late 1970's [58, 59].

Some key developments in antibacterial fibers and antibacterial/antimycotic polymers are listed in Table II. The most significant developments include the controlled release of a nitrofuran derivative from poly (vinyl alcohol fibers) [35], an inherently biostatic coating based on a polymeric silicone containing a pendant quaternary ammonium group [38], and controlled release of a zinc peroxide species from an inorganic polymer affixed to cellulosic fibers [41].

MODERN DAY ANTIMICROBIAL POLYMERS AND FIBERS

In the last decade, there have been many advances and trends in the use and evaluation of antimicrobial polymers and fibers. Bioluminescence as an early indicator of microbial infiltration and degradation of fibers has been used as a test method [60]. Fluorescein dyes containing organotin moieties were found to exhibit a wide range of antibacterial activity and are useful for doping plastics and as additives in latex paints and fabrics [61]. There have been numerous Japanese patents on the use of synthetic zeolites treated with antibacterial metal ions, especially silver, mercury and zinc. The modified zeolites have been incorporated into fabrics, and are claimed to be effective against several fungi and bacteria. The patent by Kanebo, Inc. is typical of this type of approach, and is one of the earlier reports of such an invention [62]. Vascular grafts and catheters made out of synthetic polymers have been complexed with derivatives of nalidixic acid. These modified materials are claimed to be effective against representative gram-positive and gram-negative bacteria [63].

Radioisotope techniques have been developed for measuring the adhesion of various bacteria to fabrics [64]. Some effort also has been made to relate the wettability of the surfaces of films and fabrics to initial adhesion of representative gram-positive bacteria. It was found that bacterial adhesion was greatest for hydrophobic film surfaces, but there was no clear correlation between surface wettability of fabrics and bacterial adhesion [65].

Novel chemical approaches for imparting antimicrobial activity to polymers and fibers have been shown such as incorporation of metallic phthalocyanine into socks [66], air filters made of fabrics with microbiocides such as imides and imidazole derivatives [67] and the incorporation of bisphenoxyarsines into thermoplastic nylon resins for outdoor use [68]. A recent commercial development uses a broad-spectrum phosphate antimicrobial incorporated into the poly (vinyl chloride) backing of carpets. This antimicrobial provides outstanding activity to the carpet by slow release over a period of years. Carpets so protected are particularly useful in facilities where microorganisms are readily spread (hospitals, commercial buildings and condominiums) [69,70].

What future developments and inventions will occur in these areas? Past prognostications of scientific and medical advances have not been always accurate and have not anticipated unexpected political or economic events that served as stimuli for inventions. However, speculation on some areas that look promising and others that could develop as we approach the 21st century can be made.

Rapid advances have been made in modifying the surfaces of polymers and fibers by use of plasma, excimer lasers and other discrete, high-energy sources. Such surface modification should allow for binding or incorporation of antimicrobial agents not possible with current science and technology. The present concern about the spread of insidious viruses such as hepatitis and HIV by contact with contaminated surfaces should lead to development of effective virucides in materials. Another consequence of these diseases will probably be the issuance of more stringent health and medical ordinances to have fibrous and plastic materials treated with microbiocides. Thus, the end uses for such materials will greatly proliferate and will result in opportunities for developing unique and useful products to meet these regulations.

Advances in the basic sciences of microbiology, mycology and virology should lead to development of new and effective polymeric materials that are either inherently antimicrobial or can be made microbiocidal by appropriate and new strategies. There will be a greater need for many of these synthetic biomedical devices to resist body rejection and infection as more artificial parts and organs are used as replacement parts in our body. Thus, the biological properties of the biomedical device will become increasingly important. Finally, our journeys into outer space and ultimately space colonization will require that many materials be treated to resist terrestrial and even extraterrestrial microorganisms.

REFERENCES

1. Anon., Assassination by poisoned clothing, Lancet 1: 661 (1865).
2. Lister, J., On the antiseptic principle in the practice of surgery, Lancet 2: 353-356; 668-669 (1867).
3. Ries, J. Von, Antiseptic military clothing, Weiner Klin. Wchnschr. 30: 577 (1917).
4. Davies, M., Antiseptization of military garments as a prophylactic against infection from wounds, J. Pharm. Chim. 15: 122 (1917).
5. Vigo, T. L. and Benjaminson, M. A., Antibacterial fiber treatments and disinfection, Text. Res. J. 51: 454-465 (1981).
6. EPA, "Pesticide Program: Guidelines for Registering Pesticides in the United States," Fed. Reg. 40 (123): 26802-26928 (June 25, 1975).
7. Reddish, G. F., Methods of Testing: General Review, Ch. 3, in "Antiseptics, Disinfectants, Fungicides and Sterilization," 2nd Edition, G. F. Reddish (Ed.), Lea and Febiger Publishers, Philadelphia, PA. (1957), pp. 31-92.
8. Reddish, G. F., Definition of Terms, Ch. 2, in "Antiseptics, Disinfectants, Fungicides and Sterilization," 2nd Edition, G. F. Reddish (Ed.), Lea and Febiger Publishers, Philadelphia, PA. (1957), pp. 23-29.
9. Petrocci, N. A. and Clarke, P., Proposed test method for antimicrobial additives, J. Assoc. Off. Anal. Chem. 52: 836-842 (1969).
10. Hollaender, A. (Ed.), "Radiation Biology; Vol. II: Ultraviolet and Related Radiations", McGraw Hill Book Co., New York, 1955.
11. Phillips, C. R. and Kaye, S., The sterilizing action of gaseous ethylene oxide. I. A review, Am. J. Hyg. 50: 270-279 (1949).
12. Vigo, T. L., Preservation of Natural Textile Fibers--Historical Perspectives, in "Preservation of Paper and Textiles of Historic and Artistic Value," J. C. Williams (Ed.), Am. Chem. Soc., Washington, D. C., Adv. Chem. Ser. 164: 189-207 (1977).
13. Osborn, T. G. B., Molds and mildews: their relation to the damaging of grey cloth and prints, J. Soc. Dyers Color. 28: 204-208 (1913).
14. Thaysen, A. C., The effect of the action of microorganisms on fibers and fabrics, J. Soc. Dyer. Color. 40: 101-104 (1924).
15. Thaysen, A. C. and Bunker, H. J., Studies on the bacterial decay of textile fibers. II. A preliminary study of the deterioration of artificial silk through the action of microorganisms, Biochem. J. 19: 1088-1094 (1925).
16. Thaysen, A. C. and Bunker, H. J., Studies on the bacterial decay of textile fibers. III. The occurrence of humus compounds in deteriorated fabrics and the bearing of their formation on the origin of peat and coal, Biochem. J. 20: 210-216 (1926).

17. Thaysen, A. C. and Bunker, H. J., Studies on the bacterial decay of textile fibers. IV. Note on the action of B. subtilis and B. mensentericus on cellulose, Biochem. J. 20: 692-694 (1926).
18. Burgess, R., Microbiology of wool, J. Text. Inst. 19: 315-322T (1928).
19. Veitch, F. P. and Levine, S. S., Testing the mildew resistance of fabrics, Science 49: 618 (1919).
20. Doree', C., Action of sea water on cotton and other textile fibers, Biochem. J. 14: 709-714 (1920).
21. British Dyestuffs Corp. Ltd, A. Renshaw and T. H. Fairbrother, Protecting textiles and other materials from mildew, Brit. Pat. 259,690 (July 24, 1925).
22. British Cotton Industry Res. Assoc., R. G. Fargher, L. D. Galloway and M. E. Probert, Protecting textiles and other materials from mildew, Brit. Pat. 323,579 (Oct. 18, 1928).
23. Pisarev, S. E., Impregnating of vegetable fibers to prevent rotting, Russ. Pat. 1,547 (Sept. 15, 1924)
24. Siu, R. G. H., "Microbial Decomposition of Cellulose," Reinhold, New York, 1951.
25. Howard, J. W. and McCord, F. A., Cotton quality study IV: resistance to weathering, Text. Res. J. 30: 75-117 (1960).
26. Vigo, T. L., Protection of textiles from biological attack, in "Handbook of of Fiber Science and Technology: Vol. II, Chemical Processing of Fibers and Fabrics, Functional Finishes, Pt. A, M. Lewin and S. B. Sello (Eds.), Marcel Dekker, New York, (1983), pp. 367-425.
27. Hirschmann, D. J. and Robinson, H. M., Testing impregnated fabrics for antibacterial properties, Soap. Sanit. Chemicals 17 (9): 94-119 (1941).
28. Rosenzweig, S., Treating fibrous materials comprising albumin, Brit. Pat. 543,948 (March 20, 1942).
29. Musser, D. A., Bacteriostatic cellulosic textile treated with derivatives of dihydroxyhexachlorodiphenylmethane, U. S. Pat. 2,559,986 (July 10, 1951).
30. Goldsmith, M. T., Latlief, M. A., Friedl, J. L. and Stuart, L. S., Effect of quaternary treatment under varied ratios of weight-volume concentration on the bacteriostatic property of fabrics, Appl. Microbiol. 4: 91-94 (1956).
31. Kuchle. C., Stabilization of protein solutions, Ger. Pat. 1,030,522 (May 22, 1958).
32. Parker, R. P. and Abbey, A., Textile fabrics containing neomycin, U. S. Pat. 2,830,011 (April 8, 1958).
33. Gagliardi, D. D., Antibacterial finishes, Am. Dyestuff Reptr. 51: P49 (1962).
34. Quinn, H., A method for the determination of the antimicrobial properties of treated fabrics, Appl. Microbiol. 10: 74-78 (1962).
35. Vol'f, L. S., Meos, A. I., Koteskii, V. V. and Hillers, S., Biologically active poly (vinyl alcohol) fibers, Letilan, Khim. Volonka (6): 16-18 (1963).
36. Hoch, P. E., Wagner, G. M. and Vullo, W. J., The bactericidal properties of THPC-resinated cotton fabric, Text. Res. J. 36 (8): 757-758 (1966).
37. Hopfenberg, H. B. and Tulis, J. J., Advances in antibacterial plastics, Modern Plastics (7): 110-116 (1970).
38. Isquith, A. J., Abbott, E. A. and Walters, P. A., Surface-bonded activity of an organosilicon quaternary ammonium chloride, Appl. Microbiol. 24 (6): 859-863 (1972).
39. Chu, C. C., Recent advancements in suture fibers for wound closure, in "High Tech Fibrous Materials," T. L. Vigo and A. F. Turbak (Eds.), Am. Chem. Soc., Washington, D. C., Symp. Ser. 457: 167-213 (1991).
40. Anon., Getting read to tape the bugs, Chem. Week 115: 43-44 (Aug. 21, 1974).
41. Danna, G. F., Vigo, T. L. and Welch, C. M., Permox-a hydrogen peroxide-zinc acetate antibacterial agent for cotton, Text. Res. J. 48: 173-177 (1978).
42. Moncrieff, R. W., "Man-made Fibres," Chapters VII, VIII and IX, John Wiley Publishers, New York, 1959.
43. Fels, M., Weathering of textile yarns, J. Text. Inst. Proc. 51: P648-P656 (1960).
44. Scullin, J. P., Dudarevitch, M. D. and Lowell, A. I., Biocides, in Encyl. Poly. Sci. and Tech. Vol. 2 (H. Mark, Exec. Ed.), Interscience, New York, 1965, pp. 379-401.
45. Dadd, A. H., Dagnall, V. P., Everall, P. H. and Jones, A. C., The survival of Streptococcus pyogenes on bacteriological swabs made from various fibres, J. Med. Microbiol. 3 (4): 561-572 (1970).
46. Sidwell, R. W., Dixon, G. J. and McNeil, E., Quantitative studies on fabrics as disseminators of viruses. I. Persistence of vaccinia virus on cotton and wool fabrics, Appl. Microbiol. 14 (1): 55-59 (1966).
47. Sidwell, R. W., Dixon, G. J. and McNeil, E., Quantitative studies on fabrics as disseminators of viruses. II. Persistence of poliomyelitis virus on cotton and wool fabrics, Appl. Microbiol. 14 (2): 183-188 (1966).
48. Sidwell, R. W., Dixon, G. J. and McNeil, E., Quantitative studies on fabrics as disseminators of viruses. III. Persistence of vaccinia virus on fabrics impregnated with a virucidal agent, Appl. Microbiol. 15 (4): 921-927 (1967).
49. Sidwell, R. W., Dixon, G. J., Westbrook, L. and Forziati, F. H., Quantitative studies on fabrics as disseminators of viruses. IV. Virus transmission by dry contact of fabrics, Appl. Microbiol. 19 (6): 950-954 (1970).

50. Sidwell, R. W., Dixon, G. J., Westbrook, L. and Forziati, F. H., Quantitative studies on fabrics as disseminators of viruses. V. Effect of laundering on polio-virus contaminated fabrics, Appl. Microbiol. 21 (2): 227-234 (1971).
51. Gutfrend, K. and Dravnicks, A., Odor retention in cotton and synthetic fabrics, Ill. Inst. Tech. Res. Inst. Rep. No. C 8224-6, Chicago, IL., Jan. 1973.
52. Chalmers, L., Aspects of foot care and hygiene (Part I), Dragoco Rep. 11: 215-222 (1972).
53. Berard, W. N., Gautreaux, G. A. and Reeves, Formic acid colloid of methylolmelamine as a weather and rot resistant finish for cotton, W. A, Text. Res. J. 29: 126 (1969).
54. Gip, L. and Magnusson, Impregnation of hose material with an antimycotic agent: method to counteract reinfection with foot mycosis, Nord. Med. 82 (44): 1375-1378 (1969).
55. General Aniline and Film Corp., Polyvinylpyrrolidone mixtures with halogens, Brit. Pat. 680, 968 (Oct. 15, 1952).
56. Rohm & Haas Co., Quaternary ammonium compounds, Brit. Pat. 776, 480 (June 5, 1957).
57. Buckley, A. J., Gazzard, E. G., Greenshields, J. N., Pemberton, D., Wild, J. H. and Leitch, D. S., Polymeric biguanide biocides, Brit. Pat. 1,531,717 (Nov. 12, 1974).
58. Regelson, W., The antimitotic activity of polyanions, Advan. Chemother. 3: 303-370 (1968).
59. Samour, C. M., Polymeric drugs in the chemotherapy of microbial infections, in "Polymeric Drugs, Proc. Intl. Symp. 1977," L. G. Donaruma and O. Vogl (Eds.), Academic Press, New York, 1978, pp. 161-184.
60. McCarthy, B. J., Bioluminescent assay of microbial contamination of textile materials, Int. Biodeterior. Bull. 19 (2): 53-57 (1983).
61. Carraher, C. E, Jr., Venkatchalam, R. S., Tiernan, T. O. and Taylor, M. L., Stannane polydyes as additives: structure and applications, Org. Coat. Appl. Poly. Sci. Proc. 47: 119-123 (1982).
62. Zenji, H., Hideo, O., Shigetaka, H. Saburo, N., Shunya, I., and Kenichi, T., Particle-packed fiber article having antibacterial property, Eur. Pat. Appl. EP 103,214 (Mar. 21, 1984).
63. Fox, C. J., Jr., Modak, S. and Reemtama, K., Infection-resistant materials and method of making them through nalidixic acid derivatives, U. S. Pat. 4,563,485 (Jan. 7, 1986).
64. Hsiesh, Y. L., Timm, D. and Merry, J., Evaluation of a radioisotope labelling technique for measuring bacterial adherence to fabrics, Appl. Radiat. Isot. 37 (12): 1235-1239 (1986).
65. Hsiesh, Y. L. and Timm, D. A., Relationship of substratum wettability measurements and initial Staphlyococcus aureus adhesion to films and fabrics, J. Colloid Interface Sci. 123 (1): 275-286 (1988).
66. Kamimura, M., Socks containing phthalocyanine derivatives as microbiocides, Jpn. Kokai Tokkyo Koho JP 63,112,701 (May 17, 1988).
67. Takahishi, E., Aihane, T., Shibagaki, T. and Satomi, H., Filters for air cleaners, Jpn. Kokai Tokkyo Koho JP 62,247,816 (Oct. 28, 1987).
68. Burton, W. L., Stabilization of bisphenoxyarsine and bisphenoarsazine antimicrobial agents in thermoplastic compositions, U. S. Pat. 4,769,268 (Sept. 6, 1988).
69. McIntosh, R. H., Jr. and Turbak, A. F., Biocidal delivery system of phosphate ester and method of preparation thereof, U. S. Pat. 4,908,209 (Mar. 13, 1990).
70. Interface Flooring Co., Inc., 1990 Annual Report, Atlanta, Ga.

SYNTHETIC FIBERS FROM "PURE SCIENCE": DUPONT HIRES CAROTHERS

MATTHEW E. HERMES, PH.D.
U. S. Surgical Corp., Materials Research and Development, 150 Glover Ave., Norwalk, CT 06856

INTRODUCTION

The success modern-day DuPont would achieve was assured the day DuPont hired Wallace Carothers. DuPont's Chemical Department, under Dr. Charles M. A. Stine, was in a decade-long search for new identity in the '20's. A decision -- one of many new policy directions from those years -- to introduce a "pure science" program, absent the requirement for useful technical outcome, resulted in the employment, on February 6, 1928, of Dr. Wallace Hume Carothers. Carothers was a midwesterner and a graduate of the University of Illinois chemistry program under Professor Roger Adams. Though only an Instructor in Chemistry at Harvard, his thinking was in the pattern of the inventive pioneers: of Edison, of the Wrights, and of Elmer Sperry.

Carothers led but one of several of DuPont's pure science groups. However, within three years he laid the foundation for the synthetic fiber industry through development of the theory of condensation polymerization, the discovery of methods to make high molecular weight polyesters and the observation, characterization and recognition of the importance of cold drawing.

Nylon-6,6 resulted from the early observations. This fiber transformed DuPont from the "merchant of death" of the '30's to the great industrial chemical and fiber giant of the '50's and beyond and set the table for the other designed synthetic fibers which have transformed modern life and culture.

We find the bold and acquisitive DuPont Company of the '20's strangely quiet and naive in its research and development. Growth of the Company through research and development, technical merger and licensing and other technical means was a distant second objective for the family owners. The controlling duPonts set as first priority, growth through family and Company acquisition. Their swollen coffers, full from unspent wartime dividends, bought control of General Motors, a large share of U. S. Rubber and even, for a time, a sizeable investment in U. S. Steel.

Dr. Stine's fundamental research program was not a new idea and Stine consciously forsook key elements of the successful pure science programs long established at General Electric and Bell Labs. His proposal was accepted, in part because Stine waited until the more research-minded of P. S. duPont's brothers, Lammot succeeded to the Presidency of the Company, and in part because of the emerging organizational culture of the Company which now encouraged origination of such major policy suggestions from the operating units for approval by the Executive Committee.

It is likely Stine hired Carothers on the recommendation of Roger Adams and James Conant. Adams, the Illinois Chemistry Department "Chief", carried the skills of a tight-fisted, cost-conscious New England business-man along with considerable chemical creativity, into his growing Department. He established Illinois as a conduit of scientists to the industrial laboratories and even taught his students the practical economics of the chemical business through an innovative, commercially profitable 'summer preps' program.

Stine wanted Adams first, but Roger's stage was larger; Adams cooperated with DuPont's efforts to hire one or more organic chemistry group leaders, even from the Illinois Department. But when an offer came to one of his staff, he countered with persuasive rhetoric extolling the reasons why C. S. (Speed) Marvel, then R. C. Fuson should remain in Urbana. Thus the tensions between academe and industry had an effective buffer in Adams at Illinois.

At Harvard, Professor James Conant, then Chemistry head and later President of the University, grudgingly played the same role. He heard out Louis Fieser's objections and refusal of Stine's offer, but placed no obstacle in Carothers' path to DuPont.

The months from September 1927, when DuPont first contacted Carothers until February 1928 when Wallace started in Wilmington passed with uncertainty, indecision and finally acceptance by Carothers under the steady pressure of Stine and his associates, Hamilton Bradshaw and Arthur Tanberg, to bring him on.

This chapter describes characteristics of American Research and Development and of the DuPont Company in the '20's, reveals the young Carothers, and details the courting and hiring of this uncertain genius, one of "The 100 Most Important Americans of the 20th Century".[1]

PURE SCIENCE WORK

> "By every means in our power, let us show our appreciation of pure science, and let us forward the work of the pure scientists, for they are the advance guard of civilization." J. J. Carty, Bell Labs, 1904

As the afternoon train rolled down the Hudson Valley in mid-March, 1927, Dr. Charles M. A. Stine, Chemical Director at DuPont sought advice from his research counterpart at General Electric, Dr. Willis R. Whitney. The previous December, Stine had proposed a radical departure for his Department. Searching for a name for the new budget-line item of $20,000 for the year 1927, he called the program "pure science" or "fundamental research".

Stine had proposed a series of studies, fundamental in nature, arguing that one of the largest chemical industries in the world should pursue such work for:

1. Its advertising value,
2. Its positive effect on the scientific staff,
3. The possibility of enhancing contacts with researchers outside DuPont, and secondarily,
4. For possible practical applications.

Practical application would be only a side benefit and this plan was far afield from previous DuPont research. Stine noted the existence (but not the experience) of fundamental programs at General Electric and in the laboratories of German industry. And he noted Herbert Hoover, then Secretary of Commerce, who had recently stressed the vital need for greater funding of such programs.[2]

The Executive Committee forced Stine to do further work to support his proposal and now, having cited General Electric's research as evidence for the value of his proposal, he belatedly sought Whitney's counsel. Stine went to Albany; two chairs were reserved on the train back to New York and Stine accompanied Whitney to a Research Directors meeting at the American Institute of Chemists.

The first research laboratory in American industry had been initiated at Schenectady to serve General Electric. Though the company was scarcely ten years old at the time, the technical rigors of the development of smoothly interacting power systems, of electrical generation, of transmission and of consumption, had called upon the total skills of the engineers and the scientists of the time.

The marshalling place was General Electric in Schenectady, and by 1900, industrial science began to breathe life of its own independent of its direct relationship to a technical goal. The technical triumvirate of Elihu Thomson, Charles Steinmetz and E. Wilbur Rice, along with Albert G. Davis, head of the patent department "deemed (it) wise .... to establish a laboratory to be devoted exclusively to original research."

Willis Whitney, recruited from MIT, had led the laboratory from its inception and by the time of Stine's letter seeking an audience, Whitney's laboratory had a staff of nearly 500.[3]

Perhaps Stine reviewed his "pure science" objectives with Whitney that day; he may have listed the benefits he believed would accrue to DuPont, but his initial questions dealt more with the tactics of operating the research group: when and whether to publish technical papers, exchange of information with other basic research laboratories and the handling of personnel.

Whitney could have reflected on his two decades of experience and reminded Stine that it was the original intent at General Electric to set the new laboratory apart from the rest of the operating company, to free its scientists from the daily and current problems, but in fact he knew that detachment had been impossible to maintain. Calls for help from business could not be left unanswered in the practical world. And Whitney had become more realistic in assessing the problems of attack. "It is, of course, difficult to assign a value to the results of purely scientific work. This is always relative," he wrote to Stine. "I feel .... since we are part of an industrial organization our investigations should be directed along those lines which are not only of scientific interest, but also promise to yield knowledge of importance to industry."[4]

Elmer K. Bolton accompanied Stine to New York ten days later to visit Edward B. Craft, Vice President of Bell Laboratories. The Saturday morning visit brought the two DuPonter's into this University-like environment whose identity had grown slowly since Frank B. Jewett joined with Edwin H. Colpitts to form the nucleus of laboratory in 1912. Its purpose -- fundamental scientific research. Its organization -- an independent entity, set aside in a different place from the operations of Western Electric and AT&T. Bell Laboratories had, within the last year, been incorporated under its own name and drew on a budget larger than any of the nation's research universities.

Bolton, as head of research in dyestuffs, had clear opinions on the conduct of research, and although he was invited to participate in the fact-finding mission by Stine, he never was convinced that Stine's altruistic purposes for the work were realistic.[5] But Stine forged ahead. The program was funded, with the principles and objectives Stine had proposed; with the diminished importance of practical outcome, left intact. The group would be housed in a new building at the Experimental Station, hard by the laboratories housing the chemists working for the operating departments. For those departmental staff, the building soon became known as "Purity Hall".

Lammot duPont, Stine's boss and the DuPont President, wished no outside publicity for the new adventure. It was simply left to Stine to now begin hiring staff and start the "pure science" program.

DU PONT AFTER THE GREAT WAR

What was the DuPont Company now, nine years after the end of the war?

There was no parallel; no previous experience which could prepare the DuPont Company for the transition required at the end of the World War in 1918.

In 1902, DuPont was a sleepy and segmented black and smokeless powder manufacturer. In a sea-change, orchestrated by three young members of the duPont family the Company had become, by the start of hostilities, the coordinated and unified giant of the explosives industry through merger, acquisition and modern organization. Research had played a secondary role, although the Company's Chemical Department; its central research organization into which Wallace Carothers would be hired in 1928, had been established in 1911.

Three cousins: Alfred I., T. Coleman and Pierre S. duPont bought the 100-year old explosives Company from their duPont relatives in 1902. It was Alfred's idea[6]; he needed T. Coleman's entrepreneurial skills and money and both believed they required the financial and organizational skills of P. S. Their purchase headed off the sale of the Company to competitor Laflin and Rand and within two years, as they established a DuPont monopoly, the new owners had purchased Laflin and Rand along with much of the rest of the powder capacity in the country and destroyed the Gunpowder Trade Association.

The courts tell much of the story of the second decade of this century. Alfred I. was edged out of his position as the manufacturing Vice President of the Company, ostensibly over his divorce and remarriage to Alicia Maddox. (The marriage split the DuPont family for generations.) With Alfred now pursuing other interests and the mercurial, political and big-city minded T. Coleman spending most of his time in New York, P. S. soon ran the Company.

Strong Departmental operating organizations, reporting to an Executive Committee[7] for general direction, emerged as the signature of the DuPont Company. P. S. introduced measurable concepts of return on investment as the criteria for financial commitment of the Company.[8]

P. S. was a cunning and ruthless gatherer of assets and the Supreme Court soon held DuPont's monopoly of power in explosives was not legal. But Dupont maneuvered to retain virtually all of the smokeless powder and two-thirds of the dynamite capacity of the nation after Atlas Powder and Hercules Powder were formed by partition of the monopoly.

Eventually, T. Coleman wished to sell his substantial stock in DuPont back to the Company so he could pursue construction of the Equitable building in New York. The Company board, including P. S., paused to consider Coleman's asking price of $160. per share. In his report to Coleman, P. S. misrepresented the board's consideration of the offer as a rejection, then negotiated purchase of the shares for himself. The family, and eventually once again the courts, judged P.S. had acted for his interests alone and had used the Company's credit improperly in financing the loan required for the stock. But the deed was done and could only be undone by a vote of the shareholders -- and of course they were dominated by P. S.'s interests.

P.S. shrewdly anticipated the demands for munitions as the world prepared for war. DuPont installed enormous, new manufacturing capacity, and in the years from 1914-1918 the Company's profits soared to astonishing levels on the basis of its manufacture and sale of the smokeless powder for which it was still essentially the sole supplier. Earnings in 1915 were nearly $60 million on sales of just over $130 million.

Both P. S.'s own growing personal resources and the excess cash from Dupont Company operations was directed to one major goal during and after the war -- toward purchase of a controlling interest in one of the smaller motorcar companies challenging Ford's dominance, General Motors. By mid-1919, DuPont owned 33% of the carmaker -- and P. S. duPont became its President.[9]

DuPont had pursued research along two lines in the period 1902-1920. There was a strong commitment toward the improvement of the manufacturing skills of the operating factories through the medium of research. The Eastern Laboratory reduced the amount of glycerin required for dynamite by 20% and Stine had built DuPont's first "semiworks" for development of the European process for trinitrotoluene (TNT).

Additionally, longer range work was given priority at the Company's Experimental Station on the banks of the Brandywine River, just downstream from the Company's ancestral Hagley Yard where the Company had been founded in 1803. It was more difficult to define the success of the longer range work, nevertheless there had been a good deal of patience with the concept through the end of the war.[10]

After the war Dupont was forced into a period of retrenchment and change, particularly in research. The return of the munitions and powder business to normal levels and the flash recession of 1920-1921 led to significant cutbacks. The final phases of the organizational reforms initiated by P. S. before the war gutted the Chemical Department. Whereas in earlier days this organization, responsible for the research of the entire Company, had a staff of up to 500, now, within the "new federalism" of the Company, with Departmental responsibility shifted to the Departmental General Manager, separate research organizations were established within each Department, and the Chemical Department mission became uncertain. Perhaps it would simply be as the caretaker-landlord of the Company's research site -- the Experimental Station.

To look at public perception of the duPonts and the DuPont Company in the 1920's is a curious undertaking. Both family, and to a greater extent Company played and labored in relative obscurity when compared to the decade to come. National political prominence (save for T. Coleman's few years in the U. S. Senate) for the family and the investigations of duPont as the "merchants of death" in the war were still in the future.

For the previous two decades, the three cousins and finally P. S. had purchased, defined and dominated the business. Family control of the Company would continue, of course, during the '20's, but the major players were on a larger stage, propelled by the astonishing windfall of dividends generated by the organizational preparation for the war years. Company sales fell to only 25% of the wartime peak, and although earnings were generally solid, the flow of peacetime dollars into the family had been dwarfed by the wartime dividends. P. S., his brothers Lammot and Irenee, Alfred I. and the other family shareholders had received more than 250 million dollars in dividends during the four war years from a business that now had sales of less than 100 million dollars a year.

For some, direction of the Company interfered with management of this great wealth in new ventures outside the family heritage. P. S.'s focus was General Motors, because both DuPont's and his personal wealth depended on the powerful surge of GM past Ford; Alfred I. quietly dominated real estate and railroads in Florida; many of the major family players completed the manicured estates which line Delaware Route 100 from the Pennsylvania State line southward until the road nears the Brandywine and the old Hagley Yard at the Experimental Station.

The reorganized DuPont Company began to contribute measurably to the ways of American life through the '20's through its own science and through its acquisition and development of the intellectual property of others. Among the inventory of significant achievements of the decade: tetraethyl lead, dyestuffs, rayon and cellophane, paints and finishes and shatterproof automotive glass, only the last two were generated within the Company. DuPont development expertise and manufacturing and marketing skills had been perfected to the extent that these new products began to impact American life, however awareness of the role and benefits of synthetic chemistry was for the most part hidden from a public whose technical interests lay more with the wonder of the radio and the glamour of Lindberg.[11]

CHARLES M. A. STINE

The reorganization and decentralization of the Chemical Department in 1921 seriously undermined the franchise of the Department's long-time Director, Dr. Charles L. Reese. Although Reese would remain on as titular head of the Department until his retirement in 1924, the mantle of leadership of the Research unit passed to his Assistant, Dr. Charles M. A. Stine.

In 1909, DuPont assigned young Stine, a 27 year-old PhD, two years out of Johns Hopkins, the task of developing the process for manufacture of the new, European explosive discovery, trinitro toluene(TNT). His nitration process and product refinement was best demonstrated, Stine reasoned, by the construction of an intermediate scale manufacturing unit -- a semiworks. Stine's innovative and low-cost scale-up of his laboratory work became a model for process development; Stine and others after him argued intermediate scale process development, prior to the commitment to full plant manufacture gave a margin of safety and opportunity for process development assuring efficient and productive commercialization of the full scale plant.

Later, as Assistant Chemical Director in 1921, with the Chemical Department at "semiworks" size itself, its role reduced to little more than a caretaker for the Experimental Station, Stine proposed the Department act as a contract laboratory for Research for non-competing Companies. The Executive Committee of the Company delayed action on this proposal, but Stine and Reese had narrowed their sight somewhat and were hard at work trying to negotiate a contract with General Motors, planning to supply the automaker, now controlled by the DuPont Company, with all its chemical research.

Stine particularly wished to pursue basic research in fuels to solve the inefficient combustion problems that plagued the auto industry. He never realized his grand design of a laboratory serving General Motors. Empirical testing in the GM laboratories under Thomas Midgley revealed the anti-knock properties of tetraethyl lead and DuPont contracted for the development of the additive in a separate (and in many ways, ill-fated[12]) research and development contract.

Nevertheless, Stine was persuasive with the General Managers of the operating units of the Company, who now depended on their own research units, and convinced them of the need for programs of research less closely aligned with current business interest. He revived the Chemical Department and in a broad extension of his mission, proposed his program of "pure science."

RESEARCH SCIENTISTS[13]

In the old photos, Thomas Edison works with his staff at Menlo Park; the Wright brothers, heads down, are laboring in their bicycle shop; Elmer Sperry supervises installation of his gyrocompass. The still pictures nevertheless imply action, experimentation, trial and error.

There are a few photographs of Wallace Carothers from his days at DuPont. Carothers stands at the laboratory bench, in jacket and tie, and looks down at some small object. The pattern of flasks and beakers on the wall rack is familiar; there is a distillation column, rubber tubing, a burner to heat an oil bath.

What is the creative process which marks indelibly the future? Carothers records his time at DuPont was spent "thinking, smoking, reading and talking". Is this time of contemplation and of communication important, or is discovery in the popular "Edisonian" model, simply the result of a sequence of trials or an experimental barrage?

Edison, Sperry and Orville and Wilbur Wright, along with Fessenden, De Forest, Alexander Graham Bell and many others, formed the vanguard of independent American creative genius in the period from 1875 until the time of Carothers. There are common characteristics in this diverse group even though, as inventors, they went different ways. Their success depended for the most part on their innate ability to understand that successful completion of their inventive quest required the assistance of adequate support: of model makers, laboratory assistants, specialized equipment, manufacturing capabilities and the funds necessary to continue their work.

Carothers' father reports that Wallace studied "The Life of Edison" as a boy.[14] Edison, still in mid-life, was a legend for his profound impact in lighting and electrical systems; he received roughly 1,100 U. S. Patents for his work. The "Edisonian" method of research, characterized as "hunt and try" was in fact grounded on basic principles. First, the problems chosen were significant and difficult. For Edison: understanding and controlling electricity, for the Wrights: flight, and for Sperry: feedback and control systems.

The literature was important, all the literature, carefully read and understood, searching for what had not worked in the past. Understanding of the scientists of an earlier perspective led critically toward new ideas. Edison's fine library held the work of Michael Faraday, Sperry was an eager reader of the Official Gazette of the patent office and the Wright brother's years of glider trials at Kitty Hawk, culminating in successful flight in 1903, followed extensive study and correspondence with Octave Chanute after digesting his 1894 book, Progress in Flying Machines.

There is an ability among these men, to think spatially, to divine and conceptualize, using verbal and even physical metaphors without mathematical rigor. None of them was strong in practical development through calculation as a leading step; concepts arose from patterns of three-dimensional thought; pictures not equations held primacy.

CAROTHERS

Wallace Carothers was virtually unknown on the chemical scene at the time he was hired by DuPont; he had no glamorous academic reputation, no position of authority or achievement.

He had published two short papers in the Journal of the American Chemical Society. The first, in 1923, dealt with his belief, on the basis of their physical properties, that phenyl azide and phenyl isocyanate were isoelectronic.[15] The second, published a year later, and entitled simply, "The Double Bond", was a theoretical discussion of the octet theory of electron distribution and its relationship to the reactivity of unsaturated molecules.[16] Carothers was the sole author of both manuscripts; Professor Jack Johnson of Cornell believed the paper on electronic theory of unsaturation was of pioneering stature, but little attention has been paid to it over the years.[17]

As a freshman at Tarkio College in September 1915, Carothers began his science training, studying chemistry with Dr. Arthur Pardee. At 19, he was already skilled as a secretary and earned bed and board teaching in the commercial side of the small Presbyterian college while pursuing his classes. His friends remember songs and poetry; they called him "Prof" and at the end of two years, his nickname became a reality as Carothers was assigned to assist in the teaching of English. Pardee took another position the following year and Tarkio, in the wartime summer of 1918, was left with no instructor in chemistry. So Carothers, a senior, having failed his military physical, taught the classes for the next two years, while completing his own degree.

We presume Pardee was the architect of Carothers' acceptance at Illinois to enter the graduate program in September 1920, (although perhaps a letter from Carothers, head of the chemistry department at Tarkio would have been sufficient recommendation for Carothers, the prospective Illinois student.) He completed a Masters in a year, then for the next year earned money as an instructor with Pardee, who had now relocated once more, this time at the University of South Dakota.

He could afford to return to Illinois in the summer of 1922, he completed his PhD. under Roger Adams by 1924 and stayed on at Illinois, again as an instructor until Adams arranged with James Conant for Carothers to leave the midwest and take residence in the creaky and antiquated home of the Harvard Chemistry Department, Boylston Hall.

ADAMS AND ILLINOIS

Clerk: Dr. James B. Conant. Dr. Conant please take the stand.

Atty: Dr. Conant, it was very good of you to come here on your short vacation to testify about the University of Illinois. Wasn't Adams at Harvard when you were there?

Conant: Yes, a long time ago. We don't talk much about it though.

Atty: What characteristics did he display there?

Conant: Well, he seemed to have a yery competitive spirit. Always trying to get ahead of me.

Atty.: How do you mean?

Conant: Well, I got interested in natural products, so he gets interested in natural products. I start working on chlorophyll ....
He was a hard man to get ahead of.

Atty.: How did he come to leave Harvard?

Conant: Roger couldn't stand the pace and since Elmer Bolton had gotten the only soft industrial job available, Roger decided he'd vegetate in some cow college out west. Finally caught on at Illinois, I hear.

Atty.: In your contacts with the chemical industry, have you found any evidence of domination by the University of Illinois?

Conant: Yes indeed. Illinois is the Columbia Teachers College of the chemical field. Not much quality, but you can't find a sewage treatment plant anywhere without at least one PhD from Illinois.

> Adapted from a 1953 skit performed for the celebration of 25 years of Roger Adams as a DuPont consultant.[18]

Adams, Conant, and Bolton. These three start out in Kohler's Harvard Department. Bolton goes off to Germany for a year with Wilstatter, then returns and starts with DuPont. Adams follows with Wilstatter, goes back to Harvard, then takes a position at Illinois. Conant, the youngest of the three, gets the Harvard teaching job Adams leaves behind.

Bolton will be DuPont's research Director for 17 years and will be Carothers' boss. Conant will be the High Commissioner and Ambassador to Germany after serving as President of Harvard. Adams, though, will always be the "Chief".

The Land Grant Colleges, among them Illinois, had from the beginning, the responsibility to serve public rather than private interests. The establishment of industry-related curricula in Engineering, and later in the sciences at MIT and the other technical schools provided powerful incentive for the large midwestern Universities to follow suit with science and engineering in the public weal. Engineering experiment stations at Iowa, Penn State, Kansas State and Illinois, to name a few, strove to stimulate engineering education and investigate problems of special interest to the state's industrial interests.

Departments of Chemistry, cast in this mold, followed quickly. At Illinois, Roger Adams, a Bostonian descendent of the sixth President, migrated from Harvard's tradition[19] into the newly completed Illinois Chemistry building in 1916.

In his report to the trustees of the University in 1920, Adams recounted the experience of one program he held in high regards -- "Summer Preps". Systematic preparation of Organic Chemicals, especially the particularly difficult to obtain aromatic derivatives which had traditionally been available from Germany, was carried out in the Department by student labor. To substitute for the wartime shortages, students practiced and perfected the preparations, and were paid a small hourly wage. The chemicals were sold to industrial customers; the balance sheet for the middle of 1920 placed a value of $3,000 on the stock inventory, with sales of close to $30,000 over the previous three years.

This fine chemicals business, run from the Chemistry Department, went forward to be the genesis of the important annual, "Organic Synthesis". Summer preps lasted into the '40's at Illinois and its very presence, notwithstanding the high regard with which the program was held, speaks of Adam's interests and direction.

By 1926, Adams was the "Chief"; he was Department chairman and the nickname described the authority he carried well past the end of World War II. His research interests always were of an exploratory nature, with the twist that the experience was expected to teach the student, not necessarily promote great discovery. Adams graduated 183 PhD's from Illinois over a 40-year span. At least twenty-seven of them went on to work at DuPont.[20]

CANDIDATES

"We shall have to pay between $8,000 and $15,000 a year for a man of established reputation and proven worth," wrote Stine.[21] New PhD. were earning less than $3,000 a year at the time.

It is clear that Stine had his mind set on Adams for the lead job from the beginning. Bolton, who had been the advisor to Stine and whose reputation as a successful research manager in his own right was looming larger, certainly would recommend his friend and mentor. But Adams refused the job. His salary at Illinois was $12,000 and he acknowledged the new position would offer even better compensation,[22] but the new Department chairman had center stage among the country's chemists. He properly saw his influence would better grow in the management of the Illinois Department than in the more limited role at DuPont, where he would be claimant to only a small share of the responsibility. He would not be at his friend Bolton's level in the DuPont organization; he would be untested in the direct industrial environment but he knew from years of consulting with Abbott laboratories, of the value to the University of his industrial contacts.

It appears that Stine was rapidly disabused of the idea of bringing in a first rank university man and, through Bolton, then began pursuit of three lesser known chemists: Louis Fieser, now at Bryn Mawr after his Harvard PhD., Henry Gilman who had studied under Adams at Harvard, served briefly at Illinois, but who now chose to star on his own stage at Iowa State, and Wallace Carothers.

The decision to hire a single scientist for the organic section was also amended. Negotiations with Gilman and Fieser continued after Carothers accepted, and discussions with Reynold Fuson, one of Adams junior staff at Illinois extended well into 1928 and after Carothers came to work.

Adams undoubtedly recommended Gilman, Carothers, Fuson and another member of his Department, "Speed" Marvel to DuPont. However, Fuson, in refusing DuPont's offer, noted he had spoken with Adams, who had assured him his future was secure and there would be ample opportunity if he would stay at Urbana. The "Chief" was successful in supplying DuPont with the names of prime candidates for the pure science position, while simultaneously protecting his own staff from defection to DuPont.

Conant showed no enthusiasm. Bolton asked his friend's advice on the new position and Conant recommended his former student Fieser. But the Harvard professor clearly let Bolton know that he believed less able and talented men were more suited for industry. Fieser finally refused, admitting to Conant that he felt he would not find a job attractive, whose mission was the technical applications of fundamental chemical principles.[23]

Conant, too, undoubtedly recommended Carothers.

AN OFFER FOR CAROTHERS

The early fall of 1927 found a nation prosperous, 20,000,000 automobiles on the country's three million miles of roads, efficient rail transportation, emerging commercial radio, long distance telephony and two-cent, next day mail service as a routine, up and down the East Coast. In mid-September, Wallace Carothers accepted Stine's invitation to come down from Harvard and visit the Chemical Department to talk about heading up part of the "pure science" program now underway in brand-new "Purity Hall".

It is compelling for one who has interviewed with DuPont to fashion the details of that trip. To conjure an evening train to Wilmington for Carothers, a short cab ride to the Hotel duPont; up Market Street and past the statue of Caesar Rodney, then left to the entrance on the north side of the DuPont building. One expects a small, high ceilinged room; breakfast, probably with Stine, or perhaps Tanberg[24] but certainly in the Green Room, with linen, and the Hotel duPont china and silverware.

Next, they would go to the "Station". (Perhaps Carothers would, for a moment, wonder why he would be returning to the train station.) They would ride out Pennsylvania Avenue past its fine homes, then right on Rising Sun, with the sudden dip under the railroad, down to and across the Brandywine and into the main gate of DuPont's Experimental Station.

"My dear Carothers:" wrote Stine on the 20th of September. "You no doubt have much to think about in connection with another year's work at Harvard, but I believe you will have taken time to think about your visit to Wilmington .... . I hope you that you will decide to cast in your lot with us. I suggest that an initial salary of $5,000 would appear to be suitable.

"I am enclosing herewith Company check for $34.41, to cover your expenses ..... in accordance with the itemized account which you submitted with your letter ..... of September 19th."[25]

REJECTION; ACCEPTANCE

After an exchange of letters in which Carothers asked for a few details should he plan to begin his work about February 1, 1928, Carothers refused DuPont's offer in a brief note written Sunday, October 9. He offered only an undetailed comparison with his position at Harvard as his reason. Carothers' rejection crossed a chatty letter Tanberg had sent him the previous Friday. This letter was delayed in getting to Cambridge, and its receipt on Thursday, October 13th, triggered a long and detailed explanation for the decision to remain at Harvard.

He acknowledged his enthusiasm for the basic research program as being one method of vitalizing academic research and forging an academic-industrial interface and saw his opportunity to participate as "an extraordinary piece of good fortune". Yet a "long list of scruples arose" with the requirement of making a real-time decision for or against the change.

Carothers reflected that his move from Illinois to Cambridge the previous year had been difficult and he required a clearly differentiated opportunity to make yet another move. He felt the completion of Harvard's new laboratory replacing Boylston Hall would provide the best working environment and he was not satisfied that the difference between the DuPont offer and his $3,200/year Harvard salary was sufficient to compensate for the risk as compared to the stability of his Harvard situation.

> "The most important thing to consider was the relative opportunities for scientific achievement, ..."
>
> "Some years ago, I became convinced that the important contributions of the future to Chemistry would be made almost entirely by powerful and compact research organizations; but I have since decided that discovery is too elusive an art for this to be necessarily true. Nevertheless it must be admitted that, other things being equal the amount of research which one can accomplish will depend directly up to a certain point on the number of hands one can command.
>
> "If I understood properly the intentions of your offer and correctly estimated my own abilities, it seems probable that I should have (had) more assistance at Wilmington than will be available at Harvard during the next few years. On the other hand the prospects for turning out research here seem to depend upon fewer contingencies than at Wilmington. I suppose it more or less inevitable that circumstances there might arise making it necessary to suppress the development of an investigation; and there is the question of how my ability to fit in and adapt myself to the quite different situation might affect the possibility of doing productive work. I suffer from neurotic spells of diminished capacity which might constitute a much more serious handicap there than here."[26]

His vision of scientific productivity, then, seemed to be clouded only by the fear that promises would not be kept. And he feared the resulting tension would trigger a defect that Carothers was willing to identify -- periodic depression with its numbing of the creative spirit.

But Carothers' inventory of his thinking gave Stine something to work with. The following Monday, the 17th, Stine's assistant, Hamilton Bradshaw wrote Carothers, saying he would be in Cambridge on Friday, October 21st. He asked Carothers to write back, indicating if the two could get together that morning.

On October 31, 1927, Wallace Carothers accepted a sweetened offer of $6,000/year to join the group at the Chemical Department.[27]

ACHIEVEMENTS IN PURE SCIENCE

Wallace Carothers stood on the platform of the Pennsylvania Station in Wilmington once again. As he awaited the train for New York he could reflect on the accomplishments of his first three and one half years with DuPont. It was Christmas time in 1931 and Carothers headed for New Haven; for the Sterling Chemical Laboratory at Yale where he had been invited with a few, select leaders in Organic Chemistry, to present his work at the fourth biennial Organic Symposium.

His subject was the new synthetic rubber, polychloroprene which had been discovered unexpectedly in a study of monovinyl and divinyl acetylene.

The previous September, at the Buffalo meeting of the American Chemical Society, his presentation on the design and synthesis of condensation polymers and the conversion of these polymers to fibers with properties further enhanced through the discovery of cold drawing, had been received by the attendees with "noted interest".[28]

Carothers had now published more than a score of papers from the Chemical Department and his recent review, titled simply, "Polymerization", offered a new vocabulary with the now-familiar terms: condensation polymerization, addition polymerization and monomer.[29] The paper was fashioning worldwide thinking on the mechanisms of polymerization and the expected properties of polymers.

Charles Lamb, Editor of the Journal of the American Chemical Society, named Carothers Associate editor of the journal and Carothers was serving by refereeing a dozens of submitted papers. He had responded to more than 30 such requests over the last eighteen months.

Stine could report the objectives of the pure research program were well met. DuPont through Carothers had widespread scientific visibility and intimate contact with other important research centers. The staff was growing and Chemical Department influence was rising. He could further note the discovery of an apparently useful synthetic rubber; the practical and useful could come from the area of "pure science".

Even before his acceptance of DuPont's offer, Carothers began to feed ideas and information to Stine and Bradshaw. He listed a number of areas of research interest; he continued to think about the electronic structures of simple molecules as an outgrowth of his earlier paper on the double bond; he proposed synthesis of drug intermediates, searching for convenient laboratory syntheses of compounds with pharmacological interest such as atropine and cocaine.[30]

Carothers and Stine had discussed polymerization during their interviews and Carothers later reviewed the literature for Bradshaw and Tanberg, noting the difficulty of determining the structure of polymers and the continuing arguments of Staudinger, on one side and Pummerer (for one) on the other side of the key structural question: are polymers of a massed colloidal nature, or as Staudinger believed, simply giant molecules with covalent bonds not materially different from the bonds in smaller molecules?

Carothers discussed the hydrogenation of natural rubber and confessed puzzlement at the differences in properties between the saturated, high molecular weight hydrocarbon produced thusly and the properties of the highest molecular weight hydrocarbons which had been synthesized by building up the chain length in a designed way.

The first proposals he made in his letters, for the systematic synthesis of long-chain molecules were crude and unwieldy but they set the stage for the invention of controlled methods of condensation polymerization.

It is clear from the testimony of Adams, of Conant, of Marvel and Nobel Laureate Paul Flory, Carothers was specially gifted with an intuitive feel for his creative craft. Adams, in reviewing the records of accomplishment of the nearly 200 Ph.D.'s he trained, returns to Carothers again and again as the model. He was as proficient in physics, mathematics, music, history and politics as he was in chemistry, relates the "Chief". Adams specifies an accurate and complete functional memory as an essential prerequisite for chemical 'stardom';[31] Carothers' Chemical Reviews paper on polymerization demonstrates the broad base of information from which he derived his work, and he augmented his familiarity with the science through his extensive journal editing.

Although Carothers' papers use diagrams rather than mathematics to express his conceptual thinking, he later brought into his group the "his physical chemist", Paul Flory who began his work seminal work on molecular weight distribution under Carothers. It was said in the group, "Only Carothers understood Flory."[32]

Carothers sought solutions for the difficult and challenging problem of the key to polymer structure and orderly synthesis. Even Adams questioned the wisdom of his former pupil, cautioning "the products are atrociously insoluble" and that "it is almost impossible to get a means of attack".[33]

These characteristics, and the application of them to truly important objectives marks those whom we call "Inventive Geniuses".

NYLON

We may accept that Carothers deserves nomination as a truly unique scientist, whose methods for condensation polymerization were necessary for the invention of nylon. But they were not sufficient.

In 1930, Stine was promoted to join an Executive Committee, well pleased with his work. His new position as indirect overseer of all DuPont research superceeded direct responsibility for the programs of the Chemical Department. Elmer Bolton was immediately named his successor. Bolton carried into his new responsibilities cautious skepticism for the maturing and seemingly productive fundamental research work. One of his first actions was to move quickly for the rapid commercialization of polychloroprene.

Polyester development, however, languished. Neither Carothers' group or rayon experts within the Company could modify the properties of low melting point, hydrolytic instability and solvent sensitivity; at least one of which appeared to be common to all the polyesters and limit their commercial potential as fibers. After a few experimental attempts at making useful polyamides, work in polymerization slowed to a trickle.

It was the management and vision of Elmer Bolton which, at this point, became responsible for high-volume, low-cost nylon which has generated a multi-billion dollar, 50-year earnings stream for DuPont.

Briefly told, Bolton continuously prodded Carothers to return to polymers and especially polyamides. Carothers' progressive spells of depression were becoming true barriers to consistent leadership of his group; his was a depression which deepened as Bolton redirected the fundamental research effort away from its original goals.[34]

But Carothers learned to prepare polyamides and chose for further development, nylon 5,10, from pentamethylene diamine and sebacic acid. Bolton refused to consider this material for final commercialization, reasoning that the raw materials would forever be difficult to obtain, and he insisted nylon 6,6, with both monomers theoretically available from benzene, would be the final target.

Nylon 6,6 melts far higher than the early fiber scientists were, at first, able to handle and efficient monomer synthesis from petroleum feedstocks was but a Bolton vision. However, within four years from the time Gerard Berchet made the first sample of nylon 6,6 on February 28, 1935, DuPont was in commercial production of its most important product.[35]

CAROTHERS

Wallace Carothers never saw the completion of the nylon development. He wrote, after his well-received talk at the 1931 Organic Symposium and on the dawn of the new year, "1932 looks pretty black to me right now."[36] The years became blacker still. He was hospitalized for depression in the summer of 1934, and again for a longer period after he was elected to the National Academy of Sciences in 1936.

Later that same year he tramped the Austrian Alps, part of the time alone, part of the time with Adams; his mentor monitored Carothers' mental state on a near daily basis in long letters to Mrs. Adams in Urbana.

There was no interruption of his progressive illness, however, and in the predawn darkness of April 29, 1937, Carothers drove to Philadelphia, checked into a hotel and drank a fatal potion made up of lemon juice and the contents of a small vial of cyanide his friends say he had carried around for years.

REFERENCES AND NOTES

1. LIFE, Fall 1990, Special Issue, names Carothers among the 100 Americans of this century whose contribution are most noteworthy.
2. C. M. A. Stine to Executive Committee, December 18, 1926.
3. America by Design, David F. Noble, Alfred A. Knopf, New York, 1977. pp 113-115. The activity of patent director Davis in the establishment of the laboratory is important to note. One salient characteristic of General Electric and the other technically advanced American companies of the period was the aggressive patent action. Patents, as stock-in-trade carried heavy weight in setting priority. There is a lack of patent sophistication throughout the DuPont "pure science" era, which, in the end, was extremely damaging to the Company when early publication of a comment by Carothers that caprolactam would not be polymerizable, allowed ICI to obtain patent rights to this important material.
4. Willis R. Whitney to Charles M. A. Stine, March 24, 1927.
5. Stine remembers that Bolton felt fundamental research couldn't be administered logically within an industrial organization. Later, Bolton as Stine's successor, would slowly change the nature of the "pure science" program. Charles M. A. Stine to Roger Adams, December 2, 1938.
6. Alfred I. DuPont; The Man and His Family, Joseph Frazier Wall, Oxford University Press, New York, 1990, p 180ff.
7. Edgar S. Woolard announced the replacement of the Executive Committee by a Chairmnan's office on October 1, 1990, thus ending 79 years of organizational continuity within DuPont.

8. The Visible Hand, The Managerial Revolution in American Business, Alfred D. Chandler, Jr., The Belknap Press of Harvard University, Cambridge, MA, 1977, p 438ff.

9. Pierre S. duPont and the Making of the Modern Corporation, Alfred D. Chandler and Stephen Salisbury, Harper and Row, New York, 1971, p 433ff.

10. Science and Corporate Strategy; DuPont R&D 1902-1980, David A. Hounshell and John Kenley Smith,Cambridge University Press, Cambridge, 1988, pp 11-125.

11. The Corporate State and the Broker State; The DuPont's and American National Politics, 1925-1940, Robert F. Burk, Harvard University Press, Cambridge 1990. Chapter one describes the '20's, DuPont and the duPonts.

Most of Burk's contentions as to the important societal impact of the Company's emerging technology, and its manner of doing business are certainly correct. And the family contributed more than its share of willing players in the "materialistic Eden" of the time.

I believe, however Burk's retrospective judgement could not have been made accurately in 1928 from the contemporary data. For the '20's were Fitzgerald, Hemingway, expatriates, the Jazz Age, Lindberg and Babe Ruth, not the quarrelsome yet self-protective family dynasty choosing to grow its wealth in the relative isolation along the Brandywine and the east Coast of Florida.

Yet the glamour and frolic of the duPonts did not pass entirely without notice. Portraits of the lifestyles of P. S., T. Coleman, Irenee and Henry F. duPont appeared during the decade, along with a long article discussing the curious appointment of P. S. as Delaware's multimillionaire Tax Collector, and a Times (July 30, 1924) editorial praising his gift, in the name of his deceased chauffeur, supporting hospital construction in West Chester, Pa.

Lost in this was the Company. Reader's Guide cites 19 articles on the family from 1920 through 1928; only four cite the Company. The pattern in the New York Times for the same nine years is similar -- family news outnumbers Company by about four to three.

Of about 150 references to the Company, finances and bad news dominated. We read, with no editorial notice, of worker deaths during the development of tetraethyl lead, lawsuits surrounding stock purchases, and legal problems related to the devious hiring of German dyestuff scientists after World War One but not one single reference to the impact of a technical development.

In spectacular contrast, the Times mentions General Electric in nearly 500 separate articles over these nine years. Every morsel of potential technical or scientific advance is offered. There will be radiotelephones on street railway cars, artificial lightning will have a practical use, talking movies are on the way, television(!), new searchlights will brighten airfields, radio, new materials, etc.,etc.,etc.

But the newspaper, while praising GE's technology, editorially chides its President Swope as nearsighted for anticipating technically oriented work will compensate for the "dreadful drudgery" of the new, machine age jobs. (New York Times, July 24, 1924).

12. The development of TEL in the absence of basic data on the toxicology of the compound led to chronic illness and death among many of the early workers in the development and manufacture. Nation, 62-64 (July 8, 1925), provides a short contemporary summary. Also see Science and Corporate Strategy, p. 150ff.

13. American Genesis, Thomas P. Hughes, Viking, New York, 1989.

To what do we attribute the scientific success of the pioneering inventors of the late 19th and early 20th Century? Hughes' book gives us a common pattern and establishes by inference, powerful guidelines for those of us who practice in the scientific disciplines.

14. Ira Carother to Roger Adams, Dec. 2, 1937.

15. W. H. Carother, J. Amer. Chem. Soc., 45, 1734 (1923).

16. ibid., 46, 2226 (1924).

17. Prof. J. R. Johnson to Roger Adams, November 3, 1938.

18. Dr. B. C. Anderson of Du Pont produced a copy of this skit for Prof. Carl Marvel some time ago. Marvel's copy is with his papers in the library of the Beckman Center for the History of Chemistry, in Philadelphia.

For many years, up until the mid-sixties, it was customary for those left behind to honor scientists who left Du Pont's Chemical (later Central Research) Department, with a tongue-in-cheek presentation. These celebrations generally marked the transfer or promotion of one or more of the Department staff to another Du Pont Department, and were an opportunity to harass and poke fun at the honorees, the Company and its leaders.

This night of honor for Adams, and on the same night for Prof. Marvel, indicates how closely allied the two men were with the Du Pont Company.

19. My Several Lives, James B. Conant, Harper & Row, New York, 1970, p. 76. Conant acknowledges that his aim at Harvard between the wars was "advancement of knowledge, not its application," and admits he was "oblivious" to the tradition of research which sought to serve industrial problems.

20. Roger Adams, Scientist and Statesman, D. Stanley Tarbell and Ann Tracy Tarbell, American Chemical Society, Washington, D.C., 1981. Appendix A lists 26 who worked for Du Pont; the text indicates a 27th, William Lycan also spent time with the Company.

21. C. M. A. Stine to the Executive Committee, March 21, 1927.

22. Taped interview with Roger Adams recorded at the Chemists' Club, New York City, in November 1964. Obtained from the University of Illinois Archives; Mr. Robert Chapel.

23. Science and Corporate Strategy, p. 228, 229. Perhaps neither Fieser nor Conant fully explored Stine's promise of a "pure science" program. Would Bolton, as Stine's skeptical emissary, have modified Stine's program specifications? This is not likely.

24. Arthur P. Tanberg was Director of the Experimental Station. His position was distinct from the directors of the Chemical Department; he was responsible for the overall facility housing many research units, and became involved in important personnel matters such as the hiring of the scientists for Stine's new group.

25. C. M. A. Stine to W. H. Carothers, September 20, 1927. In that time before regular air mail; before the copier or Federal Express or the fax machine era, the two-cent letter was mailed in Cambridge on Monday, received at Wilmington and delivered to Hamilton Bradshaw on Tuesday. An offer letter was prepared, a check was cut and the whole thing was sent back to Carothers the same day.

26. W. H. Carother to A. P. Tanberg, October 13, 1927.

27. Bradshaw could easily counter, in good faith, the objections Carothers had presented.

Money was no problem. In fact, the offer was far below the original specifications for the job. Bradshaw could assure Carothers of the commitment, at the highest levels, to the facilities and the staffing which would give Carothers the clear calling to the Wilmington job.

There was, in the society of the '20s, a drive to engineer narrow specifications for the hiring of personnel to very restrictive norms. Carothers' frankness about his limitations is astonishing even in our time of more open exchange of detailed experience. But there is no record of concern for this admission found in the Du Pont files covering the time of Carothers employment.

Carothers depression was, in fact, progressive and ultimately fatal.

28. Industrial and Engineering Chemistry, News Edition, september 10, 1931. Henry Gilman wrote the report of the meeting. Attendance was a record 2057.

29. W. H. Carothers, Chemical Reviews, 8, 353-426 (1931).

30. There is no evidence of any arrangement or secrecy agreement between Carothers and Du Pont from September 1927 until his employment on February 6, 1928. Apparently Stine, Bradshaw and Tanberg were willing to listen and comment and neither party endeavored to hold the discussions in a privileged way.

Three letters, Carother to Stine (September 27, 1927), to Bradshaw (November 9, 1927), and to Tanberg (November 20, 1927) give a rather complete picture of Carothers' pre-employment program ideas.

31. See Reference 22.

32. Julian Hill; personal communication.

33. R. Adams to W. H. Carother, February 18, 1928.

34. By 1932, Carothers questioned openly the course of his work and noted he would not have come to Du Pont had he been aware of the course the work would take. W. H. Carother to A. P. Tanberg, August 5, 1932.

35. Science and Corporate Strategy, pp. 257ff.

36. W. H. Carothers to W. Machetanz, January 4, 1932.

TEXTILE MILL SCALE-UP OF NYLON HOSIERY: 1937-1938*

HERBERT T. PRATT
23 Colesbery Drive, New Castle, DE 19720

INTRODUCTION

On October 27, 1938, Dr. Charles M. Stine, Vice President of E. I. Du Pont Company, speaking in New York before a meeting of 3000 women's club members, announced that Du Pont had developed from coal, air, and water a new textile fiber called nylon, having "filaments as strong as steel, as fine as a spider's web, yet more elastic than any of the common natural fibers" [1]. Dr. Stine's startling revelation came a little more than ten years after Du Pont had launched a program of pure scientific research into the fundamentals of polymers and polymerization. The research had been led by the brilliant young organic chemist, Dr. Wallace H. Carothers (Note 1), who, in March 1928, had been enticed away from a teaching career at Harvard, teamed with five or six men as capable as himself, and surrounded with every resource his heart desired.

FIBERS: A POTENTIAL MARKET

Carothers' work advanced at a rapid pace, and by the fall of 1935, the Company had determined to explore, as one end use for polyamides, the commercial possibilities of textile fibers, specifically those from the polymer based on the six carbon molecules of hexamethylene diamine and adipic acid (Note 2). In January 1936, responsibility for this fiber was assumed by the Pioneering Research Section of the Technical Division of the Rayon Department in Buffalo, New York [4] (Note 3). Thus was born the project for Fiber 66 later to be named nylon (Note 4).

By that fall, Fiber 66 had been deemed a feasible venture, and in November, Dr. G. Preston Hoff (Note 5), a research manager at Buffalo, was appointed as liaison between the Chemical Department at the Experimental Station in Wilmington that would scale up the polymer process and the Rayon Department in Buffalo that would perfect the 66 yarn.

Full-fashioned Hosiery

In casting about the textile industry for potential end uses for Fiber 66, one having particular appeal was women's full-fashioned stockings, which, at the time, were made almost exclusively from silk. Women's hosiery consumed two to three million pounds of silk annually, or about 80% of all silk used in the United States [7]. There were at least seven good reasons for targeting full-fashioned hosiery as an outlet for the new fiber: It was a sought-after, high priced product, shaped to fit the leg; the seam up the calf gave a glamorous high-fashion image not enjoyed by circular knit stockings worn every-day; Fiber 66 would not be forced to compete in existing markets for other Du Pont yarns; since a pair of silk stockings weighed only about an ounce, the output of a small Fiber

66 plant could be spread broadly across the consumer market; the physical and chemical properties of Fiber 66 were akin to those of silk; furthermore, poor yarn quality at start-up would not have as a great an impact in hosiery as in woven fabrics, since one bad yarn spoiled only a few stockings rather than many yards of fabric.

Because Du Pont had no experience with full-fashioned hosiery, the Research lab in Buffalo bought a single-head full-fashioned machine and began to learn how to use it. On February 15, 1937, Carothers filed for a patent on stockings from polyamide yarns, which he claimed would, unlike those from other synthetic fibers, cling properly and not become baggy [8].

The day after the stocking patent was filed, Dr. E. Vernon Lewis (Note 6), a research chemist at Buffalo, took a pound or two or of 50 denier (Note 7) yarn to Union Manufacturing Company, a Du Pont rayon customer in Frederick, Maryland, and over the next three days, February 16-18, knit the first full-fashioned stockings from Fiber 66. The trial did not go well at all. The yarns knitted poorly. The stockings snagged, did not absorb the dye, and were extremely wrinkled after dyeing. Bloodied but unbowed, Lewis returned to Buffalo armed with assurances from Union personnel that the problems could be solved [11].

SCALE-UP

Until the Spring of 1937, the amount of yarn available was limited to the output of a two-pound capacity polymer autoclave. When in March 1937, a new 50-pound capacity autoclave came on stream, the way was opened for a continuous supply of yarn and an extended mill trial.

Dr. Hoff now turned to the Sales Division of the Rayon Department in New York to find a suitable customer and to enlist the help of its Sales Service Section in following mill trials. The mill selected was Van Raalte Company, a household name for high quality women's underwear, hosiery, and gloves.

The Sales Service Section

The Sales Service Section, headed by Dr. J. B. Quig (Note 8), was, in technical matters, the link between Du Pont and its textile customers (Table 1). In the Fiber 66 project, it would assess the viability of the new fiber as a commercial product by adapting Van Raalte's processes to the fiber's characteristics, and, where that was not possible, it would recommend needed fiber changes to Du Pont's research and manufacturing organizations. Those to be involved directly were George J. Groh (Note 9) "coordinator of outside development" in Wilmington, and two field men, John C. Lake (Note 10) and W. Ellis Mann (Note 11). The breadth of experience of these men is telling of the kind of people selected for sales service work (Table 2). Groh, a 41-year old B.S. chemist, had spent two years in rayon research and nine years in sales service. Lake, at 34, had degrees in both the humanities and textile engineering, and mill experience in cotton testing, dyeing and finishing, and ribbon manufacture, as well as three years in acetate and rayon sales service at Du Pont. Mann, at 32, had degrees in both French literature and chemical engineering, had been both a high school principal and college teacher, and had rayon manufacturing and textile weaving experience at Du Pont.

Table 1. Some Functions of the Sales Service Section. Sales Division, Rayon Department, E. I. Du Pont Company.

- Act as the link in technical matters between customers and Du Pont manufacturing, research and sales.
- Solve customer problems related to processing and use of products.
- Adjudicate customer complains and claims.
- Monitor quality of existing products.
- Recommend product improvements needed to meet competition, to cope with changes in textile technology (new machines, higher processing speeds, etc.), and to accommodate customer process changes.
- Evaluate effect of Du Pont process changes on product performance.
- Evaluate new Du Pont products and recommend product changes.
- Teach customers how to use Du Pont fibers.
- Publish technical information on product properties and processing.
- Develop textile cost data.
- Conduct wear tests on new and improved products.

Table 2. Some Characteristics of the Sales Service Representative

- Diverse background and experience
- Technically astute
- Adept at problem solving
- Keen observer
- Adept at building customer confidence
- Aware of customer needs, production deadlines, etc.
- Practical minded and cost oriented
- Works well with others
- Does not need close supervision
- Works well under pressure
- Capable of on-the-spot decision making
- Good communicator

The Van Raalte Company Agreement

Eager to scale-up and fully believing in their mission, Hoff, Quig, and Groh met at Van Raalte's New York headquarters on May 6, 1937 with F. R. Reisenberger, vise president, a Mr. Claflin, general manager for hosiery, J. W. Crawford, superintendent of the Boonton, New Jersey hosiery plant, and a Mr. Howard. Quig's proposal was the Du Pont rent Van Raalte's equipment, services, and experience in ferreting out problems with what they described as an "improved yarn" [16]. Du Pont Sales Service would provide on-site assistance and follow-up. Seizing the opportunity to be in on the ground floor of what might turn out to be a money-maker, Van Raalte quickly agreed to the deal. They elected to start one 20 position machine, which would knit 1225 dozen pairs of stockings per week and require about 100 pounds of yarn. Du Pont would ship the necessary yarns immediately:

Leg Yarn	48 pounds	50 denier, 30 filament, 6 tpi Z, bright luster
Welt Yarn	45 pounds	75 denier, 45 filament, 6 tpi Z, bright luster
Foot Yarn	7 pounds	150 denier, 90 filament, 6 tpi Z, bright luster

Security

To ensure that information regarding Fiber 66 would not "leak out to the trade," they agreed that the yarn would be shipped in plain packing cases with no labels, Du Pont trademarks, or other identification, except denier. All paper work would to be sent to Van Raalte's New York office instead of the Boonton plant. At the plant, only Mr. Crawford would know where the yarn came from. All yarn fragments would be gathered, weighed, and held for Du Pont. Because John Lake lived in Pennsylvania, and his auto license plate would not be readily link him to Delaware and Du Pont, he would be asked to drive his personal car at Van Raalte rather than a company car [17]. Obviously, such tight-lipped security was warranted. During the meeting, Mr. Howard told the Du Pont visitors that the Celanese Corporation also was working on a new yarn for full-fashioned hosiery, and later asked "whether this yarn [Fiber 66] was the material made by one of the chemists at the Experimental Station , a good-looking middle westerner who drank heavily." No one from Du Pont replied, except to confirm that the yarn was being made both in Wilmington and in Buffalo [18].

THE FIRST TWO MONTHS

No time was wasted in getting started. On May 7, the day after the Van Raalte meeting, Lake lugged a hundred pounds of yarn to Boonton on the train [19]. His immediate objectives were two-fold: (1) to evaluate yarn performance in knitting and subsequent steps (Table 3), and (2), to assess stocking appearance, tactile hand, and elasticity. A longer term objective was to make stockings for wear tests. As a standard, Van Raalte chose their "Image" brand stocking, size 9 1/2, knit at 48 courses per inch from 4-thread, 20 turn, 13/15 silk, on a 42 gauge Reading machine running at 56 courses per minute (Note 12).

Table 3. Manufacturing Steps for Full-Fashioned Stockings From Fiber 66.

		July 1937
1.	Yarn Preparation	Twisting, twist setting, coning, etc.
2.	Topping	Knit stocking welt from 75 denier yarn
3.	Legging	Knit flat contoured leg blank from 50 denier yarn
4.	Footing	Knit foot on special machine from 150 denier yarn as continuation of leg blank
5.	Looping	Close toe and heel
6.	Seaming	Fold leg blank lengthwise and sew together to form shaped stocking
7.	Greige inspection	Cull defective undyed stockings
8.	Dyeing and finishing	Apply color and lubricant as appropriate for visual and tactile aesthetics
9.	Boarding	Dry stocking in air at 150-170°F on a form shaped like stocking
10.	Final inspection	Cull defective finished stockings

By the time Lake used his first progress report two months later, he had made only 30 dozen pairs of stockings, less than two percent of the knitting machine's rated capacity. Yarn quality had been poor, knitting and other manufacturing steps had been tedious, and the finished stockings were in no way fit for wear. Compared to silk, Fiber 66 stockings were too flimsy for easy handling in the mill, too shiny (lustrous) in appearance, too soft to the touch, lacked elasticity, were full of color variations and dye defects, and because of shrinkage during dyeing, were much smaller in size than the intended 9 1/2. Worst of all, they were full of snags and were frightfully wrinkled and creased [20].

Although solutions to these problems, ultimately, would be a team effort by the research and manufacturing and sales service, this paper emphasizes the contributions made by sales service to the body of knowledge that allowed the October 1938 decision to commercialize nylon fiber. Though problems were tackled more or less simultaneously, in this paper they are discussed individually for clarity.

PROBLEMS AND MORE PROBLEMS

Poor Yarn Quality

Quality of the first yarns was much too poor for good knitting. During Lake's first month at Boonton, knitting machine stops (pressoffs) (Note 13) were 18 times higher for Fiber 66 than for silk. The Du Pont plant had wound the yarn onto the supply cones much too tightly, causing individual filaments to adhere to one another, resulting in moment-to-moment tension variations in the yarn as it was pulled into the knitting head. These tension variations, in turn, caused the knitted stitches to be uneven in size and shape. Also, wads of broken filaments (fluff balls), snagged on yarn guides causing the yarn to break. For freer yarn delivery, Lake used Van Raalte's equipment to rewind the yarn at lower tension from the straight-edge Du Pont rayon cones onto tapered-edge pineapple cones.

Compared to silk, Fiber 66 was limber and difficult for the operator to grasp and maneuver in tying a knot. Also, it was slippery, so that once the knot was tied, it tended to come untied. Although this problem mustered the Boy Scout instinct in many, it was not resolved until months later when Ellis Mann found a particular flat knot that did not slip [21].

Snags

Unlike silk, individual filaments in Fiber 66 yarns were prone to snag on the slightest roughness, even rough hands, and pull the yarn out of its normal position in the knitting stitch. Pulled threads, as snagged filaments came to be called, were not physically damaged and did not cause a run, but gave the stocking an unsightly distorted appearance. Because intended solutions either created new problems or were too expensive to use, this seemingly simple defect would, in the months ahead, prove to be a most intractable problem. Without some improvement, Lake reported, "...66 hose would not be at all practical for normal handling through the mill [22]." Raw silk was not prone to pulled threads because it contained 20-25% sericin, a natural gum, which glued the individual filaments together. Lake set out to duplicate this effect with conventional gelatin-type protective coatings or yarn sizings used for rayon, only to find that very little sizing would adhere to Fiber 66, and the small amount that did adhere was so brittle it flaked off during knitting. After trying ten different combinations of sulfonated oils and glycerin as gelatin plasticizers, Lake tried coatings of aluminum soap and aluminum sulfate. Nothing worked [23].

Binding the filaments together by increasing yarn twist from 6 to 20 turns per inch (tpi) only swapped one problem for another. When the knitting machine stopped, the twisted yarn, now tension free, would immediately twist on itself and kink. Then, when the machine was restarted, a kind would snag some place and cause the yarn to break. Worse yet, high twist yarn caused the stocking blanks to spiral and coil like a corkscrew [24].

Highly-twisted rayon yarns could be prevented from kinking by exposing them for one or two hours to an atmosphere of 75-85% relative humidity (twist-setting), but this treatment did not work for Fiber 66. Neither did twisting the yarns while wet, nor treatment in 0.5% ammonium hydroxide solution before twist-setting. But yarns exposed to live steam for one hour did not kink [25].

In knitting though, higher yarn twist did not reduce pulled threads, and furthermore, caused a new problem. During dyeing, the knitted stitch became wildly distorted, giving the fabric a totally unacceptable lace-like appearance. But high-twist yarns did give one dramatic improvement; the stockings, as knit, appeared to be shearer and not as glossy, being more like high-twist silk. For that reason, they tried to get around twist-setting and its attendant laciness by using two-ply yarns having balanced S and Z twist. These yarns didn't kink and showed a lower tendency for pulled threads, but were too expensive to sue because of high spinning and plying costs [26]. It was November; six months had passed with no concrete progress towards reducing pulled threads.

All along, both Lake and Groh had hammered away at the need for a protective coating or sizing for Fiber 66 that would function like sericin on silk. In August, Groh had suggested that the problem "be handled by Research." In September, Lake listed the requirements for a sizing (pliable, protective, easily removable). In October, Groh reported that sales service was "firmly convinced" of the need and had again brought it to the attention of research. But no one responded; nothing happened. Finally, in his December status report, Groh flatly stated that without a yarn sizing there was no way that Fiber 66 could be handled on a commercial scale [27]. Within six weeks, twelve cones of yarn sized with three different candidates based on polyacrylic acid were ready for knitting at Van Raalte [28] (Note 14).

In a 100 stocking trial, one candidate was too sticky to knit and another gave mediocre results, but candidate MK gave only one pulled thread compared to 70 for unsized yarns [30]. Except for one drawback, MK sizing appeared to be a breakthrough. Full-fashioned knitting machines were designed for the yarn to pass through a trough of water and to be knit wet so as to obtain a more uniform stitch. MK sizing was sufficiently soluble in water that, if the machine stopped for a few minutes, or just long enough for the yarn to start to dry, the yarn became glued to contact surfaces and broke when the machine was restarted. Various lubricants added to the wetting trough to relieve sticking had no effect [31].

By late May or early June 1938, research had prepared five variations of MK sizing containing glycerin, alcohol sulfate wetting agents, and fatty softeners, additives aimed at preventing sticking. Again, all candidates stuck to the machine; but one, MK4A, was better than the others. In the mill, Lake found that adding lubricants such as sulfonated olive oil and glycerol stearate to the water trough reduced sticking in knitting, but, because the lubricants softened the sizing, pulled threads shot back to the old level [32]. He then found that coning the dry MK4A sized yarn through mineral oil both prevented sticking and pulled threads, and concluded that a yarn coating that would reduce the rate of drying "should be in the right direction [33]."

Meanwhile, knitting continued until they discovered yet another problem. The acrylic acid in MK4A was so corrosive that after only three weeks of knitting, wear to the knitting machine's needles and sinkers was equivalent to several year's of wear when knitting silk. Adding triethanolamine oleate, to the knitting machine water trough reduced corrosion, but so altered the sizing that it no longer prevented pulled threads. Suspecting that uneven application of the sizing accelerated wear, research prepared a lot of continuously sized yarns, but they were no better than skein sized.

In early August, the plant followed up on Lake's earlier success with mineral oil applied to dry MK4A sized yarn. From a field of eleven candidates, on after-treated with olive oil both reduced wear and pulled threads and caused no other problems. This yarn, by no means perfect, worked well enough to allow them to start building a badly needed stocking inventory [34].

Tedious Handling

Difficulties caused by the hosiery being too flimsy to handle were, in many ways, closely allied to pulled threads in that a good yarn sizing would prevent both. Unlike silk stockings, which, as knit, were stiff and easy to grasp, stockings of Fiber 66 were limp and slippery. As a result, when Fiber 66 stocking blanks (Fig. 1) were removed from the knitting machine, the ends and selvages curled into a tight tube. For operators to unroll the tube while forming the welt, or transferring the leg blank into the machines that knitted the foot (Fig. 2), or that closed the leg (Fig. 3), doubled the time required for each operation (Note 15). Lake improved topping to some extent by first knitting in a few courses of removable cotton yarns which stiffened the end of the fabric, and prevented the top from curling. But the extra step cost four cents per dozen stockings [36].

In March 1938, Lake reported a detailed, step-by-step, four-decimal-place comparison of manufacturing costs to "substantiate previously expressed opinions." In summary (Table 4), the study showed that costs of Fiber 66 hosiery were more than double those of silk, primarily due to the extra handling time [37] (Note 16). Anyone who had doubted before that a good yarn sizing was needed, now had to be a believer.

MK4A sizing, which became available in August and which had significantly reduced pulled threads, also improved handling dramatically. The stockings, being stiffer, did not curl and were almost as easy to top, loop, and seam as silk stockings.

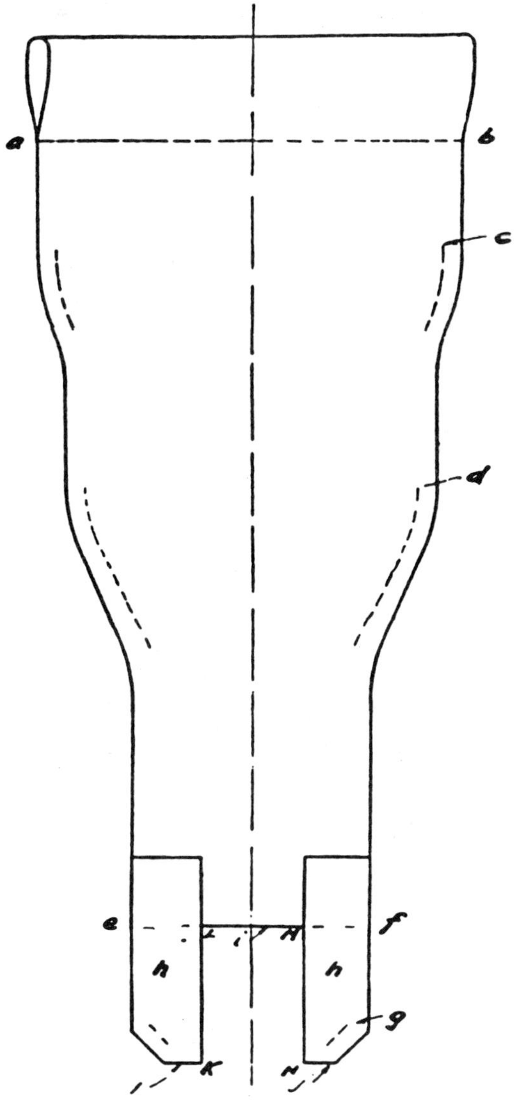

Fig. 1

Full-fashioned Leg Blank

Adapted from Miller, Ref. 35.

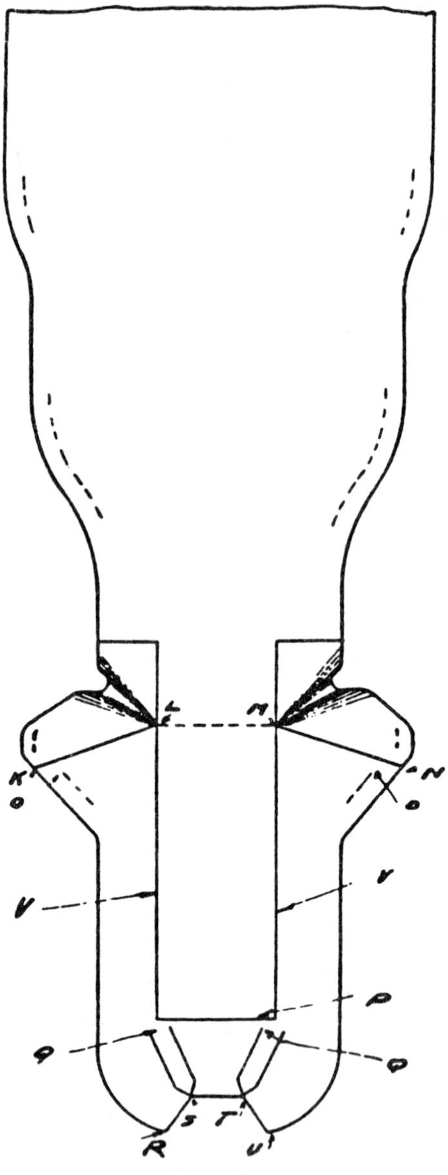

Fig. 2.

Full-fashioned Foot and Leg Blank

Adapted from Miller, Ref. 35.

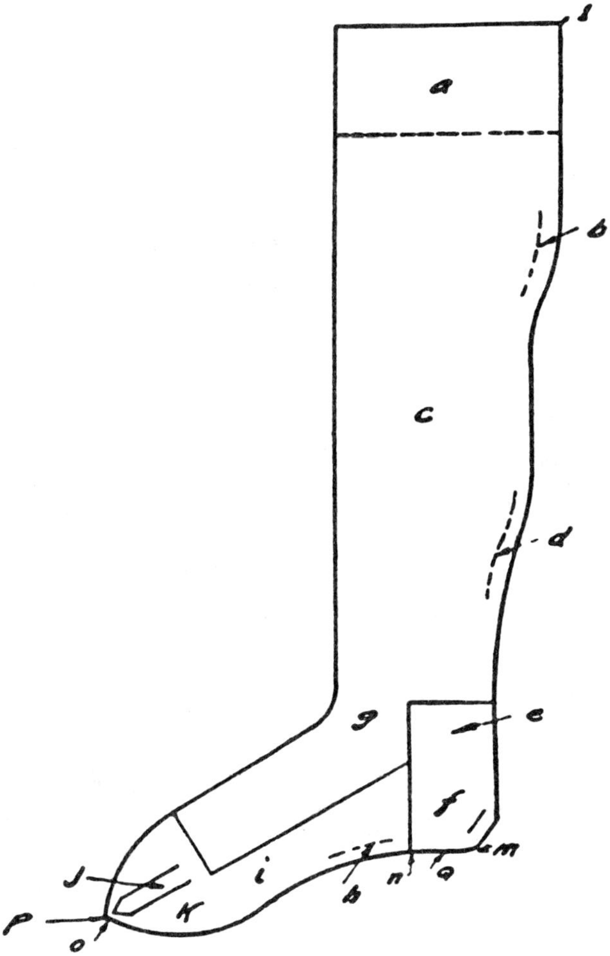

Fig. 3

Full-fashioned Stocking

Adapted from Miller, Ref. 35.

Table 4. Full-Fashioned Hosiery Manufacturing Costs.

Fiber 66 vs. Silk
March 1938

	$ per Dozen Pairs	
	4-Thread Silk	**Fiber 66**
Productive labor	.7227	2.3952
Non-productive labor	.1767	.2709
Overhead	2.6330	4.6671
Total per dozen pairs - $	3.5324	7.3332
Cost per pair - $	.29	.61

Wrinkles

From the beginning of the Fiber 66 project, wrinkles and "crowsfeet" (tiny virtually undetectable wrinkles) which formed during dyeing were the number one problem. Like silk, the stockings were added loosely into mesh bags and dyed in revolving cylinder machines, similar to washing machines used by commercial laundries. The first lots of hosiery finished at Van Raalte during May and June were too wrinkled to be of any value [38]. After rewetting and reboarding the stockings and then trying sulfonated olive oil and other finishes with no reduction in wrinkles, Lake doubted "finishing will be the answer to the problem [39]."

Launching out in several directions at once with as many as 15 different trials, by the end of August, they knew that:

(1) stockings knit from yarns preshrunk in boiling water wrinkled during dyeing;
(2)stockings dyed at 180°F instead of 212°F were not as prone to wrinkle;
(3) stockings dyed at 125°F did not wrinkle at all;
(4) stockings knit from dyed yarns did not wrinkle; and
(5) undyed stockings exposed to live steam for one hour before dyeing did not wrinkle.

But each of these possible solutions to wrinkling also had drawbacks. Preshrinking the yarns would be too expensive unless Du Pont did it as an integral part of yarn spinning. Dyeing at 125°F was impractical because little of the dye exhausted onto the fiber, and what did exhaust failed to penetrate. Dyed yarns were too expensive, and for that reason had been abandoned for silk hosiery long before. Exposing the stockings to steam before dyeing would be an expensive extra step not required for silk [40].

Since steaming, though expensive, seemed to be the best choice, by October, they were devising ways to make the operation practical by steaming stockings on regular flat boarding forms in a modified boarding machine such as used for drying silk hosiery after dyeing [41]. What sales service did to further this idea is not known, but at the end of December, Groh reported that wrinkles could be prevented by using high-twist yarns, preboarding, and applying a finish to the stockings after dyeing. As far as can be determined [42]:

The development of a preboarding operation (setting the stocking on a form with steam) has acted as a molding operation. The stockings so treated are set against shrinkage in subsequent dyeing and do not wrinkle, regardless of the temperature used in the boil-off and dyeing operation.

In the early Fall of 1937, Ellis Mann, who had been working on woven fabrics, joined John Lake at Van Raalte, and together they set out, during January and February 1938, to refine the preboarding process through detailed studies of the effects of time, temperature, and boarding form sizes, using a crude wooden steam chest equipped with a thermometer.

Their work showed that steaming the stockings for 5 minutes at 210°F on a standard sized board gave "satisfactory results" for wrinkles, while stockings boarded at 210°F and dyed at 180°F were remarkably free of wrinkles [43]. As to the design of a practical preboarding machine, they suggested that if a preboarding temperature slightly above 210°F could be attained, the time could be reduced from five minutes to three. The shorter time would reduce both the number of machines needed for a full-scale mill operation as well as the amount of floor space required. By April 1938, John Lake, had tests underway at Proctor & Schwartz Company, a Philadelphia machinery manufacturer, to determine the feasibility of building such a machine, and by October 3, an automatic preboarder that operated at 225°F had been installed at Van Raalte at a cost to Du Pont of $3,581.90 [44].

Evolution of Preboarding

Considering the overwhelming contribution that preboarding made to the commercial success of nylon hosiery, it is worthwhile to trace the chronology of known events relating to the process (Table 5).

Carothers' stocking patent, filed February 15, 1937, stated that polyamide fibers could not, like silk, be boarded in air at 70°C, but could be boarded at 100-150°C in steam. This obviously referred to boarding after dyeing, since Carothers made no claims relative to boarding before dyeing [45].

However, in another patent, filed on February 15 by Dr. J. B. Miles (Note 17), a research chemist in the Chemical Department in Wilmington, claimed that the properties of polyamide fibers "are affected in a remarkable way by steam," such that they will hold a "predetermined shape or form." By "predetermined shape," he generally meant the end result of processes such as "spinning, casting, molding or the like," but in claim 13, he included "fabrics...[and[...treating said fabrics with steam [47]." Although this patent contained the germ of the idea of preboarding, nothing practical came from it, at least immediately.

During the summer of 1937, sales service continued to wrestle with ways to overcome wrinkles at Van Raalte. By August 27, they were setting stockings before dyeing, both on regular boarding forms in air at 240°F and while lying loose in a steam box. By October 1, they had merged the two ideas and were planning an experiment to steam stockings on a regular boarding form.

On or before September 27, Dr. Hoff suggested, apparently as an afterthought to a previous directive, that E. V. Lewis in Buffalo try "boiling water and boiling salt solutions as setting agents early in the program." Lewis tried water on the 30th, but noted in his lab book, "Salt not practical [48]." In a memo to Hoff covering this work, dated October 12, Lewis concluded that setting a stocking on a form in steam for 10 minutes "is effective both in the elimination of wrinkles and of lacy stitches..." As evidence, he sent along stockings which he asked Hoff to return for the permanent file [49].

Table 5. Some Events Relating to Du Pont Developments for Preboarding Nylon Hosiery.

	1937
February 15	W. H. Carothers files patent for boarding dyed polyamide hosiery in steam at 100-150°C.
February 15	J. B. Miles files patent claiming that polyamide fibers "are affected in a remarkable way by steam such that they will hold a predetermined shape."
February 16-18	E. V. Lewis finds first commercially dyed stockings extremely wrinkled.
May-July	J. C. Lake not successful in preventing wrinkles in dyed stockings at Van Raalte Co.
August 27	J. C. Lake reports stockings either boarded on a form in hot air at 240°F or steamed loose at 210°F before dyeing are less wrinkled after dyeing.
September 27	G. P. Hoff suggests that E. V. Lewis set stockings in hot water.
October 1	G. J. Groh reports that sales service planning to steam stockings on forms at Van Raalte Co.
October 11	E. V. Lewis reports stockings steamed 10 minutes on a form do not wrinkled during dyeing.
November 10	G. P. Hoff reports to top management that nylon can be set to a predetermined shape.
December 28	G. J. Groh reports that preboarding stockings in steam molds stockings to shape and prevents wrinkles during dyeing.
	1938
January-February	J. C. Lake and W. E. Mann study preboarding conditions at Van Raalte Co.
February 24	Lake and Mann propose preboarding above 210°F to reduce time required.
April	Tests underway to determine feasibility of high-temperature preboarding machine.
June 28	J. B. Miles files patent on preboarding based on Van Raalte studies in February.
October 3	Van Raalte preboarding at 225°F on machine built by Proctor & Schwartz for Du Pont.

Thus it appears that John Kale and Vernon Lewis, independently and at about the same time, made the connection between steaming the stocking on a form before dyeing and wrinkle prevention during dyeing. In a status report for top management written November 10, 1937, Hoff reported [50]:

> The ability of this yarn to 'set', i.e., to assume a shape to which it tends to return after deformation, upon exposure to heat and moisture is most unusual. It is believed that this property will be a definite advantage rather than a detriment in producing hosiery of a quality superior to silk in

respect to uniformity of stitch, sheerness, retention of shape, and resistance to deformation.

Considering that from May through August sales service was dutifully reporting the severity of the wrinkling problem, and that research had filed for a patent in February that would solve it, one can only wonder about the effectiveness of communication between the groups involved. At least, from the tenor of their correspondence, one can infer that as late as mid-September the idea of presteaming on a form was new to Hoff, Lewis, and Groh. Perhaps the research people simply failed to make the connection between the devastating effects of steam on a stocking held to shape on a form. If sales service knew about the Miles patent, they might have been so blinded by the costs of steaming as an extra step that they failed to recognize its potential. Or, sales service might not have clearly stated the problem. In August, Lewis asked Groh to publish an illustrated glossary of defect terms after finding that different persons were attaching different meanings to the terms wrinkle, crowsfeet, and distorted stitches [51].

Such speculation, while interesting, serves no purpose, because on June 28, 1938, Dr. Miles filed for a patent on preboarding which cited as examples the work done by Lake and Mann at Van Raalte during February [52].

Shrinkage During Dyeing

To counteract abnormally high (8-10%) stocking shrinkage that occurred during the earliest dyeings, Lake had to knit the stocking two inches longer and make the foot size 11 so as to end up with size 9 1/2. Although this gave finished stockings of the right dimensions, the increased knitting time and extra yarn added 10%, perhaps, to the costs [53].

Every subsequent change in yarns or in the stocking-making process also affected stockings sizes and required empirical adjustments to the knitting stocking dimensions. For example, stocking blanks knit from unsized Fiber 66 were, must off the knitting machine, only about half as wide as silk blanks, but blanks knit from MK4A sized yarns were about the same width as silk. And preshrunk yarns having only 5% residual shrinkage gave wider stockings than unshrunk yarns having 8-10% shrinkage.

A key to dimension control came out of the preboarding studies by Lake and Mann in January-February 1938. They found that soaking the stockings for 5 minutes in a 1% solution of sulfonated olive oil before preboarding caused the stocking to cling to the form and reduced its tendency to shrink during steaming. The results was that the stockings were set to reasonably uniform dimensions [54].

By July 1938, the best and cheapest way to keep stocking shrinkage during dyeing to a uniform level was to use unshrunk yarns, to uniformly apply a yarn sizing to give stocking blanks of uniform dimensions, to wet the stockings before sliding them on the boarding form, and to preboard in steam at 210°F for five minutes [55].

Color Nonuniformities

In comparison to pulled threads and winkles, color nonuniformity was a relatively minor problem in the beginning. The Dyestuffs Division of the Organic Chemical Department had become involved in the Spring of 1937, and by Fall, had found a sufficient range of colorfast dyes available to dye any shade [56]. No one seemed overly concerned about dyeing defects such as:

(1) Dye rings or bans of one or more courses caused by denier variations [57],
(2) Glaring two-tone shade differences among the various denier yarns comprising the leg, welt, and foot [58],
(3) Short lengths and flecks (fraction of an inch) of deeply dyed, and obviously undrawn, filaments [59],
(4) Shade differences between stockings within a dye lot, and
(5) Shade differences between different lots of stockings.

The effects of rings could be broken up and minimized by using a system patented for silk, and charging stiff royalties, that drew successive knitting courses from three different yarns [60]. Also, two-ply yarns, although cost prohibitive, gave a more uniform color [61].

The dyeing procedure generally followed Van Raalte's procedure for silk (Table 6), and did not change appreciably during 1937-1938 [62]. As far as was known, sales service could do little or nothing to make the yarns dye more uniformly; only improvements in the yarn manufacturing process would make a difference.

Table 6. Dyeing Procedure for Fiber 66 Stockings: June 1937. Six Dozen Pairs (5 lbs.) - 10 Gallon Volume.

1. Scour
   - Set bath @ 120°F with
     - 5% olive oil soap (based on weight of goods)
   - Run for 15 minutes
   - Dump bath
2. Dye
   - 2.1 Prepare a new bath with
     - 2% olive oil soap (b o w g)
     - 1% Nekal BX wetting agent (b o w g)
   - 2.2 Bring dye bath to a boil and add
     - Acid dyes for Fiber 66

Du Pont Neutral Brown RS	120 grains	.34%
Du Pont Neutral Yellow GS	26 grains	.07%
Pontacyl Fast Blue 5R conc.	12 grains	.03%

     - Direct dyes for cotton sewing thread

Calcodur Fast Yellow RL	20 grains	.06%
[IG] Cotton Brown R	40 grains	.12%

   - 2.3 Continue dyeing at boil for 15 minutes and add
     - Glauber's salt [$Na_2SO_4 10H_2O$]. 4 oz.
   - 2.4 Continue dyeing for 15 minutes
   - 2.5 Compare shade against standard and add dyes as necessary to correct shade.
   - 2.6 Repeat above step as necessary, running 10 minutes after each dye addition.
   - 2.7 Dump dye bath and rinse in cold water.

Manufacturing steadily reduced yarn denier variation, so that by November 1937, it was at a value half that of silk, thereby reducing rings and making it possible to knit with a single yarn, and to save the royalties on multiple yarn machines [63]. Although every dye lot still had color defects, Hoff reported management, perhaps with too much optimism, that the dye uniformity of Fiber 66 compared favorably to that of silk [64].

But as other problems one-by-one were reduced in severity, dye uniformity assumed more importance. A study by sales service in April 1938, showed that 60% of all effects in second grade stockings were due to color nonuniformities of one kind or another [65].

Shortly after learning, in the Spring of 1938, "that non-uniform drawing has been causing much of the non-uniformity of dyeing in Fiber 66," manufacturing and research began sending a variety of test yarns to Boonton, all intended either to improve color uniformity or reduce wrinkling: in May, changes in draw ratio [66], in June, yarns treated with tetrachloroethylene [67], in July, preshrunk yarns [68], and in August, yarns passed through a steam chamber before applying the finish. Steamed yarns gave excellent dye unions between leg and welt yarns and had fewer dark dyeing defects, but the shades were 20% lighter, which increased dye costs [69]. Little progress was made, however, and a major breakthrough would not come for another six months.

Tactile Softness

Whereas silk stockings had a firm crisp hand, sometimes described as "scroopy" or "bitey", Fiber 66 stockings were soft, pliable, and slippers. Almost any change in the yarn, knit construction, or processing steps also changed the tactile properties of the stocking in one way or another. For example, increasing yarn twist from 6 to 20 tpi had given the stocking a more silk-like feel, as had preboarding. Usually there were tradeoffs, but one variable that could be changed without appreciably affecting other characteristics was denier per filament (dpf) of the yarn.

While Lake was trying to find ways to reduce pulled threads, he noted that 50 denier, 20 filament yarns (2.5 dpf) gave a stocking that was crisper than one from the standard 50 denier, 30 filament yarn (1.7 dpf). But since softness was not a high priority problem [70], they didn't explore higher filament deniers until May 1938. In a small wear trial of stockings from 3 and 4 dpf yarns, some women liked the softer feel, while others liked the crisper. Because Van Raalte's executives preferred the crisper 4 dpf, which felt more like trendy crepe silk, sales service recommended that 4 dpf be adopted [71]. And that ended the matter.

It appears that they did not consider finishes applied after dyeing as a means of modifying hand, but used whatever Van Raalte applied to silk. Finishing was more art than science and finish formulations often were put together by trial and error (Table 7). In the Van Raalte formulations, Aquapell was probably a wax emulsion. The precipitate of barium sulfate formed from barium chloride and Glauber's salt not only harshened the hand, but reduced fabric luster. Starch helped bind the barium sulfate to the fiber and formaldehyde kept down mildew [72].

Shininess

Based on a sales service recommendation in July 1937, manufacturing, in October, drastically reduced the objectionable shiny appearance of Fiber 66 stockings by adding sufficient titanium dioxide pigment to the polymer to change the official designation of the yarn from bright to dull [73]. With the cumulative effects of the changes, such as increasing yarn twist from 6 to 20 tpi, increasing dpf from 1.7 to 4.0, and using a barium sulfate stocking finish, shininess was reduced to an almost acceptable level by mid-1938.

Table 7. Van Raalte Stocking Finishing Procedures. 10 Dozen Pairs Stockings, 25 Gallon Volume.

July 1937

1. Set a bath @ 100°F and add
   5% Fancourt Aquapel water repellent.

2. Run 20 minutes. Dump bath.

3. Whiz. [centrifugal extract]

4. Postboard @ 160-170°F until dry.

February 1938

1. Set a bath @ 70°F and add	
Fancourt Aquapell water repellant	3/4 lbs.
Run 5 minutes.	
2. Raise temperature to 90°F and add	
Barium chloride	1 1/2 oz.
Run 5 minutes.	
3. Add	
Starch	1 1/2 oz.
Glauber's salt [$Na_2SO_4 10H_2O$]	1 1/2 oz.
Run 5 minutes.	
4. Add	
Scroop Oil [sulfonated fatty ester]	1 1/2 oz.
Formaldehyde	40 cc
Run 5 minutes.	
5. Drop bath, drain.	
6. Whiz. [centrifugal extract]	
7. Board @ 160-170°F until dry.	

Lack of Elasticity

In his stocking patent of February 15, 1937, Carothers claimed that polyamide fibers, like silk, were highly elastic, therefore ideal for hosiery. In practice, however, high yarn elasticity and recovery did not translate into high stocking elasticity and recovery. A stocking's ability to cling to the leg and not bag at the knee was dependent "to a large extent on the ability of the knitted stitches to slip past each other and [then] return to their normal position after the fabric had been pulled or stretched [74]."

Little attention was given to stocking elasticity until it appeared that it would be a deterrent to sales. In a small wear test made in the Fall of 1937, a third of the women commented about poor elasticity, and in December, low welt elasticity was the chief criticism of I. K. Hessberg, President of Van Raalte [75].

Starting in February 1938, rough studies by Lake and Mann showed that Van Raalte's 4-thread illusion stocking would stretch 21% more in width than its counterpart in Fiber 66 [76]. Later studies showed this difference to be as great as 30%, readily explaining shy Fiber 66 stockings would not fit stout legs [77].

Realizing that over the past year, experiments with sized yarns, plied yarns, preset yarns, yarns of higher twist levels and higher denier per filament, preboarding, and lubricants had affected elasticity little or not at all, they now began to focus on knit construction of the stocking. Under high magnification, they could see considerable free space around the knitting loops in silk stockings, whereas Fiber 66 loops were jammed together. Silk gave oval shaped stitches, while Fiber 66 stitches were more rectangular in shape. From knitting studies, they learned that as the number of courses per inch increased, the total length of yarn in the course decreased, and stretchability decreased [78]. Also, they established a relationship between leg width and the number of knitted narrowings used to shape the leg [79]. Such understanding enabled them, in due time, to make stockings which would stretch enough in width to fit almost any leg, but it still was necessary to supply three lengths - short, medium, and long - as it was with better grades of silk stockings [80].

The Watershed

By September 1938, sales service had a reasonably reproducible stocking process in hand. The color line had been expanded to all six of the Van Raalte shades, and stockings were being knit in sized 8 1/2 to 10 1/2 in anticipation of a large consumer wear test [81]. Small wear tests had already confirmed many of the anticipated advantages of Fiber 66 stockings over silk. They had a better feel, bagged less at the knees, were less prone to snag, and if they did snag, they were less likely to run. They also were easier to wash, dried more quickly, and lasted longer [82].

After the new pilot plant with its 500-pound capacity autoclave came on stream July 11, sales service was kept busy at Van Raalte evaluating the yarn, particularly for dye uniformity between spinning position and over time (Note 18). Starting September 15, knitting machines were run on two shifts so that yarns spun one day in Wilmington could be knitted the next day in Boonton [83]. As pilot-plant process changes were made, stockings once too short became too long and then too short again, requiring constant fine-tuning of stocking leg widths and lengths for the right balance of dimensions and stretch.

On October 14th, Du Pont's Executive Committee approved eight and a half million dollars for a plant that would produce four million pounds of nylon a year, and on the 19th, Seaford, Delaware, was chosen as the plant site. On the 27th, the name nylon officially replaced Fiber 66, the same day that Dr. Stine announced nylon to the world [84]. But behind the hoopla, the project still had serious problems. Research was frantically looking for a yarn sizing that would not corrode knitting machines, and 60% of all stockings being knit were second quality, mostly because of color nonuniformities [85].

1939 and Beyond

Plant construction started January 20, 1939, and shortly afterward [86], twenty-five more knitting mills were brought into the program and the number of sales service representatives was increased accordingly [87]. Stockings were shown at the San Francisco World's Fair on February 16, and starting on February 20, 5000 pairs of stockings went on sale to Du Pont employees at $1.15 per pair compared to $1.00 a pair for comparable silk stockings [88].

The two remaining major problems were resolved in the spring of 1939. In April, the Organic Chemicals Department discovered that dispersion-type acetate dyes colored nylon very uniformly compared to the acid dyes used silk [89]. And just in the nick of time, research solved the pulled thread problem with a sizing based on polyvinyl alcohol and boric acid. Continuous sizing machines, ordered in July, were installed, just two weeks before plant start-up on December 15, 1939 [90]. Full-fashioned nylon stockings, which went on sale nationwide on May 14, 1940, captured more than 30% of the market by the time World War II required every ounce of nylon production [91].

Following the war, seamless nylon stockings knit on circular machines took over the full-fashioned market. With the advent of stretch yarns in the late 1950s, the one-size-fits-all concept was born, and as hem lines rose in the 1960s, so did panty hose sales. By the 1980s, a half century after the beginning, women's hosiery was consuming more than 70 million pounds of nylon a year [92].

ACKNOWLEDGEMENTS

The author is indebted to H. I. Salmons, Jr. for leads on W. Ellis Mann, to Ellis Mann for leads on John C. Lake, to John Lake, Larry Coffin, John Zimmerman, and Dr. Stephen J. Kennedy for leads and data on George J. Groh, to John Lake and Ellis Mann for reminiscences of their work during 1937-1938, and to Hagley Library for permission to quote from the G. Preston Hoff papers.

NOTES

1. Wallace Hume Carothers was born April 27, 1896 in Burlington, Iowa. He earned a Ph. D. in chemistry at the University of Illinois in 1926 and taught there before going to Harvard in 1928. He committed suicide April 29, 1937 [2].
2. Two critical decisions had been to develop 6-6 polyamide instead of 5-4 polyamide and to melt spin the fiber rather than solvent spin it [3].
3. The Du Pont Rayon Co. became the Rayon Department of the Du Pont Company in 1936.
4. The fiber was called Rayon 66 until around November 1937 when the designation was changed to Fiber 66 because the fiber "was so distinctly different and superior" to rayon [5]. For consistency, the name Fiber 66 is used throughout this paper.
5. George Preston Hoff was born in Shelby, Ohio, on May 1, 1899, and received A.B., M.S. and Ph.D. degrees at Ohio State University, the latter in 1925. He joined the Du Pont Rayon Company, Buffalo, as a research chemist in 1926, and was promoted to manager in 1928 [6].
6. Everett Vernon Lewis, born August 3, 1907, received an S.B. at Massachusetts Institute of Technology in 1928, and a Ph.D. in physical chemistry there in 1931 [9].
7. Denier is a numbering system for expressing linear density, equal to the mass in grams of 9000 meters of yarn. For example, 9000 meters of a 50 denier yarn weighs 50 grams [10].
8. Joseph Bradley Quig, who was born in East Orange, New Jersey on January 8, 1892, received a B.S. degree at Lafayette College in 1917. After working as a chemist for American Smelting and Refining

Company at Atlas Powder Company, he received a Ph.D. at Iowa State College in 1926. That same year, he became assistant chemical superintendent at Du Pont Rayon Company, Old Hickory, Tennessee and manager of sales-service in 1928 [12].

9. George John Groh, who was born August 7, 1896, in Cattaraugus, New York, received a B.S. degree in chemistry from Colgate University in 1918. In 1924, he joined the Du Pont Rayon Company as a research chemist, and, in 1928 became one of the company's pioneer sales service representatives in New York City. He transferred to the Fiber 66 project in Wilmington, Delaware, in 1937. When World War II started, he became an expert in weaving and finishing nylon parachute cloth. Leaving Du Pont at the end of 1945, he held management positions in the development of fabrics from man-made fibers with Robbins Mills (1946-1948), Cannon Mills (1948-1951), and Burlington Mills (1951-1953). In 1954, he joined the U.S. Army Laboratory, Natick, Massachusetts, retiring as Chief, Textile Engineering Section in 1966. He died February 6, 1985, at the age of 89 [13].

10. John Caldwell Lake, the son of James L. and Lula Caldwell Lake, was born December 6, 1903, in Wake Forest, North Carolina, where his father taught physics and mathematics at Wake Forest College, a Baptist school. Lake received a B.A. degree at Wake Forest in 1923, and a B.S. in textile engineering at Clemson College in 1925. After graduation, he worked in the U.S. government cotton testing laboratory at Clemson for about six months before joining Sayles Finishing Company, Saylesville, Rhode Island. During 1930-1931, he was a sales service representative for Du Pont Rayon Company in New York, but got laid off when the depression hit full force. He was then at General Ribbon Company, Catasauqua, Pennsylvania, for two years before returning to Du Pont in acetate sales service in 1934. He transferred to the Fiber 66 project as principal investigator on hosiery in Wilmington in 1937. Lake left Du Pont in 1945 to become President and General Manager of Grenada Industries, a nylon hosiery manufacturer in Grenada, Mississippi. By the time Lake retired in 1968, this company had grown from 200 employees to 2,000. He married Julia Loukes of Lowville, New York, in 1934, and has three daughters. He still lives in Grenada, where he divides his time between travel and golf [14].

11. William Ellis Mann, the son of William Stephen and Martha Ellis Mann, was born on August 27, 1905, and spent his early years in Young Harris, Georgia, where his father taught mathematics at Young Harris College, a Methodist school. After graduating from Young Harris in 1924, Mann received a B.A. degree in French from Emory University in 1926, and then returned to Young Harris for three years to teach. In 1922, he entered Georgia Institute of Technology and received a B.S. in chemical engineering in 1933. For the year 1933-1934, he was principal of the high school in Jackson, Georgia. During his stay at Georgia Tech, he had been in a work-study program at Du Pont Rayon Company, Old Hickory, Tennessee, and in July 1934, joined the company full-time at its Buffalo, New York, plant. In August 1935, he transferred to the sales service laboratory in Chester, Pennsylvania. After weaving the first Fiber 66 fabric at Frank Ix and Sons, New Holland, Pennsylvania, in July 1937, he joined the hosiery project. In August 1939, he became a chemist in the newly crested Nylon Research Division, and during World War II assisted in developing military flack jackets and jungle boots from nylon. He transferred to the Technical Service Section of the Sales Division around 1950. For the rest of his career, Mann worked on sales-related technical problems and developments in nylon hosiery. By the time he retired in 1970, his expertise was recognized world wide. He married Gladys Austin, a Georgia school teacher, in 1936, and had a son and a daughter. He now resides in Atlanta [15].

12. A course is a series of successive loops lying crosswise the stocking. A 4-thread silk consists of four individual strands of silk, as spun by the silk worm, twisted together; 20 turn refers to the amount of twist in the yarn, measured in turns per inch. 13/15 represents the range of deniers in individual strands of silk; a 4-thread silk would range from 52 to 60 denier. Gauge, or gage, in full-fashioned knitting, is the number of knitting needles per 1.5 inches, and is a measure of stitch fineness.

13. In a press-off, the yarn fails to knit and either falls off the knitting needles or the knitted design is distorted and incomplete. Yarn breakage is a common cause of press-offs.

14. Contrary to sales service perceptions, the record shows considerable research on sizings. During early 1938, one of the two assigned chemists screened no fewer than 60 compounds, ranging from beeswax to rubber latex [29]. Apparently, communication was poor between research and sales service.

15. The stocking was knitted from the top down. The welt to which the garter attached was knitted first from 75 denier yarn; then, the contoured leg blank and part of the heel knitted from 50 denier yarn. The leg blank was then transferred to a simpler machine and knitting continued with 150 denier yarn to complete the foot. After toes and heels were closed on a looping machine, the blank was folded lengthwise and the edges sewn together to complete the stocking [35].

16. Lake estimated production labor costs were at least 15% higher than they should have been because he suspected that operators, who were being paid by the hour rather than the normal piece rate, although appearing busy, "were holding back somewhat due to orders given them by labor union

leaders..." Whether or not the stocking could be made economically was only academic at this stage of the development. The questions were, would women wear the stockings, would they like them, etc. If the answers were affirmative, means could be found for reducing costs.

17. John Blanchard Miles, born in 1908, graduated from Princeton and earned a Ph.D. there before joining Du Pont's Chemical Department in 1929 [46].
18. This was the third scale-up; a 250-pound capacity autoclave had been in operation since November 1937.

REFERENCES

1. D. A. Hounshell, J. K. Smith, Jr., "Science and Corporate Strategy, Du Pont R&D, 1902-1980," Cambridge University Press, 1988, p. 270.
2. W. W. Hassler, in "American Chemists and Chemical Engineers," W. D. Miles, ed., American Chemical Society: Washington, 1976, pp. 65-66.
3. F. Schulze, R. Soukup, "The Technical Division of the Rayon Department 1920-1951," Textile Fibers Department, E. I. Du Pont Co., Wilmington, DE, 1952, p. 160.
4. Schulze and Soukup, Ref. 3, p. 166.
5. Useage in reports, C. H. Rutledge, "The Name Nylon and Some of Its Adventures," Textile Fibers Department Production Information Bulletin, June 20, 1966, author's copy, Hounshell and Smith, Ref. 1, pp. 268-269, 668 n. 91.
6. J. Cattell, "American Men of Science," The Science Press, Lancaster, PA, 8th ed., 1949, p. 1146.
7. W. H. Carothers, "Stocking and Method for Producing Same," United States Patent 2,157,116, May 9, 1939, p. 1, col. 1, line 35; "Rayon Textile Monthly," 20(2), 38 (Feb. 1939), 20(12), 40 (Dec. 1939).
8. Carothers, Ref. 7.
9. Cattell, Ref. 6, p. 1486.
10. "ASTM Annual Book of Standards," Philadelphia, 07.01, 281 (1990).
11. E. V. Lewis, Memo, "Hosiery Trial," Feb. 22, 1937, in G. P. Hoff papers, Hagley Library, Wilmington, DE, Acc. 1905, Series II, Part 2, Box 951, folder 37.
12. Cattell, Ref. 6, p. 2003.
13. Alumni files, Colgate University, Information supplied by S. J. Kennedy formerly of the U. S. Army R&D Center, Natick, MA.
14. J. E. Kale to H. T. Pratt, conversations, Oct. 16 and Nov. 1, 1990, Feb. 27 and April 1, 1991.
15. W. E. Mann to H. T. Pratt, conversations, Oct. 2 and Dec. 24, 1990, personal knowledge.
16. G. P. Hoff, "Confidential Memorandum," May 7, 1937, Hoff, Ref. 11, Box 951, folder 38.
17. Lake, Ref. 14, Oct. 16.
18. Hoff, Ref. 16, p. 2.
19. Lake, Ref. 14, Oct. 16.
20. J. C. Lake, Progress Report No. 1, July 12, 1937. An incomplete series of 18 three to ten page progress and status reports for Sales Service Test RS 1, issued by G. J. Groh, J. C. Lake, and W. E. Mann, between July 12, 1937 and Aug. 11, 1938, were acquired by the author about 1975 from a person who had retrieved them from a trash heap. These reports form the nucleus of this paper.
21. J. C. Lake, Progress Report No. 2, Aug. 27, 1937, pp. 4-6, Ref. 20, p. 6, Ref. 14, Feb. 27, 1991, Memo, A. W. Staudt to G. D. Graves, Sept. 26, 1938, Hoff, Ref. 11, Box 951, folder 38.
22. Lake, Ref. 20, p. 10.
23. Lake, Progress Report, Ref. 21, p. 9.
24. Lake, Progress Report, Ref. 23, p. 5.
25. J. C. Lake, Progress Report No. 3, Sept. 20, 1937, p. 5.
26. Lake, Ref. 25, pp. 3-4.
27. Lake, Ref. 20, p. 7, Ref. 25, pp. 2-3, 5; G. J. Groh to J. B. Quig, Status Reports, Aug. 27, 1937, p. 2, Oct. 1, 1937, p. 2, Dec. 23, 1937, p. 4, March 25, 1938, p. 2.
28. E. W. Spanagel, "Method of Sizing Synthetic Linear Polymer Yarns," United State Patent 2,253,146, Aug. 19, 1938.
29. F. C. King, laboratory notebook 1054, Dec. 28, 1937, Hoff, Ref. 11, Box 973, Memo, G. P. Hoff to G. J. Groh, Nov. 24, 1937, Hoff, Ref. 11, Box 951, folder 38.
30. J. C. Lake, Progress Report No. 7, Feb. 25, 1938, p. 3.
31. Lake, Ref. 30, pp. 5-7.
32. G. J. Groh to G. P. Hoff, June 17, 1938, cover memo to J. C. Lake, Progress Report No. 12, June 16, 1938.
33. Lake, Ref. 32, pp. 4-7.
34. J. C. Lake, Progress Report No. 17, Aug. 11, 1938.

35. M. C. Miller, "Principles of Knitting," McGraw Hill: New York, 1931, pp. 128-147, H. R. Mauersberger, "American Handbook of Synthetic Textiles," Textile Book Publishers: New York, 1952, pp. 702-707.
36. Lake, Ref. 20, pp. 2-6.
37. J. C. Lake, Progress Report No. 8, March 23, 1938; G. J. Groh to J. B. Quig, March 25, 1938, cover memo attached to J. C. Lake, Progress Report No. 8.
38. G. J. Groh to J. B. Quig, July 12, 1937 cover memo attached to J. C. Lake, Progress Report No. 1, p. 1.
39. Lake, Ref. 20, p. 9, Lake, Ref. 25, p. 2.
40. G. J. Groh to J. B. Quig, Aug. 27, 1937, cover memo to J. C. Lake, Progress Report No. 2, Ref. 21, W. E. Mann, Progress Report No. 15, July 5, 1938, p. 4.
41. G. J. Groh to J. B. Quig, Status Report, Oct. 1, 1937, p. 2.
42. G. J. Groh to J. B. Quig, Status Report, Dec. 28, 1937, pp. 4-5.
43. J. C. Lake, Progress Report No. 6, Feb. 24, 1938, G. J. Groh, Status Report, April 28, 1938, p. 5.
44. Groh, Ref. 43, pp. 4-5, Letter, R. A. Ramsdell, Du Pont Co. to I. K. Hessbarg, Van Raalte Co., Feb. 9, 1939, Hoff, Ref. 11, Box 951, folder 38.
45. Carothers, Ref. 7.
46. Cattell, Ref. 6, p. 1703.
47. J. B. Miles, Jr., "Steam Treatment of Polyamides," United States Patent 2,157,117, May 9, 1939.
48. E. V. Lewis, Notebook 207, "Odd Notes," Sept. 27, 1937, p. 35, Hoff, Ref. 11, Box 972.
49. Letter, E. V. Lewis to G. P. Hoff, "Reduction of Laciness in Stockings," Oct. 12, 1937, Hoff, Ref. 11, Box 951, folder 37.
50. Letter, G. P. Hoff to E. K. Gladding, "Status of Fiber 66," Nov. 10, 1937, p. 6, Hoff, Ref. 11, Box 963, folder 1.
51. Memo, E. V. Lewis to G. J. Groh, Aug. 27, 1937, Hoff, Ref. 11, Box 951, folder 37, W. E. Benzinger, "Textile Color Terms and Specifications," Sales Service Section, Rayon Division, E. I. Du Pont, Dec. 3, 1937.
52. J. B. Miles, Jr., "Method of Making Fabric," United States Patent, 2,157,119, May 9, 1939.
53. Lake, Ref. 20, pp. 4-5.
54. Lake, Ref. 43, pp. 5-9.
55. Mann, Ref. 40.
56. G. J. Groh to J. B. Quig, Status Report, Oct. 1, 1937, p. 5, Memo, G. P. Hoff to I. H. Godlove, April 30, 1937, Hoff, Ref. 11, Box 977, folder 34.
57. Lake, Ref. 20, pp. 3, 9, G. J. Groh to J. B. Quig, Status Report, April 28, 1938, p. 3.
58. W. E. Mann, Progress Report No. 13, May 26, 1938, p. 1, Lake, Ref. 20, p. 9.
59. Lake, Ref. 37, p. 9.
60. Lake, Ref. 20, p. 4, Groh, Ref. 42, p. 8.
61. Groh, Ref. 41, pp. 4, 5, J. C. Lake, Progress Report No. 9, April 1, 1938, pp. 2, 4.
62. Lake, Ref. 20, p. 8, Ref. 30, p. 8.
63. Hoff, Ref. 50, p. 5.
64. Hoff, Ref. 50, p. 5.
65. Groh, Ref. 43, p. 3.
66. W. E. Mann, Progress Report No. 13, May 26, 1938, p. 1.
67. W. E. Mann, Progress Report No. 14, June 17, 1938.
68. Mann, Ref. 40, pp. 2, 3.
69. W. E. Mann, Progress Report No. 16, Aug. 11, 1938, p. 2.
70. Comment by G. P. Hoff written by item 6, p[. 2, of memo G. J. Groh to J. B. Quig, July 12, 1937, Hoff, Ref. 11, Box 951, folder 38.
71. J. C. Lake, Progress Report No. 10, May 24, 1938, p. 3.
72. Lake, Ref. 64.
73. Based on designation changes in progress and status reports.
74. Lake, Ref. 20, p. 10, J. C. Lake, Progress Report No. 11, June 14, 1938, p. 6.
75. G. J. Groh to J. B. Quig, Status Report, Dec. 28, 1937, pp. 8, 9.
76. J. C. Lake, Progress Report No. 7, Feb. 25, 1938, pp. 9-10, Groh, Ref. 43, p. 6.
77. Lake, Ref. 61, p. 9.
78. J. C. Lake, Progress Report, Ref. 74, pp. 3-6, Groh, Ref. 43, pp. 6-7.
79. Lake, Ref. 61, pp. 7-9, Ref. 20, pp. 4-5, Groh, Ref. 43, p. 7.
80. Memo, J. C. Lake to G. J. Groh, Oct. 6, 1938, Hoff Papers, Ref. 11, Box 951, folder 38.
81. Lake, Ref. 80.
82. Hoff, Ref. 50, pp. 6-7, Groh, Ref. 42, pp. 9-11.
83. Hounshell and Smith, Ref. 1, p. 267, Memos G. J. Groh to J. C. Lake, Sept. 12, 1938, A. W. Staudt to G. C. Hazelwood, et al, Sept. 12, 1938, Hoff, Ref. 11, Box 951, folder 38.

84. Hounshell and Smith, Ref. 1, p. 268, D. F. Holmes, "History of the Du Pont Company's Textile Fibers Department," privately printed, Wilmington, DE, 1983, pp. 5-6.
85. J. C. Lake, "Report of Meeting Between the Van Raalte Organization and the Nylon Division...," January 12, 1939, p. 2, Hoff, Ref. 11, Box 941, folder 38.
86. Chemical and Metallurgical Engineering, 46(12), 760-761 (Dec. 1939).
87. Customers listed on letter, G. J. Groh to J. J. Kronenberg, Wayne Knitting Mills, May 10, 1939, Hoff, Ref. 11, Box 951, folder 37.
88. Letter, K. M. Mayer to R. A. Ramsdell, March 22, 1939, Hoff, Ref. 11, Box 951, folder 37, Groh, Ref. 87.
89. Memo, G. J. Groh to R. A. Ramsdell, April 14, 1939, p. 3, Hoff, Ref. 11, Box 951, folder 37, Groh, Ref. 87.
90. Schulze and Soukup, Ref. 3, p. 169.
91. Hounshell and Smith, Ref. 1, p. 272.
92. Hosiery and Underwear, 61(1), (July 1976), America's Textiles International, 17(4), 50 (April 1988).

HISTORY OF FIBER PRODUCTION BY MELT SPIN-ORIENTATION PROCESSES

H. D. NOETHER
TRI-Princeton, 601 Prospect Avenue, P.O. Box 625, Princeton, NJ, 08542

Yarn manufacture and other textile operations are highly labor intensive. Initially synthetic fibers were spun from solutions, evaporating the solvent (dry spinning) or by regenerating the polymeric material in fiber form from a viscous solution in a well controlled precipitationbath (wet spinning). With the invention of nylons, Carothers increased the productivity of synthetic fibers by extruding a 100% polymer melt rather than a 20 to 25% polymer solution. Further he oriented the filaments or yarns to produce mechanically desirable materials for both textile and industrial applications.

It does not take much imagination to realize the advantages of combining melt-spinning and drawing processes into a single operation in which the "as spun" fiber bundles are oriented between the spinnerette and the yarn wind up device. It would save machinery and yield more product. Results were a partial success: Only yarn strengths of textiles (3-5 g/den) were obtained with considerably higher elongations than desirable.[1] (Fig.1). At the time, such behavior was completely unexpected and unexplainable by our fiber structural models.

Two advances in our knowledge of polymers occurred in the late 1950's. One was the synthesis of stereo-regular polymers of olefins using Ziegler-Natta catalysts, the other showed that polymers could form single crystals like other crystalline materials and that the polymer chains in the crystals were folded. (Figs. 2, 3). The latter discovery, (by electron microscopy, Fischer, Keller, Till)[2] originated a completely new model of morphological polymer structure and established correlations of molecular data with observed macroscopic mechanical properties. The Ziegler-Natta discovery started the commercialization of a flood of crystalline polyolefins. They had useful melting points, broad molecular weight distributions which allowed melt-extrusion over a wide temperature range, in contrast to the nylons and polyesters, which had narrow molecular weight distributions and a narrow melt spinning range but until then were the only major raw-materials to be melt processable.

In spin-orientation processes the resulting yarn properties depend on the crystallization rate:

(1) Polymers with very high rates of crystallization such as polyethylene, polypropylene and other stereo-polyolefins, polypivalolactone (Statton), polyoxymethylene (Celanese, Clark at DuPont)[3].

(2) Materials with low rates of crystallization, primarily PET(Polyethylene terephthalate). Fig. 4 shows rates of crystallization (log scale) in dependence on supercooling temperature of a few polymers. These data are critical since spin-orientation fiber formation is a non-isothermal and very fast process.

Polypropylene demonstrates spin orientation results of a fast crystallizing material. The polymer melts at 165°C and is extruded in the range of 200 to 350°C. Slow extrusion followed by several drawing steps yields a strong industrial fiber.[4] Its x-ray diffraction pattern (wide and small angle) show a highly oriented crystalline structure (Fig. 5, 6) with a fibrillar arrangement of the polymer chains. Spin-orientation at 500 to 3000 m/min take-up speed also produces a highly oriented crystalline fiber, but its morpholgy and mechanical properties are vastly different.[5] Fig. 7 is the x-ray data and Fig. 8 is a comparison of tensile properties.

This fiber is relatively weak, it has an extension at break of 300 to 500% and when extended to about 100%, the extension is highly recoverable (Fig. 9). On annealing it becomes more crystalline and has improved elastic characteristics (about 95% immediate elastic recovery). At the time this behavior was not known and therefore we tried every physical method available (x-ray diffraction, microscopy, electron-microscopy, tensile and elastic evaluations) to understand the elastic mechanism. Microscopic observation showed that the fiber did not contract laterally on extension as would be the case for rubber-elastic behavior. Wide angle x-ray did not indicate a change in orientation on extension (Fig. 10). Small angle x-ray showed that the original lamellar structure of the fiber expanded proportional to the degree of extension and that the scattering intensity increased markedly, pointing to increased density differences in the system. The pattern was reversible on relaxation. Electron microscopy showed the formation of gradually increasing voids between the lamellar crystallites (Fig. 11). There are fibrillar connections between the crystallites, highly oriented during extension which collapse on relaxation, closing the voids. Further studies show the process to have both energy-elastic and rubber-elastic components (Muller & Goritz).[6]

The formation of the lamellar crystals proceeds as follows: Oriented chains in the stressed extruding melt form long nuclei on cooling; the lamellar crystals grow epitaxially onto these nuclei, forming a lateral lamellar morphology. Annealing produces a more uniform lamellar system which is more resistant to local irreversible deformations. The epitaxial crystallization process can be interrupted at the nucleation stage by quenching during the spinning process. Such a yarn is highly oriented, poorly crystalline (smectic?) and shows fibrillar morphology[7] (Fig. 12).

What can one do commercially with such a material? Elastic applications (Spandex?) seem possible. However the fiber is much too stiff for normal elastic uses. At present the commercial applications make use of the void formation. The voids are all accessible. Fibers or film are heat-set while extended and form microporous materials useful in permeable membrane operations. An electron microscope picture of a membrane is shown in Fig. 13. The holes are stabilized by crystalline fibrils crossing the voids (as can be ascertained by small angle x-ray diffraction) (Fig. 14).[8] Later research showed that the take-up velocity of 2000 to 3000 m/min is optimum for elastic properties; at higher spin-orientation velocities the elastic characteristics declined continuously.[9]

Polyester (polyethylene terephthalate) presents a somewhat different picture. This polymer had already been produced in every industrial country and was an important synthetic fiber. High speed extrusion also had been tried. In 1952 Hebeler[10] of DuPont received two patents, one to prepare a wool-like highly crimped PET yarn by spinning at take-up speeds between 3000 to 5200 yds/min and heat treating the product to crimp it, the other with take-up velocities above 5200 yds/min that gave a yarn of only 2-4% shrinkage on boil-off. These yarns were found useful as textile products after voluminizing, bulking or crimping processes.

In the 70's a new series of patents and publications appeared from DuPont, ICI, Celanese, Hoechst, Rhone-Poulenc, Akzo and somewhat later a book on spin-orientation by A.Ziabicki and H.Kawai.[11,12] It was shown that spin-orientation gave amorphous, slightly oriented fibers (POY, partially oriented yarns) and that above 3500 to 4500 m/min TUS (take-up speed) crystalline, highly oriented fibers were available (HOY, highly oriented yarns). These are the two types of yarns, originally proposed as shrinkable, wool-like and non-shrinkable. But even at 6000 m/min TUS, the "as spun" yarns only have textile properties, and have to be drawn further for high tenacity industrial products.[13] Today it is understood that the superstructure of the spin-oriented yarns is formed in a temperature range close to the melting point of the polymer, crystallization occurring preferentially in the highly oriented polyester chains; the remainder has no chance to crystallize in the short time to solidification. Thus while the orientation of the crystallites is as high as in the drawn yarns, (Figs. 15, 16, 17) the amorphous regions have low orientation; therefore they have low tensile properties, but also a non-shrinking structure up to the melting range of the yarn (Figs. 18, 19). In contrast, in a spun and subsequently drawn yarn the orientation is similar in both the crystalline and the amorphous regions. Here, drawing starts slightly above the glass transition temperature and involves several drawing steps at stepwise increasing temperatures and is followed by annealing to stabilize the drawn structure. The spin-orientation process for PET is equivalent to that in polypropylene, but because of the slow rate of crystallization and large supercooling, the process is stopped at the fibrillar nucleation stage. An additional equivalence is covered in a recent DuPont patent[14] which produces an amorphous, oriented structure at and above 5000 yds/min TUS by quenching the yarns at the nucleation stage. These yarns have a high boil-off shrinkage and are especially useful for false-twist texturing; except for the higher spinning speed, these yarns are similar to those spun earlier below 5200 yds/min to give wool-like properties.

High speed spinning demands extensive machinery design and engineering, both for experimental studies and commercial production. Wind-up machinery has to be accurate and safe; temperature control along the spin line is essential. New measuring devices had to be developed to observe process conditions (temperature, yarn speed, etc.) without touching the yarn along the spin line.

Melt spin-orientation processes resulted in new types of yarn products and forced us to develop better models for correlating molecular structural features with macroscopic mechanical and thermal properties.

References

1. W. Griehl and H. Versaumer, Faserforsch. u. Textiltechnik, 9, 226 (1958).
2. E. W. Fischer, Z. Naturforsch. 12a, 753 (1957); A. Keller, Phil. Mag., 2, 1171 (1957); P. H. Till, Jr., J. Polym. Sci., 24, 301 (1957).
3. Summary reference to polymers with high rates of crystallization which may yield "hard elastic" materials: H. D. Noether, Poly. Eng. and Sci., 19, 427 (1979).
4. H. D. Noether and R. W. Singleton, U.S. Patent 3,161,709; to Celanese, Dec. 16, 1964.
5. H. D. Noether and W. Whitney, Kolloid Z., 251, 991 (1973); I. K. Park and H. D. Noether, Colloid Polym. Sci., 253, 824 (1975); B. S. Sprague, J. Macromol. Sci. Phys., B8, 157 (1973); S. L. Cannon, G. B. McKenna and W. O. Statton, Macromol. Reviews, 11, 209 (1976).
6. D. Goritz and F. H. Muller, Coll. Polym. Sci., 252, 862 (1974); D. Goritz and F. H. Muller, Coll. Polym. Sci., 253, 844 (1975).
7. W. H. Sheehan and T. B. Cole, J. Appl. Poly. Sci., 8 2359 (1964); H. D. Noether, Prog. Colloid & Polym. Sci., 66, 109 (1979); H. D. Noether, Poly. Eng. &. Sci., 19, 427 (1979).
8. H. D. Noether and I. L. Hay, J. Appl. Cryst., 11, 546 (1978).
9. J. Shimizu, N. Okui and Y. Imai, Sen-I Gakkaishi, 36, 166 (1980).
10. H. H. Hebeler, U.S. Patent 2,604,657; U.S. Patent 2,604,689; to Du Pont, July 29, 1952.
11. PET spin-orientation publications: E. Liska, Koll. Z. u. Z. Polymere, 251, 1028 (1973); H. M. Heuvel and R. Huisman, J. Appl. Poly. Sci., 22, 2229 (1978); C. Lecluse and G. Perez, Man-made fiber conference at Dornbirn, Austria, June 20-22, 1979; W. D. Dietrich, G. Reichelt, H. Renkert, Chemiefasern u. Textilindustrie, Sept. 1982, 612. A. Ziabicki and H. Kawai: High Speed Fiber Spinning, Science and Engineering Aspects, John Wiley & Sons, Inc. Publishers: March 1985.
12. PET spin-orientation patents: H. R. E. Frankfort and B. H. Knox, U.S.Patent 4,134,882, Jan. 16, 1979, U.S. Patent 4,195,051, March 25, 1980, to Du Pont.
13. H. L. Davis, M. J. Jaffe, H. L. LaNieve, E. J. Powers, U.S. Patent 4,195,052 to Celanese March 25, 1980.
14. G. Vassilatos, U.S. Patent 4,425,293 to Du Pont, Jan. 10, 1984.

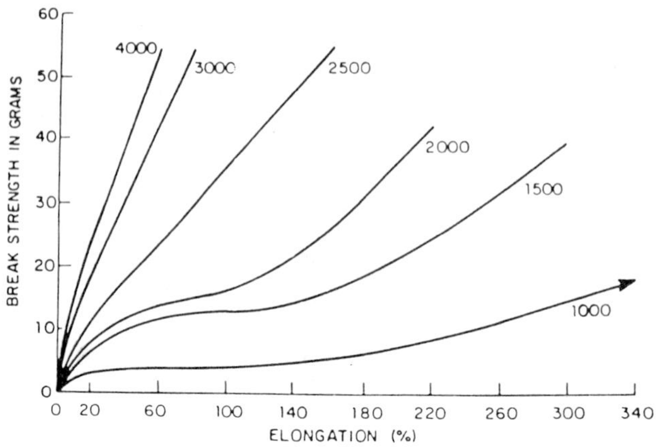

Figure 1. Stress-strain curves of spin-oriented Nylon 6 as a function of take-up velocity (in m/min).

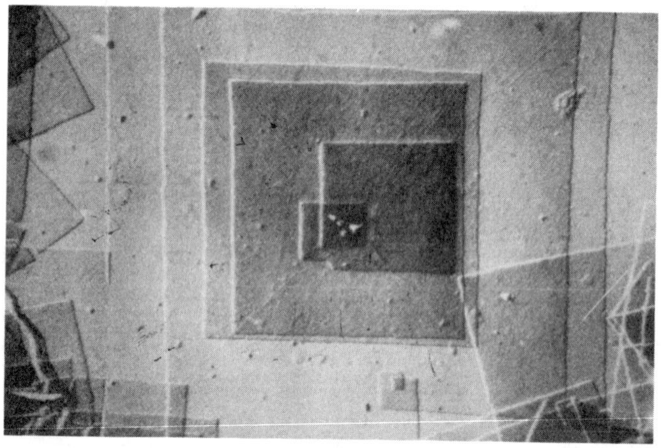

Figure 2. Single Crystals of 4-Methyl Pentene. B.C. Basset, F. R. Dammont and R. Saloway, Polymer, 5, 579 (1964).

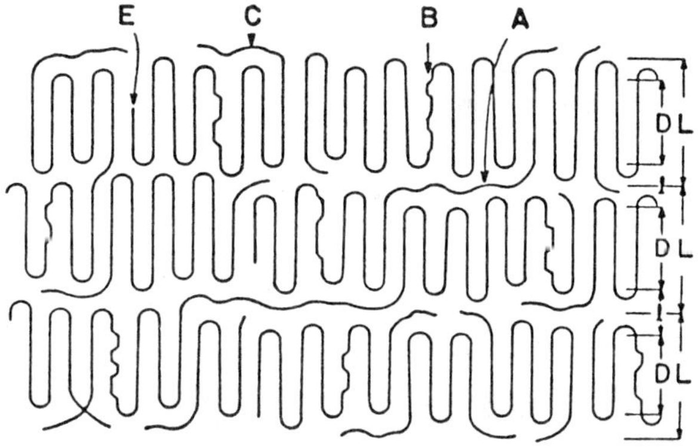

Figure 3. Molecular model of a stack of parallel lamellae of the spherulitic structure; A. interlamellar tie molecule; B. boundary layer between two mosaic blocks; C. chain end in the 'amorphous' surface layer (cilium); D. thickness of the crystalline core of the lamella; E. linear vacancy caused by the chain end in the crystal lattice; L. long period; l. thickness of the 'amorphous' layer (Anton Peterlin)

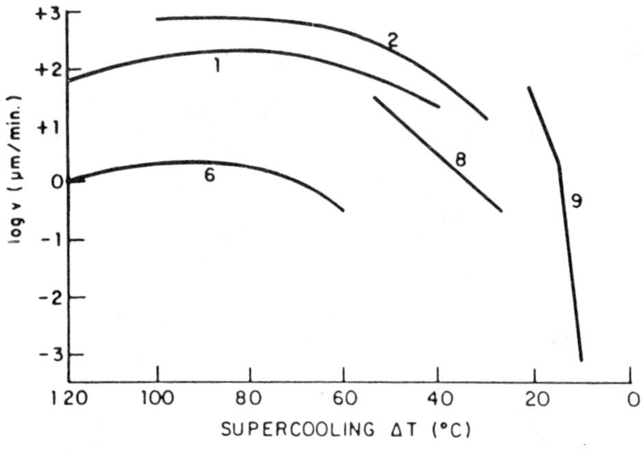

Figure 4. Crystal growth rates as a function of supercooling: (1), Nylon 6; (2), Nylon 6,6; (6), poly(ethylene terephthalate); (8), polypropylene; (9), polyethylene. H. D. Noether, Intern. J. Polymeric Materials, 7, 57-82 (1979).

Drawn Polypropylene Fiber X-ray Pattern

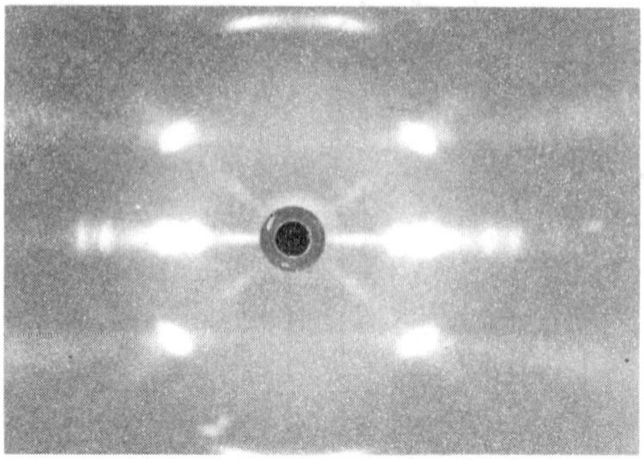

Figure 5. Wide Angle X-ray

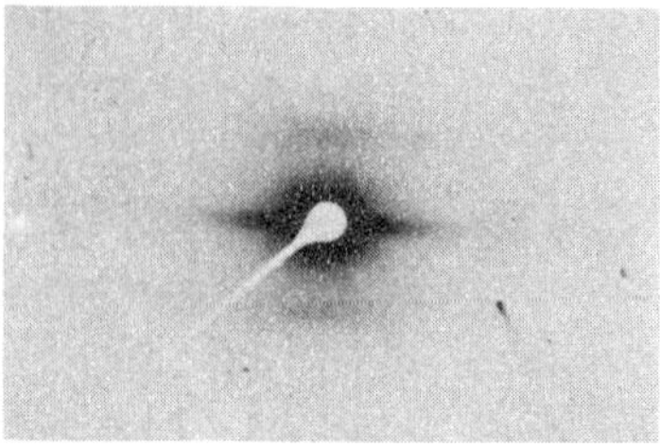

Figure 6. Small Angle X-ray

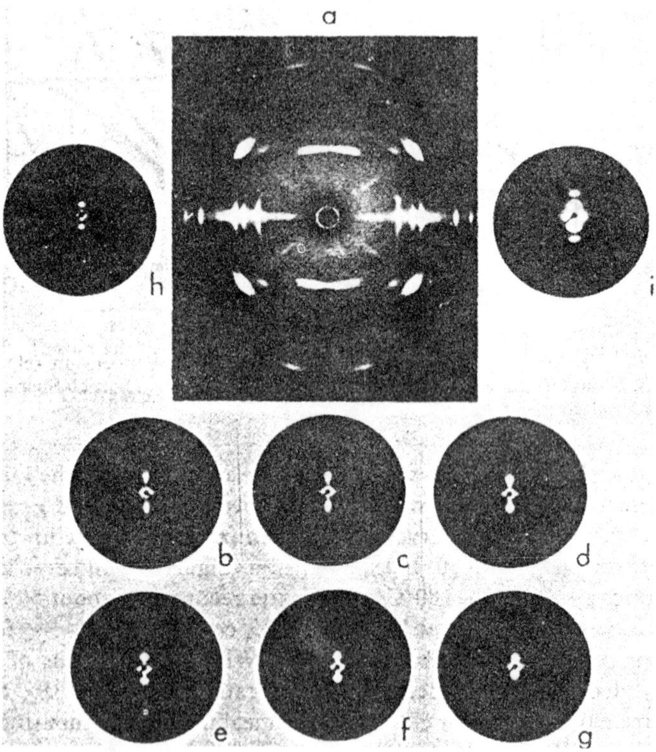

Figure 7. WAXR and SAXR of polypropylene as a function of annealing temperature. (a), WAXR (150°C/1 hr). SAXR data: (b) 70°C/1 hr; (c) 90°C/1 hr; (d) 120°C/1 hr; (e) 140°C/1 hr; (f) 155°C/1 h4; (g) 160°C/1 hr; (h) 150°C 1/hr, SAXS exposure 1/2 hr; (i) 150°C/1 hr, SAXS exposure 16 hr. H. D. Noether, Ref. for Fig. 4.

Figure 8. Tensile Properties of Drawn and Spin-oriented Fibers.

	Modulus g/den	Tenacity g/den	Elongation %
PET (Drawn)	140-150	9-12	11-15
PET (Spin Or.)	60-100	3-5	30-60
PP (Drawn)	160	10-12	10-12
PP (Spin Or.)	20-40	1-2	300+

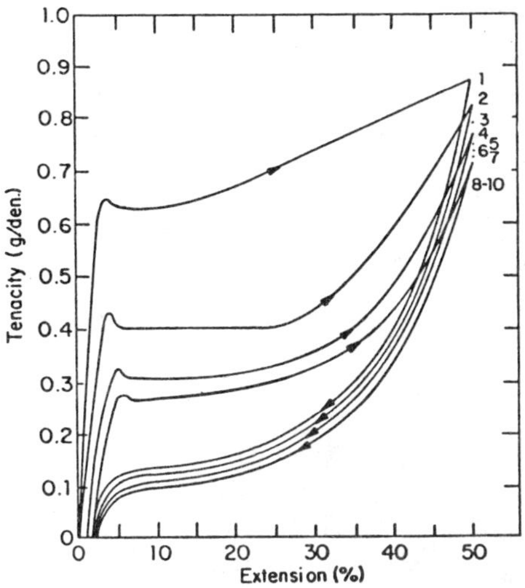

Figure 9. I. K. Park & H. D. Noether, Coll. & Polym. Sci., 253, 824 (1975). Annealed elastic polypropylene stress-strain behavior on cycling to 50% extension (10 cycles). Extension rate: 100%/min.

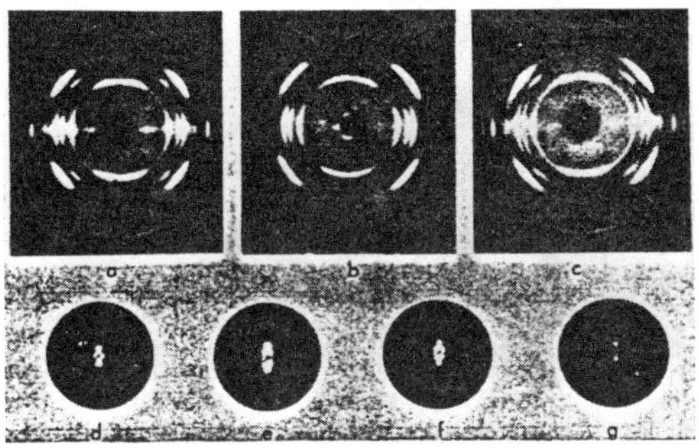

Figure 10. H. D. Noether & W. Whitney, Koll. Z. & Z. Polymere, 251, 991 (1973). WAXR patterns of polypropylene on extension to 73% and relaxation (from left-to-right: 0% extension, 73% extension, back to 0% extension). SAXR patterns (from left-to-right): 0%, 50%, 100%, and back to 0% extension.

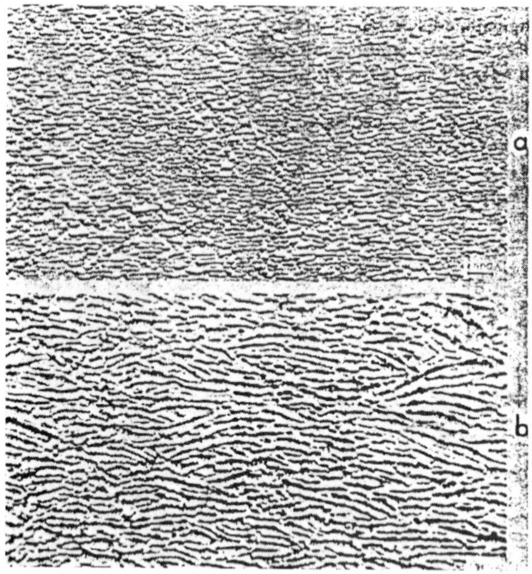

Figure 11. Electronmicrograph of polypropylene annealed at 144°C. (a) 0% extension, (b) 80% extension. Direction of stretch: vertical. Same Ref. as Fig. 10.

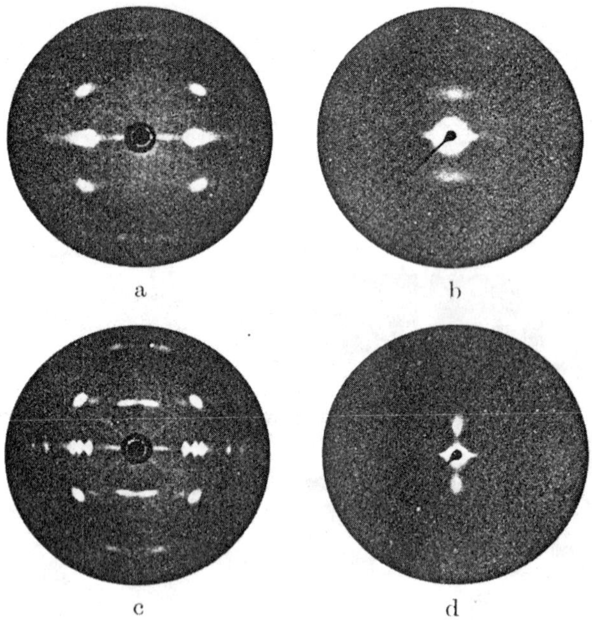

Figure 12. H. D. Noether, Progress. Coll. & Poly. Sci., 68, 109 (1979). X-ray diffraction pattern of smectic and crystalline polypropylene. Smectic structure: 1a. WAXS; 1b. SAXS. Crystalline structure: 1c. WAXR; 1d. SAXS.

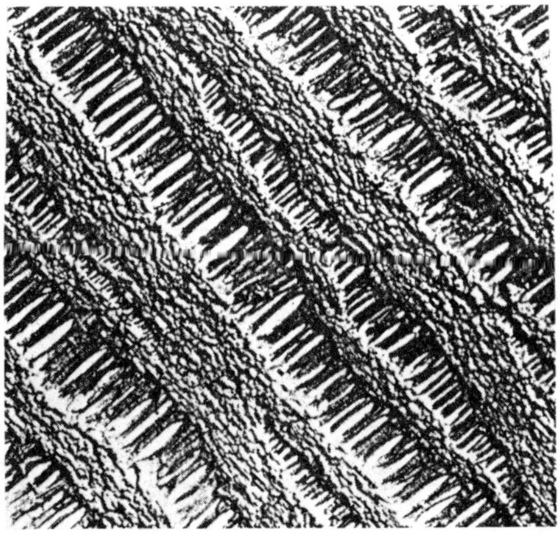

Figure 13. H. D. Noether & I. L. Hay, J. Appl. Cryst., 11, 546 (1978). Electronmicrograph of Celgard microporous material. The thickness of the lamellae in the unextended parts is about 180-200 Å. (Courtesy of Dr. I. Hay.)

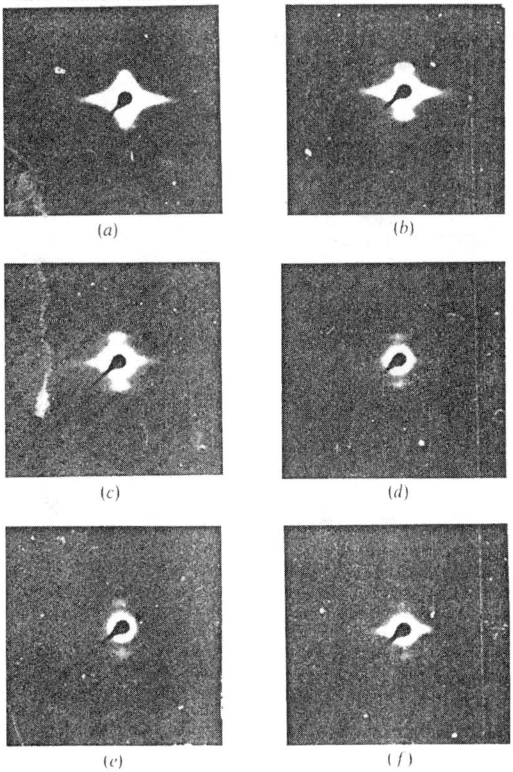

Figure 14. Same Ref. as Fig. 13. SAXS pattern of Celgard with solvents of various electron densities. (a) air, (b) heptane, (c) methanol, (d) decalin, (e) cyclohexanol, (f) methylsalicylate.

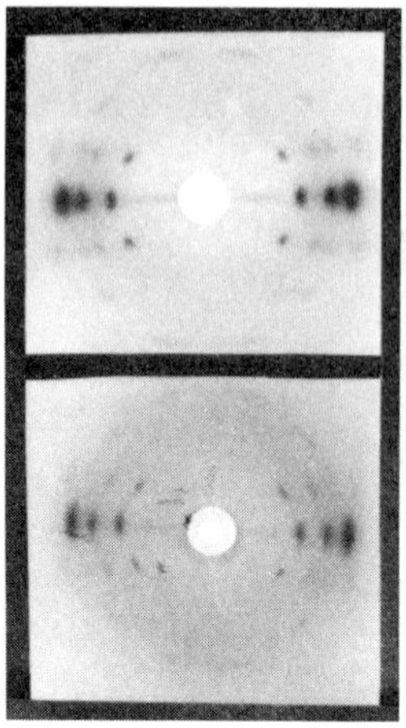

Figure 15a

Figure 15b

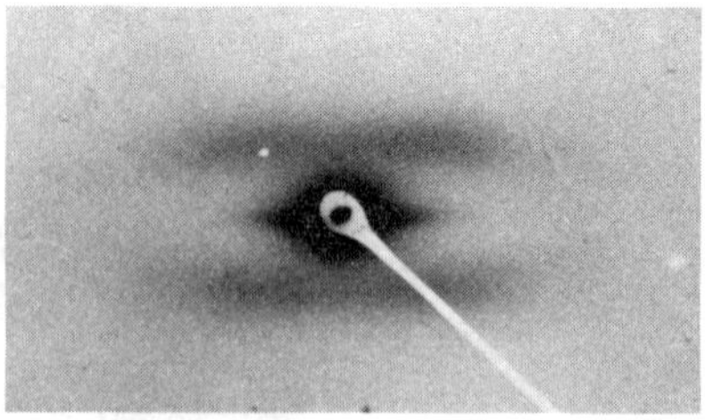

Figure 16

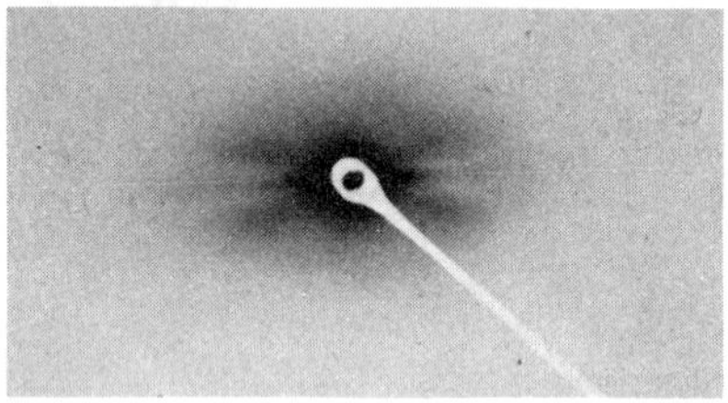

Figure 17

Figure 15-17. X-ray pattern of polyethylene terephthalate fibers. Fig. 15- wide angle pattern, (a) drawn fiber, (b) spin-oriented fiber; Fig. 16 - small angle pattern, drawn fiber; Fig. 17 - small angle pattern, spin-oriented fiber.

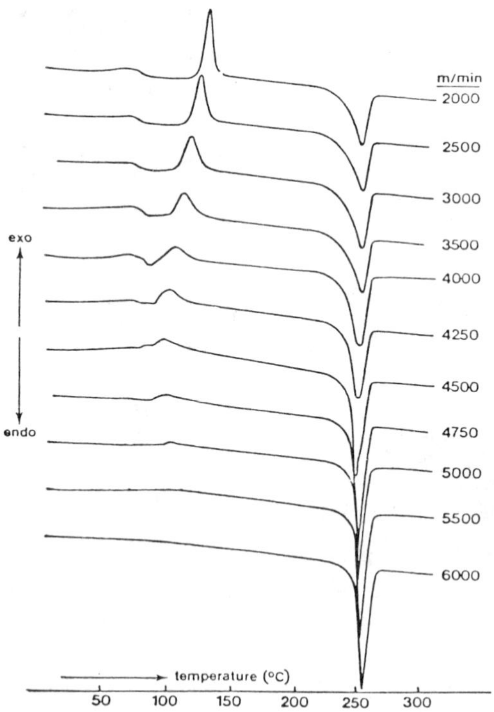

Figure 18. H. M. Heuvel & R. Huisman, J. Appl. Poly. Sci., 22, 2229 (1978). Spin-oriented PET. DTA curves as function of take-up velocity.

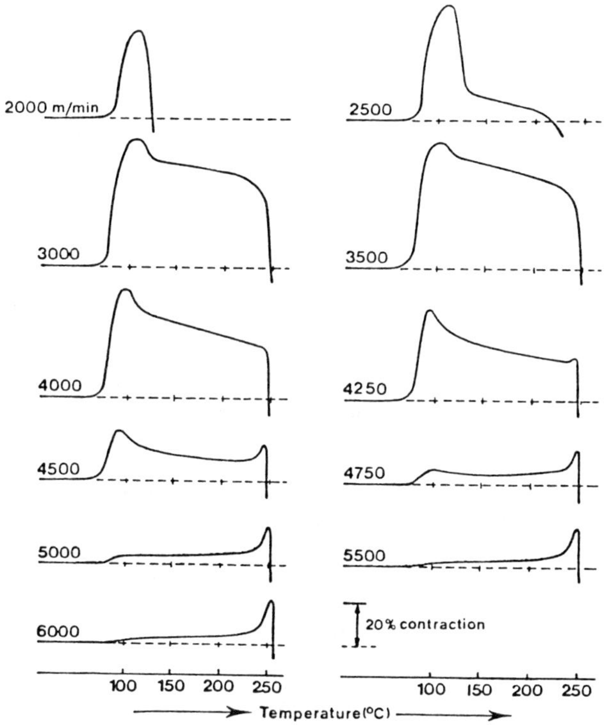

Figure 19. Same Ref. as Fig. 18. Spin-oriented PET. TMA curves as function of take-up velocity.

HISTORY OF SPANDEX ELASTOMERIC FIBERS

A. J. ULTEE
Research Associate, E.I. DuPont De Nemours and Company, Waynesboro, VA 22980

INTRODUCTION

Elastomeric fibers are used as the minor component in stretch garments to provide stretch with recovery. The main categories of elastomeric fibers are rubber and spandex. Spandex fibers, elsewhere also called elasthane fibers, by definition consist for at least 85% of segmented polyurethane. Segmented polyurethanes contain alternating "soft" and "hard" segments linked by urethane bonds. Figure 1 shows an example of such a composition.

Spandex fibers were invented in the nineteen fifties to improve on the process and properties of rubber fibers, and became commercial around 1960. While the history of spandex fibers is therefore short, it is much longer if one considers that their discovery was only possible by combining knowledge of rubber technology, organic chemistry, polymer chemistry and synthetic fiber spinning technology, each of which had a long history going back to the nineteenth century.

CONTRIBUTORY HISTORIES - RUBBER TECHNOLOGY

To start with the contribution of rubber technology to spandex, this came mostly in the form of understanding the process and product deficiencies of rubber thread rather than adopting rubber technology for spandex. The history of rubber technology starts with the invention of plasticization by Hancock in 1820 and of vulcanization by Goodyear in 1839. To convert Rubber into useful articles like rubber thread, it was necessary to first break the molecules down to a tractable size for compounding and shaping, then vulcanize by a slow batch process. In case of cut rubber thread, vulcanization took place in the form of thin sheets, which then were sliced to threads of rectangular cross-section. A better process is used for latex rubber thread [1]. Latex can be compounded with vulcanizing ingredients without plasticization and extruded into a coagulating bath to form continuous threads with round cross-section. Accelerated vulcanization recipes allow a continuous vulcanization process. Nevertheless rubber latex thread manufacture is still a slow process in comparison with synthetic fiber spinning and finishing.

The property deficiencies of rubber thread are low recovery force from stretch and sensitivity to atmospheric degradation. The recovery force (called "power" in textile uses) can be increased by reinforcement with carbon black, but this cannot be used with rubber thread for textiles because of color. The sensitivity of rubber to atmospheric degradation can be traced to the double bonds of its polyisoprene composition, which are especially prone to ozone attack.

In summary, the inventors of spandex learned from rubber experience to look for more rapid network formation than vulcanization, preferably during fiber spinning; for an increase in "power" by some built-in reinforcement without color; and for compositions without oxidation-prone double bonds in the polymer chain.

CONTRIBUTORY HISTORIES - ORGANIC CHEMISTRY

Spandex is made from segmented polyurethane, and to synthesize polyurethanes in an economical fashion, isocyanate chemistry is required. Isocyanates were discovered by A. Wurtz in 1848 [2]. The basic reactions by which isocyanates can for urethanes as well as ureas are shown in Figure 2. These reactions take place in high yields without forming volatile byproducts, a special advantage for commercial polymerization processes. Isocyanates are therefore predominantly used to put the urethane bonds in the segmented polyurethanes for spandex. In addition, the reaction forming ureas is commonly used to build urea hard segments for spandex. Thus, spandex development was very much dependent on isocyanate raw materials. The long-known phosgenation reaction of amines to form isocyanates became the basis for commercial processes to provide diisocyanates for spandex.

In addition to isocyanates, organic chemistry supplied the necessary ingredients for the soft segments of spandex polymers. Diols and dicarboxylic acids are starting materials for polyester soft segments, while tetrahydrofuran is the preferred starting material for polyether soft segments.

Finally, organic chemistry contributed by the synthesis of various stabilizers which are used as additives in spandex.

CONTRIBUTORY HISTORIES - POLYMER CHEMISTRY

The history of spandex fibers is closely interwoven with the history of segmented polymers which are block copolymers with alternating segments with greatly different properties. After the discovery of nylon by Carothers of DuPont, more work followed in the field of alternative polyamides from different diacids and diamines. Wittbecker [3] and coworkers showed, as illustrated in Figure 3, that copolyamides from a mixture or primary and secondary diamines in a certain composition range were elastomers. Apparently the combination of high-melting, hydrogen-bonded and low-melting, non-hydrogen-bonded polymer segments somehow produced a network structure with elastic properties. This did not necessarily require hydrogen bonding for the high melting segments. M.D. Snyder [4] showed shortly afterwards that copolyesters from ethylene glycol with mixtures of stiff terephthalic acid and flexible sebacic acid produced polymers with an elastic region. It was then found that for both polyamides and polyesters more "blocky" copolymers with better properties could be made by starting with the high- and low-melting homopolymers and melt blending these at a temperature where amide or ester interchange occurs. But the interchange has to be interrupted before a random copolymer results. While such compositions could be used for melt-spun elastomeric fibers, it would be hard to control properties with ongoing interchange reactions. J.C. Shivers (Figure 4) then showed [4] that copolymers of non-interchanging poly(ethylene ether) and ethylene terephthalate in a certain composition range were stable elastomers with high melting points, as illustrated in Figure 5. These copolymers had previously been investigated by Edgar and Hill [6], but only in the low poly(ethylene ether) range which did not give elastomers. Further improvement of elastic properties was obtained by using lower-melting polyethers for the soft segment, in particular poly(tetramethylene ether) which could be obtained by polymerization of tetrahydrofuran as shown by H. Meerwein [7]. The significance of Shivers' work for spandex is that it laid the groundwork for understanding the requirements for segmented polymers useful for elastomeric fibers, that is

1) soft segments of low-melting polymers forming the major part of the segmented elastomers;

2)hard segments consisting of blocks of high-melting polymers held together by hydrogen bonding or other forces;

3) stability of hard and soft segments during the synthesis and processing of the segmented polymers; and

4) no covalent cross-links required.

There were many possible compositions for these elastomers: for example, for the soft segments polyethers, polyesters and polylactones; for the hard segments, polyurethanes and polyureas in addition to polyamides and polyesters. For coupling hard and soft segments, however, urethane links formed by isocyanate chemistry became the preferred route. For this we have to go back to the developments in polyurethane chemistry.

As mentioned before, the reaction of isocyanates with hydroxyl groups proceeds in high yields without formation of volatile byproducts. It was a major contribution when O. Bayer (Figure 6) of IG-Farben (later Bayer) applied this reaction to diisocyanates and diols to produce polyurethanes by his polyaddition process [8]. This opened up a whole new area of useful polymers. The first fibers made of polyurethane were nylon-like "hard" fibers with names like "Perlon-U" and "Dorlon". At Bayer softer polyurethanes were also obtained by reacting diisocyanates with low-molecular weight, hydroxyl-terminated aliphatic polyesters to give highly elastic materials called Vulcollans [9]. Best properties were obtained by reacting the polyester glycols with an excess of diisocyanate and a limited amount of water. As shown in Figure 7, water reacts with two isocyanate groups to give urea linkages which then can react with more isocyanate to give biuret cross-links. The Bayer workers also studied the effect of polyester composition and molecular weight on elastomer properties, such as crystallization or admixture with poly(1,2-propylene adipate) was the chosen way to lower the melting point.

CONTRIBUTORY HISTORIES - SYNTHETIC FIBER TECHNOLOGY

Spandex fiber being a latecomer in the synthetic fiber world has profited from the various techniques that had been developed for other synthetic fibers. The main choices for fiber spinning were:

1) Melt-spinning as developed for nylon and polyesters - here the main limitation for spandex was one of melting point. For good elastic properties, a high-melting hard segment was required, but the reversibility of isocyanate reactions put an upper melting point limit on melt-spinnable spandex because of degradation of urethane and urea bonds.

2) Dry-spinning as developed for cellulose acetate and acrylic fibers - Here the main requirement was solubility of the segmented elastomers in a suitable dry-spinning

solvent and the stability of such solutions. It ruled out any significantly cross-linked compositions.

3) Wet-spinning, using techniques similar to those developed for viscose rayon and latex rubber - this also required use of a solvent for the polymer, although concentration requirements were less stringent than for dry-spinning.

4) Reaction- or chemical-spinning, a form of wet-spinning where the polymer is formed during the spinning operation. This is applicable to spandex compositions by spinning a prepolymer of soft segment with isocyanate endgroups into a bath containing ingredients for hard segment formation.

BIRTH OF SPANDEX FIBERS

Two large companies, DuPont and Bayer, were mainly responsible for the successful development and commercialization of spandex fibers. Contributions of other companies will be discussed only where they led to successful new types of spandex.

A. DuPont

It was an exciting time in the early fifties in DuPont fiber departments' pioneering research laboratory. After the compositional requirements of segmented elastomers for spandex had been defined, a large group of chemists and engineers was assigned the job of coming up with the best composition as well as the best fiber process for commercialization. Segmented polyurethanes were just one group of compositions included in this investigation. For polyether soft segments, poly(tetramethylene ether) replaced poly(ethylene ethers) which were to hydrophilic. Fortunately, at another department of DuPont, a commercial process for polymerization and termination of tetrahydrofuran to give highly bifunctional poly(tetramethylene ether) glycols had been developed [10]. For the hard segments, the emphasis wa on building the most strongly cohering hard segments, by selecting the highest melting and still processible from the classes of polyesters [11], polyamides [12], polyurethanes [13], Polyureas [14] and polyureylenes [15]. While segmented elastomers with hard segments of the first three groups were melt-spinnable, the strongly hydrogen-bonded polyureas and polyureylenes would not melt without decomposition and had to be spun from solution by dry- or wet-spinning, or formed during spinning by reaction spinning. All these methods were explored [14,16,17]. The final selection of the composition for commercialization was a segmented elastomer with polyureylene hard segments based on the work of P.E. Frankenburg and A.H. Frazer (Figures 8 and 9). As shown in Figure 10, the polymer was made by reacting poly(tetramethylene ether) glycol with an excess of diphenylmethane diisocyanate (MDI) to give a "capped" glycol with isocyanate end groups, which was then "chain-extended" with hydrazine to form the segmented elastomer. Solutions of the polymer in dimethylformamide were dry- and wet-spun into spandex fibers, using modifications of known spinning methods. The new spandex fiber was introduced from pilot-plant production as "Fiber K" in 1959 and commercialized from plant production as "Lycra" in 1962.

B. Bayer

Developments at bayer in the nineteen fifties toward creation of spandex fibers moved along somewhat different lines. Already in 1949, E. Windemuth [18] had produced a fiber with spandex composition by extruding a isocyanate-capped prepolymer into an aqueous piperazine solution, but properties were poor because of hard segment composition. More importantly, W. Brenschede [19] showed in 1951 that the supposedly cross-linked, water-extended vulcollans could be dissolved in dimethylformamide or dimethylacetamide and wet-spun into elastic fibers. However, the time was not ripe in post-war Germany to commercialize an expensive fashion fiber, so research was directed to more conventional uses of the new elastomers. This changed in 1958 after the announcement of spandex fibers by DuPont and US Rubber in America. A group was organized under H. Rinke (Figure 11) to develop fibers from linear segmented elastomers with high-melting urea-type hard segments. Such compositions would have to be dry- or wet-spun from a solvent like dimethylformamide or dimethylacetamide.Bayer, like DuPont, had dry-spinning technology from their acrylic fiber. As far as composition goes, they had already explored polyester soft segments, so the main job was to design the proper hard segment. Rinke's coworkers H. Oertel and W. Thoma (Figures 12 and 13) came up with a hard segment from MDI diisocyanate and a dihydrazide [20], as shown in Figure 14, for the initial version of "Dorlastan" which was commercialized as a dry-spun spandex in 1964.

C. Other Producers

The first spandex fiber made by reaction spinning was developed by the US Rubber Company [21] and commercialized as "Vyrene". In their process, a soft-segment prepolymer with aromatic isocyanate end-groups was extruded into an aqueous diamine bath to produce filaments with a skin of spandex polymer having urea hard segments. The polymerization inside the filaments was then completed by chain-extension and/or crosslinking with water under pressure. Other rubber, chemical and fiber companies followed the US Rubber Co. lead by developing reaction-spun spandex. The most successful of these was the Globe Rubber Company, which hired A.F. Cacella (who had worked at DuPont) to help develop a process [22] based on extrusion of a polyester diol capped with tolylene diisocyanate into a diamine bath. Their products "glospan" and "cleerspan" still are a sizable part of world spandex production.

The next spandex developments were of wet-spun products. T.V. Peters, who had worked on spandex for DuPont, developed a wet-spinning process [23] which was picked up by several chemical and fiber companies around the world. The most successful of these was Fujibo of Japan. Another wet-spun product was developed by Union Carbide Company on the basis of a polycaprolactone soft segment [24]. This was later commercialized by Filatice in Italy as "Lynel".

In dry-spun spandex, the first after DuPont was Toyobo of Japan [25], which produced "ESPA" as early as 1962. Their polymer had urea hard segments from MDI diisocyanate and 1,2-propylene diamine chain-extender. Bayer as we have discussed followed soon afterwards with "Dorlastan", and another dry-spun spandex, "Roica", was commercialized by Asahi in 1970 [26].

Finally, melt-spun spandex with its limitation of low-melting hard segment was initially not successful. Products made by B.F. Goodrich and Bayer/Mobay based on thermoplastic polyurethane elastomers "estane" and "texin", respectively, did not have the required properties. However, Nisshinbo of Japan [27] later developed and commercialized a melt-spun spandex "mobilon" which has found a place in the market.

STRUCTURAL UNDERSTANDING

In discussing history of spandex, the progress of understanding spandex structure has to be included as it had a major impact on spandex developments. The early concept of what the structure of an elastomeric fiber should be was that of a network formed by crosslinking a low-melting polymer, like the network of rubber formed by vulcanizing polyisoprene. During and after development of the first spandex fibers, repeated attempts were made to cross-link spandex compositions in the belief that covalent crosslinks would improve properties like recovery from stretch. It only slowly became recognized that "virtual crosslinks" [28] based on hard-segment interaction by hydrogen bonding could give spandex properties superior to those of covalent crosslinks. An early concept of spandex structure - Figure 15 - showed the hard segments connected in pairs; it was based on models consisting of pieces of rubber hose (soft segments) connected by magnets (hard segments). This model, however, failed to explain superior properties, especially the higher modulus of spandex versus rubber of equal crosslink density. Rinke [29] suggested that hard segments formed crystalline regions which somehow were more effective than sulfur crosslinks in rubber, and we proposed [30] that such regions could have the same reinforcing effect as fillers such as carbon black in rubber. A.V. Tobolsky and his school [31] proposed that in segmented polyurethane elastomers like in other segmented elastomers hard segments associate in sufficient number to form "domains". which may or may not be of a crystalline nature. The domains of other segmented polyurethanes with higher hard segment content, have resisted detection by electron microscopy. Yet from overwhelming indirect evidence there can be no doubt that spandex has a structure of "islands" of hard segment domains in a "sea" of soft segment matrix, as shown illustrated in Figure 16.

Meanwhile, R. Bonart of Bayer [32], following suggestions by Oertel [33] about a possibly crystalline order of hard segments held together by hydrogen bonding, had started wide-angle x-ray studies of early spandex fibers. Only after stretching and hot-wet annealing did these compositions show reflections indicating a periodicity which could be attributed to layers in paracrystalline hard segment domains. Based on these observations as well as model studies, configurations of parallel hard segments with maximum hydrogen bonding were postulated which could explain the variation in properties of compositions with different chain-extenders. While such relatively high order probably has no time to develop during spinning of spandex fibers, the importance of hydrogen bonding capability for process and properties cannot be denied and was a large consideration in developing improved fibers.

IMPROVEMENTS OF THE SIXTIES AND SEVENTIES

As a result of better understanding structure-property relationships in spandex, many improvements were made in the nineteen sixties and seventies. To improve physical properties, the soft segments had to be made softer and the hard segments harder, usually translated as lower glass transition for soft segments and higher melting points for hard segments. With polyester soft segments, this was accomplished by copolymerization using a mixture of glycols [34], where the use of branched glycols such as neopentyl glycol [35] gave the added advantage of increased hydrolytic stability and decreased crystallization under stress [36]. Branched glycols can also be incorporated in the soft segment by adding these glycols used by both Bayer [37] and Toyobo [38] for the additional advantage of providing dye sites for acid dyes.

With the hard segments, considerations of oxidative and thermal instability led both DuPont and Bayer to abandon hydrazine and hydrazide type chain extenders. Highest melting hard segments with MSI diisocyanate were obtained with simple ethylene diamine chain-extender. Because of extremely high reaction rates in chain extension, Toyobo chose 1,2-propylene diamine chain extender from the beginning [25], and Bayer developed a process using ethylene diamine carbonate [39]. However, another problem with MDI-ethylene urea hard segments was that their tendency to hydrogen-bond could be too strong, leading to gelation of spinning solutions and some undesirable yarn properties like reduced heat-settability. The gelation problem could be reduced by adding a secondary amine chain terminator forming unstable urea endgroups which would decompose with secondary chain-extension during spinning [40]. The general approach was to back off on the crystallization tendency of the hard segments by copolymerization with other diamine chain-extenders [41].

Other improvements from understanding the structure occurred in processes of wet- and reaction spun-fibers. In wet-spinning processes, where viscous drag may cause unwanted orientation, productivity can be increased by a coupled hot-annealing after treatment, taking advantage of the heat-settability of spandex with its virtual cross-links [42]. With reaction-spun fibers, the skin-core structure can be reduced by spinning the prepolymer into a non-aqueous diamine bath and diluting the prepolymer with an inert solvent like toluene [43]. This allows deeper penetration of the diamine with more skin formation as the fiber forms.

Spandex history would not be complete without mentioning the developments in additives to the polymer which serve to change fiber appearance and/or protect it from discoloration and degradation. The first "lycra" was protected with T102 pigment against UV degradation and a polymeric amine against discoloration from atmospheric fumes [44]. Knowledge of the domain structure helped formulate a host of other stabilizers protecting the hard-segment domains or the soft-segment matrix as needed for specific uses. Reviewing the further developments in this still ongoing stabilizer work falls outside the scope of this paper.

SUMMARY

The history of spandex elastomeric fibers is characterized by an almost explosive development some forty years ago, which led to the establishment of a new and highly successful class of synthetic fibers. Credit for this belongs first to the scientists at companies like DuPont and Bayer who with support of their management set out to create what was needed for a superior elastic fiber. But on closer inspection, it appears that their knowledge of other, older disciplines was essential for their success.

I want to thank Dr. Seymour for the invitation to give this paper, the DuPont Company for their support, and Dr. Oertel and the Bayer Company for cooperation on the contents.

REFERENCES

In this bibliography, patent years are those of first application because of greater historical significance than year patents were allowed.

1. Weiss, G. H. R., Nat. Rubber Technol. 10, 80 (1979).
2. Wurtz, A., Compt. Rend., 27, 242 (1848); Ann. chem., 71, 328 (1849).
3. Wittbecker, E. L., Houtz, R. C., and Watkins, W. W., Ind. Eng. Chem., 40, 875 (1948).
4. Snyder, M. D. (to DuPont), U.S. Patents 2,623,031 and 2,623,033 (1950).
5. Charch, W. H. and Shivers, J. C., Text. Res. J., 29, 536 (1959).
6. Edgar, O. B. and Hill, R., J. Polymer Sci., 8, 1 (1952).
7. Meerwein, H., Delfs, D., and Morschel, Angew. Chem., 72, 927 (1966).
8. Bayer, O., Angew. Chem., 59, 257 (1947).
9. Bayer, O., Mueller, E., Peterson, S., Poepenbrink, H. -F., and Windemuth, E., Angew. Chem., 62, 57 (1950).
10. Hill, F. B. (to DuPont), U.S. Patent 2,751,419 (1954).
11. Schaefgen, J. R. and Shivers, J. C. (to DuPont), U.S. Patent 3,044,987 (1958).
12. Koller, C. R. (to DuPont), U.S. Patent 2,929,801 (1955).
13. Katz, M. (to DuPont), U.S. Patent 2,929,802 (1955).
14. Frazer, A. H. (to DuPont), U.S. Patent 2,929,803 (1955); Steuber, W. (to DuPont), U.S. Patent 2,929,804 (1955).
15. Frankenburg, P. E. and Frazer, A. H. (to DuPont), U.S. Patent 2,957,852 (1955).
16. Blomberg, R. N. (to DuPont), U.S. Patent 2,965,437 (1958); Boyer, C. (to DuPont), U.S. Patent 3,047,909 (1959).
17. Steuber, W., U.S. Patent 2,813,775 (1955), Reissue 24,691 (1958); Koller, C. R., U.S. Patent 2,813,776 (1955), Reissue 24,689 (1958).
18. Windemuth, E. (to Bayer), Ger. Patent 826,641 (1949).
19. Brenschede, W. (to Bayer), Ger. Patent 888,766 (1951).
20. Oertel, H., Rinke, H., and Thoma, W. (to Bayer), Ger. Patent 1,123,467 (1960).
21. Kohrn, R. C., Slovin, D. G., and Bliven, F. L. (to U.S. Rubber Co.), U.S. Patent 2,953,839 (1957).
22. Cacella, A. F. and Killen (to Globe), W. R., U.S. Patent 3,115,384 (1962).
23. Schneider, A. D. and Peters, T. V. (to Polythane Corp.), U.S. 3,180,854 (1961); Peters, T. V. and Schneider, A. D. (to Polythane Corp.), U.S. 3,267,192 (1961); Peters, T. V. (to Polythane Corp.), U.S. Patent 3,161,706 (1961).
24. Horn, C. F., Magnus, G., and Vineyard, H. (to Union Carbide), U.S. Patent 3,509,102 (1963/1964).
25. Toyo Spinning Co., Jap. Pat. Appl.14,673 (1961) and 12,858 (1964).
26. Asahi Kasei Kogyo, FR. Patent 1,361,789 (1964).
27. Kimura, K., Katayama, S., Hirai, T., Shiniada, S., and Ishoguro, H. (to Nisshin Spinning Co.), Jap. Patent 69 07960 (1965).
28. Schollenberger, C. S., Scott, H., and Moore, G. R., Rubber World, 137, 549 (1958).
29. Rinke, H., Angew. Chem. Int. Ed., 1, 419 (1962).
30. Ultee, A. J., Z. Ges. Textil -Ind., 67, 453 (1965).
31. Cooper, S. L. and Tobolsky, A. V., J. Appl. Polym. Sci., 10, 1837 (1966); Koutsky, J. A., Hien, N. V., and Cooper, S. L., J. Polym. Sci. B, 8, 353 (1970).
32. Bonart, R., J. Macromol. Sci. Phys., 2, 115 (1968).
33. Oertel, H., Bayer Farbenrev., 11, 1 (1965).
34. Schilit, R. J. (to DuPont), U.S. Patent 3,097,192 (1958).
35. Mueller, E. and Schmidt, K. L. See Oertel, H., Rinke, H., and Thoma, W., (to Bayer), Ger. Patent 1,290,659 (1962).
36. Morbitzer, L. and Bonart, R., Kolloid-Z., Z. Polym., 232, 764 (1969).
37. Oertel, H., Rinke, H., Moosmueller, F. and Rosendahl, F. -K. (to Bayer), Ger. Patent 1,495,830 (1964.
38. Toyobo, Jap. Pat. Appl. 21440/1974(1970) and 40158/1974(1970).
39. Thoma, W., Rinke, H., and Oertel, H. (to Bayer), DB Patent 1,122,253 and 1,223,154 (1963).
40. Bleasdale, J. L. and Sandquist, C. L. (to DuPont), U.S. Patent 3,557,044 (1967).
41. Disclosures in Ref. 39 (Bayer) and in Booth, M. H. (to DuPont), U.S. Patent 3,379,683 (1963/1964), DuPont Brit. 1,102,819 (1966), and Bottomley, C. G. and Hunt, O. L. (to DuPont), U.S. Patent 3,555,115 (1968).
42. Peters, T. V., U.S. Patent 4,002,711 (1971/1974).
43. Cahill, J. W., Powell, J. L., and Gartner, E. A. (to Globe), U.S. Patent 3,387,071 (1963/1964).
44. Arvidson, H. C. and Blake, N. (to DuPont), U.S. Patent 2,999,839 (1958).

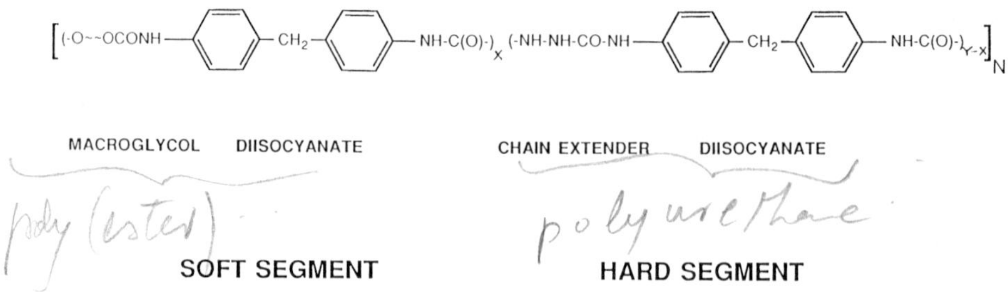

SEGMENTED POLYURETHANE

Figure 1. General Formula of Spandex

$$\text{R-NCO} + \text{R'-OH} \rightarrow \underset{\textit{URETHANE}}{\text{R-NH-CO-O-R'}}$$

$$\text{R-NCO} + \text{R''-NH}_2 \rightarrow \underset{\textit{UREA}}{\text{R-NH-CO-NH-R''}}$$

$$\text{R-NCO} + \text{H}_2\text{O} \rightarrow \text{R-NH-COOH} \rightarrow \text{R-NH}_2 + \text{CO}_2$$

$$\text{R-NCO} + \text{R-NH}_2 \rightarrow \underset{\textit{UREA}}{\text{R-NH-CO-NH-R}}$$

Figure 2. Basic Isocyanate Reactions

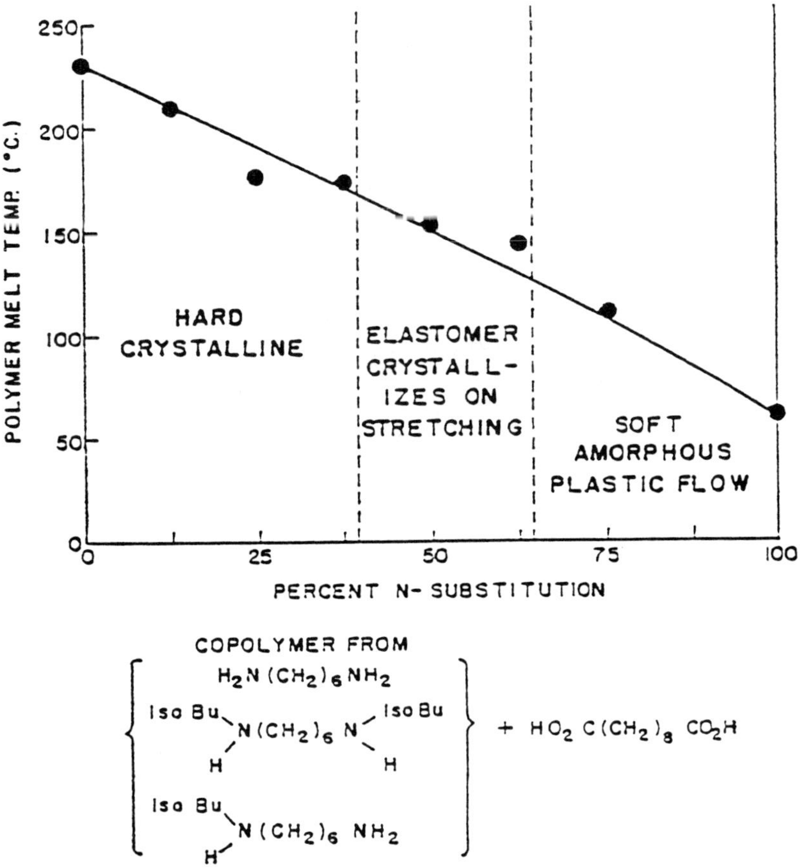

Figure 3. Polyamide Elastomers (from Ref.5)

Figure 4. J.C. Shivers

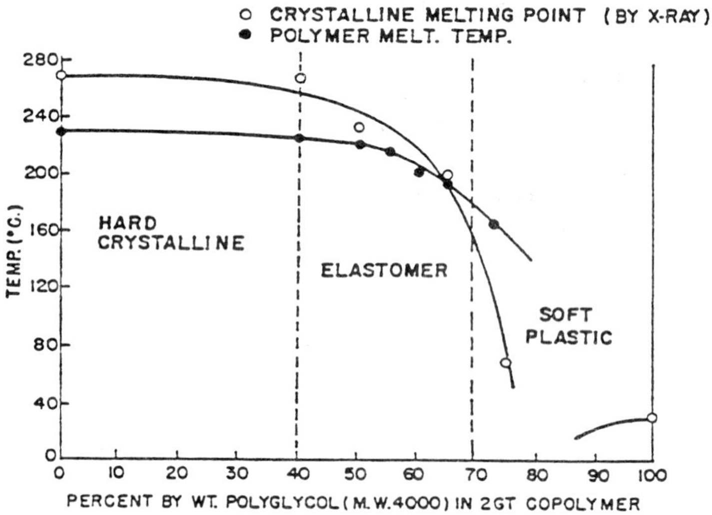

Figure 5. Polyester Elastomers (from ref.5)

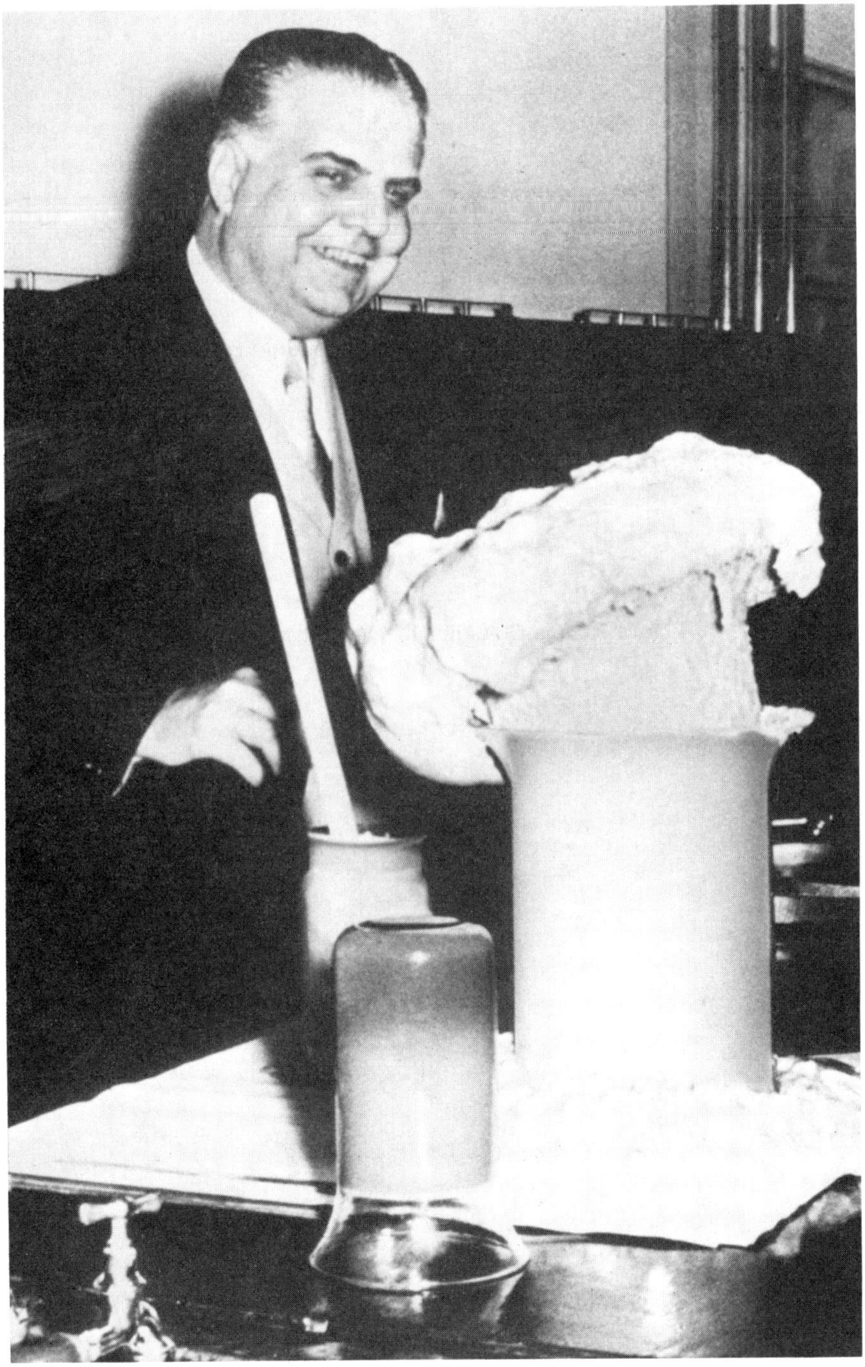

Figure 6. O. Bayer

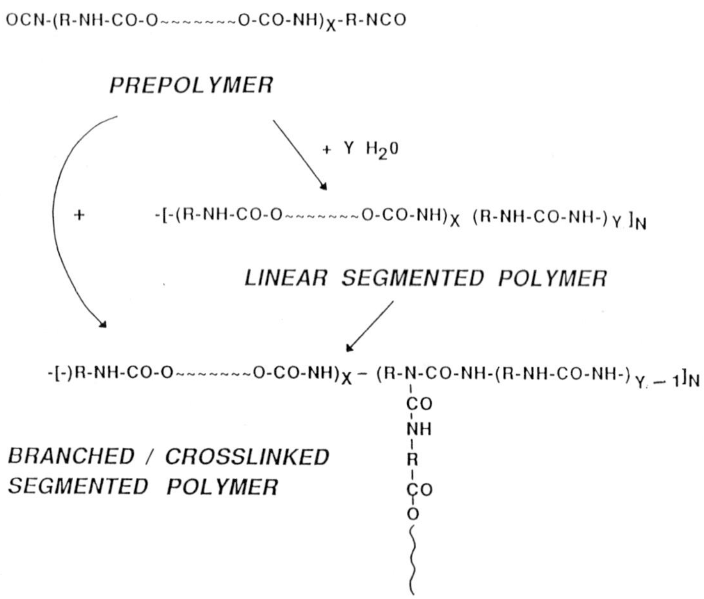

Figure 7. Urea Cross-linking (Vulcollan Chemistry)

Figure 8. P.E. Frankenburg

Figure 9. A.H. Frazer

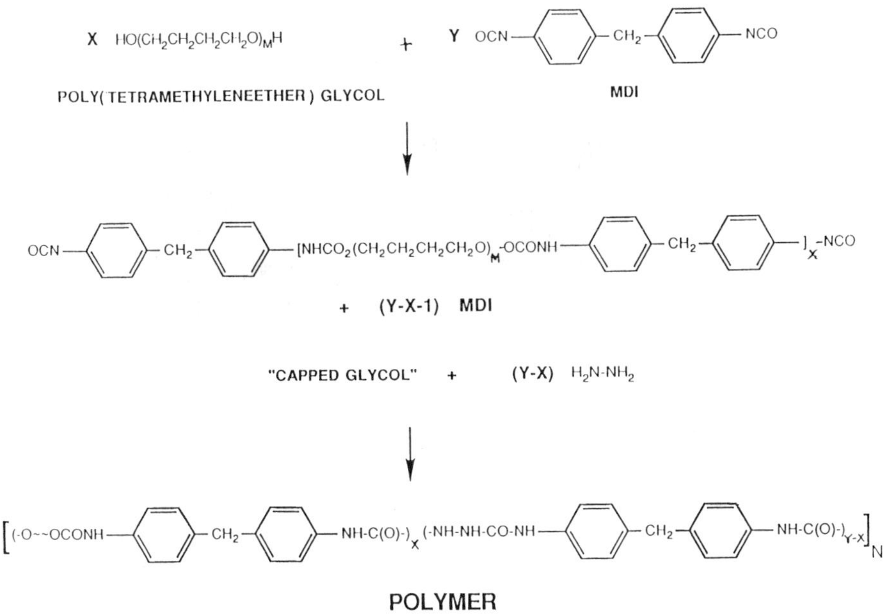

Figure 10. T-124 "Lycra" Chemistry

Figure 11. H. Rinke

Figure 12. H. Oertel

Figure 13. W. Thoma

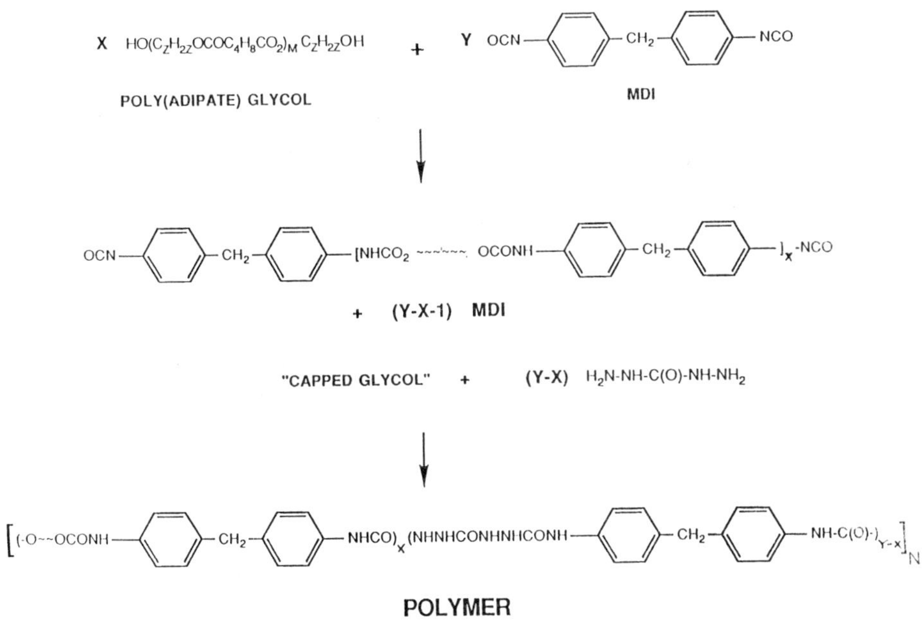

Figure 14. First "Dorlastan" Chemistry

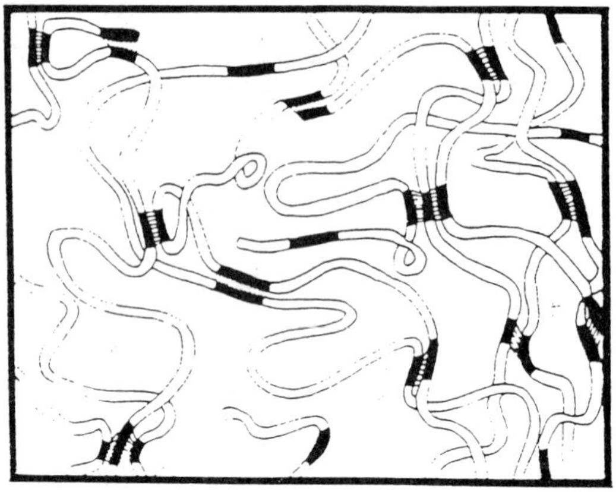

Figure 15. Early Spandex Model (from Ref.30)

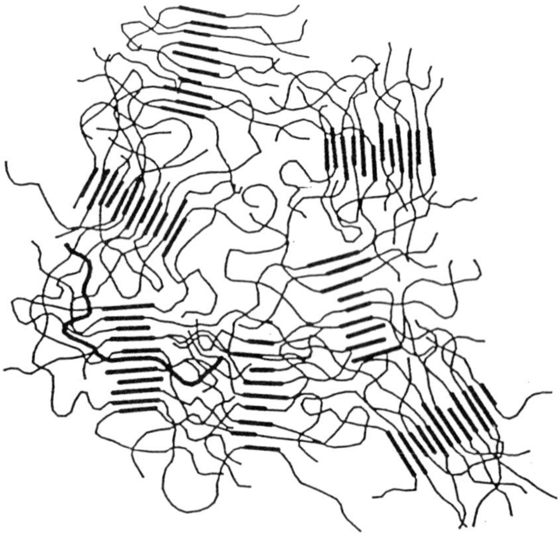

Figure 16. Later Spandex Model

DEVELOPMENT OF POLYETHYLENE FIBERS

ROGER S. PORTER
Polymer Science & Engineering Department, University of Massachusetts, Amherst, MA 01003
TETSUO KANAMOTO
Department of Chemistry, Science University of Tokyo, Kagurazaka, Shinjuku-ku, Tokyo 162, Japan

INTRODUCTION

The development of polyethylene fibers and their applications has been one of the mst recent and most surprising developments in the polymer and fiber fields. Yet the potential for such fibers had long been recognized, but the fraction of all research on polyethylene fibers has been insignificant compared to the huge, long standing research efforts on the range of other products developed from this world's largest volume plastic.

The potential of polyethylene as a quality fiber comes from its inherent features of structure that include:

- the high strength of carbon-carbon bond,
- the small cross-sectional area of the chain ~18A^2,
- the high crystallinity for close molecular packing, and
- the long and regular chain for uniform and high mechanical loading on the polymer backbone.

The origins of ideas and efforts to develop polyethylene fibers have been highly diverse. By this multifaceted development, it is not always easy to identify the origin of ideas and the seminal experiments for each aspect of development. Nonetheless, by timing and by intensity of effort, several groups in the world can be identified as having been major contributors in the development of polyethylene fibers and their applications. The following sections chronicle these developments from the several different lines of research that have been employed. For brevity of presentations, a review of the fundamental concepts is left to the cited primary references.

The following section centers on solution and gel processing. The gel spinning process is thus detailed in Section II, as invented at DSM in the Netherlands for the preparation of high-modulus, high-strength fibers. In the 1960s, two other routes were being developed for the preparation of chain-extended polyethylene as precursors of fibers. These are the solid-state drawing of single-crystal mats that are prepared from a dilute solution of polyethylene and the stirred-induced crystallization of polyethylene from solution. These approaches have led to important contributions towards a better morphological understanding of high-strength and high-modulus polyethylene fibers. Up to now, they have not been shown to be technologically feasible processes. An additional approach for mention and evaluation is the direct draw of the reactor product of polyethylene. Possible production of fibers by melt processing is also to be discussed. References for each process are given by chapter section.

The approaches for the development of polyethylene fibers have not also always been distance from each other. It is thus appropriate to discuss deformation of polyethylene single crystals in the following section on solution processing. This is due in part to the recognition that folded-chain crystals, as possible in both solution and in single crystals, are now known to b desirable to achieve high draw and thus high tensile properties.

The highest strength presently reported for polyethylene fiber is 9.9 GPa:[1] Marichin et al have measured only once this unconfirmed value. The test was made on a sample made by the surface growth technique in combination with solid-state post-drawing for the production of the structures of ultrahigh molecular weight polyethylene, UHMWPE. Kanamoto has reported his highest modulus of 250 GPa at -100°C and 230 GPa and a strength of 6 GPa measured at ambient.[2] It is an interesting question whether polyethylene fibers can be produced with consistently higher tensile strengths. Theoretical calculations and experiments vary widely, yet show that the strength of linear polyethylene may be as high as 19025 GPa [3-6] with a corresponding modulus of about 300 GPa.

REFERENCES

1. Marichin, V. A., Mjasnikova, L. P., Zenke, D., Hirte, R. and Weigel, P., Polymer Bulletin, 12, 287 (1984).
2. Kanamoto, T. and Porter, R. S., "Integration of Fundamental Polymer Science and Technology-3," Lemstra, P. J. and Kleintjens, eds., Elsevier Appl. Sci.: London, 1989, p. 168.
3. Mark. H. F., "Polymer Science and Materials," Mark, H. F. and Tobolsky, A. V., eds., Wiley-Interscience: New York, 231 (1971).
4. Boudreaux, D. S., J. Polym. Sci., Polym. Phys. Ed., 11, 1285 (1973).
5. Termonia, Y., Meakin, P. and Smith, P., Macromolecules, 18, 2246 (1985).
6. Prevorsek, D. C., 2nd Ed., Encyl. Polym. Sci. Eng., Supplement, 803 (1989).

SOLUTION PROCESSING OF POLYETHYLENE

In approximate order, the origin and the development of the various solution and gel spinning processing routes are to be reviewed. Especially addressed are contributions which seem to be of historical interest for the commercial development of high-modulus, high-strength polyethylene fibers. Excellent reviews on solution and gel spinning of polyethylene have been published by Barham and Keller[1] and subsequently by Lemstra, Kirschbaum, Ohta and Yasuda[2].

In 1956 Jurgeleit[3] patented the spinning of a conventional high density polyethylene (HDPE) from a 10-18 wt% solution in paraffin oil. After extraction of the solvent with a non-solvent, the fibers could be extended to a draw ratio of up to 9. For a molecular weight of 150 kg/mol, a maximum tensile strength of 1.2 GPa was obtained. Similar values for the tensile strength were later claimed by Sato and Hirai [4] for HDPE spun with a solvent using highly concentrated (40-50%) solutions and with subsequent drawing of the extrudate. Blades and White[5] introduced in 1962 the so-called flash-spinning technique. A solution of linear polyethylene was extruded at high speeds and temperatures (200°C). The fibrillated "strand" was subjected to draw after evaporation of a halogenated solvent. The maximum tensile modulus and strength were 20 and 1.4 GPa, respectively.

In 1965 Zwick,[6] who was primarily concerned with high-speed solution spinning (800 m/min), was able to process UHMWPE with napthalene as a solvent. The fibers produced were likely of low tensile strength, due to both the absence of a drawing step and the inhomogeneous fiber morphology caused by liquid-liquid phase separation. Solid-state drawing of the fibers was not investigated, possibly because the low denier fibers were fragile.

An alternate approach, the solid-state deformation of single-crystal mats, has received attention since the work of Statton[7] and by Ishikawa et al[8] and Maeda et al[9]. These studies were mainly concerned with the morphology of drawn single-crystal mats, rather than for the mechanical properties achieved. The principal observations of interest were that single-crystals mats of HDPE are remarkably ductile at temperatures just below the melting temperature (100-120°C) and that the lamellar crystals readily unfold in the direction of the applied stress. Using conventional molecular weight polyethylene, Statton[7] obtained a maximum tensile modulus of 10.3 GPa and a tensile strength of 0.67 GPa for a uniaxially drawn single-crystal mat. Maeda et al[9] reported a modulus of approximately 20 GPa and a strength of 0.5 GPa for a draw ratio of 23.

Later studies by Barham and Keller[10] describe the solid-state drawing of HDPE single crystal mats up to x40 with ultimate moduli of about 50 GPa. The tensile modulus increased with draw ratio. Fibers with a modulus of 60 GPa were produced by Kanamoto, Sherman and Porter by SSE (solid-state extrusion) of HDPE single crystal powder or by using the split-billet solid-state extrusion technique, originally developed by Porter and coworkers, to deform single-crystal HDPE mats. The maximum achievable draw ratio and consequent modulus were higher for single-crystal mats than for spherulitic melt-crystallized samples of HDPE.[11]

Studies of stirred-induced crystallization began in 1963. Mitsuhashi[12] first attempted inducing chain extension directly in solution by using a Couette type of apparatus, shown in Figure 1. He reported the formation of fibrous "string-like" HDPE structures upon stirring of linear polyethylene in xylene. However, the mechanical properties of the fibrous crystal mats precipitated on the rotor surface were not reported.[12]

The above observations remained almost unnoticed until in 1970, when independently, Pennings, van der Mark and Kiel[13] used a similar set-up for the fractionation of polyethylene. They discovered that polyethylene crystals with a so-called shish-kebab morphology can be grown. A typical example of a shish-kebab morphology, as observed from electron microscopy is shown in Figure 2a.[13]. A shish-kebab is a central core consisting of more or less extended polyethylene molecules (shish) surrounded by platelets of folded chain crystals (kebabs), as illustrated schematically in Figure 2b. The molecular arrangement within a shish-kebab-type polyethylene is far from ideal, because of the existence of non-stress-carrying parts (kebabs) that make no significant contribution to stiffness and strength.

The work on stir-induced crystallization was continued by Pennings and his group[13-18] with different molecular weights, including UHMWPE: Various processing conditions were used such as the free growth and the surface growth techniques. Draw of the gel state was also evaluated by Pennings and in Canada by St. John Manley. In 1976 Meihuizen, Pennings and Zwijnenburg [17-18] used very dilute solutions of UHMWPE and succeeded in preparing crystals with a tensile modulus of 100 GPa and strength of 3 GPa. These advanced results were mainly due to the higher crystallization temperature,[18] leading to a micro-shish-kebab morphology, as visualized by Schaper et

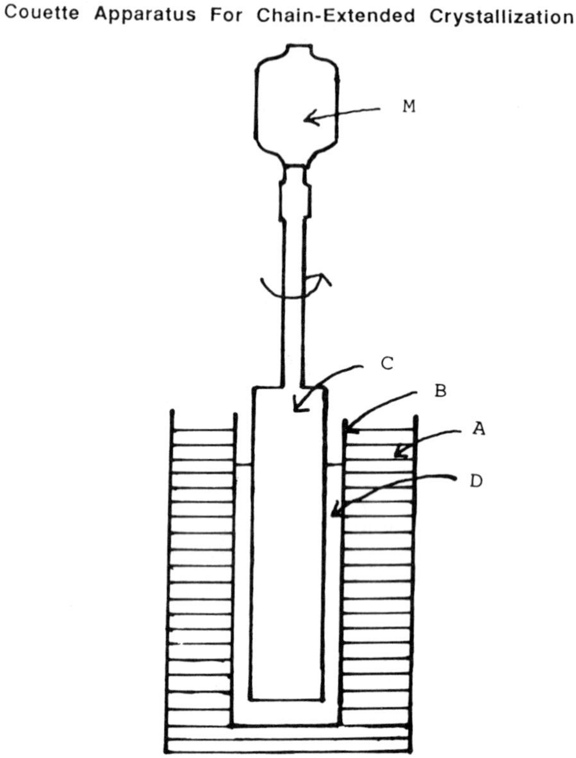

Figure 1. M, motor; A, oil thermostat; B, cylinder with PE solution; C, rotor.

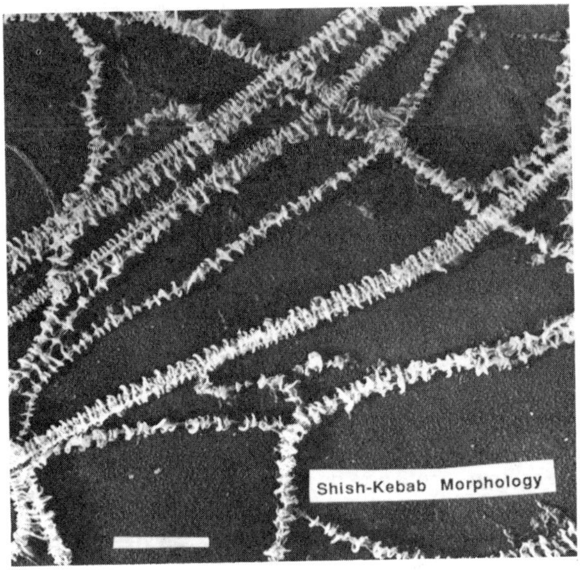

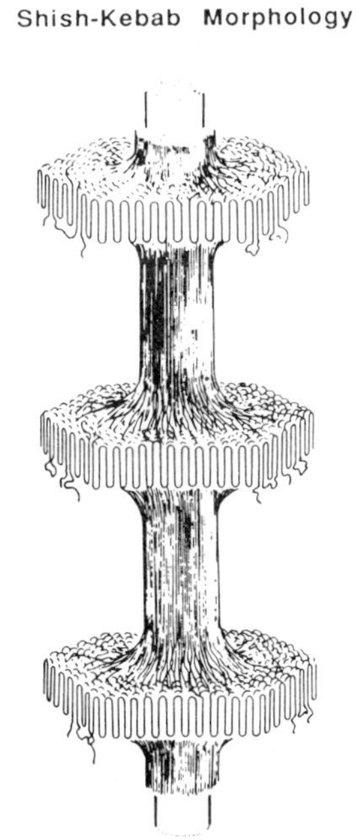

Figure 2. a. Electron Micrograph, scale 1μ b. Schematic Representation.

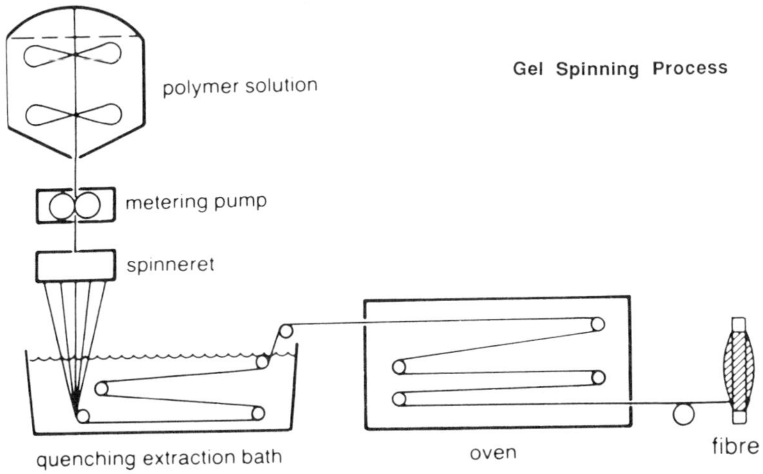

Figure 3.

al[19] using transmission electron micrographs of stained sections of high-modulus surface-grown UHMWPE.

In the mid-seventies, the usefulness of HUMWPE for the production of stiff and especially of strong polyethylene was demonstrated using the surface-grown technique. The main limitations of this technique were the low production speed and multifilament scaling-up and the lack of controllability of crucial product specification parameters such as strength, modulus, and cross section, as required for fiber production.

In 1979, more than two decades and more than one decade, respectively, after the inventions of Jurgeleit[3] and Zwick[6] in solution spinning of polyethylene, a breakthrough in the areas of both UHMWPE processing and continuous production of high-performance polyethylene fibers was achieved by Smith and coworkers at DSM.[20-21] They described the so-called "gel spinning process" shown schematically in Figure 3. A solution of UHMWPE is spun into a water bath or a gas-controlled solidification chamber in which crystallization takes place upon cooling, resulting in a gel of UHMWPE. This gel is sufficiently stable mechanically to be transported via conventional roller systems into an oven where ultra-drawing is performed at elevated temperatures. The solvent can be removed before, during or after this drawing.

High-performance polyethylene fibers could now be produced in quantity for which the crucial product specifications could be controlled by relatively simple processing parameters like molecular weight of the feedstock, polymer solution concentration, die geometry and winding speeds. Using this processing route, high-modulus (90 GPa) and high-strength (3 GPa) UHMWPE fibers were obtained.[20-21] Indeed, these fibers are on the market under various trade names, as discussed alter.[22-23] The fibers have a smooth, fibrillar morphology as observed from a scanning electron micrograph (Figure 4) and a single crystal-like orientation revealed by the "dot-like" wide-angle x-ray spots.[24] Smith and Lemstra[20,21,25] observed a linear relationship between the modulus and the draw ratio (Figure 5a) and between the tensile strength and the draw ratio (Figure 5b). The modulus-draw ratio relationship is similar to that shown for solid-state drawing of melt-crystallized high-density polyethylene by Capaccio with Ward[26] and for drawn single-crystal mats.[8]

A surprise of the gel spinning process is that, even after complete removal of the solvent from as-spun fibers,[25] solution-crystallized UHMW PE can be ultra-drawn in the solid-state ($\lambda_{max}>30$). This contrasts with the natural draw ratio of melt-crystallized UHMW PE ($\lambda_{max}<10$). Hence, the constraints limiting the drawability can be removed by spinning from solution. In other words, the crystallization of UHMW PE from solution introduces a morphology which is favorable for solid-state drawing in contrast to the morphology of melt-crystallized UHMW PE. The strongly enhanced drawability of solution-crystallized UHMW PE has been explained by Smith et al[27] in terms of the controlled removal of entanglements which act as crosslinks on the time scale of the drawing experiment.

Because of the high rate of transformation from a draw-resistant into an ultra-drawable polyethylene and vice versa, Lemstra et al[28] recently suggested that, apart from rheological concepts, viz entanglements, the arrangement of chain segments in the crystals may also drastically affect the drawability.

Since the invention of the gel spinning process, variations on the same theme have been described in several patent applications, with the original principles maintained, see Table 1. Stamicarbon is the licensing subsidiary of DSM and Dyneema is the trade name

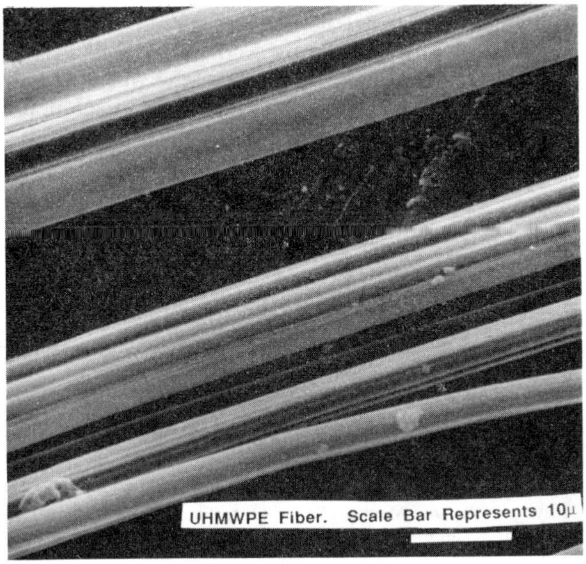

Figure 4. Gel-Spun and Solid-State Drawn UHMWPE Fibers.

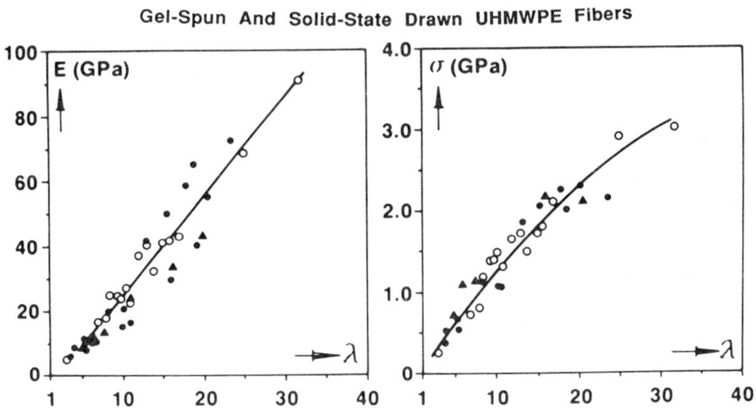

Figure 5. Tensile Modulus (a) and Tensile Strength (b) of Gel-Spun and Solid-State Drawn UHMWPE Fibers versus Draw Ratio, λ.

for the DSM high-modulus, high strength UHMWPE fiber. In 1982 Allied patented a process[29] in which paraffin oil is used as a solvent for UHMWPE. From 1983 Allied has been issued a range of patents on the polyethylene fibers and their applications. Mitsui filed a patent application for a process[30] in which a solution of UHMWPE in paraffin is spun, quenched to induce gelation, and subsequently solid-state drawn. Apart from minor changes in processing parameters the same concepts appear to be involved.

REFERENCES

1. Barham, P. J. and Keller, A., J. Mater. Sci., 20, 2281 (1985).
2. Lemstra, P. J., Kirschbaum, R., Ohta, T. and Yasuda, H., in "Developments in Oriented Polymers-2," Ward, I. M. ed., Elsevier Appl. Sci. Publ: New York, Ch. 2, (1987).
3. Jurgeleit, W., U.S, Patent 3,048,465 (priority June 16, 1956).
4. Sato, H. and Hirai, T., Japanese Patent Sho 37-9765 (priority May 15, 1959).
5. Blades, H. and White, J. R., U.S. Patent 3,081,519 (Priority Jan. 31, 1962).
6. Zwick, M., NL Patent 1,100,497 (priority Feb. 1, 1965).
7. Statton W. O., J. Appl. Phys., 38, 4149 (1967).
8. Ishikawa, K., Miyasaka, K., and Maeda, M., J. Polym. Sci., Polym. Phys. Ed., 7, 2029 (1969).
9. Maeda, M., Miyasaka, K. and Ishikawa, K., J. Polym. Sci., Polym. Phys. Ed., 8, 355 (1970).
10. Barham P. J. and Keller, A., J. Mater. Sci., 11, 27 (1976).
11. Kanamoto, T., Sherman, E. S. and Porter, R. S., Polym. J., 11, 547 (1979).
12. Mitsuhashi, S., Bull. Text. Res. Inst., 66, 1 (1963).
13. Pennings, A. J., van der Mark, J. M. A. A. and Kiel, A. M., Coll. Polym. Sci., 237, 336 (1970).
14. Zwijnenburg, A. and Pennings, A. J.,Coll. Polym. Sci., 253, 452 (1975).
15. Zwijnenburg, A. and Pennings, A. J., Coll. Polym. Sci., 254, 858 (1976).
16. Pennings, A. J., J. Polym. Sci., Polym. Symp., 59, 55 (1977).
17. Meihuizen, C. E., Pennings, A. J. and Zwijnenburg, A., U.S. Patent 4,137,395 (priority May 20, 1976).
18. Zwijnenburg, A. and Pennings, A. J., J. Polym. Sci., Polym. Lett. Ed., 14, 339 (1976).
19. Schaper, A., Walenta, E. and Schulz, E., Progr. Coll. Polym. Sci., 78, 183 (1988).
20. Smith, P., Lemstra, P. J. and Pennings, A. J., U.K. Patent 2,040,414 (priority Feb. 8, 1979).
21. Smith, P. and Lemstra, P. J., U.K. Patent 2,051,667 (priority June 27, 1979).
22. Kirschbaum, R., Yasuda, H. and van Gorp, E. H. M., Chemiefasern/Textilindustrie, December, 1986, T134-T139.
23. Kirschbaum, R. and van Dingenen, J. L. J, in "Integration of Fundamental Polymer Science and Technology-3," Lemstra P. J. and Kleintjens, eds., Elsevier Appl. Sci.: London, p. 178 (1989).
24. Smith, P. Lemstra, P. J., Pijpers, J. P. L. and Kiel, A. M., Coll. Polym. Sci., 259, 1070 (1981).
25. Smith, P. and Lemstra, P. J., J. Mater. Sci., 15, 505 (1980).
26. Capaccio, G., Crompton, T. A. and Ward, I. M., J. Polym. Sci., Polym. Phys. Ed., 14, 1641 (1976).
27. Smith, P., Lemstra, P. J. and Booij, H. C., J. Polym. Sci., Polym. Phys. Ed., 19, 877 (1981).
28. Lemstra, P. J., van Aerle, N. A. J. M. and Bastiaansen, C. W. M., Polym. J., 19, 15 (1987).
29. Kavesh, S. and Prevorsek, D. C., U.S. Patent 4,413,110 (priority March 19, 1982).
30. Motooka, M. and Ohno, T., E.P. Patent 0,115,192 (priority Dec. 28, 1982).

MELT PROCESSING

There has been research on melt extrusion that parallels the development of the solution spinning technology for the production of high molecular weight PE fibers of exceptionally high strength and modulus.

A highly oriented extrudate can be obtained by extruding polyethylene through a capillary rheometer with a conical die at temperatures close to the melting point.[1] Initial work led to the development of transparent and fibrous polyethylene extrudates.[2] These were obtained by extruding HDPE through a conical die from the molten state (132-136°C) above a critical shear rate in a capillary rheometer. As a result of the rising extrusion pressure, the polymer undergoes supercooling, which combined with the (extensional) elongational velocity gradient, leads to massive crystallization in the conical die. Because of this rapid secondary crystallization, it has been difficult to make this process continuous and to achieve polyethylenes with extreme tensile moduli.

Professor I. M. Ward of Leeds University developed a melt extrusion process that involves a multiple stage drawing for conventional molecular weight polyethylene.[3,4] He has also done interesting related research on hydrostatic extrusion and die drawing. The commercial development of melt spun and drawn fibers by the methods of Ward are being carried out by Snia Fibre who is developing in the fiber in Italy under the trade name Tenfor for applications such as ropes, cords, fishing lines, nets, lightweight composites, etc. The fibers are marketed through the British Technology Group.[8] For interest in the corresponding composites and fibers with a strength of 1.0-1.5 GPa and modulus (40-70 GPa) for polyethylene, references may be consulted.[5-8]

The recent research of Professor I. M. Ward on polyethylene fiber has been on radiation crosslinking to reduce the creep of polyethylene fibers and for fiber treatment by oxygen plasma to improve bonding in composites.

Also to be noted are Baranov's experiments to produce high-performance polyethylene fibers by solid-state deformation of melt-crystallized polyethylene: in 1975 Baranov[9] reported a tensile strength of 4 GPa with a modulus of 60 GPa for HDPE. This linear polyethylene had a molecular weight below 300 kg/mol (melt index = 0.2). For his measurements he used one fibril with a length of about 10 cm, which he obtained after rupture of the sample.

REFERENCES

1. Bowman, M., U.S. Patent 3,382,220 (1968), see also van der Vegt, A. K. and Smith, P. P. A., S.C.I., Monograph, 26, p. 313.
2. Southern, J. H. and Porter, R. S., Chem. Eng. News, 45 (Sept. 22, 1969); J. Macromol. Sci., Phys., B4, 541 (1970); J. Appl. Polym. Sci., 14, 2305 (1970).
3. Capaccio, G. and Ward, I. M., Polymer, 15, 233 (1974).
4. Capaccio, G., Gibson, A. G. and Ward, I. M., in "Ultra High Modulus Polymer," Ciferri, A. and Ward, I. M. eds., London Appl. Sci. Publishers, Ch. 1 (1979).
5. Ladizesky, N. H. and Ward, I. M., Compos. Sci. Technol., 26, 129 (1986).
6. Ladizesky, N. H., Sitepu, M. and Ward, I. M., Compos. Sci. Technol., 26, 169 (1986).
7. Ladizesky, N. H. and Ward, I. M., Compos. Sci. Technol., 26, 199 (1986).
8. Ward, I. M., High Performance Textiles, 9(11), 10 (1989).
9. Baranov, V. G., Krenev, V. V. and Frenkel, Y., Sov. Phys. Solid State, 17, 1018 (1975).

SOLID-STATE FORMING

Solid-state extrusion (SSE) has been investigated extensively for the deformation and evaluation of uniaxial molecular chain extension for a wide range of semicrystalline and amorphous thermoplastics. The method is the successor to the melt extrusion processes of Porter described in the prior section. The conditions of SSE are extrusion through tapered dies below the melting point of semicrystalline polymers or just above the glass transition temperature for amorphous polymers. High density polyethylene (HDPE) has been most widely studied by SSE, in order to understand the major morphological and properties transformations that occur. By subsequent drawing steps,

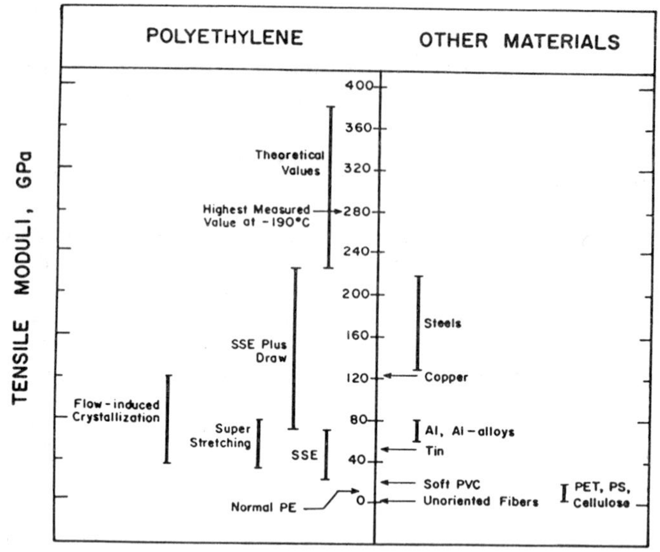

Figure 6. Polyethylene Moduli by Processing and by Comparison with Other Materials.

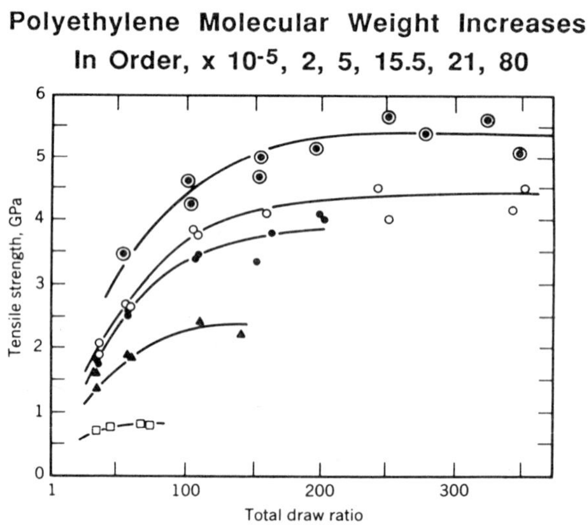

Figure 7.

polyethylene prepared by SSE of selected premorphologies can now be drawn into fibers of the highest specific tensile moduli and strengths.[1-9]

Early attempts to produce high modulus fibers were initiated by direct tensile drawing of linear polyethylenes at elevated temperatures. In a typical tensile-drawing experiment, dumbbell or rod samples of melt-crystallized morphologies are mounted and drawn on a tensile testing machine. In addition to the molecular weight and molecular weight distribution of the polyethylene, the draw temperature and extrusion rate must be selected carefully for maximum effective draw. Polyethylene fibers with a tensile modulus of 35 GPa (5 x 10^6 psi) were produced[10] by drawing melt-extruded films. Linear polyethylene fibers of draw ratio 20 with a modulus of 27 Gpa (3.9 x 10^6 psi) were obtained.[11] Polyethylene fibers were prepared at a draw ratio greater than 15 with a dynamic tensile modulus of 38 GPa (5.5 x 10^6).[12] Thus, linear polyethylenes had been drawn by several groups of workers up to a draw ratio of 30,[13-16] with moduli up to 70 GPa (10 x 10^6 psi), which is perhaps 25% of the estimated ultimate PE modulus.[4,5] Additional SSE experiments and a scale of moduli attained for PE has been reported, see Figure 6, after Figure 1, ref. 6. The high value given in the figure at 190°C is consistent with the low temperature value of 262 Gpa measured by Barham and Keller at a frequency of 2.5 Hz.[16]

The SSE technique has also been sued for drawing polyethylene powders,[17] which may be as fine as single crystals. Calendaring is a recent simplifying advance for the SSE process.[18] single-crystal aggregates of ultrahigh molecular weight polyethylene (UHMW PE)[19] grown from dilute solutions generally exhibit much higher extensibility than compacted powders or melt-crystallized HDPE.

A combination of solid-state coextrusion followed by tensile drawing give a high and efficient draw. HDPE molecular weight and entanglements strongly affects drawing conditions, drawability, and draw efficiency. The maximum attainable DR is markedly higher for higher MW and lower entanglements, see Figure 7.[20] These values extend to a higher limit the strength values in Figure 5. Also provided is additional background on the SSE process.[20]

The demonstrated ductility of PE reactor powders has led to interesting related studies at Du Pont on the direct draw of "never processed" polyethylene.[21] Independently, two-state process[17,18] is being applied by Nippon Oil of Japan in a commercial process for drawing of UHMWPE reactor powder.[22] This process involves the direct draw of reactor powder and without the requirement of solvent as in the gel spinning.

Kanamoto, Porter and coworkers[19,20] have used solid-state extrusion and subsequent drawing of UHMWPE single-crystal mats to obtain samples with tensile moduli up to 220 GPa for a corresponding draw ratio of about 250. Kanamoto with coworkers[23] have shown that UHMWPE can be solid-state drawn up to a draw ratio of 800.

REFERENCES

1. Mark, H., Trans. Faraday Soc., 32, 144 (1936).
2. Meyer, K. A. and Lotmar, W., Helv. Chim. Acta., 19, 68 (1936).
3. Dulmage, W. J. and Contois, L. E., J. Polym. Sci., 28, 275 (1958).
4. Sakurada, I., Ito, T. and Nakamae, K., Bull. Inst. Chem. Res. Kyoto Univ., 42, 79 (1964).

5. Sakurada, I., Ito, T. and Nakamae, Macromol. Chem., 75, 1 (1964).
6. Sakurada, I., Nukushira, Y. and Ito, T., J. Polym. Sci., 57, 651 (1962).
7. Sakurada, I. and Kaji, K., J. Polym. Sci., Part C., 31, 57 (1970).
8. Britton, R. N., Jakeways, R. and Ward, I. M., J. Mater. Sci., 11, 2057 (1976).
9. Black. W. B., J. Macromol. Sci. Chem., A7(1), 3 (1973).
10. Takayanagi, M., Urabe, M. and Fukuda, T., Rep. Prog. Polym. Phys. Japan, 8, 107 (1965).
11. Meinel, G., Peterlin, A. and Sakaoku, K., in Analytical Calorimetry, Porter, R. S. and Johnson, J. F., eds., Plenum Press: New York, p. 15 (1968).
12. Fischer, E. W., Goddar, H. and Piesczek, W., J. Polym. Sci., 32, 149 (1971).
13. Capaccio, G. and Ward, I. M., Nature (London) Phys. Sci., 243, 143 (1973).
14. Capaccio, G. and Ward, I. M., Polymer, 15, 233 (1974).
15. Capaccio, G. and Ward, I. M., Polym. Eng. Sci., 15, 219 (1975).
16. Berger, W., Grafe, F. and Kammer, H. W., Acta Polymerica, 35, 504 (1984); Barham, P. J. and Keller, A., J. Polym. Sci. Polym. Lett., 17, 591 (1979).
17. Kanamoto, T., Ohama, T., Tanaka, K., Takeda, M. and Porter, R.S., Polymer, 28, 1517 (1987).
18. Wang, L. H. and Porter, R. S., J. Polym. Sci., Polym. Phys., 28, 2441 (1990).
19. Kanamoto, T., Tsuruta, A., Tanaka, K., Takeda, M. and Porter, R. S., Polymer J., 15, 327 (1983).
20. Southern, J. H., Capiati, N. J., Kanamoto, T., Porter, R. S. and Zachariades, A. E., 2nd Ed., Encycl. Polym. Sci. Eng., 15, 346 (1989).
21. Smith, P., Chanzy, H. D. and Rotzinger, B. P., Polym. Comm., 26, 258 (1985); J. Mater. Sci., 22, 523 (1987); U.S. Patent 4,769,433 (1988).
22. Sano, A., Iwanami, Y., Matsuira, K., Yohoyama, S. and Kanamoto, T., Pres. IUPAC Macro-90, Session 2.2.6, July 8-13, 1990, Montreal, Canada.
23. Tovmasyan, Y., Takahashi, K. and Kanamoto, T., Reports Prog. Polym. Phys., Japan, 32, 147 (1989).

APPLICATIONS OF POLYETHYLENE FIBERS

This latest entry into the field of high performance fibers is clearly the interesting high tenacity-high modulus polyethylene fibers. Many years of research activities at universities and in industry have resulted in an appreciable technical and patent literature.[1-3] The Dyneema SK60 and SK66 fibers produced by DSM are commercially available in Europe. In the U.S. fibers are available from Allied Signal under the trade names of Spectra -900 and Spectra -1000. This latter grade, for the ballistics markets, exhibits, in the form of multifilament yarn, a strength of 3.0 GPa and modulus of 172 Gpa.

The outstanding tensile properties of PE fibers are now established. On a weight basis, they are the strongest and almost the stiffest materials available. High resolution electron microscopy of Spectra polyethylene fibers reveals (Figure 8) a fibrillar morphology with the microfibrils consisting of long single crystal sections of crystals and small "amorphous" domains. For this and the following illustrations the authors are indebted to Allied Signal, and particularly to the help and writings of Dusan Prevorsek and Sheldon Kavesh.

Commercial fibers are produced from solution using UHMWPE. The tensile strength of Spectra -1000 is higher than that of aramid (Kevlar) fibers below ~100°C and falls below that of aramid fibers above this temperature. It is thus inappropriate to use the polyethylene fibers in applications where significant loads are applied for an appreciable time at temperatures exceeding 80-90°C.

Attention could be directed to several features of polyethylene fibers that make them special. For example, they readily buckle in compression, due to the weak lateral forces in polyethylene. This feature can be a plus as DSM has shown that Dyneema fiber can be knotted without fibrillation and failure.

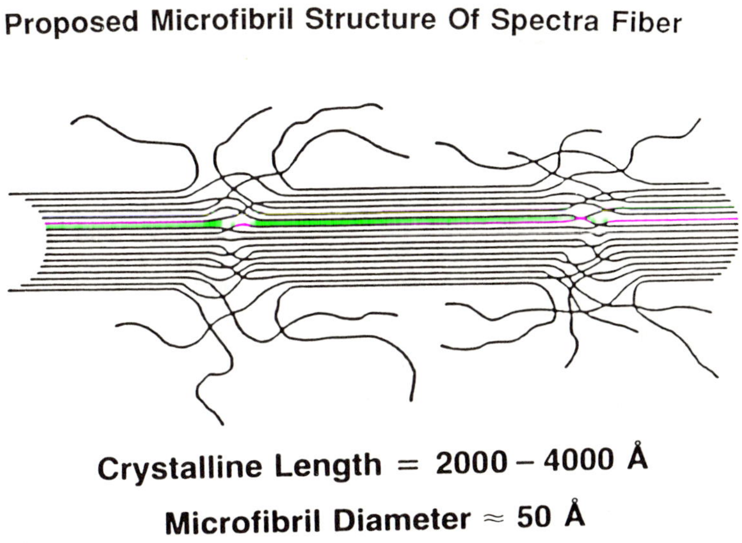

Figure 8.

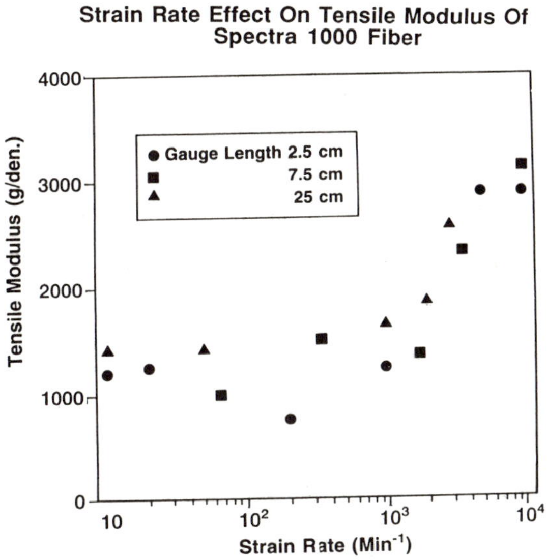

Figure 9.

Of many other features of distinction, addressed here are only the remarkable properties of polyethylene fiber under load. The following sections thus address Creep and Impact Applications.

Creep Under Load

Dimensional stability under load has been a notable problem with all polyolefin fibers. industrial technology has made important steps towards developing a dimensionally stable product. Currently these fibers are successfully utilized in many conventional industrial applications. Nevertheless, the designer must be aware of their tendency to creep under constant load, which is much higher in polyethylene than in aramid fibers. This property has led to a major area application which will be featured in this review. A range of interesting alternate applications are detailed in Kaleidoscope, a bulletin published by Allied Corp., High Performance Fibers[4] for DSM in Chemical and Engineering News, March 25, 1991, pp. 20-21.

Impact Applications

In comparison with all other high performance fibers, the thermoplastic polyethylene fibers show by far the largest capacity to withstand impact. The reason for this is the low glass transition temperature of polyethylene which is low, but varies widely, depending on the type of measurement.[5] As a result, polyethylene fibers retain their mechanical strength even at ballistic rates of deformation and at low temperatures. For example, the impact resistance of Spectra composites is 5-10 times higher than that of an aramid control, depending on construction.

At high rates of deformation, the adiabatic heating tends to bring fibers to their melting point or to high temperatures at which the breaking stress is very low.[5] Microscopic examinations of armor constructed with aliphatic polyamides showed indications of fiber melting after impact. However, ballistic experiments with ultrastrong PE fibers and composites show that these conclusions do not apply to polyethylene fibers. Based on it smelting point, PE armor offers better ballistic protection than aliphatic and aromatic polyamides. The ballistic event is far too short for the heat transfer from hot projectiles to be a factor in polymeric armor performance. Moreover, it has also been established that uniaxial polyethylenes become even stronger, stiffer and tougher with increasing rate of deformation. Theoretical work relative to strain dependence of uniaxial polyethylene has been carried out by Termonia et al.[6] These authors predicted that an increase in the deformation rate from 10^{-1} min^{-1} to 10^{2} min^{-1} should lead to six-fold increase in fiber strength. Testing at strain-rates achievable by standard laboratory equipment indicates large modulus increases with increasing rate of deformation.[7] Because of their favorable strain-rate sensitivity, Spectra and Dyneema fibers are gaining acceptance in applications concerned with the survivability of equipment, apparatus and people. For example, fencing suits made of Dyneema were first used in the 1988 Olympic Games.

Numerical simulations have been developed to understand Spectra composites that are highly anisotropic and viscoelastic.[8] The approach considers that the properties of Spectra fibers are highly strain rate dependent as shown in Figure 9.[5] A general finite element code, Marc, developed by Marc Analysis Research Corporation has been used. Marc is capable of handling three dimensional dynamic problems, as well as anisotropic materials. Strain rate dependence of properties was handled in a stepwise manner.

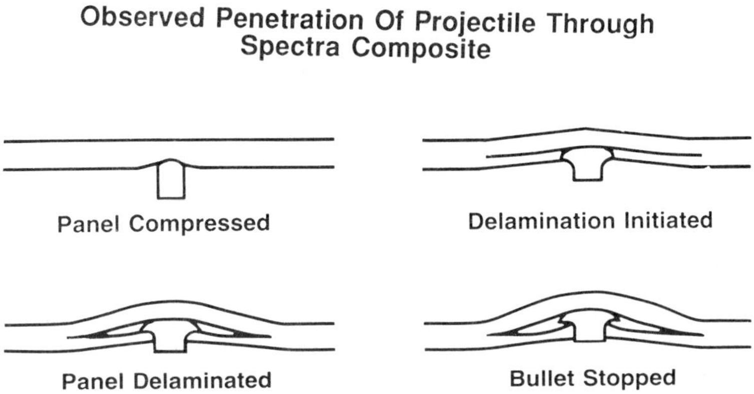

Figure 10.

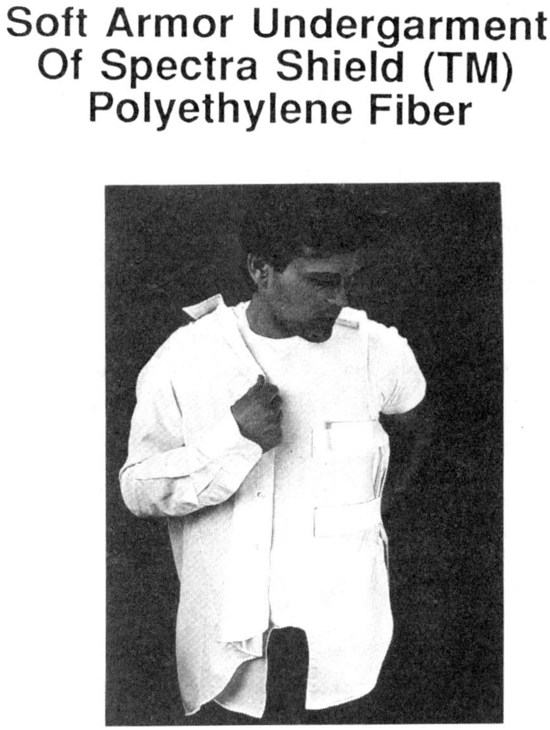

Figure 11.

The analytical procedure was applied to calculate C_s, C_l and E from the photographic records for transverse impact on the fibers. C_s is the velocity of the transverse wave. The value of E is a dynamic modulus for the strain rate of 1.2 x 10^2 sec^{-1} (i.e., 7.2 x 103 min^{-1}). The dynamic value of E ≈ 320 Gpa for Spectra is very high, partly due to the frequency of measurements. This value can be compared with that for steel of 210 GPa.[5] Since polyethylene has a low density of near one, the velocity of longitudinal stress wave, C_l is very high. The value of ca 18000 m/s is much higher than that of steel (i.e. ca 5500 m/s) and is comparable with that of diamond (also about 18000 m/s depending on the crystallographic direction).

For evaluation, Spectra composite samples were prepared by laminating unidirectional prepreg of Spectra Shield™ type using [0/90] fiber angle stacking sequence. The matrix polymer of the Spectra prepreg was Kraton D1107, a tri-block copolymer of polystyrene, polyisoprene and polystyrene produced by Shell Chemical Company. The fiber content was 70% by weight.

Spectra Shield™ is a new concept that utilizes a unidirectional fiber and resin matrix system. Hundreds of fiber bundles are laid side by side in a 'unidirectional' aligned beam. The beam is drawn through a bath of thermoplastic polymer to produce a flexible composite layer. The layers are then cut, crossplied (in a 0, 90° orientation) and pressed to produce a 'single ply.' The plies can be stacked to form a soft pliable material for soft body armor or pressed together to form a more rigid 'board' type structure.

The phenomena of high speed impact on Spectra composites has been analyzed by examining the Spectra composite panels subjected to ballistic impact. It was observed that the failure process of Spectra composites could be divided into four stages. The Spectra composite target was first locally compressed upon contact of the projectile and then several layers of Spectra fiber were broken at the impact point by high stresses developed at the contract area with the projectile (bullet). Then the projectile penetrated through several layers causing interfacial delamination. While penetrating the Spectra composite, the projectile was severely deformed into a mushroom shape, which reduced the stress at the contact area and finally the projectile came to a stop. Figure 10 illustrates a projectile deformation and partial penetration of Spectra composite as observed by high speed photography. Figure 11 is a photograph of a soft armor undergarment made of Spectra Shield™ polyethylene fiber.

REFERENCES

1. Prevorsek, D. C., in "Recent Advances in High Strength Fibers and Molecular Composites in Polymer Liquid Crystal," Ciferri, A., et al, eds., Ch. 12 (1982).
2. Pennings, A. J. and Menninger, K. E., in Ultra High Modulus Polymers, Ciferri, A. and Ward, I. M., eds., London, Appl. Sci., p. 117 (1979).
3. Smith, P. and Lemstra, P. J., J. Mater. Sci., 15, 505 (1980).
4. Allied Corp., High Performance Fibers, P. O. Box 31, Petersburg, VA 23804-9990, USA.
5. Prevorsek, D. C., Chin, H. B., Kwan, Y. D. and Field, J. E., J. Appl. Polym. Sci., Symp., 47, 45 (1991).
6. Termonia, Y., Meakin, P. and Smith, P., Macromolecules, 18, 2246 (1985).
7. Prevorsek, D. C., Kwan, Y. D., Harpell, G. A. and Li, H. L., "Spectra Composite Armor: Dynamics of Absorbing the Kinetic Energy of Ballistic Projectiles," 34th Int., SAMPE Symp. Prepr., Sec. 6A, (May 8-11, 1989).
8. Zukas, J. A., Nicholas, T., Swift, H. F., Greszczuk, L. B. and Curran, D. R., "Impact Dynamics," John Wiley and Sons: New York, 367 (1992).
9. Allied-Signal Kaleidoscope, Vol. III, No. 1, 1988, p. 3.

HISTORY OF POLYPROPYLENE FIBERS

MARVIN WISHMAN AND JACK SCRUGGS
Phillips Fibers Corporation, P. O. Box 66, Greenville, SC 29602

Introduction

About forty (40) years ago, scientists at Phillips Petroleum Company showed that a fraction of a polypropylene product, made with a chromium ion promoted silica-alumina catalyst, was a unique white solid: crystalline polypropylene. Paul Hogan and Bob Banks, the inventors of crystalline polypropylene, found that this material could be melted and drawn into strands of fiber. Initial commercial production of polypropylene resin began in 1957 and fiber production began several years later. Growth (See Figure) in traditional filament and staple product areas was slow and did not surpass ten (10) million pounds until 1965. In 1989, total fiber production reached 1.639 billion pounds. 286 million pounds of filament yarn and 377 million pounds of staple fiber were shipped, in addition to 305 million pounds of spunbond and monofilament and 635 million pounds of slit film. As you can see, growth accelerated in the 80's. How and why polypropylene fiber grew constitutes the History of Polypropylene Fibers. In the early 1960's, I worked for a man who said of polypropylene, "Looks like wax, feels like wax-always did and always will." He could not have been more wrong - so - let's take a look.

Polypropylene fiber, as initially produced, was not dyeable, had poor thermal stability above room temperature and was very sensitive to UV light. More than any other synthetic fiber, polypropylene fibers owe their commercial utility to polymer additives that enable the fiber to perform in a manner impossible for fiber made with pure polymer. Because the fiber remained undyeable from a commercial point of view, despite a great deal of research effort, pigments are added to provide color. Stabilizers are added to help (1) melt flow stability; (2) long term thermal stability and (3) to protect against sunlight. As stabilizers improved, pigment quality had to be improved in order to provide colors with life characteristics capable of matching the degradation stability which was continuously improved. Polymerization technology also contributed to a changing polypropylene fiber picture. Higher yields in terms of pounds of resin per pound of catalyst, reduced atactic polymer content and modification of polymer molecular weight distribution all had an effect on product cost and processing performance. Fiber technology changed, permitting special products to be produced. From an initial start in industrial end uses which did not require coloration, polypropylene fibers have entered virtually every market area providing increased room for growth. Initially, polypropylene was thought of only as an inexpensive, low end substitute for other fibers and efforts were made to convert it to a dyeable product. A large number of methods were developed and patented but none became commercially viable. Every method added to the cost and, at the same time, eliminated some of the good properties such as resistance to soiling. For instance, one process converted polypropylene into an acid dyeable fiber-actually a very poor substitute for nylon-with few of nylon's assets. Real growth of polypropylene in fibers came with success in finding markets which could capitalize on polypropylene's attributes rather than by eliminating all of its deficiencies. Let's examine this growth.

Discussion

In this discussion I will concentrate on filament and staple fibers although large quantities of polypropylene resin are used in slit film, monofilament, melt blown and spunbond products. Melt blown volumes are not included in this summary. Slit film is used to produce carpet primary and secondary back. Spunbond products are used in a variety of end uses for nonwoven fabrics, competing with staple products converted to fabrics via the carded route. Slit film is primarily used in carpet as primary and secondary back. The growth until 1984 was primarily in primary back, i.e., the tufting medium into which carpet yarn is inserted. The large jump since 1984 resulted from the replacement of jute as carpet secondary back. Spunbond products have shown a less spectacular but steady growth over the past twenty-five (25) years.

Staple

Polypropylene staple expanded very slowly and did not reach ten (10) million pounds until 1965. It was initially produced on slow conventional spinning equipment for the needlepunch nonwoven industry which made a variety of fabrics including so-called indoor/outdoor carpet. UV stability was initially provided by absorbers which are not very effective on fibers and later by nickel containing stabilizers such as Irgastab 2002 and Cyasorb 1084 which were the UV stabilizers of the 70's. Growth spurted after 1976 due to use of polypropylene staple in carpet secondary back. Polypropylene in slit film had begun to take over the carpet backing market as previously indicated. Staple enhanced the characteristics by providing color as capcoat on primary back and improving appearance of secondary back. Improved stabilizers based upon hindered amine and hindered phenol technology led to improved UV and thermal stability permitting polypropylene staple to become the preferred fiber for exposed areas in automobiles. Use of polypropylene staple in outdoor products was also made possible by these improved stabilizers. In other words, staple fiber growth was made possible by the development of properties which could not easily be matched by other fibers. The increase in the use of polypropylene fibers was based upon developments which reduced the price/performance ratio. However, the improvement was due to better performance rather than lower price, which had been the objective of much effort in polypropylene's early fiber history.

The increase in use of polypropylene staple since the early 1980's is related to the development of civil engineering fabrics and the development of polypropylene staple for use in disposable diapers. The latter area, disposable diapers, also marked the ability of polypropylene to be used in lower dpf products. Polymer improvements have been continuous throughout polypropylene's history. Reduction of atactic content, higher polymerization yields per pound of catalyst which resulted in reduced contaminants and improvements in resin cleanliness were the resin producers contributions to the ability to make low dpf products. Preparation of a fusion bonded diaper liner from polypropylene was the result of finding that the resin could be converted to a fiber which could be pattern fused without destroying the strength characteristics of the fiber near the fused area. The unique melting characteristics of polypropylene permitted this product to be developed. Modern fiber making equipment permitted the low dpf staple products to be made economically. Better, cleaner resins contributed by allowing very low dpf products to be processed.

The growth of staple in civil engineering products, like Petromat[R] nonwoven fabric, Supac[R] nonwoven fabric, etc., is moderated some by competition from spunbond and woven slit film products. Fabrics for road construction, railroad road beds, weed control, roofing,etc., will continue to grow dramatically and all three routes to these fabrics will participate in the growth during the 1990's.

Yarn

Growth of polypropylene filament yarns was steady but not spectacular until the 1980's. Initial use in industrial yarns was due to its low specific gravity, excellent abrasion resistance and reasonably high strength. Use in home furnishings (as upholstery and carpet fiber) was the result of use of pigments which provided excellent color stability at reasonable cost. Development of controlled rheology (CR) resins (resins with narrower molecular weight distribution) permitted higher processing speeds of relatively low dpf yarns to be produced.

The major growth of polypropylene yarns came in the 1980's with a major thrust by Amoco into the carpet market with BCF (bulked, continuous filament) yarn. The development of fiber making hardware which started with pigments and polymer and, in one continuous process, resulted in BCF yarns permitted a number of carpet companies to begin production of BCF yarn. While this equipment did not necessarily make a product equal to that from the major producers, it was relatively easy to operate, made BCF yarn suitable for the mill's own use and in some cases the yarn was good enough to sell. Backward integration has reached such proportions that Beaulieu has even entered the polymer business with Epsilon Products coming on stream in 1991. (Some of the sharp increase at the end of the 80's was due to changes in reporting. However, growth from somewhat over 100 million to 400 million pounds did occur during the decade.)

In summary, the success of polypropylene fibers during the 30 years since its introduction has been due to continuous improvements in fiber and polymer technology, stabilizers and coloration which have resulted in entry into every textile area. Polymerization rate increases have helped to moderate increases in resin cost and have resulted in cleaner resins which can be spun with higher productivity to produce fibers of lower denier (lower diameter).

Fiber making equipment changes helped to moderate costs by providing more pounds of fiber per operator, either by higher speeds or by one step processes which are run at relatively low speeds but at high filament count. Since denier times speed equals pounds of production, productivity can be and has been improved by increased speed or by a much larger number of fibers being extruded at lower speed.

By combining hindered phenols, hindered amines and synergists, the industry can now make fibers which can withstand thousands of hours at temperatures well above 100°C, will provide good performance for three to five or more years of outdoor exposure, and can sit in hot caustic or acid for hundreds of hours without deterioration. Interior fabrics in automobiles need not degrade due to heat and sunlight even with the increased use of glass.

Because of improved stabilizer technology and the ability to produce low dpf fibers, greater demands have been placed upon pigments with regard to performance during as well as after fiber making. Colors which last 3-5 and more years, without undesirable color change, are available. Pigments which permit preparation of fibers as small as 10-15 microns in diameter are available.

The Future

It is always tempting, when discussing history, to attempt to predict the future. In addition to those developments already discussed, polypropylene fiber use will continue to expand due to (a) development of fibers such as Alpha^R olefin with improved ability to retain appearance, i.e., improved resilience, which will allow polypropylene to be used in tufted floor carpets for automotive interiors; (b) development of fiber which can be radiation stabilized with excessive loss of strength or color development; (c) greater recognition of the outstanding soil resistant characteristics of the products, particularly carpet (name another carpet fiber that can be rejuvenated if required by treatment with chlorine bleach); (d) use of polypropylene fibers in all outdoor areas including awnings, casual furniture, carpet, etc.; (e) expanded use of polypropylene's capability to be fusion bonded; (f) use in apparel where comfort characteristics are of significance; (g) a growing spread between the cost of polypropylene fibers and other fibers, i.e., improvement in cost/performance comparisons because the cost of other fibers will increase faster.

Polypropylene has grown from about 5% of the fiber market in 1977 to over 20% in 1990. Greater recognition of polypropylene's properties and additional improved products should help polypropylene to continue to grow faster than the general synthetic fiber market.

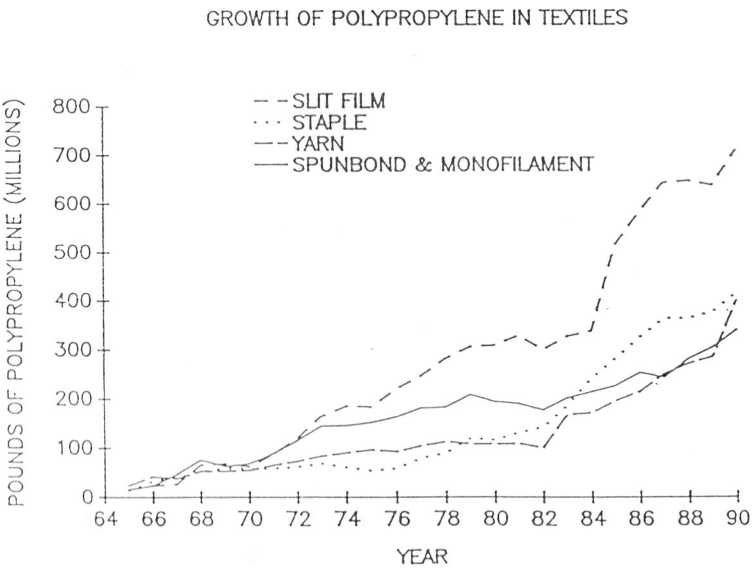

HISTORY OF ARAMID FIBERS

S. L. KWOLEK AND H. H. YANG
Du Pont Fibers, E. I. Du Pont de Nemours & Co., Inc., Richmond, VA 23234

'Aramid' is a generic definition for a polyamide with at least 85 mole % of its amide linkages directly connected to two aromatic rings. Aramid polymers were historically recognized as having excellent thermal resistance. Nomex[a] aramid fiber, a thermally resistant fiber based on a meta-oriented aramid, was first commercialized by the Du Pont Co. in 1962. A technological breakthrough occurred in the middle 1960s when Kwolek discovered the liquid crystal behavior of para-oriented aramids in solution. This discovery allowed the development of a novel fiber spinning process for anisotropic solutions by Blades. As a result, a high strength, high modulus Kevlar[b] aramid fiber was commercialized by the Du Pont Company in 1972. This was followed by the synthesis of numerous aromatic polyamide compositions and the development of a number of commercial aramid fibers such as Twaron[c], Technora[d], and SVM[e]. New aramid fibers are expected to appear in the high performance fibers market in the future.

DEFINITION OF ARAMID

The word 'aramid' is a generic term for "a manufactured fiber in which the fiber-forming substance is a long chain synthetic polyamide in which at least 85% of the amide linkages are attached directly to two aromatic rings" as defined by the U. S. Federal Trade Commission. This generic definition distinguishes aramids from conventional polyamides, such as nylon which contain mostly aliphatic and cycloaliphatic units in the polymer main chain.

By definition, therefore, the aramid family will cover Kevlar, Nomex, Technorac Teijinconex, and Twaron fibers.

EARLY AROMATIC POLYAMIDES

In the 1960s, research on low temperature processes for the preparation of condensation polymers in the Textile Fibers Pioneering Research Laboratory of the Du Pont Company led to the preparation of high molecular weight polymers that were unmeltable and/or thermally unstable.[1] The low temperature solution polymerization process, that employed fast reacting complimentary intermediates in a single organic liquid or a mixture in which the polymer may or may not be soluble throughout the reaction, was especially significant. This non-aqueous system allowed the use of a new range of solvent media, such as amides or amides with salts, and was especially applicable to the preparation of aromatic polymers.[2]

[a]NOMEX - a registered trademark of E. I. Du Pont de Nemours & Co., Inc.
[b]KEVLAR - a registered trademark of E. I. Du Pont de Nemours & Co., Inc.
[c]Twaron - a registered trademark of N. V. Akzo, The Netherlands
[d]Technora - a registered trademark of Teijin Ltd., Japan
[e]SVM - an aramid fiber produced in the Soviet Union

Early aromatic polyamides were usually characterized by high melting point, non-linear chain conformation, reduced crystallinity and moderate to good solubility. As seen in Table I, these polyamides typically contain meta-oriented or mixtures of meta- and para-oriented phenylene ring units.[3-11] In addition, some contain solubilizing groups such as sulfonyl, isopropylidene, etc.

From this work resulted the commercialization of Nomex aramid, poly(m-phenylene isophthalamide), in 1962.[12] Fibers from this polymer have textile properties in the range of conventional fibers, but because of their high temperature stability have been used largely in thermal protective apparel, high temperature filtration and industrial paper for electrical insulation.

Laboratory success with the preparation of these more tractable aromatic polyamides, served as a driving force for future research effort being directed at high performance fibers with superior heat resistance and stiffness equivalent to glass. These polymers would be the extended chain para-oriented polyamides which were noted for their intractability.

DISCOVERY OF ANISOTROPIC ARAMID POLYMERS

The development of aromatic polyamides took a dramatic change in direction with the discovery of lyotropic liquid crystalline aramids. This discovery occurred about 1965 when Stephanie Kwolek, a Du Pont research scientist, synthesized the rigid chain, poly(p-benzamide) in high molecular weight and with limited crystallinity to promote the dissolution of this seemingly intractable polymer in amide salt solvents.[13]

The polymer solutions at 10-15% concentration were very fluid, hazy, and stir-opalescent, very much unlike the clear, viscous solutions of isotropic polymers. There was considerable skepticism that these "watery" solutions would spin and that the fibers would hold together. However, these solutions were dry-spun directly, without drawing, to highly oriented strong fibers with tensile modulus greater than 400 grams per denier, which was stiffer than glass fiber. Both fiber tenacity and modulus were further increased about 2 to 3 fold by a very brief heat treatment under slight tension but with no drawing in the conventional sense. Repeated determinations of tensile properties of these fibers confirmed the high tenacity and high modulus numbers.[14-15] Subsequent spins gave fibers with tenacities up to 20 grams per denier and modulus up to 1000 grams per denier, after heat treatment.

With this evidence in hand of the potential of extended chain para-aramids, a research and development program was initiated to develop greater understanding of the behavior of liquid crystalline polymer solutions based on both practical and theoretical considerations (latter refers to Flory's theoretical examination of rod-like polymers in solutions [16]) and to evaluate possible candidates for commercial development.

Many anisotropic homo- and copolymamides were prepared and spun by a number of coworkers [17-23]. The polymers contained a variety of stiff structures but the common feature was that the components exhibited essentially coaxial or parallel extension of the chain forming bonds. Some of these anisotropic aramids are presented in Table 2. Of these, poly(p-benzamide) (PBA) and poly(p-phenylene terephthalamide) (PPD-T) are the best known.

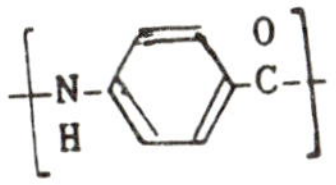

PBA

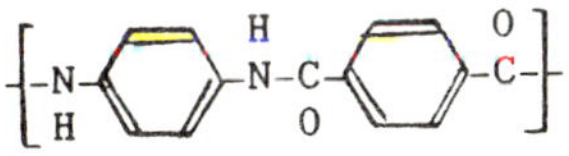

PPD-T

While the development of poly(p-benzamide) was continued, much effort was also directed to the study and development of poly(p-phenylene terephthalamide) which had the advantages of being prepared from intermediates that were readily available and less costly. Research related to the preparation of this polymer and its wet-spinning from an oleum solution (~100% sulfuric acid) at moderate polymer concentrations (10-12%) is discussed in patents by Thomas I. Bair and Paul W. Morgan.[22,23] However, even this polymer composition was not acceptable for commercialization until further breakthroughs occurred in the spinning process.

Herbert Blades found that by increasing the polymer concentration to 20% in 100% sulfuric acid, he unexpectedly formed a crystalline PPD-T/sulfuric acid complex which melted around 70°C and was stable to heating.[24,25] In addition, he spun this polymer solution by an air-gap wet-spinning method at considerably higher speeds than before possible. This improved product became the first Kevlar product.

DEVELOPMENT OF KEVLAR ARAMID FIBER

These discoveries led the Du Pont Company to undertake the commercialization of aramid fiber. The first aramid fiber spun from an anisotropic solution was coded Fiber B. It initially represented poly(p-benzamide) and later poly(p-phenylene terephalamide). A task force of researchers and technicians was formed to work on the Fiber B project. It took countless man hours of hard work and huge investment to get it into production. The first plant was built in Richmond, VA in 1971 with an annual capacity of 5 million pounds. The commercial fiber product was formally introduced under the trade name of Kevlar aramid. Initial shipment of Kevlar aramid was made in less than a year. Since then, rapid growth in the use of Kevlar aramid fiber has resulted in a further increase in its production. The Du Pont Company first increased the production capacity of Kevlar aramid to 15 million lb/year in 1978, and then to 45 million lb/year in 1982. Investments for the plant expansion have amounted to hundreds of millions of dollars. To facilitate distribution, Kevlar aramid fiber is now produced in the United States and England. It will also be produced in Japan by 1991. Thus, Kevlar aramid fiber has truly become a global business.

Preparation of Polymer

Kevlar aramid fiber is based on poly(p-phenylene terephthalamide) (PPD-T), which can be prepared by a number of synthetic routes.[14-23]. The classical synthesis involves the low temperature polycondensation of p-phenylenediamine (PPD) and terephthaloyl chloride (TCl) in an amide solvent:

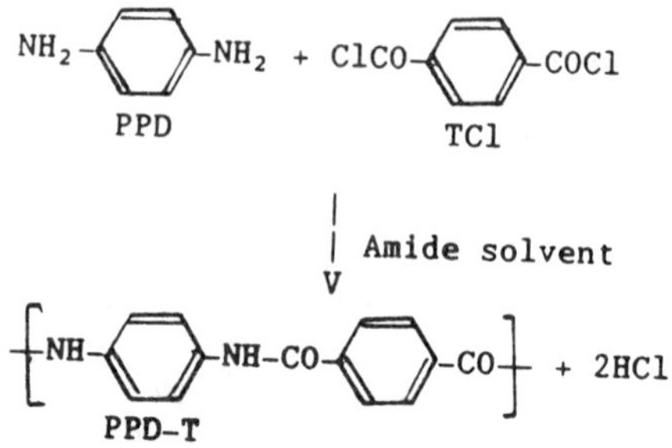

The amide solvents for this synthesis include hexamethylphosphoramide (HMPA), N-methyl-2-pyrrolidone (NMP), and dimethylacetamide (DMAc). See Figure 1. These solvents are used in single or mixed system often in the presence of an organic salt such as lithium chloride and calcium chloride. The polymer from this reaction system is isolated by precipitation with water, neutralized and subsequently washed and dried.

PPD-T polymer is generally prepared in high molecular weights to ensure high fiber properties. The inherent viscosity of PPD-T, is normally in excess of 4 dl/g. A typical number average molecular weight (Mn) of PPD-T is on the order of 20,000 which corresponds to a degree of polymerization of 84 and a chain repeat length of 108 nm. This compares well with the molecular weight and degree of polymerization of nylon 66 for industrial fiber. The molecular weight distribution of PPD-T in terms of Mw/Mn is on the order of 2-3.

Alternate Polymerization Solvent

In the history of Kevlar fiber development toward commercialization, a major hurdle occurred in the late 1970s. The toxicity of the polymerization solvent HMPA came under question.[26a]

HMPA was a well-known material for many years, but its behavior on large scale use as in the manufacturing of Kevlar aramid was unknown. There were no reports of unusual toxic effects at that time. Therefore, Du Pont initiated a lifetime exposure study with rats to determine any carcinogenic potential and other health hazards. Results of that study showed that HMPA was an experimental animal carcinogen.[27] Immediate steps were taken in the handling of HMPA to ascertain that there was no hazard to the workers, the community, or the customers. A crash technical program was mounted to find a replacement for HMPA.

The chemistry for an alternate solvent turned out to be relatively straightforward.[11] The combination of N-methylpyrrolidone and calcium chloride was selected as the alternate polymerization solvent. The polymerization of PPD-T in NMP/CaCl2 proceeded smoothly and polymer of high inherent viscosity was obtained without difficulty. Extensive internal and trade testing clearly demonstrated that the Kevlar fibers made by the two polymerization systems were indistinguishable.

The program for an alternate solvent took an estimated 40 man-years of technical effort. There was an underlying confidence in the multi-disciplinary team throughout the process scale-up phase.

Anistropic Solution

A study of the behavior of para-aramids in solution and the conditions under which they form anisotropic phases was initiated shortly after Kwolek's discovery. As seen in Figure 2, PPD-T has a rigid rod-like molecular conformation. Its solution behavior is affected by the polymer structure, molecular weight, and temperature.

Figure 3 shows that the dilute solution of an extended chain polymer is isotropic, as is the dilute solution of a flexible chain polymer. As the polymer concentration is increased, the rod-like PPD-T molecules are packed closer together. When the polymer concentration is increased beyond a certain critical limit, the molecules will adopt an ordered arrangement in small domains for more efficient packing. In a quiescent state, the domains are randomly oriented with respect to each other. This solution is in the nematic state. When the nematic solution of a low molecular weight liquid crystalline compound is cooled below a transition temperature, it is known to transform to a smectic state where its molecules form a 2-dimensional layered structure. In like manner, the nematic solution of PPD-T undergoes subcooling behavior and forms a cholesteric phase. These phase transitions are characteristic of liquid crystalline polymer solutions. In comparison, solutions of flexible polymers will not undergo such phase transitions; the molecules will remain in a random state regardless of concentration and temperature changes.

PPD-T polymer, like most rod-like aromatic polyamides, is insoluble in conventional organic solvents. It is soluble in strong acids such as concentrated (fuming) sulfuric acid, chloro- and fluorosulfonic acid, nitric acid, and hydrochloric acid. In sulfuric acid of 98% of greater concentration, PPD-T forms liquid crystalline (anisotropic) solutions at room temperature over a wide range of molecular weights. With anisotropic phase formation ,PPD-T solutions exhibit stir opalescence, optical birefringence, and decrease in solution viscosity as observed previously for anisotropic poly(p-benzamide) solutions.[17,22] Figure 4 illustrates a typical relationship of solution viscosity versus PPD-T concentration in 100% sulfuric acid. For this combination, the polymer concentration for the formation of anisotropic phase occurs at about 12%. At the left, solution viscosity rises in the normal manner for isotropic solutions, and then at the critical concentration (~12%) the system will accommodate more polymer only if an ordered anisotropic phase can form. This is followed by decreasing viscosity and increasing amount of ordered phase. At the minimum point (20%), the system is wholly anisotropic. Addition of more polymers results in viscosity increase and eventually the formation of a solid phase.

When an anisotropic solution is subjected to shear or elongational flow, its domains of ordered phase will deform and orient along the direction of flow. This behavior can be readily observed on a hot-stage polarizing microscopy. Figure 5 is an optical micrograph of PPD-T/H2SO4 solution at rest in a cross polarized light field. Liquid crystal domains of different colors and micron sizes can be seen in a random arrangement. Figure 6 shows the same PPD-T/H2SO4 solution after slight shearing. The liquid crystal domains become elongated and oriented along the direction of shear. Under high shear and elongational flows, the boundaries of liquid crystal domains may be partly destroyed and the domains may link up to form the fibrillar structure of coagulated fiber.

Fiber Spinning

The dry-jet wet spinning process discovered by Blades [24] to spin the anisotropic solution of an aramid polymer is shown schematically in Figure 7. The key feature of this process is that an anisotropic solution is extruded from a spinneret through an air gap into a coagulating bath at very high spinning speeds. The coagulated filaments are washed, neutralized and finally dried. This process produces a fiber with a tenacity of almost 23 grams per denier and a modulus of about 550 grams per denier. Heat treatment further increases the modulus.

The mechanism of polymer molecular orientation in an air-gap wet spinning process is shown as a conceptual diagram in Figure 8.[28] When an anisotropic solution is extruded through a spinneret capillary, the capillary shear will cause the liquid crystalline domains to orient along the direction of flow. At the capillary exit, some deorientation of liquid crystalline domains occurs as a result of solution viscoelasticity. However, the deorientation is quickly overcome by filament attenuation in the air gap under spinning tension. The attenuated filaments will retain the highly oriented molecular structure when they are coagulated. This give rise to a highly crystalline, highly oriented fiber structure.

Crystal Structure of PPD-T Fiber

PPD-T fiber spun by the air-gap wet spinning process has a unique three-dimensional fine structure. Figure 9 shows the unit cell structure of PPD-T as determined by wide angle X-ray diffraction.[29] The extended chain PPD-T molecules are aligned with the fiber axis. The close packing of the adjacent rod-like polymer chains enhances the intermolecular hydrogen bonding. The hydrogen-bonded molecules then form sheet-like structure across the fiber in the radical direction. This structure results in an almost perfect radial as well as longitudinal order. Such level of three-dimensional order had never been observed before.

The unique structure of PPD-T fiber gives rise to its truly outstanding mechanical properties.

Commercial Kevlar Aramid Products

Kevlar aramid fiber was first introduced as Fiber B continuous filament yarn into the high performance fibers market. A high modulus version of Fiber B was later introduced as PRD-49 fiber. These code names were later replaced by Kevlar 29 and Kevlar 49, respectively, after commercialization. Since the initial offerings, there have been several new filament yarns: Kevlar 68, Kevlar 129, Kevlar 149, and colored Kevlar 100; and new product forms: Kevlar staple, Kevlar fabrics, Kevlar pulp, and Kevlar M/B masterbatch. Figure 10 shows various forms of Kevlar aramid fiber products. They include continuous filament yarn, staple, engineered short fiber, pulp, spun yarn, needle punched felt, paper, woven fabric, cord, and narrow webbing.

Table 3 presents the typical properties of several filament yarns of Kevlar aramid fiber. Figure 10 compares the stress-strain curves of several commercial fibers.[26c-26e,30,31] In general, Kevlar aramid fiber has a high breaking tenacity that is several times that of steel wire, industrial nylon and polyester yarns. It also has a much higher tensile modulus than steel wire, fiberglass, nylon and polyester fibers. It has a low elongation at break that is comparable to that of steel. It has a lower density than steel and glass. Thus, most Kevlar-reinforced structures are of lighter weight for a given strength and stiffness. As an aramid, Kevlar fiber is inherently stable at relatively high temperatures. It has a very low shrinkage at elevated temperature, low creep, and a rather high glass transition temperature. In addition, it is corrosion resistant, non-conductive, and resistant to most chemicals except strong acids and bases. These outstanding properties make Kevlar fiber products useful for many industrial and civilian applications.

THREE DECADES LATER

The years, since the discoveries of lyotropic aramids, high strength and high modulus fibers there from and the novel air-gap wet spinning process, have been witness to many significant advances in related polymer chemistry. The latter have included new aramid compositions and high strength, high modulus fibers, new synthetic methods for condensation polymers and new spinning methods.

Synthesis Methods

A number of new synthetic routes for the preparation of aramid polymers have been reported in the literature. These involve new monomers and polymer compositions and amide solvents with salts. Table 4 presents a variety of routes to preparing PPD-T at high and low temperatures and in the gaseous, liquid and solid states.[11,22,30,32-44]

Spinning Methods

The air-gap wet spinning method has been used widely for fiber preparation from both anisotropic and isotropic solutions. This method has generally produced fibers with high strength and high modulus. In the case of isotropic solutions, post drawing of as-spun fibers at elevated temperature and high draw ratio has been practiced to achieve a high degree of molecular orientation and crystallinity.

Other Aramid Fibers

Numerous aramid compositions and fibers have been reported in the literature.[30] The general characteristics of these starting polymers can be summarized as follows:

- They are rigid or semi-rigid polyamides containing aromatic, cycloaliphatic and/or heterocyclic rings. The components exhibit essentially coaxial or parallel extension of the chain forming bonds.

- They can be polymerized to high molecular weights.

- They have a high molecular chain length/diameter ratio.

- They form melts or solutions in suitable solvents which permit processing below their decomposition temperatures and can be oriented with reasonable ease to form highly crystalline, highly oriented fibers.

Aromatic polymers with the above structural features generally exhibit high melt temperatures, poor solubility in common solvents and excellent thermal stability.

Several aramid fibers have been commercialized or are in the development stage. Table 5 presents a summary of commercial aramid fibers at the present time. It can be seen that the interest in aramid fibers is worldwide and is growing steadily.

Other Aromatic High Strength Fibers

In addition to aramid fibers, there has been much progress in the development of new classes or aromatic polymers and fibers in recent years. Most notable among these polymers are aromatic polyesters, polyazomethines, polyimides and heterocyclic polymers such as polybenzobisoxazoles and polybenzobisthiazoles. Most of these polymers exhibit anisotropic behavior in solution or melt form. Their fibers are characteristically of high strength and high modulus. Table 6 shows several selected polymers and their fiber properties.[30] While the end use performance of these fibers may vary, their physical properties have set new records and are indeed impressive.

SUMMARY

The history of aramid fibers, as shown in Table 7, covers about 30 years to date. It began about the time of the synthesis and commercial production of Nomex aramid fiber in 1962. A major step forward occurred in the middle 1960s with the discoveries of anisotropic solutions of aramid polymers and high tenacity, high modulus fibers there from, and the unique air-gap spinning method in 1970. These were followed by the difficult but rapid scale-up of Kevlar aramid fiber to meet the needs for advanced materials. With outstanding tensile and physical properties, Kevlar fiber has not only set new standards for synthetic fibers, but also has demonstrated excellent performance in many end uses such as composites, protective apparel, soft and hard armors, tires and mechanical rubber goods, ropes and cables, friction products, gaskets, masterbatches, thixotropes, etc.[31]

The development of aramid fibers has had a considerable technological impact on polymer and fiber sciences. In the last two decades, there was developed a greater understanding of liquid crystalline polymers. More study is needed along this line. We also have seen new polymer syntheses and fiber spinning methods, new aramids and other high performance aromatic polymers and fibers.

We are enthusiastic about the future of aramid fibers. Our confidence rests on the good balance of tensile and thermal properties, demonstrated end use performance, competitive cost, and availability of monomers. Thus, we look forward to the continued growth of aramid fibers throughout the world.

REFERENCES

1. P. W. Morgan, "Condensation Polymers: By Interfacial and Solution Methods," Polymeric Reviews, Vol. 10, Interscience, New York (1964).
2. S. L. Kwolek, P. W. Morgan, and W. R. Sorenson, U. S. Patent 3,063,966 (1962).
3. J. Coleonge and E. Fichet, Bull. Soc. Chim. France, 412 (1955).
4. M. Katz, Belg. Patent 569,760 (1958).
5. C. W. Stephens, British Patent 901,159 (1962).
6. W. A. H. Huffman, R. W. Smith, and W. T. Dye, Belg. Patent 620,511 (1962).
7. H. F. Mark, S. M. Atlas, and N. Ogata, J. Polym. Sci., 61, S49 (1962).
8. J. Preston and R. W. Smith, Bel. Patent 637,260 (1964).

9. R. A. Dine-Hart, B. J. C. Moore, and W. W. Wright, J. Polym. Sci. B, 2, 369 (1964).
10. J. Preston, J. Polym. Sci., A-1, 4, 529-539 (1966).
11. W. B. Black and J. Preston, 'Fiber-Forming Aromatic Polyamides,' in "Man-Made Fibers," Vol. 2, H. Mark, ed., 297-301, Interscience, New York, 1968.
12. W. Sweeny, U.S. Patent 3,287,324 (1966).
13. S. L. Kwolek, Am. Inst. Chem., The Chemist, Washington, DC, 57(11), 9 (1980).
14. S. L. Kwolek, U.S. Patent 3,600,350 (1971).
15. Ibid., U.S. Patent 3,888,965 (1975).
16. P. J. Flory, Proc. Roy. Soc. (London), Ser. A, 234, 6, 73 (1956); J. Polym. Sci., 44, 105 (1961).
17. S. L. Kwolek, U.S. Patent 3,671,542 (1972); reissued U.S. Patent 3,819,587 (1974).
18. Ibid, U.S. Patent 3,819,587 (1974).
19. P. W. Morgan and T. C. Pletchner, U.S. Patent 3,869,419 (1975).
20. L. W. Gulrich and P. W. Morgan, U.S. Patent 3,836,498 (1974).
21. P. W. Morgan, U.S. Patent 3,827,998 (1974).
22. T. I. Bair and P. W. Morgan, U.S. Patent 3,673,143 (1972).
23. Ibid, U.S. Patent 3,817,941 (1974).
24. H. Blades, U.S. Patent 3,767,756 (1973).
25. Ibid, U.S. Patent 3,869,429 (1975).
26. E.I. Du Pont de Nemours & Co., Inc., product information:
a) "Kevlar - from Laboratory to Marketplace through Innovation," H-16261.
b) Kevlar aramid: The fiber that lets you re-think strength and weight, E38532.
c) Bulletin K-2, Characteristics and Uses of Kevlar 49 Aramid High Modulus Organic Fiber, 1978.
d) Bulletin-K4, Properties of Industrial Filament Yarns of Kevlar Aramid Fiber for Tires, Hose & Rubberized Belting, 1979.
e) Preliminary Information Memo, Spun Yarns of Kevlar Type 970 Aramid Staple, June 27, 1980.
27. K. P. Lee and H. J. Trochimowicz, "Pulmonary Response to Inhaled Hexamethylphosphoramide in Rates," J. Toxicol. & Appl. Pharm., 62, 90-103 (1982).
28. H. H. Yang, 'Aramid Fibers,' in "Fibre Reinforcements for Composite Materials," A. R. Bunshell, ed., Elsevier, Amsterdam, 249-329 (1988).
29. M. G. Northolt and J. J. Aartsen, J. Polym. Sci., Polym. Lett. Ed., 11, 333 (1973).
30. H. H. Yang, "Aromatic High-Strength Fibers," Wiley-Interscience Publishers, New York (1989).
31. P. G. Riewald, 'Kevlar Aramid Fibers - Properties and Industrial Applications,' Adv. Text. Mat. Conf., Clemson Univ., Greenville, SC, April 5-6, 1988.
32. H. Shin, U.S. Patent 4,009,153 (1977).
33. H. Mera, Y. Makagawa, M. Yamaguchi, and M. Ohno, U.S. Patent 4,172,938 (1979).
34. V. Rochina and J. Sacco, U.S. Patent 4,297,478 and 4,297,479 (1981).
35. L. B. Sokolov, V. M. Savinov, and V. S. Petrukhin, U.S. Patent 4,159,932 (1979).
36. J. Preston and W. L. Hofferbert, J. Polym. Sci., Polym. Symp. 65, 13 (1978); J. Appl. Polym. Sci., 24, 1109 (1979).
37. F. Higashi, M. Goto, Y. Nakano, and H. Kakinoki, J. Polym. Sci., Polym. Chem. Ed., 18, 1099 (1980).
38. F. Higashi, Y. Nakano, M. Goto, and H. Kakinoki, J. Polym. Sci., Polym. Chem. Ed., 18, 1099 (1980).
39. F. Higashi, M. Goto, and H. Kakinoki, J. Polym. Sci., Polym. Chem. Ed., 18, 1711 (1980).
40. Asahi Chemical Industry Co., Japanese Patent 33,591 (1979).
41. F. Higashi and T. Nishi, J. Polym. Sci., Part A, Polym. Chem., 26, 3235 (1988).
42. R. Cuidard, B. Giroud-Abel, and V. Rochina, U.S. Patent 4,579,895 (1986).
43. W. Heitz and Strohriegel, U.S. Patent 4,598,143 (1986).
44. J. D. Hodge and K. S. Lee, U.S. Patent 4,684,409 (1987).

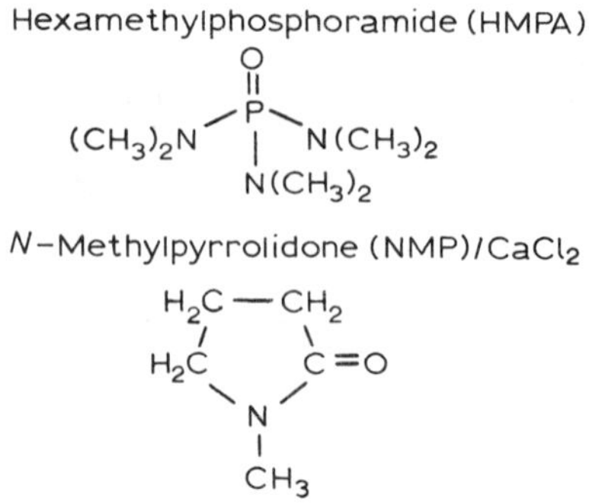

Figure 1. HMPA and NMP

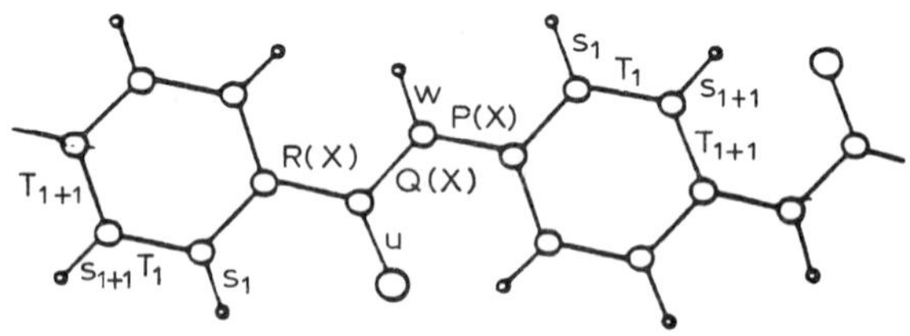

Figure 2. Molecular Model of PPD-T

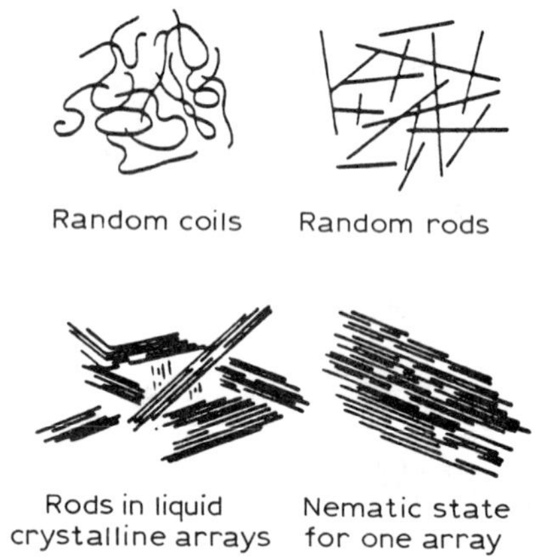

Figure 3. Solution Behavior of Stiff Chain Polymer

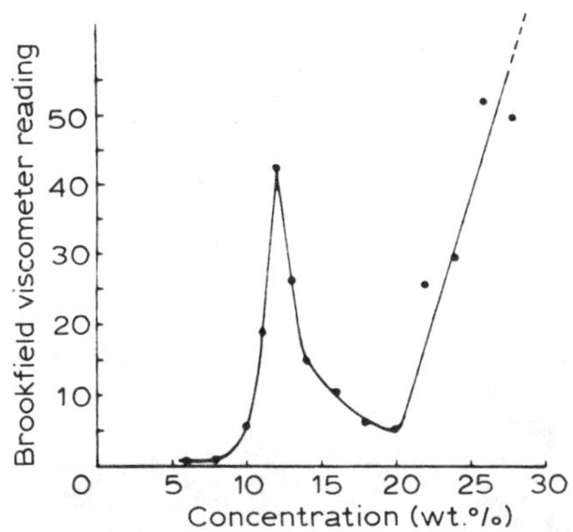

Figure 4. Bulk Viscosity vs. Concentration of PPD-T/H_2SO_4 Solution

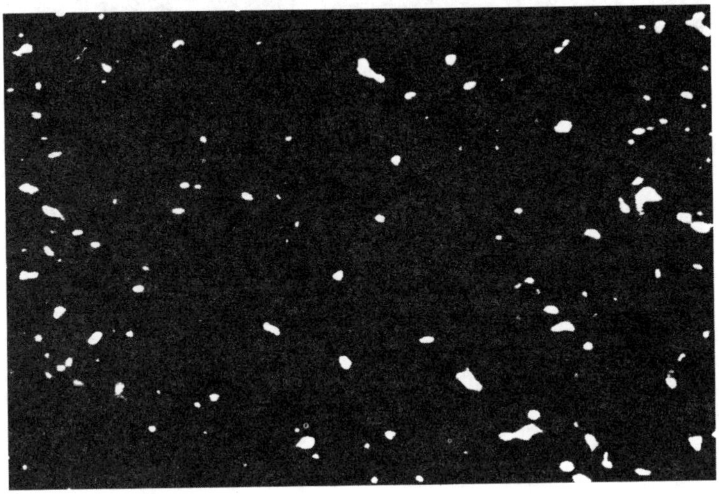

Figure 5. PPD-T/H_2SO_4 Solution at Rest

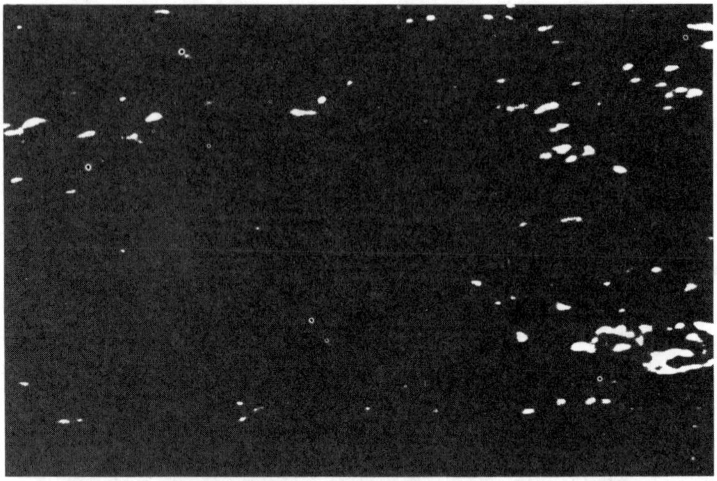

Figure 6. PPD-T/H_2SO_4 Solution After Shear

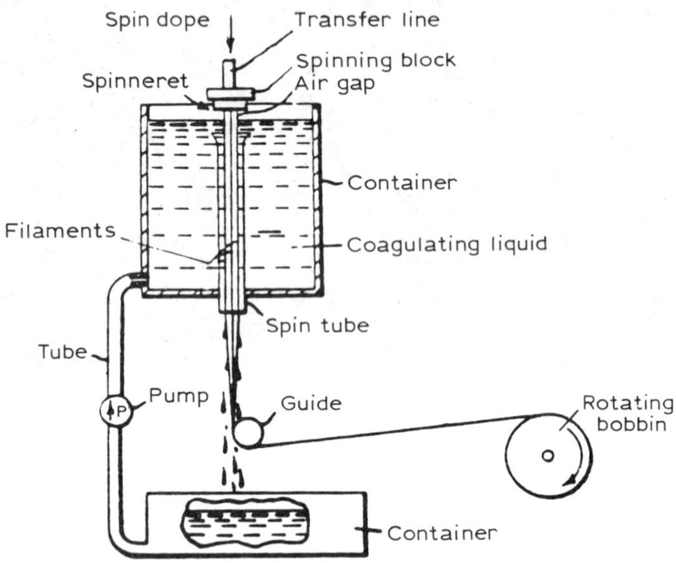

Figure 7. Air Gap Spinning Process

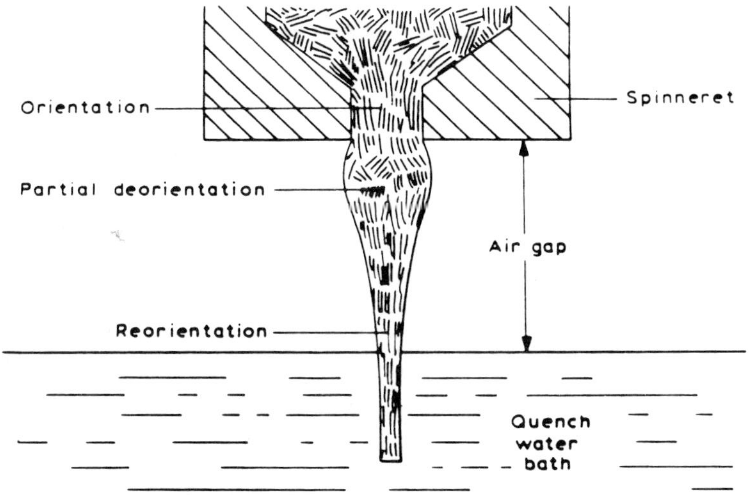

Figure 8. Molecular Orientation During Spinning [28]

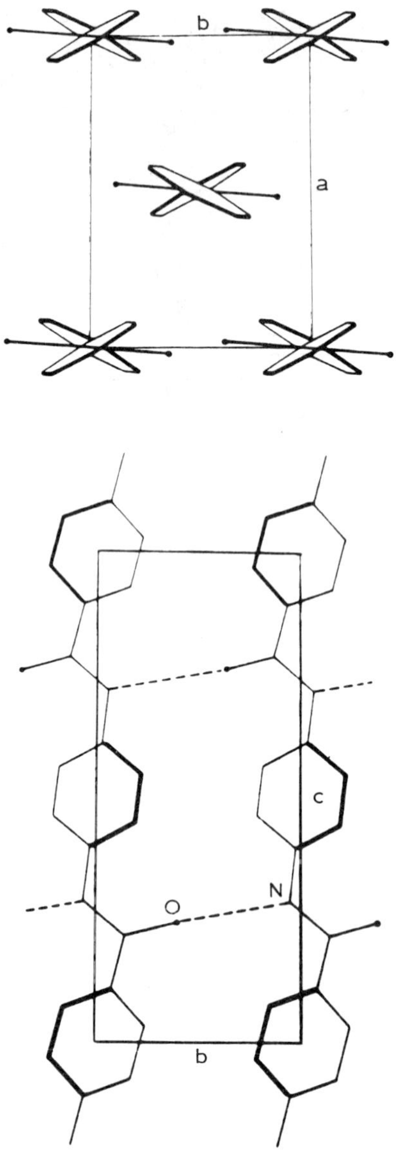

Figure 9. Crystal Structure of PPD-T Fiber [29]

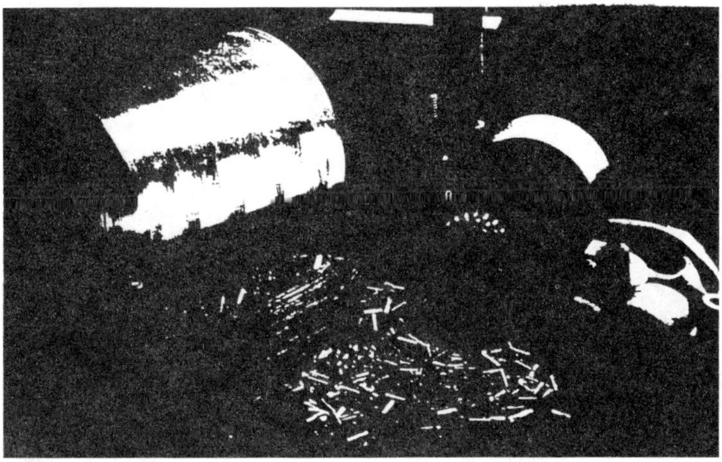

Figure 10. Various Product Forms of Kevlar Aramid [26b]

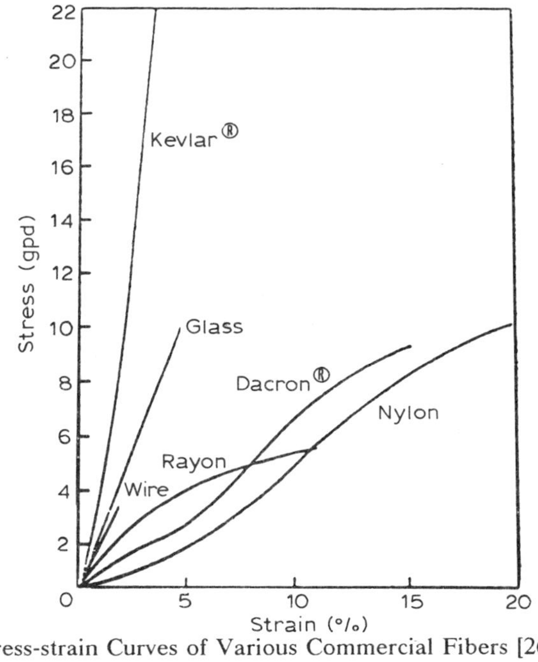

Figure 11. Stress-strain Curves of Various Commercial Fibers [26c]

Table 1. Early Aromatic Polyamide Compositions

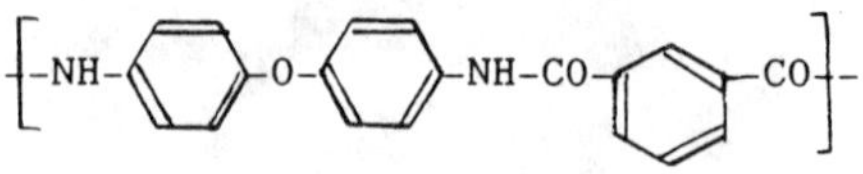	4,4'POP-I
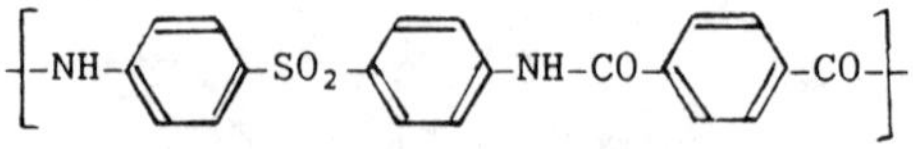	4,4'PSO_2P-T
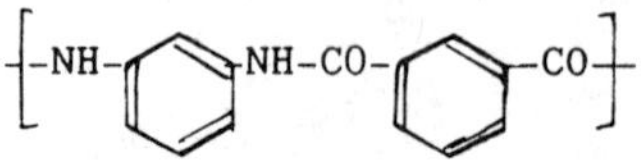	MPD-I (NOMEX)
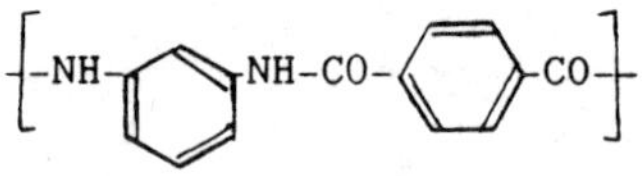	MPD-T
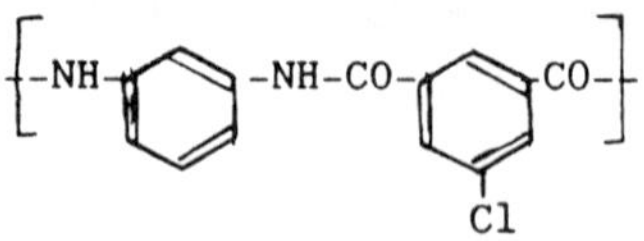	MPD-ClI

Table 2. Some Polyamides Yielding Liquid Crystalline Solutions

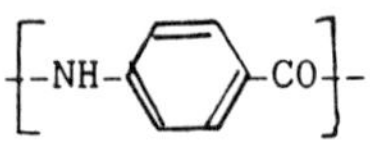	PBA
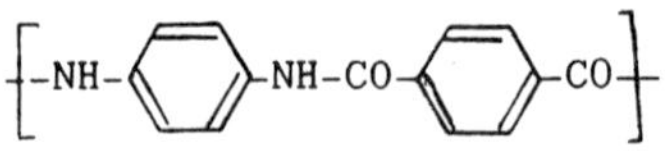	PPD-T
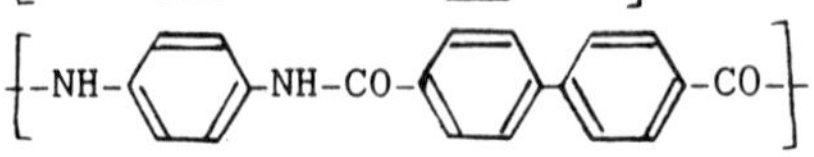	PPD-BB
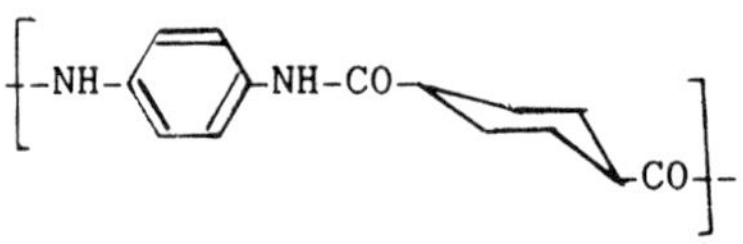	PPD-HT
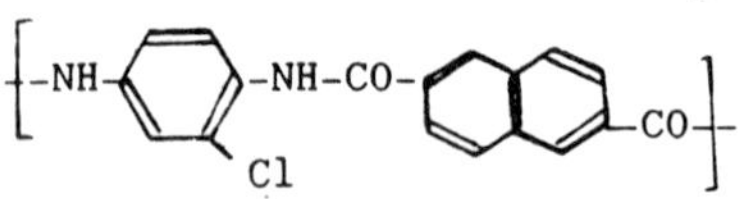	ClPPD-2,6N
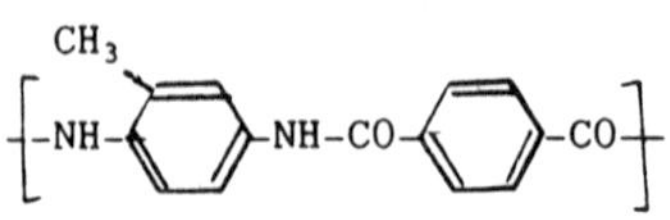	MePPD-T
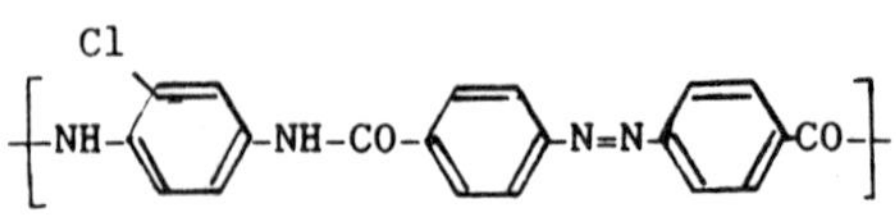	ClPPD-4,4'BN=NB

Table 3. Typical Yarn Properties of Kevlar Aramid

Yarn Properties	Kevlar & Kevlar 29	Kevlar 49	Kevlar 68	Kevlar 119	Kevlar 129	Kevlar 149
Tensile strength,[1]						
gpd	23.0	23.0	23.0	24.0	26.5	18.0
Kpsi	420	420	420	440	485	340
Initial modulus,						
gpd	550	950	780	430	750	1100
Mpsi	10.3	17.4	14.4	8	14	21
Elongation, %	3.8	2.8	3.0	4.4	3.3	1.5
Epoxy impregnated strand[2]						
Tensile strength, Kpsi	525	525	525	610 est.		500
Initial modulus, Mpsi	12	18-19	16	16 est.		25-26
Elongation, %	4.4	2.9	---	---		1.9
Density,						
g/cc	1.44	1.45	1.44	1.44	1.45	1.47
lb/ft^3	89.9	90.5	89.9	89.9	90.5	91.7
Moisture regain, % 25 C, 65% RH	6	4.3	4.3	---	---	1.5

[1] Yarn properties determined on 10° twisted yarns (ASTM D-885)
[2] Strand properties determined on untwisted epoxy-impregnated yarn (ASTM D-2343)

Table 4. Synthesis Methods for PPD-T [17,22,30,32-44]

Author	Monomers	Solvent System	Polymer in Solvent (%)	Salt in Solvent (%)	η_{inh}
Kwolek	PPD + TCl	HMPA/NMP	10-12	0	3.4
Bair and Morgan	PPD + TCl	HMPA/NMP	10-12	0	5.5
Akzo	PPD + TCl	NMP/excess CaCl	6-9	12	4-5
Mera et al	PPD + TCl	NMP/$CaCl_2$		4	5.5-6.5
Rochina and Sacco	PPD + TCl	NMP/$CaCl_2$	5-16	>0.7	1.2-7
Sokolov et al	PPD + TCl	DMAc/LiCl/pyridine	6	2.5	5.0
Preston	PPD + TCl	DMAc/LiCl			0.28
Higashi et al	PPD + TPA	NMP/LiCl/ $P(OC_6H_5)_3$/pyridine			0.53 0.74
Higashi et al	PPD + TPA	NMP/LiCl/ $P(OC_6H_5)_3$/P4VP/ pyridine			0.45
Higashi et al	PPD + TPA	NMP/LiCl/ $P(OC_6H_5)_3$/PEO/ pyridine			0.6-1.1
Shin	PPD + TCl	None (vapor phase reaction)	---		2-3
Asahi	PPD + TCl	None/tertiary amine (vapor phase reaction)	---		
Higashi et al	PPD + TPA	NMP/LiCl/$CaCl_2$/ $P(OC_6H_5)_3$/pyridine			4.51
Cuidard et al	PPD + TCl	NMP/$CaCl_2$/pyridine	8.5	7.0	3.65
Heitz et al	PPD + TA	Pyridine/$SiCl_4$			0.60
Hodge and Lee	PPD + TCl	NMP/MTBAC/TBA	11	25.9	3.8

Table 5. Worldwide Production of Aramid Fibers

Fiber Product	Base Polymer	Producer	Est. Capacity, MM lb/yr
Nomex	poly(m-phenylene isophthalamide)	Du Pont Co., USA	10-20
Teijinconex	poly(m-phenylene isophthalamide)	Teijin Ltd., Japan	
Kermel	poly(m-phenylene isophthalamide)	Rhone Poulenc, France	
Fenilin	poly(m-phenylene isophthalamide)	USSR	
Kevlar	poly(p-phenylene terephthalamide)	Du Pont Co., USA	40
Twaron	poly(p-phenylene terephthalamide)	Akzo Chemicals BV, The Netherlands	10
Technora	poly(p-phenylene/3,4′-diphenylether terephthalamide)	Teijin Ltd., Japan	1
SVM	Unknown	USSR	
New Aramid	Unknown	Hoechst, Germany	Market Development
New Aramid	Unknown	Corlon Co., S. Korea	Market Development

Table 6. Aromatic High-Strength Fibers [20]

Polymer	Fiber Type	Density g/cc	Ten. gpd	Mod. gpd.	T_g, °C	T_m, °C	T_d, °C
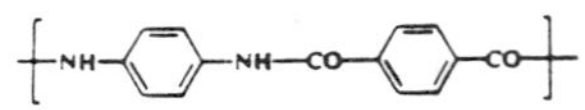 Aromatic polyamide	Kevlar 29 Kevlar 49 Kevlar 119 Kevlar 129 Kevlar 149	1.43 1.45 1.44 1.45 1.47	23 23 24 26.5 18	580 950 470 750 1100	> 375	550	555
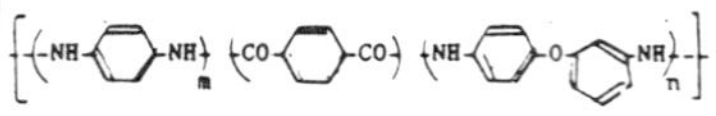 Aromatic copolyamide	Technora	139	25	550		500	
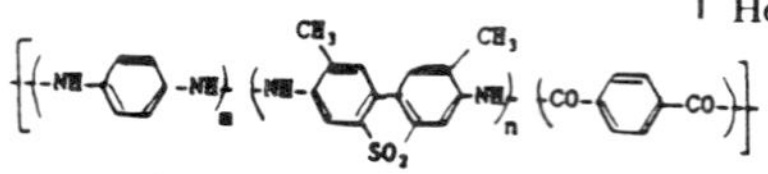 Aromatic copolyamide	As spun Heat treated		24 31	490 875			455
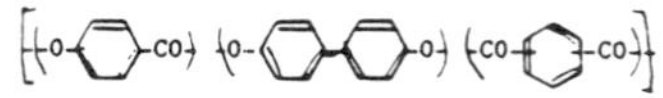 Aromatic copolyester	Ekonol	1.4	31	1100		-350	
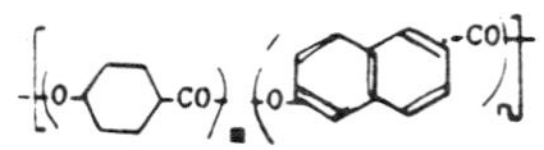 Aromatic copolyester	Vectra	1.47	22-25	600-700		-250	

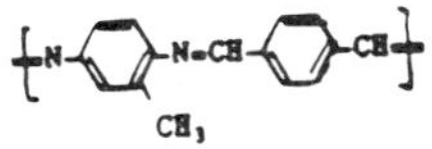 Aromatic polyazomethine		1.39	38	1012		250	
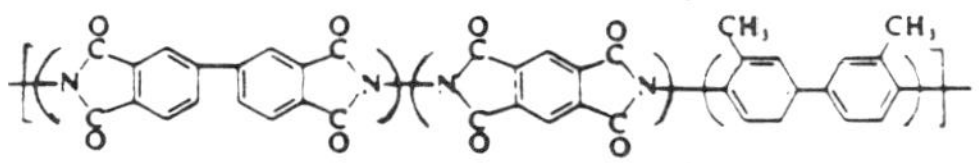 Aromatic copolyimide		1.34	27	1659			
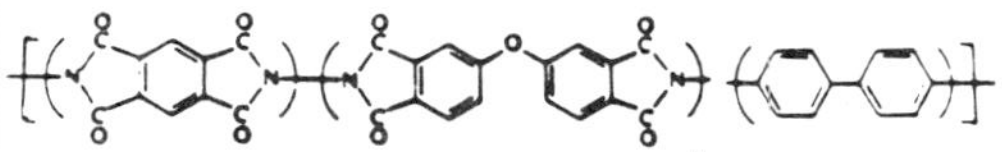 Aromatic copolyimide		1.47	22	1600			
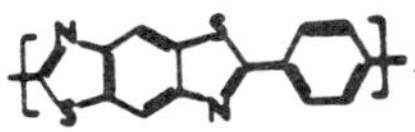 Aromatic heterocyclic polymer		1.57	25	2690			
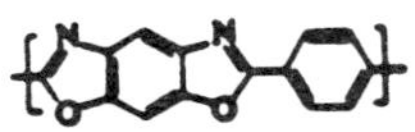 Aromatic heterocyclic polymer		1.57	25	2900			

Table 7. History of Aramid Fibers

Year	Event
1938	Commercialization of nylon
1962	Introduction of Nomex aramid fiber
1965	Discovery of anisotropic polymers
1970	Discovery of air gap spinning
1971	Scale-up of Fiber B
1972	Commercialization of Kevlar aramid fiber
1976	Introduction of SVM fiber
1978	Development of Arenka aramid fiber
1983	Expansion of Kevlar Plant
1987	Introduction of HM-50 (Technora) aramid fiber
1988	Introduction of Twaron (formerly Arenka) aramid fiber
1990	Announcement of Hoechst aramid fiber

THE OTHER ARAMID FIBERS

KAREL F. MULDER
Management and Assessment of Technology, Department of Chemistry, University of Groningen, Nyenborgh 16, 9747 AG Groningen, The Netherlands.

Introduction

When analyzing the history of technology, it is very easy to make the mistake of what historians call 'Whig History', i.e. omitting all the mistakes that were made and dead-end streets that were followed. However, science is a human activity and development of new products is not only a scientific activity but also a technological and commercial activity. Strictly scientific criteria are therefore not applicable to the acceptance of the results of the work of scientists. Often great scientific achievements in the laboratory do not fit to the social or economic context in which the corporation operates.

Moreover, it is not always original research that is carried out in corporate laboratories; trying to duplicate the tricks of the competitors is often an important and time consuming activity of corporate scientists. In my view, these features point at the most interesting point in the history of fiber development, the interaction between science and business strategy. If one wants to understand this interaction, successes and failures have to be analyzed both. Therefore, I will present to you the histories of aramid fiber R&D carried out by some corporations. Since not only scientific developments are of importance, I will try to describe the interplay of business, science and technology. Sometimes the R&D was highly original, sometimes it was mere duplication. Du Pont's pioneering research on aramid fiber technology which lead to Nomex and Kevlar, is part of another chapter. Therefore, I will only briefly mention them.[1]

Strictly speaking, the fibers that I will deal with are not all 'aramids' according to the definition of the United States Federal Trade Commission of 1974 (at least 85 % of the amide linkages have to be attached directly to two aromatic rings) (C&EN, 21-1-74). However, the R&D efforts in the various laboratories had been very competitive. In their research, most laboratories switched very easy between polymers that would later be classified as aramids, and polymers that wouldn't.

A New Polymerization Technology

In 1950, Emerson Wittbecker of Du Pont found a new polymerization technology, 'interfacial polycondensation'. Features of this technology could be found in the Allied reports on German synthetic fiber research during the war (Smith, 1946). At Du Pont's Experimental Station, the method was further elaborated. Paul Morgan developed the single phase system method, 'solution polycondensation'. In 1958, Du Pont scientists first published the interfacial polycondensation method and its main applications (Wittbecker/Morgan, 1959). In the meantime, they were developing the heat resistant fiber HT-1, which later was

---

[1] Many people have been supporting my study by giving information etc. Unfortunately not all of them can be mentioned here. I would like to thank especially the magnificent cooperation of Dr.B.von Falkai, Mr.J.Fitzgerald, Dr.V.S.Matveev, Prof.Dr.J.E.McIntyre, Dr.J.Preston, and Ir.H.Weyland.

commercialized under the Nomex trademark. In the sixties, Du Pont scientists found a way to make high strength/high modulus (HS/HM) wholly aromatic polyamide fibers. In 1970, the commercialization of a HS/HM fiber was announced by Du Pont (C&EN, 9-2-70, 2-2-70, Hannell, 1970). Some later, it became clear that Du Pont had improved its Fiber B (Lee et al., 1971). This fiber turned out to be based on poly-p-phenylene-terephtalamide (PPD-T).

In the fifties chemists of General Electric, Bayer and Gevaert had started to use methods to make polycarbonates which were similar to the ones developed by Du Pont (Morgan, 1985, Schnell, 1964) Even before Du Pont scientists published on their work, Chemstrand and Celanese scientists noticed that Du Pont was working on heat resistant fibers, based on aromatic polyamides that were made by the new polymerization method. Some of this information was exchanged by the US Air Force which evaluated heat resistant fibers for use in drogue chutes which were needed for supersonic jets.

In the following paragraphs, I will describe the work of some corporations which have been involved in aramid fiber R&D. This will not be a complete overview. I will describe the efforts of Chemstrand (Monsanto Textiles Division), Bayer, the Soviet institutes VNIIV and VNIISV, ICI and AKZO. The R&D efforts of other corporations will only be mentioned briefly.

Chemstrand

The Chemstrand company was established as a joint venture of Monsanto and American Viscose in 1952. Monsanto had developed Acrilan acrylic fiber and found in Avisco a partner to market it. Acrilan was no commercial success. However, Chemstrand was the first licensee of Du Pont's nylon proces, on which it made a lot of money in the fifties (Forrestal, 1977). Chemstrand was operating independent from its parents. Its fashion oriented New York salesmen sometimes boasted 'we are no chemists' and talked about Monsanto headquarters in Saint Louis as 'the guys out in the prairies'. However, Chemstrand's independence was endangered since Monsanto was trying to buy out Avisco. To prevent the Monsanto board from transferring Chemstrand's laboratories from Decatur to Saint Louis, Chemstrand decided to establish a new laboratory in the Research Triangle Park in Durham (NC). The scientists were transferred to this new laboratory just before Monsanto took over Chemstrand in 1961.

In 1958, Chemstrand scientists got a sample of Du Pont's HT-1 from the Air Force. They were able to reproduce this fiber. The Chemstrand scientists started working in various directions to develop a heat resistant fiber with comparable or better properties than Du Pont's HT-1, which could be produced at comparable or lower cost and without infringing on Du Pont's patents. However, it could only be estimated what Du Pont would have patented, since there still were no patents issued on HT-1. Various research groups of the Exploratory Department of the Durham laboratory were working on the subject following various lines:

I) W.A.H.Huffman's research group started to make an analogue of Du Pont's HT-1 (AB order instead of HT-1's AABB order). One finished the work on this AB type polymer successfully (Huffman et al., 1965). The fiber had most properties of HT-1. However, later it turned out that the polymer was covered by DuPont patents.

II) J.Preston and R.Smith of W.B.Black's exploratory group managed to synthesize a wholly para aromatic polyamide, poly-p-benzamide (Preston/Smith, 1964).

However, one was unable to form a fiber of this polymer because it was intractable.

III) M.Lilyquist's research group worked phenylene triazole polymers which were studied by C.S.Marvel before (Abshire/Marvel, 1961). This work was done under Air Force contract. One had to improve the process developed by Marvel to reach the high molecular weights that were necessary for spinning fibers. However, Lilyquist's group discovered that these polymers were not very appropriate for fibers.

The spinning of these resistant fibers was greatly improved by Herbert Morgan, a laboratory technician. He elaborated on an method which he studied to spin acrylic fibers. This air gap spinning method, or dry-wet-jet method[2], worked very well for aromatic polyamides and gave superior properties.

A different approach

When these three projects had failed, the idea came up to use diamines with preformed amide bonds as an intermediate[3]. The advantage of this method was that parts of the polymerization were already done when this diamine was made. Therefore, there was lower heat and less hydrochloric acid as the byproduct to get rid of when the actual polymer was made. In this way one could also polymerize with terephtaloyl chloride which was cheaper than isophtaloyl chloride. However, the order of the polymer would be less and therefore the fiber properties would probably be not as good as those of HT-1.

The product

Various polymers were made in this way (Preston et al., 1967). A polymer which was called M3P[4] had almost identical properties to HT-1. The intermediates would be some more expensive than those of HT-1, but production would be less complicated. The project to develop a fiber based on M3P got the name X-101. In 1964, some of the properties of this polymer and some related polymers were published (Preston/Dobinson, 1964). The diamine preparation was scaled up in Dayton. Polymerization on a large scale was started at Monsanto's Pensacola site. The product was evaluated by the Air Force and they founded it to be better than Du Pont's Nomex.

Monsanto worked with Hooker chemical company. Hooker agreed to make the diamine with the preformed amide bonds for the price of $ 1.10-1.15 per pound. The Textiles Division had formulated a rule that no intermediate that cost more than $ 1 per pound should be purchased. Hooker did not want to deliver for that price. This was one of the main problems to continue the project. When market crisis broke out in 1967, Monsanto panicked and wanted to cut the fiber research budget considerably. At one day, a quarter of the people working at the Durham research center were fired. X-101 was terminated.

Durette

---

[2] This is basically a combination of dry and wet spinning: The spinneret is not immersed in the bath, but an air gap of about one inch is left open. In the bath, a jet stream is used to increase the speed of spinning.
[3] Generally, diamines which contained two or three aromatic rings connected by amide linkages.
[4] M3P: **three m**eta linkages in the preformed diamine polymerized with **one p**ara, terephtaloyl-di-chloride. The resulting polymer had the ABAABABB order.

Monsanto had spent considerable efforts to improve the inflammability of Nomex and X-101 by chemical treatment. Some of the results of this project were presented at a symposium in 1968 (Hirsch et al., 1969, C&EN, 23-9-68). When the Textiles Division was hit by the 1967 crisis, this activity was taken over by Monsanto's New Enterprise Division. In 1970, Monsanto produced this improved Nomex as so called X-400 fiber at a pilot plant. Later the fiber got the trademark Durette. In 1970, it was pushed on the market (C&EN, 18-5-70). Monsanto just bought Du Pont's Nomex, another company was weaving the fabric and Monsanto ran it through a reactor. The New Enterprise Division did not find enough market for it because it still had some flaws, light stability was rather bad and it darkened up. It could not be dyed.

A HS/HM fiber

The scientists in Durham were scouting for fibers with even better heat resistance than X-101. In 1963/64, Huffman's group started studying all para wholly aromatic polyamides. They succeeded in making a fiber based on such a polymer. This fiber had a remarkable high modulus. The scientists proposed to spin this polymer from sulphuric acid but it was refused for it was to expensive and a manager believed that Celanese had tried this before without success. Moreover mechanical properties were not of primary importance. Huffman had died in 1963 and Tom Tolbert who succeeded him left rather soon. Afterwards, no one was pushing this research and it was therefore terminated. In 1966/67, the scientists could publish these results. It were the first publications on all para, wholly aromatic polyamide fibers (Preston et al. 1966, Preston et al. 1969).

In the beginning of 1966, Black and Preston discussed to make fibers containing oxadiazole units by a new method. They would first make polymers containing hydrazides, then form a fiber and heat the fiber to convert the hydrazides to oxadiazoles.[5] When the first spinning of a poly-hydrazide-amide polymer was done in September 1966, the as-spun fiber showed an extremely high modulus and strength[6]. Black and Preston agreed to forget about heat resistance, the name of the game would be HS/HM fibers. The project they set up got the name X-500.

However, the director of the basic research department they were in at that moment, did not want to pay for new spinning experiments which were done by the exploratory department. These experiments were very expensive. Although the fiber could probably be produced rather cheap, it was estimated that the combination of strength and inelasticity, which was needed for a tire cord, could not be reached. Therefore, the Monsanto Fibers Department management did not want to invest in the X-500 development. This decision was also affected by the fact that business was becoming worse at the end of 1966. Budgets were cut and some scientists were looking for other jobs. E.g. J.J.Hermans, the most distinctive polymer scientist at the laboratory, left for the State University of North Carolina.

When the Monsanto Textiles Division was hit by the crisis of 1967, the New Enterprise Division took over part of the research of the laboratory. Engineered composite systems were one of the five main activities of this division. It was decided that X-500 would be taken to development as one of their products. The pilot plant for X-500 was built in Pensacola. At

---

[5] This idea was patented almost simultaneously by Ashland Oil & Refining Co (Culbertson, 1968). I am aware of rumors concerning the cause of this simultaneity, but since they cannot be substantiated, I will not elaborate on this point.

[6] after some heat stretching the strength was 15-16 g/denier and the modulus was more than 500 g/denier.

the end of 1967, it began to make large poundage of the fiber. A lot of the evaluation was done internally. Prepregs were made and tested. According to tests of the US Army Natick Laboratories, the X-500 fiber could not be used for ballistic protection (Published as: Laible et al., 1973, Laible et al., 1974). The maximum strength reached with X-500 (X-500G) was about 19,1 g/denier (about 2/3 of the maximum strength of the later Kevlar 49) The maximum modulus was 828 g/denier (about 82 % of the Kevlar 49 modulus) (Preston, 1975).

Opposition

Gradually, there was a growing opposition to X-500 within Monsanto.
"Our marketing people said 'How can we possibly market this product ? There is nothing like it on the market." This made our (the scientists) mouth drop open, how could anybody be so stupid. It is exactly what you want, 'nothing like it on the market' (Preston, 1989)

The Monsanto Textiles Division marketeers just did not know how to develop a new market, they were just used to follow Du Pont. Scientists that were also working on budgets of the New Enterprise Division in Monsanto's Dayton research center did not like the competitors in Durham and Pensacola. In Dayton, they were working on a carbon fiber. Daytonians used the argument that glass fiber was much cheaper than X-500 and carbon fiber was, although much more expensive, much better then X-500. So there would be no need for X-500 which was in between. A salesman went out to Boeing in Seattle and came back with the report that they were not in the least interested in X-500.[7] He therefore proposed to drop the project. The scientists of the Nylon Department in Pensacola were also opposed to X-500. Relations between the various exploratory and basic research groups had been bad in Durham, and had not improved when some left for Pensacola.

In September 1968, one was ready to announce X-500 as a new product. A trademark had been picked: Fibron 500. The bulletins were ready. Then all of a sudden, it was decided to cancel the whole project. X-500 died October first 1968. Main argument for this decision was the negative report on Boeing's interest. The scientists knew from patent literature that Du Pont was interested in HS/HM fibers. Afterwards they estimated that if they had been able to show that Du Pont would come up with Kevlar 49, they would probably have been able to keep the project going.

Monsanto did not publicly declare that they had ended the X-500 project because some exploratory R&D on HS/HM fibers was continued in Pensacola. The budget of the New Enterprise Division was transferred to the carbon fiber project, which was terminated in 1971.

In April 1972, there was a special ACS symposium on wholly aromatic fibers. Du Pont did not allow its scientists to present papers. Therefore, it became an all Monsanto symposium. Black and Preston, presented a lot on X-500 (Black/Preston, 1973). The project,

---

[7] 'Much later, I went to this manager in Seattle and asked him if he really was not interested in X-500. He answered "of course I was interested, but this man came into my office and said 'we have this ivory tower research group which came up with a fiber, but you do not really want any of it, do you ?', I wanted to throw this man out of my office because we were dealing with Du Pont's PRD-49 and we would love to have a competitor because then we would not be dependent on one material. But he came in so negative, so we said no, based upon the attitude towards his own product." ' (Preston, 1989)

however, had been dead for about three and a half years. The press called commercialization of X-500 fiber 'quite likely' (C&EN 17-4-72).

In Pensacola, one made polymers from terephtaloyl-di-hydrazide (TDH) and/or oxalic-di-hydrazide (ODH) polymerized with terephtaloyl-di-chloride (TDC) in di-methyl-acetamide containing 5 % LiCl. This project was called X-702. The fibers were called H-20 (ODH-TDC), H-22 (TDH-TDC) and H-202 (copolymer ODH-TDH-TDC). The H-20 and H-22 polymers were rather intractable and hard to spin. Therefore, development concentrated on H-202. A fiber with tenacity of 21.6 g/denier and a modulus of 443 g/denier could be formed, which was some better than X-500 (Dobinson/Pelezo, 1973/74/74/75). The fiber based on the copolymer was tested for usage as a tire cord in the belts of radial tires. Some samples for the tire industry were made (Dobinson et al, 1979). The modulus and strength of the fiber were sufficient but it was much to thick to give good flex resistance to a tire. Moreover, the manufacturing proces was not very economic because of the expensive LiCl that was used in the polycondensation reaction.

After this new failure one attempted to make poly-oxadiazole fibers. In 1971/72, copolymers were made based on phenylene oxadiazole and N-methyl hydrazides. The fibers based on this polymer had good properties. However, Furukawa Electric appeared to have patents on the process which was used to make the polymer.[8]

The last attempts at Pensacola to develop a HS/HM tire cord were made by Hartwig Bach. He was spinning a pretty good fiber from terephtalic acid, di-methyl-terephtalate and hydrazide. Later, one was able to reach a tenacity of about 16-18 g/denier and more than 400 g/denier modulus (Bach, 1978, 1980). These fibers were evaluated as tire cords (Bach et al., 1979). To improve their fatigue resistance, Bach decided to replace para linkages by meta linkages. In this way, he was able to make fibers of higher tenacity and modulus (Bach et al., 1981). These good properties could only be obtained when very dilute solutions were used. Commercial production was therefore not feasible. Finally the work on tire cord was stopped in 1975. Once again a crisis had hit the fiber industry. "**Probably no one in Monsanto's Saint Louis headquarters does ever want to hear of a high modulus fiber project again**" (Preston, 1989).

The VNIIV and VNIISV Laboratories

The NPO Khimvolokno in Mytishchi (near Moscow) and the VNIISV, in Kalinin, are the federal institutes for R&D on fibers in the Soviet Union. They are subordinate to the Fibers Section of the Soviet Ministry of Chemical Industry (MCI). MCI is the responsible authority for all chemical production (except fertilizers, drugs and crude oil products) in the Soviet Union. It issues orders for production, construction, R&D etc. to all facilities working in this area.

In 1975, the NPO Khimvolokno was formed by a merger between the R&D institute VNIIV[9] and the project planning & engineering institute for the fiber industry, Giproiv. The VNIIV was the oldest institute for fiber R&D in the Soviet Union. In the fifties, it developed processes to produce the new synthetic fibers that were developed in the West before. Also

---

[8] Sekiguchi/Sadamitsu, 1977, describes the polycondensation of this polymers using terephtalic acid, its di-methyl-ester and hydrazine sulfate.

[9] VNIIV: Vsesoyuznyi Nauchno-issledovatel'skii Institut Iskusstvennogo Volokna, the All Union Research Institute for Artificial Fibers.

some original work was done like the development of 'Enant' (nylon 7) fiber. However, the R&D was clearly lagging behind US and Western European corporations in those years.

In May 1958, a special Plenum of the Central Committee of the Communist Party was organized on the acceleration of the expansion of the chemical industry and the production of synthetic materials. Kruschev announced that the chemical industry would become most important for the development of the Soviet economy. This became known as the 'chemicalisation drive'. Technology would be a key factor for its success. The 22nd Congress of the Party confirmed this policy in October 1961 (Denisova, 1982).

As a result of this new policy, fiber production and fiber R&D had to be enlarged. The VNIIV could not expand at its site. Therefore, it was decided to establish a new research institute in Kalinin, 170 kilometers north of Moscow. The new institute, VNIISV[10], would deal with the new synthetic fibers, the VNIIV would study production of rayon and special fibers. The VNIISV grew very rapidly from about 100 employees at the end of 1959 to 2000 employees in 1970.

The fiber plants could propose R&D projects to be carried out by VNIIV and VNIISV. However, the production plants did not show much interest in R&D, mainly because it was virtually impossible to transform R&D results into new production technologies. As a result of the chemicalisation drive, the institutes were growing so fast that there were often not enough R&D project proposals at MCI. Therefore, the scientists at the institutes very often suggested the R&D orders to the planners at MCI themselves.

Western contacts were virtually absent. In fact the only information on Western developments, that reached the scientists, were Russian translations of foreign patents. Since almost none of the scientists understood any foreign language, much Western publications that were not translated, were completely unknown.[11] In the seventies, more contacts with Western scientists were established by regularly organizing an international symposium.

Although the VNIIV and VNIISV were catching up with the West technologically, the Soviet fiber industry was far lagging behind. In the sixties and seventies, rayon fibers were by far the most important artificial fibers in the Soviet Union. New synthetic fiber plants were often based on Western technology and constructed by Western firms. An important reason for this was that high quality equipment was often only made by plants under military control and therefore not available for civil industry and civil R&D. Equipment therefore often had to be imported from Western countries.

Fenilon

In the Soviet Union, research on aromatic polyamides was started by Prof.L.B.Sokolov. Sokolov worked in Vladimir, 200 kilometers East of Moscow, at the Institute of Synthetic Resins[12], and was professor at the local Polytechnic. In 1958, he carried out his first experiments using interfacial polycondensation (Sokolov/Gudakova, 1959). He continued this work during the sixties and seventies and became the most respected Soviet expert on low temperature polycondensation. E.P.Krasnov, deputy director of the VNIISV, had been a

---

[10] Vsesoyuznyi Nauchno-issledovatel'skii Institut Iskusstvennogo Sintetitseskogo Volokna, the All Union Institute for Scientific Research of Synthetic Fibers.

[11] e.g. Morgan's book of 1965 was not translated and therefore unknown.

[12] Nauchno-Issledovatel'skii Institut Sinteticheskikh Smol, Scientific Research Institute for Synthetic Resins, under MCI authority.

student of Sokolov. Sokolov proposed to Krasnov to consider the possibilities of developing fibers based on the aromatic polyamides that he had made[13]. Sokolov had given these polyamides the name Fenilon.

Du Pont R&D on aromatic polyamides (Nomex) became gradually known at the VNIISV. This raised the interests of the scientists. The military authorities showed some interests in a fiber with increased thermal resistance too. The VNIISV scientists suggested to MCI to order a R&D project for aromatic polyamide fibers which it did. In this way the project showed up at the VNIISV and not at the VNIIV where it probably belonged more.

At the VNIISV, two groups started working on the project: A group headed by Prof.A.S.Chegolya studied the synthesis of the polymer, another group studied fiber spinning. In 1964/65, the scientists at the VNIISV decided to try to develop a polymerization process for MPD-I without the use of salts in the solution. Salts caused corrosion and affected the quality of the solvent. Since corrosion-free equipment was very hard to get for civil industry in the Soviet Union, prevention of corrosion was an important aim for R&D. Moreover, if a salt-free process could be developed, one would not have to infringe on Du Pont patents.[14] This attempt succeeded and, although the new process was not advantageous in every respect, it was developed further. Since filing for foreign patents was too expensive, patents were only applied for in the Soviet Union, and most of them were not published[15].

In 1966, the VNIISV was 'ordered' by MCI to build a pilot plant for Fenilon fiber. It was constructed at the Kalinin site. In 1969, it came on stream. The new polymerization process was applied in the pilot plant.

Sulfon T

After the R&D on synthetic fibers had been transferred to Kalinin, scientists at the VNIIV in Mytishchi were mainly engaged in R&D on the various rayon spinning processes. This gradually caused some discomfort amongst the scientists. Various new aromatic and heterocyclic polymers were synthesized at several institutes and the VNIIV scientists wanted to get involved in these developments.

In 1964, a group of about 10 scientists, headed by Prof.G.I.Kudryavtsev started R&D on spinning fibers based on these polymers. They did not concern much about the application of these fibers. One knew that some ministries on military production were interested. Since the project on aromatic polyamide fibers in Kalinin was well on its way, the VNIIV scientists chose to study especially polymers containing combinations of aromatic and heterocyclic units.

The polymers that were obtained from other research institutes were often insoluble and could therefore not be spun. This caused the VNIIV scientists to start making polymers themselves. One tried to improve the solubility of the polymers by inserting specific functional groups (Kudryavtsev et al., 1968). Thousands of polymers were made[16] and spun to fibers using dozens of solvents. On many occasions, one cooperated with specialists of

---

[13] One of these polymers was MPD-I that was used by Du Pont to make HT-1 fiber, the later Nomex fiber.
[14] If a foreign corporation refuses to licence Soviet corporations and the product cannot be purchased by the Soviet Union, then Soviet corporations have the right to copy foreign patents.
[15] One patent was published: (Gribov et al., 1968)
[16] The main ones were aromatic polyamides, polyimides, polyoxadiazoles, polyphenylene--triazoles, and polybenzimidazoles.

institutes of the Academy of Science to synthesize the polymers or the intermediates. From 1966 onward, one started further developing a fiber based on an aromatic polymer which contained a sulfon group to enhance solubility.[17] The fiber was called Sulfon T (Shchetinin/Kudryavtsev, 1968, Shchetinin, 1969). The properties of this fiber were almost identical to that of Fenilon which was developed by the VNIISV in Kalinin[18].

The research group of Kudryavtsev grew in these years to become a separate department of the VNIIV. It developed even more heat resistant fibers. Of these fibers, 'Lola' became best known. It was based on a ladderpolymer and had a very good thermal and chemical stability. When specially treated it could reach an LOI of 70 (VNIIV, 1975). Lola was used to make the suits for the Russian as well as the American crew of the combined Apollo-Saliut space project of 1974. The VNIIV is still proud of this achievement. The suits are exhibited at the VDNCha[19]. Lola is probably still not produced in large quantities. A polyimide fiber by the name of Arimid, that was also developed by Kudryavtsev's department, was also not produced in large quantities.

Only Oksalon fiber was made in larger quantities. This aromatic poly-oxadiazole fiber was also developed at the end of the sixties. It was later patented in the West (Volokhina et al., 1975). Its properties were not as good as those of the other heat resistant fibers but it was cheaper. Larger quantities of Oksalon were made at the NPO Khimvolokno branch facility at Svetlogorsk since the end of the seventies.

The Fenilon-Sulfon struggle

VNIIV and VNIISV were both working on heat resistant fibers and each had picked out a fiber, Sulfon T and Fenilon, to be further developed for large scale production. The properties of these fibers were almost identical. However, it was clear that MCI could not dispose of enough resources to bring both products to production on an industrial scale: commercialization of newly developed products was difficult anyway in the Soviet Union. In the beginning of the seventies MCI choose to continue the development of Fenilon and terminated the Sulfon T project.

The main advantage of Fenilon was that the intermediates could probably be manufactured easier than those of Sulfon T. However, the good relations of the VNIISV management were also an important factor. Prof.A.S.Chegolya, which had lead the R&D group for synthesis of the Fenilon polymer, had become director of VNIISV in the beginning of the seventies. His predecessor as a director, V.P.Yunitskii was promoted to MCI. In these functions both man lobbied much to stimulate the development of applications for Fenilon and to get the industrial production of Fenilon, its intermediates and its applications into the economic plans for industrial production. At the end of the seventies, Chegolya was also promoted to MCI. The VNIIV did not have political relations to compensate.

Further development of Fenilon

The R&D for end products based on Fenilon fiber was not carried out by the VNIISV. In the planning system, this R&D was the responsibility of the branches of industry that would manufacture the end-products like e.g. the knitwear industry and the aircraft industry. The

---

[17] The polymer was para-di-phenyl-sulfon polymerized with terephtalic acid.
[18] The LOI value was a little better, mechanical properties some worse. (Budnitskii, 1981)
[19] The National Exhibition of the Achievements of the Soviet Economy in Moscow.

Central R&D institute for staple fibers in Kalinin[20] played an important role in developing textile Fenilon applications. The research institute of the paper industry in Leningrad carried out R&D on the production Fenilon paper for high voltage insulation[21].

Further R&D at the VNIISV concentrated on achieving an even higher heat resistance. Fibers based on Fenilon copolymers were studied which contained varying fractions of meta and para linkages. The copolymers were often made by Sokolov and his assistants in Vladimir. One of the intermediates for Fenilon, isophtaloyl-di-chloride, was not available in large quantities in the Soviet Union. Therefore a production process had to be developed. This task was assigned to the state research and design institute for the chlorine industry in Moscow[22]. This institute also developed a process for the production of terephtaloyl-di-chloride which was needed for SVM and Terlon fibers, which will be dealt with in the next paragraph. In the meantime, the VNIISV bought the IDC that was needed for the pilot plant, in the West.

The contacts of VNIISV director Chegolya within the party and MCI, and the support of former director Yunitskii, were of great importance to get industrial scale Fenilon production into the plans of the various ministries. Decisive for the question whether to built a full scale industrial plant for Fenilon was of ban on Nomex exports by president Reagan. The Fenilon plant was built in Kustanay, Kazachstan. As one expected environmental problems by the emission of di-methyl-acetamide, this remote town was chosen. The environmental problem had become one of the main tasks of the ecological research group that was established at the VNIISV in 1975. Industrial scale Fenilon production started in 1985. Fenilon insulation paper production started in the city of Malena (Shetomir district).

A HS/HM fiber

In 1969, the group of Kudryavtsev at the VNIIV came up with an HS/HM fiber. The high strength and modulus were found by surprise. The polymer that was used had both aromatic and heterocyclic rings[23]. The fiber was called Vniivlon.

The intermediates for the polymer were synthesized by the Lensoviet Technologic Institute in Leningrad and Sokolov's institute in Vladimir. In 1971 a statement on the properties of the fiber was published in Khimicheskie Volokna, the Soviet Journal on synthetic fiber research. The synthetic fiber had unequalled mechanical properties, some of which were released: Its tenacity was 2-2.5 GPa and its modulus 110-120 GPa. Vniivlon could be applied in cables, tire cord, nets and composites (VNIIV, 1971).

The VNIIV management considered its fiber as a major achievement. One estimated that Vniivlon could probably be exported. Therefore, one applied for patents in Western countries (e.g. Tokarev et al., 1974). Usually, one did not apply for Western patents because these required too much hard currency. Some Western corporations objected to these applications. This forced the VNIIV to spend even more hard currency.

[20] Tsentral'nyi Nauchno-Issledovatel'skii Institut Shveinoi Promysel, Central scientific research institute for the sewing industry.

[21] Vsesoyuznyi Nauchno-Issledovatel'skii Institut tsellyulozno-bumazhnoi Promyshlennosti, (VNIITsP).

[22] GOSNIIchlorproekt.

[23] The polymer was based on terephtalic acid and 5,4-di-amino-2-phenyl-benzimidazole.

After the patent applications had been filed, the military Ministry of aviation industry (Minaviaprom) showed interests in the application of Vniivlon in composites. Therefore, export abroad became undesirable. The VNIIV decided to drop the patent case and to keep further results of R&D on Vniivlon secret. Minaviaprom took over control of this part of the VNIIV R&D. The project was 'closed', i.e. no results were allowed to be published. Research orders for Kudryavtsev and his group were no longer issued by MCI but by Minaviaprom. There had been only one scientific publication in which the polymer was described on which Vniivlon was based (Kudryavtsev et al., 1970). Only some fundamental scientific articles were published on this research. Some articles on liquid crystal polymers were authored by the remarkable S.P.Papkov[24], in age almost the Soviet equivalent of Herman Mark.

The properties of Vniivlon could be improved in the beginning of the seventies. Therefore one decided in 1974 to change the name of the fiber to SVM[25] (Kudryavtsev et al., 1974). The intermediates for SVM were rather difficult to produce and therefore expensive. Since R&D groups working under Minaviaprom had access to various facilities, and the price of the fiber did not matter, this problem was not important.

Kudryavtsev and his department also carried out R&D on another HS/HM fiber called Terlon. This fiber was basically a copy of Du Pont's Fiber B (Kevlar). The polymer for Terlon was made by Sokolov's institute. Sokolov and his group were not under control of Minaviaprom. Therefore, they were allowed to publish. Some of their publications were of great importance.

The R&D on Terlon fiber was also carried out under responsibility of Minaviaprom. However, SVM became the primary target of the R&D efforts. It was seen as an original piece of R&D of the VNIIV to be proud off. The application of SVM in composites was especially developed further by Minaviaprom institutes. Tire cords and bullet proof vests were also developed. As president Bush is wearing a Kevlar vest, president Gorbachov is probably wearing SVM. Recently, an even stronger Fiber was announced: Armos. Its tenacity would be 4.5-5.2 GPa and its modulus 145-160 GPa (Avrorova et al., 1989).

By the reforms of Gorbachov, the SVM and Terlon R&D was no longer under military control. The VNIIV, now working by the name of NPO Khimvolokno, was allowed to export SVM which it produced itself on a semi commercial scale. Only now, the high production costs of the fiber have become an important barrier.

Bayer

In 1960, the Bayer fiber division established a new scientific fiber laboratory, 'das Wissenschaftliches Labor' (WL) at Dormagen. Dormagen was the main production site of the fiber division. It was 25 kilometers North of Leverkusen, Bayer's home town. In 1964, the fiber division added the 'Organisch Wissenschaftliches Labor' (OWL) to the existing laboratories in Dormagen. This laboratory studied fiber chemistry.

The Bayer fiber division only produced textile fibers. The reasons for this were historic: The main IG Farben factories which produced industrial yarn had been located in the East part of Germany. These factories had been lost for the IG descendants after the war.

---

[24] e.g. (Papkov et al, 1974)

[25] Sverchprogibks Visokomodulsyi Material: super strong, high modulus material.

Therefore, it did not seem to be very attractive for Bayer to try to develop a high performance industrial yarn.

However, as one of the worlds largest producers of dyes, Bayer had a lot of aromatic diamines at its disposal which could be used as intermediates for aramids. Therefore, the Bayer scientists chose to experiment with a huge mass of aromatic diamines to see whether they could produce a new high performance textile fiber.

In 1964, the OWL of the Fibers Department in Dormagen started studies on aromatic polyamides. A great number of polymers were made. The information about HT-1, and from 1967 its marketing as 'Nomex' by Du Pont, excited the marketing and sales group in Dormagen. Positive reactions came from fiber processing corporations. The scientist reacted rather skeptical to this enthusiasm of the marketeers:

> **"wenn anderen mit etwas neues kommen sagt die Vertrieb immer zu uns 'warum haben wir das nicht' und wenn wir mit etwas neues kommen sagen Sie 'was sollen wir damit ?'"** (Künzel, 1989).

Because of the interest of the sales group, the work on aromatic/heterocyclic polyamides was intensified in 1966. As an aim for the project was set: to create a fiber for textile use with an increased heat and flame resistance. One was hoping to find a fiber with good thermal and textile properties which was easy to manufacture. Because Bayer had no interests in industrial fibers, one did not include the industrial use of heat resistant fibers as an important aim. In 1966, the patent office in Leverkusen carried out a extensive literature search, which reinforced the decision not to aim at polymers to be used in industrial fibers for it was estimated that there were to many patents already on the most interesting polymers for this aim.

The choice for textile fibers resulted in important limitations for the scientists: Textile fibers had to be comparatively cheap. Therefore, the manufacturing process had to be simple. This meant that one had to be able to spin a fiber from the solution in which the polymerization had taken place. Moreover, expensive salts could not be used in the polymerization reaction.

In the period till 1968, four scientists and their assistants at the OWL synthesized about 600 polymers. As intermediates, they could use various diamines which were made by Bayer's dyestuffs department. The polymers were spun by the WL. The total number of people working on the project was about 20 to 25, 4 or 5 of them were scientists. Because this work was rather explorative, the patent office covered all of the results with patents.

Some fibers seemed to be some more interesting. One evaluated their properties further. This further evaluation was done by ATA Textil[26] which was part of the sales group. At the end of 1966, a potentially very interesting fiber was made.[27] The most interesting property of this fiber was its moisture regain; it was almost cotton like. It had two important advantages compared to polyester: it could be white washed and it did feel 'cotton-like'. The flame resistance of the fiber was extremely good, which made it a good material for children's pajamas etc. However, the fiber could not be spun directly from the polymerization solution. Since this last point was a clear disadvantage, exploratory research was continued for a while.

---

[26] Anwendungs Technische Abteilung Textil, Department for textile application technology.

[27] One made a polymer of 3-para-aminophenyl-7-chinazolin-2,4-dion and isophtaloyl-di-chloride. (von Falkai, 1974)

The sales-group of Bayer was very positive on the fiber. It estimated that this fiber could replace polyester textile fiber in some applications when the price would be about 20 % above the polyester fiber price. Moreover it could also compete with Du Pont's Nomex. Therefore, it was decided to develop this specific fiber further and approach customers to evaluate the properties. It got the name ATF 2000.[28] In the workshop of the WL, one made some hundred kilograms ATF 2000 in 1970/71. ATA Textil was carrying out the application research of the fiber. Besides as a protective garment, ATF 2000 could be used as a fiber for industrial filter cloth for hot gases. The fiber was spun by several methods. At last one selected dry spinning because in this way one could use di-methyl-formamide as a solvent, which was also used for Dralon acrylic fiber production in Dormagen.

In its textile application there were some important problems left. The fiber was hard to dye, which was no problem in industrial applications, but was a major problem for textile. By copolymerization Dr.G.D.Wolf succeeded in 1969 to improve the dye-ability of the fiber (Wolf, 1971). One succeeded even in making white ATF 2000 fabrics. Laboratory personnel tested the garment itself. One was unanimous on the great comfort of it. Textile manufacturers which were approached were very positive on the fiber.

The end of ATF 2000

Of major importance for the project was the price of the intermediates. The price of intermediates would have to come down to a level that would make it possible to use the fiber in high quality textile. Since the diamine was the main intermediate of the polymer, its price would be of key importance. The Hauptlabor in Leverkusen studied the synthesis. It succeeded in bringing the estimated industrial scale production costs of the diamine back from DM 50 per kilogram in 1967 to DM 12 per kilogram at the end of 1969. However, in 1970, the price of the polyester fiber, with which ATF 2000 would have to compete as a textile fiber, had decreased considerable. Since the polyester fiber price dropped that much, it became no longer realistic to try to catch up with this price level. However, one could try to set up a much smaller production in order to market the fiber for protective garment only. Since Nomex's textile properties were rather bad, one could market it at a somewhat higher price.

In 1969, the production costs of ATF 2000 were estimated at DM 40-50 per kilogram which could probably be reduced. The price of Nomex was at time about $ 11 per kilogram but was reduced to about $ 8 per kilogram. It was gradually further reduced to $ 4.5-5.5 in 1975. The pursuit race of the Bayer scientists to catch up with the Nomex price level, became after some initial successes, gradually pointless. Not only was Nomex getting cheaper, the DM/$ exchange rate changed in an unfavorable way for ATF 2000. Moreover, fiber business was not as profitable as it had been in the sixties. Initially, decisions on further development of ATF 2000 were postponed. In the beginning of 1973 work on ATF 2000 was terminated. When a new crisis hit the fiber industry, the Dormagen laboratories were shut down in 1977.

A HS/HM fiber

In 1969, a serendipitous event caused the scientists of the OWL to start, besides heat resistant fibers, a second line of development: When they were looking for polymers with a-symmetric structures to see if one could find polymers with enhanced solubility, a polymer was found which formed a rather strong fiber. This was most remarkable because the polymer chain was not all para but contained some coils. Its straight equivalent, which was also

[28] ATF means Anwendungs Technische Forschung, Applied Technical Research.

synthesized, was far less strong. The scientists had no explanation for this fact. However, it appeared to be very interesting for further elaboration.

Bayer did not sell industrial fibers like tire cords. However, since strength became a property of increasing importance in the fiber business, and it seemed that Bayer scientists had made a unique discovery, the scientists were able to convince the marketing and sales group and the fiber division management that this was an important development. The R&D on this fiber was therefore continued and some batches were made. Bayer applied for patent protection in 1971 (Künzel et al., 1973). Especially the reinforcement of plastics (composites) seemed interesting. The corporate department Anwendungtechnik Neue Gebiete[29] in Leverkusen started developing composites using this fiber. However, soon afterwards, when Du Pont announced semi commercial production of Fiber B in the beginning of 1972, the marketing and sales group lost interest. The fiber was not as unique as one had believed initially. Moreover, unlike Du Pont, Bayer did not have relations in the market of industrial fibers.[30]

ICI

The ICI Fibers department has its headquarters together with its research facilities in Harrogate, North England. Its main product were polyester fibers. In 1964, ICI acquired the Courtaulds stock in British Nylon Spinners (BNS) by which this joint venture became a 100 % owned ICI subsidiary (Coleman, 1980, Knight, 1974). ICI also acquired the BNS laboratory at Pontypool. This laboratory initially remained the place for polyamide fiber research. Polyester and polypropylene fiber research were carried out in Harrogate. At the end of the sixties there were a couple of hundred scientists working in each laboratory.

In the early seventies, the ICI Fibers Division decided that the two laboratories should be integrated. This could facilitate and stimulate communication. The divisional management chose Harrogate as the preferred site for it could also facilitate communication with other divisional departments. The Pontypool laboratory was closed down between 1973 and 1975. The scientists were gradually transferred to Harrogate.

All of the fibers that ICI produced, were melt-spun and therefore the ICI scientists were unexperienced in other spinning technologies. The end uses of ICI's fibers were often industrial: polyamide and polyester were often used as tire cords and polypropylene in cables and ropes.

a4

One heard of Du Pont's Fiber B in 1969. Everybody was extremely interested since this was a major new development which could affect ICI's position in the industrial fibers market. By 'friendly relations in the rubber industry' one found out that Fiber B was based on PBA. A report was made on the probable production costs of a PBA fiber.

A research project by the name A4 was initiated which was carried out at Pontypool (since it was a polyamide). Soon, one realized that it was unthinkable to reach the properties, that Du Pont had reached, by the technology that was preferred by ICI, melt spinning.

---

[29] Application Technology New Areas.

[30] Bayer's concept did not lead to an inferior product: It was later successfully elaborated by the Japanese corporation Teijin, which started commercial production of Technora aramid fiber in the eighties. See: (S.Ozawa, 1987)

Therefore, one had to start using wet spinning procedures. Quite a lot of resources were diverted to the A4 project. Geoffrey J.Tyler and his section of about 20 people were working on it. Duplicating Du Pont's results was not easy. In part one tried to see if a HS/HM fiber could be made without infringing on Du Pont patents: So there were efforts aimed at making various aromatic copolyamides. After the scientists found out that the improved Fiber B of 1971 was based on PPD-T, the major part of the work done was aimed at this polymer.

a5

At ICI, another R&D project was carried out which was later seen as having also potential to result in a HS/HM fiber:

In 1969, Donald Brydon had started studying aromatic polyoxadiazoles. He had become very interested in this chemistry when he stayed in 67/68 with Prof.Olah at the Case-Western Reserve University in Cleveland (Ohio). Back at ICI, Brydon studied these polymers more or less just by himself. He achieved some successes and a project called A5, started in Pontypool. This project initially aimed at heat resistance. The project grew when Brydon developed an important improvement: In the beginning of the seventies, it became clear that bromination of the polymer resulted in a self extinguishing material (ICI, 1973, 1975).

The A5 project reached a second important breakthrough: First the aromatic polyoxadiazoles could not be hot drawn to improve their mechanical properties. However,

> **"Don had given Norman Wolfendale a set of brominated fibers and Don had said: 'just keep going up with the temperature and at each temperature, try to achieve the highest draw ratio.' Norman, as a very quiet and systematic chap, suddenly reached a draw ratio of 15 with a certain fiber. That was a crucial point in the whole story, because the properties jumped by a factor of 2. At that moment it became interesting."** (Tomka, 1990).

The patent application mentioned a draw ratio of 18 which resulted in a strength of 9.75 g/dtex and a modulus of 240 g/dtex (ICI, 1976). It was hoped that this could be doubled again when a proper pilot plant was used. The fiber might have a dual purpose as HS/HM fiber and as a heat resistant/flame retardant fiber (McIntyre/Tomka, 1990).

The divisional patent department in Harrogate made an assessment on an ongoing basis of the patent situation. From the early patents, it looked as if there could be scope for A4. When Du Pont's Blades patents were published, the situation became worse. In 1973/74, there were contacts between ICI and Du Pont executives about licensing Du Pont's HS/HM fiber process. The message that came back to the people working on A4 was that Du Pont was not interested in licensing others.

ICI continued its work on A4 until late 1974. The assessment of the patent department was at that time that, in spite of the novel things that had been worked, it would be extremely difficult to produce a competitive fiber without infringing on Du Pont patents. Since A5 could perhaps also produce a HS/HM fiber, the laboratory management decided to stop A4 and divert the means A5.

Fibers were produced on a semi technical scale. There was a peculiar thing in these years on the development of aromatic polyoxadiazoles. Inventa, a Swiss research corporation, acquired a patent by which they claimed to make very good fibers with the unhalogenated aromatic polyoxadiazole homopolymer (Inventa, 1975). However, the ICI scientists were never able to reproduce that result.

One of the main problems of the A5 project was what to do with the sulphuric acid which was produced by the process. Since one could only use a dope strength of 7-8 %, (as compared to about 20 % for Kevlar) the amount of sulphuric acid produced was enormous. Some research was aimed at achieving the maximum dope strength (ICI, 1980). However, there were sharp limits to it. One could not think of a better solution than turning it into gypsum and dump it in a mine.

> **'Some people think this sulphuric acid problem is trivial, certainly in the lab stage, but when it comes to large scale manufacture it becomes a very important factor'** (McIntyre, 1990).

A5 was a group of products, a very small modification would result in a fiber for the high combustion segment. That made it attractive, more sectors could be covered with very little change in technology. One member of the A5 family had excellent in tire stability, it would beat aramids because there were no hydrolyzable bonds in that polymer. A5 did not reach the pilot plant stage, so application research did not become really involved.

The project went on until the summer of 1976. At that time, the technical/commercial evaluation was not sufficiently promising to continue the project. It was a high modulus and high tenacity product but not in the Kevlar range. The cost calculations showed that the fiber would not be cheaper than Kevlar. The bromine made the fiber more expensive and introduced problems of corrosion. The total financial position of the Fiber Division, which was considerably affected by the market crisis, influenced that decision as well. Over the years 1973-1979 the average trading margin of the Fibers Division was - 1.2 % (Pettigrew, 1985).

> **'The company had reached a stage where the Fibers Division would have great difficulties persuading the main ICI board to put a large amount of capital into the project'** (McIntyre, 1990).

Moreover, much people started to lose faith in another break through which was necessary. A5 had to be stopped or had to be moved into a different gear, a decent pilot plant.

In 1976, after a two hour meeting with the project staff, the management decided to stop the project. Scientists were very demoralized after the negative decision on the project. A 'wake' was organized for A5. In a market village not far from Harrogate there were about 12 pubs around the marketplace. The participants were to visit these pubs and drink beer from the jars in which the spinning dope had been prepared. Because it looked to be still useful, not much of the ICI research on this subject has ever been published. Only Brydon published some articles on the chemistry of substances containing 1,3,4-oxadiazole units (Brydon, 1980, 1984, 1989).

AKZO

In 1969, the large European fiber conglomerate AKU[31] took part in a merger by which a new corporation was established: AKZO. The fibers department of AKZO became the largest European fiber producer. Its position was especially strong in the field of industrial fibers. The AKU headquarters at Arnhem (Netherlands) became the new corporate headquarters. The fiber scientists at Arnhem became the neighbors of the new corporate board.

---

[31] AKU: Algemeene Kunstzijde Unie, General Artificial Silk Union. This corporation operated often under the names of its affiliates: Enka, Glanzstoff, Italo-Ollandese, La Seda, American Enka, British Enkalon etc.

In the sixties, AKU had decided on a strategy which would allow them to supply every major industrial fiber to their customers. Besides viscose, nylon and polyester, AKU had started to manufacture glass fiber and steel wire and it had acquired polyacryl fiber plants. As many other fiber manufacturers, AKZO expected at the end of the sixties that the market share of rayon and nylon fibers would substantially decline in the near future (See e.g. Van Krevelen, 1967). Therefore, AKU had established a joint venture with General Electric to develop fibers based on the phenylene oxide polymers that were made by the oxidative coupling method developed by GE. This project was terminated at the end of 1969 for no good fibers could made using these polymers.

The Fiber B announcement

In the beginning of 1970, AKZO scientists noticed the rumors on Du Pont's Fiber B. The properties that were published, seemed very interesting for an industrial fiber. Dr.Leo Vollbracht, who was assigned to this subject, concluded from published patents that Du Pont probably would have succeeded in making a fiber based on a wholly para aromatic polyamide. AKU had considered developing aromatic polyamides in the sixties, but had rejected it. Vollbracht picked out PPD-T as the most promising polymer, based on the availability of intermediates. In December 1970, he succeeded in making PPD-T of good quality.

The laboratory management was aware that one was far behind Du Pont on this development. Therefore, Vollbracht was urged not to go on any sideways but duplicate Du Pont's results first. In January 1971, salesmen of the fibers department laid hands on a Fiber B sample. The polymer turned out to be PBA. For this reason, Vollbracht had to study PBA instead. His argument against PBA, the high cost of the intermediate, did not convince the laboratory management: costs could not be that high, otherwise Du Pont would not have chosen this polymer. In the summer of 1971, AKZO scientists noted the publication which mentioned improved Fiber B (Lee et al., 1971, n43). In the autumn of 1971, a representative of the AKZO Fibers Department reported that the improved Fiber B, that Du Pont had shown at the tire symposium in Akron, was yellow. Vollbracht concluded that Du Pont was developing a PPD-T fiber, because PBA was not yellow. As a result, he returned to PPD-T. Shortly afterwards, AKZO salesmen brought in Du Pont samples which confirmed this. The AKZO patent office reported that a PPD-T fiber could be made without insurmountable patent objections. When Du Pont announced that it would start semi-commercial production of Fiber B, the support for Vollbracht's work was considerable reinforced, for it was estimated that if Du Pont would commercialize the fiber, it would be more than a laboratory curiosity. As a result, more scientists were assigned to the project.

There were some contacts between AKZO and Du Pont at rather high level. Du Pont's Mitchell visited Arnhem on which occasion he showed reports on Fiber B's properties. In Arnhem, one was not able to reach the same results. Despite the knowledge that was disclosed in patents, it was not easy to make a fiber as good as Du Pont's Fiber B. Only after the patent applications of Blades were disclosed at the end of 1972, AKZO scientists succeeded in reaching the same properties. When the scientists finally succeeded, the fiber was shown to corporate executives when they visited the laboratory by hanging a large pile of water at a very thin thread. This fiber was further developed as Fiber X.

The Fiber Department board tried to speed up the researchers as much as possible in order not to loose too much ground to Du Pont. A project coordinator, Weyland, was appointed who started to make a business plan for the production of Fiber X. Based on fiber properties and the number of plies needed in the belt of tires, he supposed that Fiber X could be used very well to reinforce tires for light weight trucks or heavy passenger cars. Since the steel reinforced radial had conquered Western Europe already, he did not aim at the large market of commodity passenger car tires.

In February 1971, the AKZO patent office had reported that Du Pont had no patents on the PBA fiber which could stop competitors from manufacturing the same product. Later they also cleared the situation for a PPD-T fiber. However, could AKZO also get around the Blades patents ? After a very intensive study, the corporate patent office reported that they were willing to take the responsibility of eventually fighting these patents in court. However, it was possible that Du Pont would sell a licence. Anyway, every improvement which AKZO scientists could make on Du Pont's process would be very important to give the corporation a stronger position if it had to negotiate with Du Pont.

Process development

At the end of 1973, Vollbracht started looking for an alternative solvent to make the PPD-T polymer. The main disadvantages of the HMPT/NMP[32] solvent mixture, that was also used by Du Pont, were that HMPT was only manufactured at a very limited scale in Europe, that it was subject to degradation during the process, and that it seemed hard to recover after it had been used. Moreover, if an alternative solvent was found, it could reinforce AKZO's position vis-à-vis Du Pont's patents. By a trick, Vollbracht was able to test an enormous amount of solvent/salt combinations. He found that NMP/ $CaCl_2$[33] could be used as well to make PPD-T (Vollbracht/Veerman, 1975). In the meantime, the process research group had designed a pilot plant which was based on the use of HMPT/NMP. The process design had to be done all over again if the solvent was changed. They saw no need for a change of design. Therefore the alternative was not used in the design of a pilot plant. The construction of the pilot plants started at the end of 1974. In April 1975, one publicly announced that the fiber would be further developed as 'Arenka'.

In 1973, another process improvement was found. It was very hard to make a good spinning dope of PPD-T and sulphuric acid. AKZO's Pharmaceuticals Laboratory advised that substances mix better in the solid state. Therefore, one tried to mix frozen sulphuric acid with PPD-T and then heat the mixture again. This gave very good results.

Crisis

The crisis of the (European) fiber industry which started in 1974, struck AKZO severely. It was said that the AKZO Fibers Department lost Fl 1 million a day. In 1975, AKZO lost Fl 440 million[34]. In September 1975, drastic measurements had to be taken. There were mass redundancies and the corporate board forbade all employees to order any goods from outside suppliers. In the following years there were again massive redundancies. At the end of the seventies, AKZO's financial position was very weak. For several projects, the corporation tried to get support from Dutch or German government.

As a result of the financial crisis of AKZO, the Arenka project was cancelled in September 1975. The engineers of AKZO's Engineering Department terminated their work on the pilot plants. However, the scientists were not assigned to other projects and therefore they had nothing else to do than to continue their work on the Arenka process.

---

[32] Hexa-methyl-phosphor-tri-amide/N-methyl-pyrolidon.
[33] N-methyl-pyrolidon with at least 5 % Calcium-chloride.
[34] 1 Fl was by then about $ 0.35.

Just a few days after the AKZO board had banned external ordering, Du Pont's Fiber Department executive Mitchell called AKZO to tell that the HMPT solvent was carcinogenic. Since further construction of the polymerization pilot plant was hold up anyhow, the process research group could redesign it to the use of the alternative solvent NMP/$CaCl_2$. The construction of the pilot plant was continued 'unofficially' since the project was cancelled. Some executives knew that one continued the project in the laboratory, but the Fibers Division board as well as the AKZO board were not officially informed. Scientists and technicians had to use rather uncommon means sometimes to continue the construction of the pilot plant.

In 1976, the ban on external ordering was lifted. The corporate executives were divided on the matter of continuing Arenka. A compromise was reached: Arenka was allowed to continue but only if the Fibers Division succeeded to get considerable government support for the project. In December 1976, the Arenka pilot plants came on stream. Pressure came from marketeers to start manufacturing fibers to be able to start developing end products. Moreover, sales of fiber would turn in some money to finance the pilot plants.

Patent struggle

By the patent on the alternative solvent, AKZO's patent office thought to have a good position to bargain with Du Pont. Until April 1977, there were discussions between the corporations but no compromise was reached. After May 1977, Du Pont started to oppose to AKZO's patent applications on the alternative solvent. After eight rejections and two refilings, a US patent was issued to AKZO in December 1981. This decision was annulled some years later (US Court for the Federal Circuit, 1987). In other countries, both corporations also opposed to one another's patent applications.

In 1978 AKZO started preparations for building full scale aramid polymer and spinning plants. At this point the struggle became harder. AKZO's financial situation was still not very good and therefore the company applied for support from various government agencies. Du Pont warned the Dutch Ministry of Economic Affairs that they would lose loans given to AKZO. The Ministry ordered an independent study on the patent situation. Although difficulties for AKZO could be foreseen, it was estimated that Du Pont could not legally stop AKZO from producing aramid fibers.

In November 1979, Du Pont filed complaint at a London court. Shortly afterwards it also filed complaints in France. AKZO reacted by contesting Du Pont's patents in a US court. In the following years, the patent struggle spread all over the world. In almost every industrialized country AKZO and Du Pont fought each others patents. There were victories for both. Unsurprisingly they both won the legal battle in their home country; In 1986 AKZO's aramid fiber was banned from the US market by decision of the International Trading Commission. The conflict was becoming more and more a US-European trade conflict: The European Commission contested the procedures used by the ITC at the GATT. President Reagan send over his special representative Clayton Yeutter to Europe to calm down the political turmoil that followed the ITC decision.

In June 1987 AKZO's aramid fiber, which was renamed 'Twaron', was banned from the UK, but the ban was lifted in April 1988. Du Pont's Kevlar was banned by court decision from the Netherlands and Germany. In April 1988, AKZO had a major legal victory in the UK: the ban on Twaron was lifted and Kevlar was banned instead. Du Pont was just finishing the construction of a Kevlar plant in Northern Ireland. Therefore, Du Pont was forced to sign a compromise.

Others

Japan

The main aramid fiber research in Japan was carried out by Teijin. The history of their research could not be studied. Therefore, I will only present some facts that have been published.

Teijin was a comparatively small corporation with only 14000 employees in 1970 (Reuben/Burstall, 1973). In October 1969, Teijin launched an aromatic polyamide fiber, called 'Conex' (C&EN, 13-10-69, 20-10-69). In 1972, the Conex production capacity was 14 tons per month. It appeared to be virtually the same as Nomex. However, it was manufactured somewhat different. Conex was wet spun from Di-methyl-formamide (Mohajer/Ferguson, 1976). In the beginning of the seventies, Teijin studied the production of a HS/HM fiber based on PPD-T. Later, one started to work on the method found by Bayer scientists to make HS/HM fibers (Künzel et al. 1973). The fiber that was developed by Teijin was initially named HM 50. The monomers that were used were besides PPD and TDC also 3,4'-di-amino-di-phenyl-ether. In 1981, Teijin's Kiyoshi Kazama announced that the tenacity was 26 g/denier and elongation at break 4.2 % (CE, 2-11-81). The manufacture of the fiber was less complicated than the manufacture of Kevlar, but its intermediates were more expensive. HM 50 was for some period withdrawn from the market to make some improvements, especially to reduce interfilament adhesion. In the summer of 1987, Teijin opened a 500 tons per year production unit. The trademark became Technora (NRC-Handelsblad, 6-9-88).

In the eighties, the Japanese corporation Unitika started manufacturing a Nomex like product which it called Apyeil. Together with Mitsubishi Paper, Unitika produced Apyeil insulation paper. Total production reached 200 tons in 1987 (Nuttall et al, 1987).

Asahi had also invested much in aramid research. At least halfway the seventies it was carrying out serious efforts. Like Teijin, Asahi was planning to introduce a fiber to the Japanese market at the end of the eighties. Asahi worked on a PPD-T aramid fiber (Nuttall et al., 1987).

Europe

Rhone Poulenc developed a polyamide-imide fiber which was produced under the trademark Kermel. The fiber was very heat resistant but rather expensive. In 1971 it was introduced to the market (McIntyre/Mohajer, 1977). In 1974, Pigeon en Allard published the fiber properties (Pigeon/Allard, 1974). In 1990 the fiber was still produced especially for yarns to be used in protective garment. Production capacity is probably small. In the early seventies, Rhone Poulenc presumably also studied the production of a PPD-T based HS/HM fiber (Allard/Sacco, 1978). Kermel is still produced on a small scale, especially for protective garment.

The Austrian corporation Lenzing markets a heat resistant polyimide fiber by the name P84. This fiber was initially developed Upjohn under the name polyimide 2080. It was sold to Dow Chemical in 1974 (CE, 1-4-74), which sold it again in 1982. Recently, Hoechst announced that it would start pilot plant production of para-aromatic polyamide fibers at its Kelheim site in 1991 (ECN, 1-5-89). The proces is probably related to that of Teijin.

The development of high performance organic fibers has not come to an end[35] and therefore final conclusions on the success or failure of these projects are still hard to draw. However, what is revealed by the history of high performance fiber development is that the road which is thought to lead from scientific success to commercial success, often turns out to be a dead end street. Scientific achievements often did not fit the corporation in which they were reached, or just came at the wrong moment. This means that a lot of scientific accomplishments were actually wasted. An industrial scientists can be a genius, but if (s)he is not able to make colleagues and management believe that (s)he creates the answers to the corporate problems, the project will fail. By these social processes, not only good R&D can fail but also projects which are only moderately interesting, are sometimes kept going to long.

[35] see e.g. Dow's announcement of PBT fiber (ECN, 12-03-90).

LIST OF VARIOUS FIBERS AND POLYMERS MENTIONED BEFORE

AKZO	- poly-p-phenylene-terephtalamide	Fiber X	Arenka/Twaron	HS/HM
Asahi	- poly-p-phenylene-terephtalamide	?		HS/HM
Bayer	- 3-p-amino-phenyl-7-chinazolin-2,4-dion + isophtaloyl-di-chloride	ATF 2000		HR textile fiber
	- 3-p-aminophenoxy-2'-methyl-4'-amino-benzanilide + terephtaloyl-di-chloride		HS/HM	
Du Pont	- poly-m-phenylene-isophtalamide	HT-1	Nomex	HR
	- poly-p-benzamid	Fiber B		HS/HM
	- poly-p-phenylene-terephtalamide	Fiber B	Kevlar	HS/HM
Hoechst	- probably similar fiber to Teijin's Technora ?		HS/HM	
ICI	- poly-p-phenylene-terephtalamide	A4		HS/HM
	- brominated aromatic 1,3,4 oxadiazole	A5		dual use
Lenzing (Upjohn, Dow)	- benzo-phenone-tetra-carboxylic-acid-di-anhydride, toluene-di-isocyanate (20 %) and 4,4'-di-phenyl-methane-di-isocyanate (80%)	P84		HR
Monsanto	- all meta diamine of BABBAB order + terephtaloyl-di-chloride	X-101		HR
	- treated poly-m-phenylene-isophtalamide	X-400	Durette	HR
	- p-amino-benzhydrazide + tereph-taloyl-di-chloride	X-500	Fibron	HS/HM
Rhone Poulenc	- aromatic poly-imide-amide based on tri-mellit-anhydrid-chloride and di-amino-di-phenyl-methane	Kermel	HR	
Teijin	- poly-m-phenylene-isophtalamide	Conex		HR
	- copoly-terephtalamide of 3',4-di-amino-di-phenyl-ether and p-pheny-lene-diamine	HM 50	Technora HS/HM	
Unitika	- poly-m-phenylene-isophtalamide		Apyeil	HR
VNIISV	- poly-m-phenylene-isophtalamide		Fenilon	HR
VNIIV	- p-di-amino-di-phenyl-sulfon + terephtalic acid		Sulfon-T	HR
	- poly-aromatic 1,3,4 oxadiazole		Oksalon	HR
	- ladderpolymer		Lola	HR
	- 5,4-di-amino-2-phenyl-benzimid-azole + terephtaloyl-di-chloride		Vniivlon/SVM	HS/HM
	- poly-p-phenylene-terephtalamide		Terlon	HS/HM
	- ?		Armos	HS/HM

References

Some of the references refer to interviews:
*Preston, 1989, interview Dr.J.Preston, Raleigh, November 10-12, 1989.
*Tomka, 1990, interview Dr.J.G.Tomka, Leeds, 14 June, 1990.
*McIntyre, 1990, interview Prof.J.E.McIntyre, Leeds, 14 June, 1990.
*Kunzel, 1989, interview Dr.H.E.Kunzel, Dormagen, 2 May 1989.

Some news magazines are abbreviated: C&EN, Chemical & Engineering News, CE, Chemical Engineering, ECN, European Chemical News.

*C.J.**Abshire**, C.S.**Marvel**, 1961, Die Makromolekulare Chemie 44-46, pp. 388-397.
*P.**Allard**, J.**Sacco**, 1978, French Patent 2,354,356, application 1975, assigned to Rhone Poulenc Textile SA.
*L.V.**Avrorova**, A.V.**Volokhina**, V.B.**Glazunov**, G.I.**Kudryavtsev**, R.A.**Makarova**, Z.G.**Oprits**, A.V.**Tokarev**, A.S.**Semenova**, 1989, Khimicheskie Volokna, nr. 4, pp. 21-26.
*H.C.**Bach**, 1978, US Patent 4,115,503 to Monsanto, application: 26 November 1971.
*H.C.**Bach**, F.**Dobinson**, K.R.**Lea**, J.H.**Saunders**, 1979, Journal of Applied Polymer Science, 23 (2), pp. 2125-2131.
*H.C.**Bach**, 1980, US Patent 4,202,962 to Monsanto, application 26 November 1970.
*H.C.**Bach**, H.E.**Hinderer**, 1981, Polymer Preprints, 22(1), pp. 31-32.
*W.B.**Black**, J.**Preston** (edit), *High modulus wholly aromatic fibers*, New York, 1973.
*D.**Brydon**, 1980, Journal of the Chemical Society, Perkin Transactions II, pp. 773-777.
*D.**Brydon**, 1984, Journal of Chemical Research (m), pp. 3485-3497,
*D.**Brydon**, 1989, Polymer 30, pp. 619-627.
*G.A.**Budnitskii**, 1981, Chemical Fibres, (English translation of Khimicheskie Volokna), pp. 81.
*D.C.**Coleman**, 1980, *Courtaulds, An Economic and Social History* Volume III, *Crisis and Change, 1940-1965*, Oxford University Press.
*B.M.**Culbertson**, 1968, US patent 3.389.122, application 18 May 1966, assigned to Ashland Oil and Refining Co.
*M.V.**Denisova**, G.V.**Talaeva**, K.N.**Ushakova**, 1982, *Perventsy Khimitseskich Volokon, stranitsy, istorii, vseso-yuznogo nautsno-issledovatelskogo i projektnogo instituta iskusstvennogo volokna*, Moscow.
*F.**Dobinson**, C.A.**Pelezo**, 1974, US patent 3.787.348, application 20 September 1972, assigned to Monsanto.
*F.**Dobinson**, C.A.**Pelezo**, 1974, US patent 3.842.028, assigned to Monsanto.
*F.**Dobinson**, C.A.**Pelezo**, 1975, US patent 3.926.887, application 19 July 1972, assigned to Monsanto.
*F.**Dobinson**, C.A.**Pelezo**, 1973, US patent 3.748.298, assigned to Monsanto.
*F.**Dobinson**, C.A.**Pelezo**, W.B.**Black**, K.R.**Lea**, J.H.**Saunders**, 1979, Journal of Applied Polymer Science, 22(1), pp. 2189-2195.
*D.J.**Forrestal**, 1977, *Faith, Hope and $ 5000, the trials and triumphs of the first 75 years*, New York.
*R.N.**Gribov**, G.A.**Kurakov**, V.A.**Ivanov**, V.S.**Zakharov**, 1968, USSR inventor certificate 218,371, application 5 July 1966.
*J.W.**Hannell**, 1970, Polymer News, 1, pp. 8-13.
*S.S.**Hirsch**, J.R.**Holsten**, 1969, Applied Polymer Symposia 9, pp. 187-194.
*W.A.H.**Huffman**, R.W.**Smith**, W.T.**Dye**, jr., 1965, US patent 3.203.933 assigned to Monsanto.
***ICI**, 1973, German Offenlegungsnummer 2,222,640, first application 14 May 1971.
***ICI**, 1975, German Offenlegungsnummer 2,447,519, first application 4 October 1973.
***ICI**, 1976, German Offenlegungsnummer 2,541,482, first application 17 September 1974.
***ICI**, 1980, European Patent application 5,913, first application 26 May, 1978.
***Inventa** AG fur Forschung und Patentverwertung, 1975, German Offenlegungsnummer 2,439,948, first application August 22nd, 1973.
*A.**Knight**, 1974, *Private Enterprise and Public Intervention, the Courtaulds Experience*, George Allen & Unwin Ltd.
*G.I.**Kudryavtsev**, A.V.**Tokarev**, L.V.**Chikurina**, 1968, USSR inventor certificate 194.229, application 17 November 1965.
*G.I.**Kudryavtsev**, L.F.**Balakleitseva**, A.M.**Shchetinin**, L.V.**Chikurina**, 1970, Vysokomolekulyarnye Soedenii, Serie A, 12(10), pp. 2205-2210.
*G.I.**Kudryavtsev**, A.V.**Tokarev**, L.V.**Avrorova**, V.A.**Konstantinov**, 1974, Khimisheskie Volokna, (6), pp. 70-71.
*H.E.**Kunzel**, G.**Blankenstein**, F.**Bentz**, G.**Nischk**, 1973, German Offenlegungsnummer 2,144,126, application September 3, 1971, assigned to Bayer AG.
*R.C.**Laible**, F.**Figucia**, W.J.**Ferguson**, 1973, Journal of Macromolecular Science-Chemistry, A7(1), pp. 295-322.
*R.C.**Laible**, F.**Figucia**, B.H.**Kirkwood**, 1974, Applied Polymer Symposium, nr. 23, pp. 181-191.
*Z.S.**Lee**, L.**Skolnik**, C.Z.**Draves** jr., 1971, Rubber World, July, pp. 41-47.

*J.E.**McIntyre**, A.A.**Mohajer**, 1977, paper Annual Conference of the Textile Institute, Manchester, 61, pp. 89-105.

*A.A.**Mohajer**, W.J.**Ferguson**, 1976, Textile Progress, 8, pp. 102.

*P.W.**Morgan**, 1965, *Condensation polymers: by Interfacial and Solution Methods*, Interscience Publishers, New York.

*P.W.**Morgan**, 1985, Journal of Polymer Science, C72, pp. 27-37.

*H.S.**Morgan** jr., 1968, US Patent 3,414,645, assigned to Monsanto, application 19 June 1964.

*S.**Nuttall**, I.**Guy**, M.**Nixon**, 1987, *Japanese New Materials Yearbook 1986/1987*, Oxford.

*S.**Ozawa**, 1987, Polymer Journal 19, nr. 1, pp. 119-125.

*S.P.**Papkov**, V.G.**Kulichikhin**, V.D.**Kalmykova**, A.Ya.**Malkin**, 1974, Journal of Polymer Science, Physics 12, pp. 1753-1770.

*A.**Pettigrew**, 1985, *The Awakening Giant, continuity and Change in ICI*, Basil Blackwell.

*R.**Pigeon**, P.**Allard**, 1974, Angewandte Makromolekulare Chemie, 40/41, pp. 139-158.

*J.**Preston**, R.W.**Smith**, 1964, Belgian patent 637,260, assigned to Monsanto, first application in USA, 11 September 1962.

*J.**Preston**, F.**Dobinson**, 1964, Polymer Letters 2, pp. 1171-1174.

*J.**Preston**, R.W.**Smith**, 1966, Polymer Letters 4, pp. 1033-1038.

*J.**Preston**, R.W.**Smith**, C.J.**Stehman**, 1967, Journal of Polymer Science, part C, no. 19, pp. 7-15.

*J.**Preston**, R.W.**Smith**, W.B.**Black**, T.L.**Tolbert**, 1969, Journal of Polymer Science, part C, nr. 22, pp. 855-865.

*J.**Preston**, 1975, Polymer Engineering and Science, March 15, nr. 3, pp. 199-206.

*B.G.**Reuben**, M.L.**Burstall**, 1973, *The chemical economy*, Longman.

***Rubber World Staff Report**, 1970, Rubber World 62-63, pp. 54-72.

*H.**Schnell**, 1964, *Chemistry and Physics of Polycarbonates*, Wiley-Interscience, New York.

*H.**Sekiguchi**, K.**Sadamitsu**, 1977, Japanese patent 77 30,555, assigned to Furukawa Electric Co. Ltd., application February 22nd, 1969, (equivalent to US patent 3,886,251, May 27th, 1975).

*A.M.**Shchetinin**, G.I.**Kudryavtsev**, 1968, Khimicheskie Volokna, nr. 1, pp. 22-25.

*A.M.**Shchetinin**, 1968, Khimicheskie Volokna, nr. 5, pp. 50-52.

*L.H.**Smith**, 1946, *Synthetic fiber development in Germany*, Textile Research Institute, New York.

*L.B.**Sokolov**, T.V.**Gudakova**, 1959, USSR inventor certificate 122,279.

*A.V.**Tokarev**, L.V.**Chikurina**, G.I.**Kudryavtsev**, V.N.**Kulakov**, S.S. **Gitis**, V.N.**Aleksandrov**, I.Ya.**Kvitko**, V.I.**Sorokina**, I.M.**Nikiforov**, 1974, German Offenlegungsnummer 2,208,811, assigned to VNIIV, application 24 February 1972.

***United States Court for the Federal Circuit**, 1987, nr. 86-1327, appendix pp. 4-5.

***D.W.Van Krevelen**, 1967, Plastica 20, pp. 9.

***VNIIV**, 1971, Khimicheskie Volokna, nr. 1, in translation in Chemical Fibers, 1972, pp. 106-107.

***VNIIV**, 1975, Khimicheskie Volokna, nr. 3, pp. 36-37.

*L.**Vollbracht**, T.J.**Veerman**, 1976, German Offenlegungsnummer 2,605,531, 2 September, first application 21 February 1975, assigned to AKZO GmbH.

*A.V.**Volokhina**, M.N.**Bogdanov**, N.P.**Okromchedlidze**, G.I.**Kudryavtsev**, L.A.**Plyashkevich**, T.V.**Kravchenko**, B.A.**Batikyan**, A.B.**Raskina**, G.N.**Freidlin**, et al., 1975, German Offenlegungsnummer 2,408,426, application, 21 February 1974, assigned to VNIIV and the VNII for monomers.

*B.**Von Falkai**, 1974, Lenzinger Berichte, Folge 36, pp. 48-55.

*E.L.**Wittbecker**, P.W.**Morgan**, 1959, Journal of Polymer Science XL, pp. 289-297, main applications interfacial polycondensation method: Journal of Polymer Science XL (1959), pp. 299-406, XLII (1962), pp. 33-58.

*G.D.**Wolf**, 1971, German Offenlegunsnummer 2,000,927, application 9 January 1970, assigned to Bayer AG.

*G.D.**Wolf**, 1971, German Offenlegunsnummer 1,953,358, application 23 October 1969, assigned to Bayer AG.

HIGH PERFORMANCE FIBERS

H. H. YANG
Du Pont Fibers, E. I. du Pont de Nemours & Co., Inc., P.O. Box 27001, Richmond, VA 23261.

INTRODUCTION

This paper presents a brief review of high performance fibers for industrial applications. The term 'high performance' is not well defined, but is loosely used in conjunction with highly advanced materials. From a product point of view, it apparently dictates certain acceptable performance under some extraordinary conditions. In that sense, it encompasses end uses which involve high temperatures, high tensile or compressive load, hostile environments, dimensional stability, lightweight, or high electrical conductivity. Consequently, high temperature fibers which were developed in the 1950-60s are not considered as a part of high performance fibers. In addition to high temperature applications, synthetic fibers have now expanded into high performance applications because of advances in material properties. New fibers which have been developed since 1970s offer high strength, high modulus, thermal stability at very high temperatures, chemical and solvent resistance, low weight, and a number of other properties. These fibers are rapidly entering into new materials markets. They are not used in forms of pulp, short staple or continuous filaments for reinforcement in many industrial, military and aerospace applications.

The new fibers have been about by several technological discoveries. These include the inventions of aromatic high strength fibers, gel-spun polyethylene fiber, graphite and inorganic fibers. All of these fibers are specialty products aimed at certain small segments of industrial, military and aerospace applications which require high performances. They are characteristic of relatively small volume of business, high prices, and complicated production methods. With rapid development of advanced materials, material scientists and fiber producers have projected a steady increase in the demand for high performance fibers. Therefore, research and development efforts in this area are very extensive worldwide.

Table 1 presents a summary of high performance fibers presently produced in the United States based largely on a previous survey by Relsch.[1] Table 2 compares the physical properties of selected high performance fibers. It can be seen that the tensile strengths of most high performance fibers are on the order 20-35 gpd and the tensile moduli are in the range of 500-2500 gpd, both several times those of conventional fibers. Although these summaries are by no means complete, they do reflect the current advances in fiber technology towards high strength, high modulus materials. Some general comments on each class of high performance fibers are given below.

AROMATIC POLYMER FIBERS

High strength, high modulus fibers have been prepared from aromatic polymers.[2] Of our major interest are aromatic polyamides, aromatic polyesters, aromatic polyimides, and aromatic heterocyclic polymers.

Table 1. High Performance Fibers Produced in the United States[1,2]

Fiber Type	Producer	Trademark	End Uses
Aromatic Fibers			
Aramid	Du Pont	NOMEX	Thermal protective apparel, electrical insulation, filter
Aramid	Du Pont	KEVLAR	Protective apparel, armor, composites, rubber goods, ropes and cables, asbestos replacement
Polyester	Hoechst-Celanese	Vetran	Thermoplastic composites
Polybenzimidazole	Hoechst-Celanese	PBI	Protective apparel
Heterocyclic polymers	Dow	PBT, PBO, AB PBO	Composites, military
Polyolefin Fibers			
Polyethylene	Allied Signal	Spectra	Armor, ropes, composites
Bicomponent	Avtex	ES	Nonwoven fabrics
Carbon Fiber			
PAN based	Hercules	Magnanite	Composites
	Hysol-Grafil		Composites
	BASF	Celilon	Composites
Pitch based	Amoco	Thornel	Composites
	Ashland	Carboflex	Composites
Inorganic Fibers			
Boron	Avco		Aerospace structures
SiC	Avco		Metal matrices
Al_2O_3-B_2O_3-SiO_2	3M	Nextel	Ceramic reinforcement
Glass	Owens-Corning	Fiberglass	Composites, armor
Poly(phenylene sulfide)	Phillips	Ryton	Chemical-resistant filtration

in a rope form was used to tie down insulating pieces on a space shuttle. More recently, Ube Company and Toray Company of Japan have succeeded in modifying aromatic polyimides and achieving high strength, high modulus fibers by sophisticated spinning and hot drawing methods. These fibers will undoubtedly compete in the high performance fiber market in the next few years.

Aromatic Heterocyclic Polymers

Aromatic heterocyclic polymers include poly(m-phenylene bensimdazole) (PBI), poly(p-phenylene benzobisthiazole) (PBI), poly(p-phenylene benzobisoxazole) (PBO), and Ab type polybenzoxazole (AB PBO).

Polybenimidazole (PBI) was mentioned earlier (cf. Table I-4) as a high temperature polymer. PBI fiber was developed under government research program for aircraft landing chutes and manned spacecraft uniforms. Celanese Corp. commercialized it in 1983. It has excellent chemical and solvent resistance, and does not burn in air. For account of these properties, PBI fiber is used in woven fabrics as aircraft seat fire blocker and protective coats for firemen. It can be blended with KEVLAR fiber to enhance the physical properties and cost effectiveness. PBI fiber has the potential of competing in some high performance markets where KEVLAR and NOMEX aramid fibers are well established. However, it is presently priced at $40/lb. which is fairly expensive.

PBT, PBO, AB PBO and other heterocyclic polymers belong to a group of rigid rod polymers investigated by SRI Research International under the U.S. Air Force Ordered Polymer Program. PBT fiber was jointly developed by Celanese, Du Pont, and several universities. PBO and AB PBO fibers are more recent developments. All of these fibers exhibit very high modulus and high strength in addition to other properties characteristic of rigid-chain aromatic polymers. Dow Chemical Co. has recently acquired a license from Commtech International Management to develop these polymers. They plan to commercialize them as a family of PBZ fibers.

High Strength Polyethylene Fibers

Allied Signal Inc. is currently marketing a high performance polyethylene fiber under the trademark of Spectra[6]. This fiber is spun from an isotropic dilute solution of polyethylene in a suitable solvent. After solvent extraction, the as-spun fiber is drawn at very high draw ratio to achieve high degree of crystallinity and orientation. Spectra fiber offers high strength, high modulus and light weight. it is therefore finding inroads into rope and cordage markets. It has the potential for reinforcement of boat hulls, automotive parts, and ballistic structures. It has a melting point of 150°C and cannot be used in high temperature applications. Spectra fiber products are currently sold at $22-28/lb. in the United States.

CARBON FIBER

Carbon fiber, often referred to as graphite fiber, is produced from a precursor fiber such as rayon, polyacrylonitrile (PAN), or mesophase pitch. The precursor is thermally decomposed, followed by heat treatment at high temperatures to convert it to carbon. For example, PAN precursor fiber is first oxidized at 250°C, then carbonized at 800°C, and finally graphitized by heating to 1400-2500°C. The PAN-based carbon fiber generally gives higher mechanical properties than the pitch-based fiber. However, both types of fibers give outstanding tensile modulus, and high tensile and compressive strengths. It is also thermally stable at temperatures up to 400°C. These good properties have made carbon fiber attractive for use in advanced composites. The major benefit of carbon fiber in composite is the savings of up to 30-40% in structural weight, which is a critical design factor in many weight-sensitive applications.

The business outlook for carbon fiber in 1990s, according to on estimate, will reach $3 billion. Many fiber producers are therefore pursuing the carbon fiber market. Hercules, Hysol-Grafil, BASF, Amoco and Ashland are the U.S. producers. There is now a worldwide carbon fiber capacity of 3.9 million lb. per year with about 50% idled. With complicated processing and heavy investments, carbon fiber is presently priced at $30-40/lb. Few producers are reported to have turned in a real profit. Thus, the carbon fiber market is growing steadily but will remain extremely competitive for the next few years.

INORGANIC FIBERS

There are several kinds of inorganic fibers, i.e., boron, boronoxide, aluminum oxide, and silicon carbide. The oxide and carbide fibers are sometimes called ceramic fibers on account of their high temperature resistance.

Boron Fiber

Avco produces boron fiber by vapor deposition on a heated tungsten wire. It has an annual capacity of 35-50,000 lb. and sells at $350-400/lb. Boron fiber offers high specific strength and specific stiffness. In composites, it offers higher compressive strength than a carbon/epoxy matrix. This permits the use of boron fiber in composite structures to save weight, particularly in aircraft designs.

Silicon Carbide Fiber

Avco also produces silicon fiber which is heat resistant up to 1200°C and offers similar reinforcing characteristics. Metal matrices of silicon carbide can be more readily fabricated. The present production volume of silico carbide is only 2000 lb./yr. at $1000/lb.

Alumina Fiber

Alumina fiber has excellent compressive strength, moderate tensile strength, high stiffness, and thermal stability in composites of nonferrous metals, ceramics, epoxy and polyimide resins. Potential applications for fiber FP include composites, automotive parts, batteries and helicopter engine housings.

Alumina-Boria-Silica Fiber

Nextel[7], an alumina-boria-silica fiber manufactured by 3M Co., also offers high temperature utilities up to 1200-1370°C. It is used in thermocoupling devices, wire insulation, high temperature seals and gaskets. It is priced at $40-100/lb.

Glass Fiber

Glass fiber was commercialized by Owens-Corning Fiberglass CO. in 1939. It has found numerous household and industrial applications on account of its low cost, good chemical and thermal resistance. For this reason, it is well known as a household item under the trade name of fiberglass.

There are several types of glass fibers which are produced from variations in ingredients and process conditions. For industrial applications, we are primarily interested in E glass and S glass. E glass has a low-alkali composition of alumino-borosilicate. It offers excellent electrical insulation properties, and constitutes the majority of textile fiberglass products. S glass has a composition of magnesium aluminosilicate. It offers approximately 40% higher filament strength than E glass. It also retains significantly higher strength than E glass at high temperatures. It is used to reinforce composite materials for aircraft, aerospace, and ballistic applications.

Poly(phenylene Sulfide) Fiber

Phillips Fibers commercialized a Ryton[8] poly(phenylene sulfide) (PPS) fiber in 1983. Ryton is principally used as high temperature filtration fabrics up to 190°C. It offers good chemical and hydrolytic resistance. It is presently priced at about $8/lb.

COMPARISON OF HIGH PERFORMANCE FIBERS

A cursory survey of high performance fibers in the United State shows that inorganic,aramid, carbon and glass fibers have dominated the advanced materials market in recent years. However, there is now rapid infusion of highly competitive new fibers such as the high-strength polyethylene fiber and ceramic fibers. From the preceding discussions, it can be noted that these fibers fall into three groups: (a) high-temperature textile fibers such as NOMEX aramid, PBI, and PPS; (b) inorganic specialty fibers such as boron, silicon carbide; and (c) high strength, high modulus reinforcing fibers such as aromatic fibers which include KEVLAR aramid, carbon, polyethylene, and glass fibers. In many end sues, this classification permits us to simply compare several high performance fibers within the same group. There are no simple ways to compare these fibers. One may attempt to draw some overall assessment according to their end uses.

Carbon fiber has the best thermal stability with a melt temperature of about 4000°C, but moderate specific strength and modulus because of its high density. High-strength polyethylene fiber has excellent specific strength and modulus, but very low melt temperature, 147°C. Aramid fiber provides good thermal stability and specific properties which are intermediate between carbon and polyethylene fibers. The costs of these fibers, based on specific properties, are fairly competitive. For maximum cost effectiveness, a recent trend in advanced composites is to use the hybrid structures of two different fibers such as KEVLAR/carbon or KEVLAR/glass. We should expect this trend to continue. Thus, all major high performance fibers will find their respective places in the market place.

Aramid fibers appears to have a much broader range of applications than other high performance fibers. They are used for structural reinforcement while offering wear, friction, ballistic resistance in both rigid and flexible structures. Carbon and glass fibers are largely limited to the reinforcement of rigid composites. High-strength polyethylene fiber is finding applications in rigid and flexible structures. Although with some processing difficulties in molding and laminating, it can be used as a reinforcing fiber for composite materials.

For applications at extremely high temperatures, inorganic fibers, and carbon fiber, are clearly the choice materials. Fibers of aromatic polyamides, polyimides and heterocyclic polymers, are suitable for the temperature range of 200-250°C. High-strength polyethylene fiber is useful for only low temperatures below 150°C. These use temperatures impose practical limits on end use environment where a high performance fiber can be applied successfully.

FUTURE OUTLOOK

As we look into the future, we can see the continued growth of high performance fibers for various composites, advanced structures, electronic and biomedical applications. The market shares of each type of high performance fibers may shift on account of advances in product properties. We are convinced that the fiber technology will not only have a place in the science of high performance materials, but also continue to become increasingly sophisticated.

It is predictable that future technological development will probably dictate further increase in fiber mechanical and thermal properties. New and improved fiber products will continue to be developed. Judging from the most recent development, aromatic polyimide and heterocyclic polymer fibers may emerge as the upcoming high performance fibers in the foreseeable future. We may also see new aromatic polymers of rigid or semi-rigid chain structures for high strength, high modulus fibers. There are now demonstrated process routes to convert such polymers. Even some of the early polymers, which were impractical to convert into fibers,may one day reappear as high performance fibers.

REFERENCES AND NOTES

1. Relsch, M. S., Chem. Eng. News, (Feb. 2, 1987).
2. Yang, H. H., "Aromatic High-Performance Fibers," Wiley-Interscience: New York (1989).
3. Bunsell, A. R., Centre des Materiaux, Ecole Nationale Superieure des Mines de Paris, France, "Fibre Reinforcements - Past, Present, and Future," (1987).

Note: [1] NOMEX, KEVLAR - registered trademarks of E. I. Du Pont de Nemours & Co., Inc.
[2] Twaron - a registered trademark of Akzo BV, The Netherlands
[3] Technora - a registered trademark of Teijin Inc., Japan
[4] Kermel - a registered trademark of Rhone Poulenc Company, France
[5] Vectra, Vectran - registered trademarks of Hoechst-Celanese Inc.
[6] Spectra - a registered trademark of Allied-Signal Inc.
[7] Nextel - a registered trademark of 3M Co.
[8] Ryton - a registered trademark of Phillips Fibers Co.

MAJOR ADVANCES IN POLYESTER 2GT TECHNOLOGY 1941-1990

N. W. HAYMAN AND F. S. SMITH
ICI Fibres Research Centre, Brockworth, Gloucester, England GL3 4HP

INTRODUCTION

In the 50 years from its invention Polyethylene Terephthalate (Polyester 2GT) had developed into the world's largest synthetic fibre with sales world-wide in 1989 of 8.4 million tonnes [1] which represents 21% of the total of all fibre sales. This was exceeded only by the natural cotton at 18.2 million tonnes. If the present annual rate of growth continues, one can predict that polyester will overtake cotton in sales by 2010.

The reason for this spectacular success is the cost-effectiveness of the fibre. Simpler intermediate chemistry than its main synthetic rival the polyamides results in a lower cost base, and this is combined with an excellent spectrum of product properties. This product profile has enabled it both to replace existing fibres and to develop new end uses. Polyester is not however perfect in every respect for every market segment and more specialist fibres such as polyamides and elastanes and even lower cost base fibres such as polypropylene will continue to happily co-exist in a world of ever increasing volume demand.

Given the enormous sales volume, it is not possible in one chapter to go into much depth of detail on the historical development of specific products - what we have tried to do is to pull out the key technology advances, a broad brush approach with the advantage of half a century of hindsight. No development, especially in its early days proceeds in a strictly logical order - we have ignored the deviations as also we have omitted most reference to the influence of individuals and political decision making. This is in no way to belittle such influence, they are the key to success or failure, but in the space available in this will be primarily a scientific historical review. We will also concentrate solely on 2GT - there are a range of aromatic polyesters of which 4GT (polybutyl terephthalate - PBT) is the second most important, but all are of relative insignificance compared to the molecule with the $2(CH_2)$ link in the chain from the 2 carbon Glycol.

The history of polyesters substantially starts with Carothers' papers whilst working at Du Pont on Condensation Polymerization 1929-1931. In the early 30s many essentially aliphatic polyesters were synthesized but were discarded as unsuitable for fibre spinning primarily because of low melting point. |In the enthusiasm to develop polyamide 66, polyethylene and several other fibres in the late 30s it was perhaps not surprising that Du Pont research in polyesters halted [2]. What is somewhat more surprising is that in 1940, two chemists J. R. Whinfield and J. T. Dickson commenced pioneering research on polyesters at the Accrington UK Laboratories of the Calico Printers Association - surprising inasmuch that CPA were not in fibre spinning and the allocated resource was modest. However, their faith was amply rewarded, for very rapidly in July 1941 a patent [3] was filed for Polyester 2GT, named 'Terylene', the innovative step by Whinfield and Dickson being to use a symmetrical aromatic unit in the polyester polymer chain which significantly raised the melting point and at the same time gave a fibre forming polymer. The use of a para oriented aromatic molecular to encourage linear configuration of the molecule and to thus raise the potential for crystallization compared with an ortho- or

Figure 1. Whinfield and Dickson - 1941

meta-oriented molecule was original thinking which previous researchers had missed, see Fig. 1.

In the environment of the second World War, the Terylene patent, as with many others of that time, was declared secret by the British Ministry of Supply. Fortunately, the then Controller of Chemical Research decided that the material was of sufficient potential importance to warrant further work and as the CPA had no suitable facilities, research was transferred to the British Government Department of Scientific and Industrial Research at Teddington. Over the next two years a few pounds of polymer were successfully synthesised and X-ray analysis of undrawn and drawn fibres confirmed a semi-crystalline material similar to nylon. Resource was not available at Teddington to take the investigation further and at this stage the invention was offered to Imperial Chemical Industries (ICI) to develop. Work started within ICI in December 1943 and in February 1944 the British government withdrew, leaving ICI to negotiate with CPA an agreement for the development of TERYLENE. This agreement signed in February 1947, gave ICI the worldwide exclusivity rights (with the exception of the USA) in exchange for a royalty payment. In July 1944, in accordance with the then current agreement between ICI and Du Pont, ICI disclosed information on Terylene in a letter to Du Pont and subsequently gave them samples of the new fibre in February 1945. Du Pont then conducted separate negotiations with CPA for the USA patent rights which were acquired in August 1946, the name of the polyester for the USA was changed, and so Dacron was born [4,5].

Intensive development then took place on both sides of the Atlantic on Intermediates manufacture, Polymerisation, Fabrication and Dyeing and Finishing, with ICI having the advantage of a year's start and Dyestuff expertise, and Du Pont the advantage of more expertise in fibre production and resources not so badly effected by the ravages of war. Free exchange of information on Polyester 2GT between the two companies continued in a spirit of constructive rivalry until July 1948 when the US threat of indictment under the new US Sherman Act halted the cooperation. On the marketing side an initial foray by Dacron into the Tyre market in September 1946 was a temporary failure and the first recorded sale of Terylene was in October 1948 as brass-bobbin yarn for lace-making. Trials in all fabric end uses were conducted in the late 1940s with a concentration on wool staple blends in the early 1950s.

It became evident in the early 1950s that Terylene was a winner and that ICI could not hope to supply world demand outside the USA. In 1953 and 1954 ICI therefore licensed Rhodiaceta, Montecatini, Farbwerke Hoescht AG, Glanzstoff-Fabriken and AKU for manufacture in France, Italy, Germany, Germany and Netherlands, respectively. Today, more than 200 companies in over 50 countries [6] of the world manufacture Polyester 2GT and many different detailed technologies have emerged for so doing. What follows is our interpretation of the key advances that have caused today's situation to happen.

BASIC BUILDING BLOCKS - THE INTERMEDIATE CHEMICALS [7,8,9]

To meet present day fibre quality standards for customer satisfaction, particularly in terms of colour, it is necessary that polyester intermediates must be of a purity higher than is common-place amongst the pharmaceuticals industry. Attempts have been made to specify tolerable impurity levels [7] but these are often at the limits of analytical techniques and it is also in practice impossible to obtain agreement as to what are the harmful species present, so that in the end judgement is based mainly on customer response.

The required monomer intermediates for the polycondensation reaction to polyethylene terephthalate (2GT) are ethylene glycol (EG) and either dimethylene terephthalate (DMT) or terephthalic acid (TA). In the 1940s ethylene glycol was already essentially a commodity chemical having been developed in the 1930s as an antifreeze for engines, however dimethylene terephthalate and terephthalic acid substantially did not exist apart from in the experimental laboratory. The path of development of industrial processes for TA and DMT manufacture was therefore initially governed by the availability of possible sensible starting materials, then by the stringent purity requirements and finally with the growth of polyester into a major chemical, the challenge of low cost production. The end point today bears little resemblance to the start in 1940 and considering the intervening years it is doubtful whether the development of the DMT based route would have taken place at all if pure TA had been readily available. However, TA is a particularly intransigent chemical to purify and the DMT route which is now less cost effective was essential up to the 60s to produce acceptable quality. DMT based 2GT capacity is still in existence but with modern plant being TA based, it is gradually dwindling away as old installations are replaced apart from being retained for some specialist polyester polymers.

Ethylene Glycol

The simplest chemical route to envisage and the largest method of production world-wide is based on a two stage route involving ethylene oxide (Laporte process 1937).

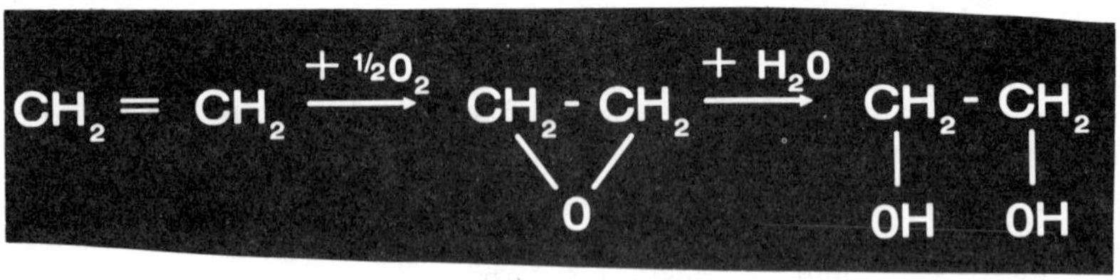

Laporte Process 1937.

The oxidation by air or by now more favoured oxygen is carried out over a silver oxide catalyst at around 250°C. However yields are only of the order of 70% because of the concomitant reaction simply burning to carbon dioxide and water, so research continues on improving catalyst selectivity.

The hydrogenation reaction can be conducted with basic or acid catalyst, but the preferred rate is under high pressure at neutral pH and 100°C with a large excess of water under which conditions conversion to the desired monoethylene glycol is around 90%, byproducts diethylene glycol and other polyglycols being minimised. Ethylene glycol can then be purified by classical methods, specifically distillation to a quality suitable for 2GT production.

With yields of the two stages at 70% and 90% it is not surprising that alternative technologies have been developed for the manufacture of ethylene glycol from ethylene. Two such are the Oxirane process using acetoxylation and the Teijin catalytic oxychlorination route but whether one or either of these routes should be preferred is a matter of detailed cost analysis.

Terephthalic Acid

In 1944, TA was produced within ICI on a small scale as a dyestuffs intermediate, but a new route was required to make the necessary volume for a polymer and fibre development programme. p-toluic acid was the chosen initial building block made by the Wyler process from an hydrolysable intermediate made from toluene, ethyl aniline and phosgene and scaled up in the late 40s to 20 t/a [10]. This material was subsequently oxidised by potassium permanganate to yield TA of excellent whiteness and this hopelessly uneconomic process maintained the momentum on 2GT development until 1952 by which date an alternative preferred route based on p-xylene oxidation had been put into production.

p-xylene from petroleum was foreseen to be the most likely source for a practical production route to TA, the alternative p-cymene from wood pulp being quickly abandoned on technical and economic grounds. These decisions required excellent visionary judgement as in the 40s there was no certain way of separating xylenes from crude oil, separating p-xylene from its isomers or converting p-xylene to TA. However, in the UK, by the end of the decade 3,000 tonnes of xylenes had been received from Iran, the p-xylene separated by freezing out and converted to TA by pressure oxidation using dilute nitric acid.

TA produced by this method was not pure enough to give satisfactorily white fibre and it was also slow to react with glycol with the then known catalyst systems. Therefore as foreseen in the 1942 Whinfield/Dickson patent a preferred route for manufacturing the intermediate DMT from the impure TA was developed by ICI. A number of similar routes or routes based on direct production of DMT from p-xylene were thus preferred until, in the 1960s, a process was commercialized that produce TA of sufficient purity for direct esterification. This process (Mid-Century, Amoco) essentially is still state of the art and has two main sections, an air oxidation followed by a purification stage.

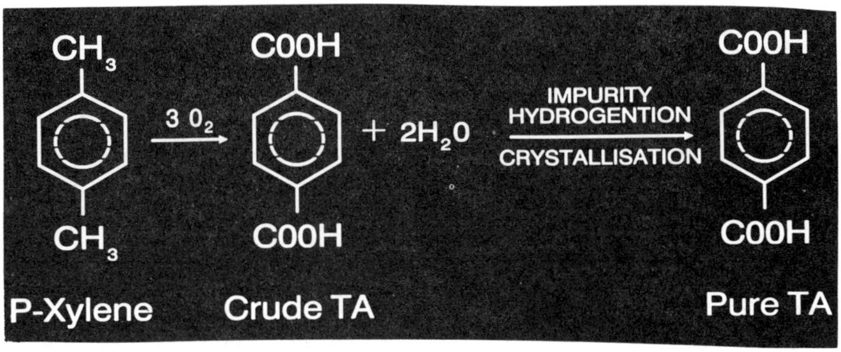

Mid-Century, Amoco 1958.

The air-oxidation is conducted in acetic acid liquid phase at temperature (ca 200°C) and pressure (ca 300 psi) using metal catalyst (eg Co) and Br_2 as a free radical source.

The purification step involves hydrogenation of the main impurities in TA (especially 4-formyl benzoic acid), in which the impure TA is dissolved in water at temperate (250°C) and pressure and still as a liquid passed over a noble-metal catalyst where it is mixed with hydrogen. Temperature is then dropped and pure TA crystalises out, the impurities remaining in the mother liquid from which the crystal is centrifugally separated. As for the other monomers there are alternative routes today to form TA, the one described above is the most favoured.

Dimethylene Terephthalate

The reason for the development of DMT in the 1940s was that it could be reasonably easily purified to high quality by the then available techniques. ICI's original route was to make crude TA and then esterify with methanol using sulphuric acid as catalyst. A high pressure low concentration route at 275°C was developed to aid the dissolution of the TA in methanol. Conversion to DMT is limited by equilibrium to the acceptable figure of 96%. The product was then recrystallised. However, in the 1950s a four step direct route from p-xylene was patented by Hercules/Dynamit Nobel which became the dominant world-wide technology.

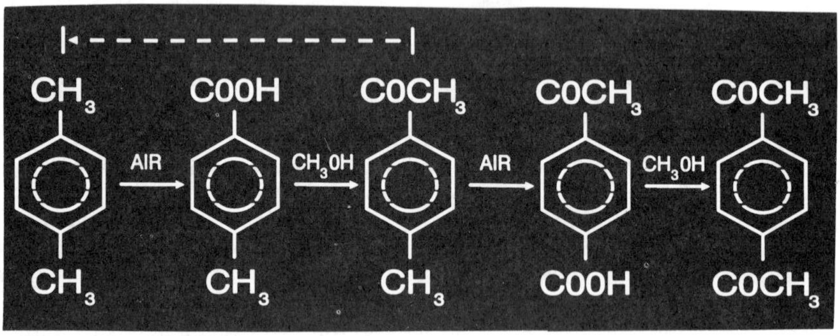

Hercules/Dynamit Nobel 1952.

A mixture of p-xylene, air and catalyst are fed to the oxidation vessel where they are joined by recycled methyl p-toluate from stage 2, oxidation occurs at 160°C and 90 psi. Methanol is added in stages 2 and 4 at 225°C and at a pressure sufficient to maintain a liquid phase. Conversion to DMT based on p-xylene is 87% the product being purified by a sequence of distillation, crystallisation from methanol and a second distillation.

POLYMERISATION [9,10]

The objective of this process is to produce a colour free polyethylene terephthalate polymer with very low levels of impurities and possessing the normal MW distribution for a polymer produced by a condensation reaction. The low level of impurities is required not only to give acceptable colour but to give the necessary extremely high levels of mechanical quality at fibre processing, particularly at warping where modern day standards will permit only 5 fractures of a filament of a threadline bundle (typically 20 filaments) per 10^8 metres of threadline.

Impurities which can cause problems may exist in the starting monomeric chemicals, result from catalyst residues, be formed in side reactions that can produce cyclic oligomers or higher glycols or be formed in degradant reactions that produce crosslinked high molecular weight gels. The extent of these undesirable reactions depend on a number of process and equipment parameters - gels are promoted by holding polymer for too long a time at too high a temperature and also by air (oxygen) ingress into the reactor which incidentally must be made of carefully chosen timeless steel to avoid catalytic degradation of polymer on its surface. Avoidance of dead spots in equipment is achieved not only by detailed engineering design but by the development of appropriate agitators whose job is not only to encourage bubbles of gaseous by-product of the condensation reaction to break free from the polymer melt, but to ensure that polymer is not stagnant and always on the move. In the latter stage of polymerisation vacuum is used to encourage production of high MW products and the combination of a rotating shaft to the agitator and prevention of air ingress, has resulted in the development of new specialised gland technology.

In theory, given the high purity of terephthalic acid and dimethylene glycol achievable in 1991, it should be possible in a single stirred batch reactor, operating firstly on a pressure and then on a vacuum cycle to produce excellent polyester from the two monomers using a catalyst(s) to promote rapid reaction and other additives to inhibit undesirable side reactions.

The basic chemical reaction would be:

Basic Polyester Condensation Reaction

and the -COOH and -OH ends of monomer and dimer would then continue to react, steadily building up MW with water being eliminated. A number of secondary species are also generated, for example a central terephthalate moiety ester capped at both ends with glycol. The relative rates of these competing reactions and their reversal is governed by the reaction environment in terms of pressure, temperature, etc. To control the order of these reactions, thus encouraging the basic condensation polymerisation to occur, and secondly because of historical evolution in which DMT was so important and also thirdly for advantageous process economies, a series of vessels rather than a single reactor is used. These can either work on an all batch principle or as more common today a combination of continuous and batch reactors or all continuous reactors. The final vessel is always well agitated to encourage liberation of remaining gas (by this stage predominantly glycol rather than water due to the kinetics of the various side reactions) from the viscous high MW melt where diffusion of bubbles is very slow.

Whinfields's first product was made at the 100 gm scale in glass equipment. Process and equipment development involving stirred metal reactors then continued through the 1940s and by the early 50s had reached 1000 tpa at ICI's first polymers plant at Huddersfield, see Fig. 2. A typical batch route to polyester chip polymer starting from DMT is schematically, in Fig. 3.

In the first vessel, solid DMT is melted under a N_2 blanket and transferred at 150°C to the Ester interchange vessel where glycol is added at 2:1 mole ratio and a

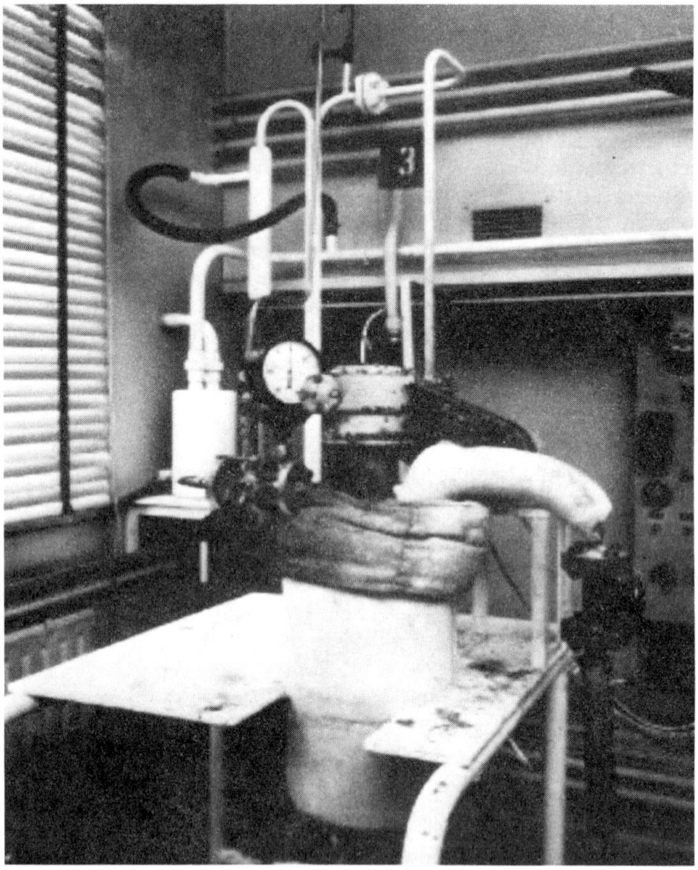

Figure 2. Early Polymerisation Vessel

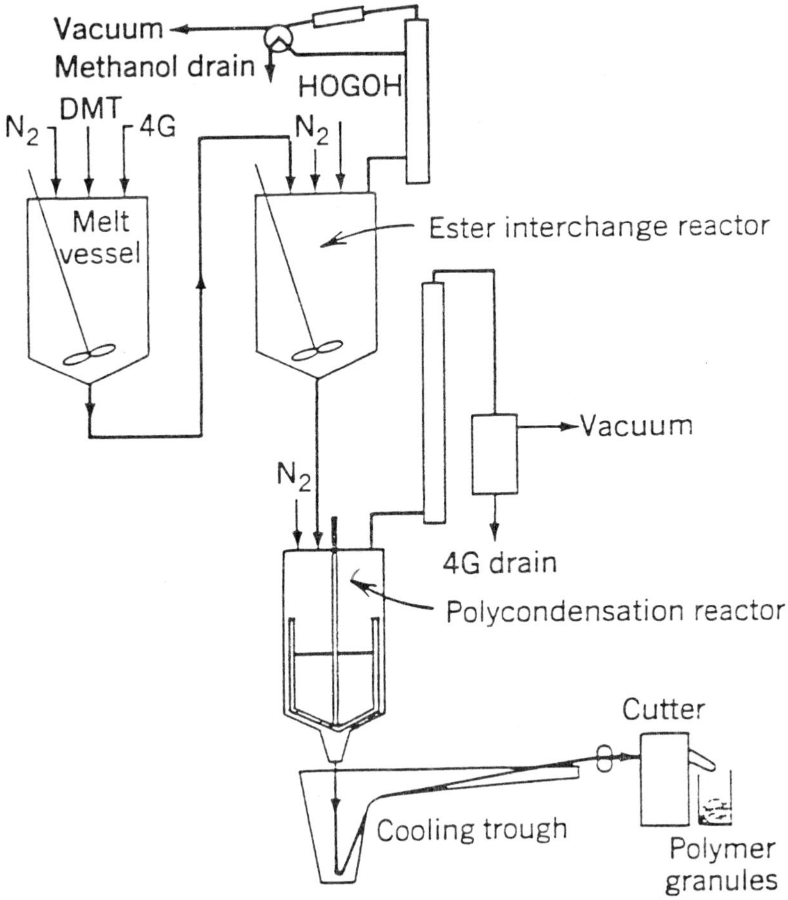

Figure 3. Batch Polymerisation - from Vol. 12 of Encyclopedia of Polymer Science and Engineering, Mark et al., Wiley Science, p. 44. Courtesy of Elsevier Science Publishers Ltd.

suitable catalyst also added (eg acetates of zinc, magnesium or manganese - many others are mentioned in the Whinfield patents). The temperature is steadily raised and methanol is evolved to travel up through a reflux column and then away to condensation. Great care however must be taken with the temperature profile, according to catalyst activity, to prevent column blockage by the twin problems of the vessel contents surging into the reflux column or DMT subliming. By the time temperature has reached 210°C, a CE of 99.99% has been achieved.

The ester is then transferred to the polymerisation vessel where by applying heat to the stirred ester to which a catalyst (typically Sb_2O_3) is added, the condensation polymerisation reaction occurs and glycol is liberated which is removed for recovery by an applied vacuum of around 1 mm Hg. Final temperatures are of the order of 280°C and the reaction is terminated when melt viscosity reaches the desired value (as measured by stirrer torque) say after 1 hour. The molten polymer is then extruded under N_2 pressure as ribbon or lace into a cold water trough and fed to a granulating cutter.

So far we have concentrated on Batch Polymerisation, but in the early 1950s Du Pont had seen the possibility of the exciting concept of Continuous Polymerisation for polyester. With rapid development, their first production plant at Kinston started up on CP in 1953, a notable triumph of Chemical Engineering prowess. A more modern continuous polymerisation route starting from TA is illustrated in Fig. 4.

In the first vessel a pumpable paste is prepared of TA and glycol (mole ratio 1:1.16) with a little water and this is forwarded to the first stage polymeriser which operates at 250°C and 40 psig. Water is driven off in the condensation reaction, and glycol recovered from a reflux column, although some is lost. After three hours around 90% of the free TA has reacted and the mixture is passed out to second stage polymeriser where it is held for a further two hours at 245°C and atmospheric pressure. At this stage virtually no free TA or glycol remains and the average length of the polymer chain is around 3 units. Catalyst (eg Sb_2O_3) is added and the mixture passed into the third stage polymeriser which is a multistage vessel where the temperature is progressively raised from 255-275°C and the pressure dropped from 70 to 20 mm Hg. The polymer chain now contains an average of 30 units and is fed to the fourth and final stage polymeriser (E) which has a wiped wall, operates at 295°C and 1 mm Hg and produces a product of typically 150 repeat units. This is either fed direct to a spinning machine or extruded as for the batch process, into a water bath and hence to a cutter [11].

A number of additives are put into either batch, continuous/batch or continuous processes at various reaction stages. Catalysts have already been mentioned but also phosphorous containing compounds (eg triaryl phosphates) are added to prevent precipitation of insoluble terephthalate catalyst cation salts and for DMT based polymers manganese or cobalt salts are added for light stability. Other proprietary knowledge trace chemicals can be added to discourage undesirable side reactions such as the formation of diethylene glycol. However, the most common additive is titania, TiO_2 added at various levels up to 2% to change the optical appearance of the polymer as desired - increasingly giving a delustred effect. The TiO_2 is added in glycol slurry and the technology of its addition to the continuous polymerisers is considerably more difficult than to the batch route to obtain good dispersion, even though dispersant aids, such as coating the TiO_2, are regularly used.

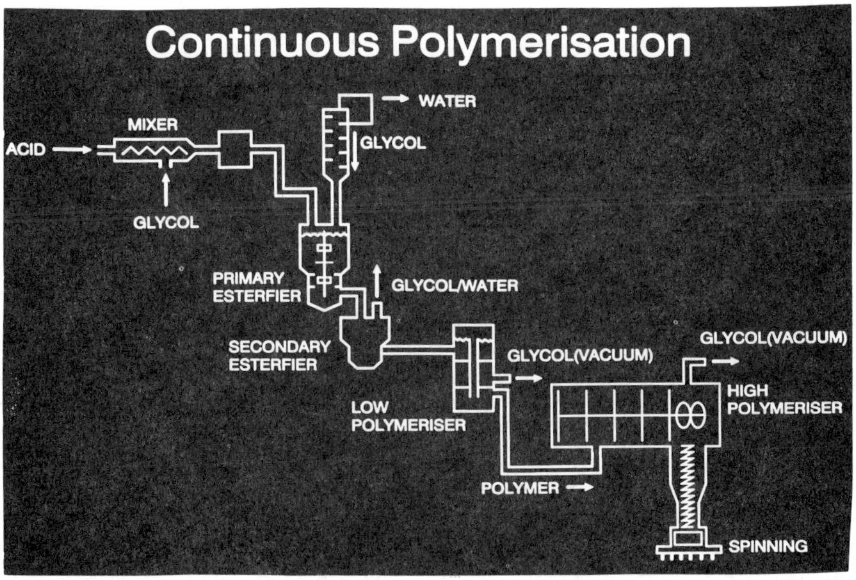

Figure 4.

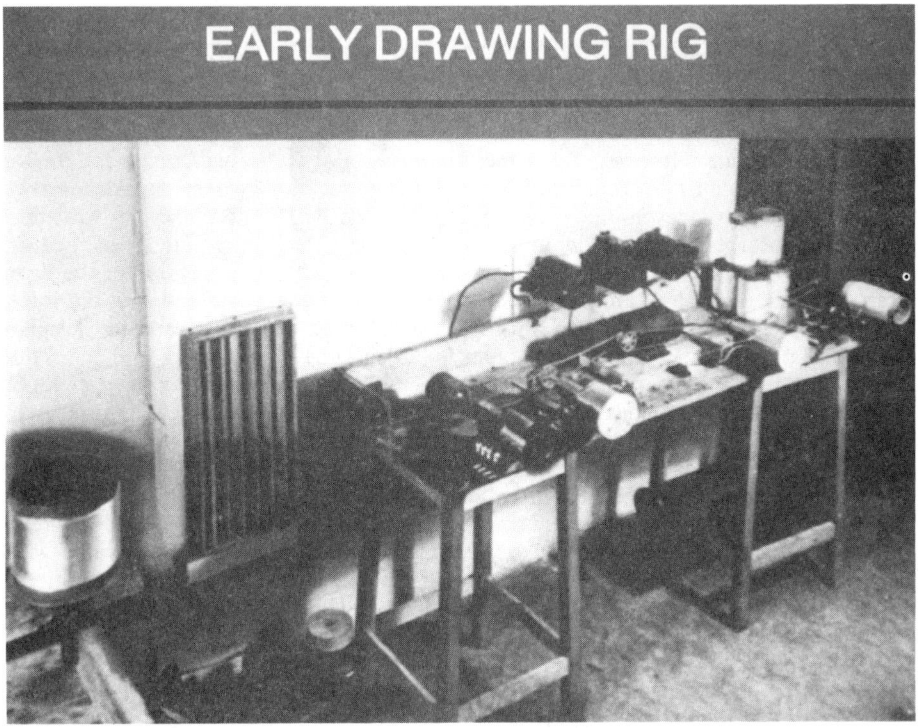

Figure 5.

EXTRUSION/SPINNING AND DRAWING

Originally, the basic process for turning polyethylene terephthalate into fibres was based on that already developed for nylon. Having made the polymer in the form of granules or chips they were melted and then extruded downwards through a multi-orifice spinneret to form filaments. The filaments were cooled by passing through a duct provided with a cross flow of air, and were then wound up at speeds of about 200 m/min onto large bobbins. At this stage, although the solid filaments had been formed, the molecular orientation was low, and they had no real strength. A further stretching process, known as drawing, was necessary to align and crystallise the molecules, so as to impart useful textile properties to the fiber, see Fig. 5.

Although melt spinning technology had been developed and brought to a reliable practical operating level for nylon, the adaption of this technology for polyester turned out to be more complicated and to require more ingenuity than was at first expected.

Two serious problems had to be overcome. The first was the high viscosity of the polyester melt. Whereas the melt viscosity of polyamide 6.6 in the fibre spinning process is something like 100 Ns/m, polyester is nearer 300 Ns/m, and this caused difficulties in obtaining adequate flow through the melter. The second problem was the sensitivity of polyester to moisture. Even small amounts of water in the molten material caused severe depolymerisation, and so polyester chip had to be thoroughly dried before melting, and the melt itself had to be protected against moisture ingress from the atmosphere.

Polyamide melters used at the time relied on gravity feed of polymer chip onto a grid which was heated using biphenyl vapour passing through channels within the grid. The heat transfer from grid to chip was not particularly good, but the relatively low viscosity of the polyamide melt allowed adequate drainage, so that reasonable melt rates could be achieved. However, these melters turned out to be inadequate for melting polyester at the melt rates required. The alternative technology of screw extruders was at an early stage of development and in experiments by ICI at Welwyn it was found difficult to control these extruders with polyester chip, due to irregularity of feed and poor temperature control [9]. The breakthrough came with the invention of the Screw Pressure Melter, which retained the heated grid of the old polyamide melter, but forced the polymer chip into contact with it using a power driven screw. By carefully designing the drainage holes, giving them a fine taper to increase the pressure by a wedging action, good melt rates could be achieved, even with the high viscosity of the polyester melt [12]. A system for drying the polymer chip in rotary driers and then transferring it in sealed hoppers to the melters was devised. The steam blanketing of the melters used for polyamide was replaced by dry nitrogen purging, and since the melt was always maintained completely dry, no equilibration of water content with steam was required as was the case in nylon machines, and so the melter design after the grid could be greatly simplified. Further simplifications to the spinning machine were also possible, since polyester does not need the steam conditioning tube used to promote crystallisation in nylon. This is a result of the higher glass-rubber transition temperature of PET, which ensures that no serious morphological changes can occur at room temperature on the wound bobbin, unlike polyamide 6.6 which is unstable on the bobbin unless crystallised before winding up. Fig. 6.

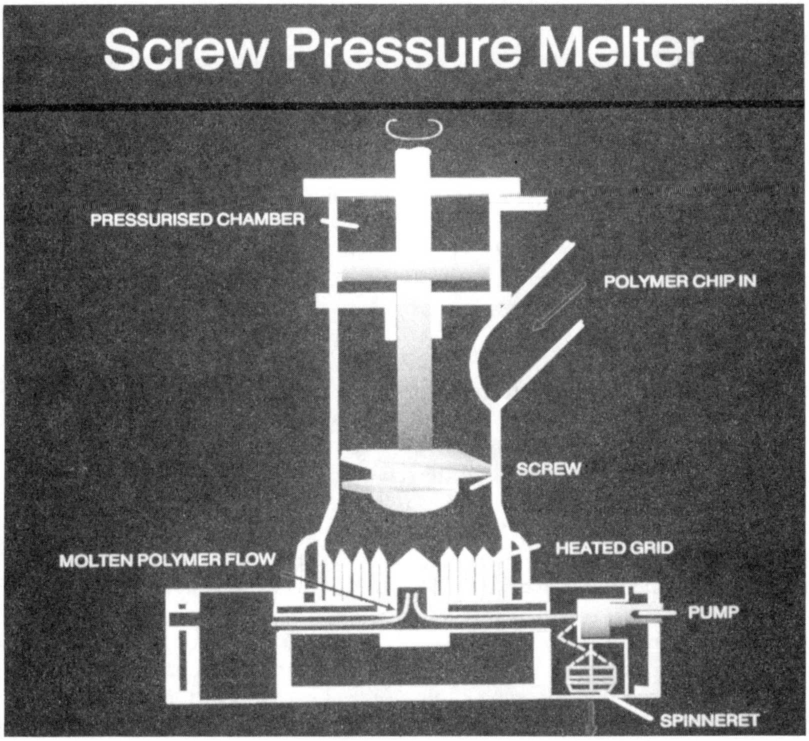
Screw Pressure Melter
PRESSURISED CHAMBER
POLYMER CHIP IN
SCREW
HEATED GRID
MOLTEN POLYMER FLOW
PUMP
SPINNERET

Figure 6.

In the 1960s screw extruder engineering developed significantly and extruders gradually replaced screw pressure melters. This move was accompanied by the modification of driers to a two stage process (a) an initial rotary section where most of the water was driven off, and sintering of chip due to crystallisation was avoided by the tumbling action followed by (b) a vertical plug flow gas column where the chip was dried to a carefully controlled low level.

Whilst this melter development for batch chip had been occurring Du Pont had followed the alternative path of continuous polymerisation followed by direct spinning. This technology is particularly suited for high throughput single product lines such as cotton-type staple. The batch route is today reserved for specialist polymers and fibres, most modern plant is based on continuous polymerisation.

The higher glass-rubber transition temperature of polyester made the development of the drawing process more difficult. Polyamide was drawn cold by passing the yarn continuously over a stationary peg between rolls, with the second roll rotating about 3.5 times faster than the first. The yarn then stretched by this amount as it ran over the peg. It was soon found that although polyester could be cold drawn by hand, a continuous process was not stable unless the yarn was heated to the region of this glass-rubber transition temperature [13] which is about 80°C. This necessitated first of all the development of a greater understanding of drawing in polymers, [14] and then the modification of polyamide draw frames to incorporate various heating devices [9].

The understanding of the way in which molecular orientation was produced in the spinning threadline was crucial in the development of high tenacity yarns for industrial use. It was discovered that increasing the molecular weight of the polymer gave better fatigue properties in the finished yarn, but the higher melt viscosity led to difficulties in spinning. In particular, the large increase in viscosity as the filaments cooled on leaving the spinneret caused frequent breaks to occur. By retarding the cooling of the filaments using a heated shroud round the spinning chimney, satisfactory processes for high molecular weight polyester fibres were developed [15].

It was also recognised very early that the molecular orientation of the yarn in the spinning threadline increased with wind-up speed, and that if a high enough wind-up speed could be reached, then further orientation by drawing would be unnecessary, and a yarn suitable for textile use could be made directly in a single step. As early as 1950, Du Pont was patenting a process for winding polyester at speeds in excess of 4700 m/min [16]. Unfortunately, this idea could not be developed commercially at the time, because of the lack of a reliable winder which could operate at the speeds required. Indeed it was more than twenty years before single stage processes could be run commercially at speeds of 4000 m/min and upwards.

However, the boom in polyester fibres at the end of the 1950s prompted large investments in production equipment based on separate spinning and drawing machines. In a bid to reduce costs and improve yarn quality, the major manufacturers introduced spin-draw machines, where the yarn passed directly from a conventional spinning stage to a draw stage on the same machine. However, these machines were complicated, and it was difficult to increase the speeds at which they ran, not least because of the need to heat the yarn for the drawing stage.

The start of the 1970s saw the development of draw-texturing machines which reduced the amount of handling that yarn packaged had to undergo [17]. Now, instead of three separate stages of spinning, drawing and texturing, the last two could be combined on one machine. However, the advantage of this system could at first only be realised if the fibre produced textured his own yarn, since the spun polyester yarn had a short shelf life and the unoriented yarn was very sensitive to damage during shipment. To enable the independent texturer to partake of the technical benefits of draw-texturing, which included better uniformity and improved yarn characteristics, [18] an additional development in technology was needed. This was the provision of yarn with a higher degree of orientation than conventional spun yarn, but which was still suitable for further drawing.

By the early 1970s, machine manufacturers were able to supply high speed winding machines which would run reliably at speeds up to about 4000 m/min [19]. This enabled fibre manufacturers to make partially oriented yarn (POY), which could then be shipped safely and processed economically by texturisers. A new era in polyester usage had started. However, fibre manufacturers could not produce fully oriented yarns in a single stage process at these POY speeds. The development of winders which would run at the required speeds of 6000 m/min and higher needed much more effort. However, an ingenious adaption of the spinning process was invented in the early 70s at the Celanese Corporation which enable the fibre maker to produce almost fully oriented yarns in a single stage, yet still winding up at less than 4000 m/min. This was the hot tube spinning process [20], see Fig. 7.

By arranging the right amount of heating in combination with the tension due to air drag in different parts of the spinning threadline, it was found possible to increase the molecular orientation produced in a single stage process above that normally obtained at such speeds. Although not possessing the high degree of orientation required for industrial uses, such yarns were adequate for direct textile use, and were put into production by Fiber Industries Inc. and ICI on both sides of the Atlantic.

TEXTURING

Although continuous filament polyester yarns were excellent for making fabrics which were crease resistant, yet retained their shape and held permanent pleats, they suffered from the same lack of aesthetic as polyamide yarns. The way forward, as for polyamide, was to texture the yarn in some way. This had been first achieved in 1948 on staple lines where the staple tow was fed through a crimper head in which the filaments were fed through a pair of nip rolls into a confined space under pressure and steam injection. This produced a random zig-zag crimp in each filament which gave a yarn with greater bulk more like natural fibres. However, a more attractive process for light decitex synthetic continuous filament yarns is fibre-twisting. Thermoplastic yarns, such as nylon and polyester, lend themselves to texturing by such processes as false twisting, where the yarn is twisted, heated to set in the twist, and then untwisted. By this process the individual filaments which are originally straight become deformed, and this imparts a bulkiness to the yarn which gives a more pleasing feel and appearance, rather like a spun wool or cotton yarn.

By the latter half of the 1950s bulked polyamide continuous filament yarns were taking the world by storm. False-twist processes were well established, and the stretchiness of the yarns produced by these means made them ideal for socks and other garments where stretch fit was required such as leotards and swimwear. They were not suitable, however, for knitted outerwear, which was an area seen by many as one of potential growth for synthetic fibres [21].

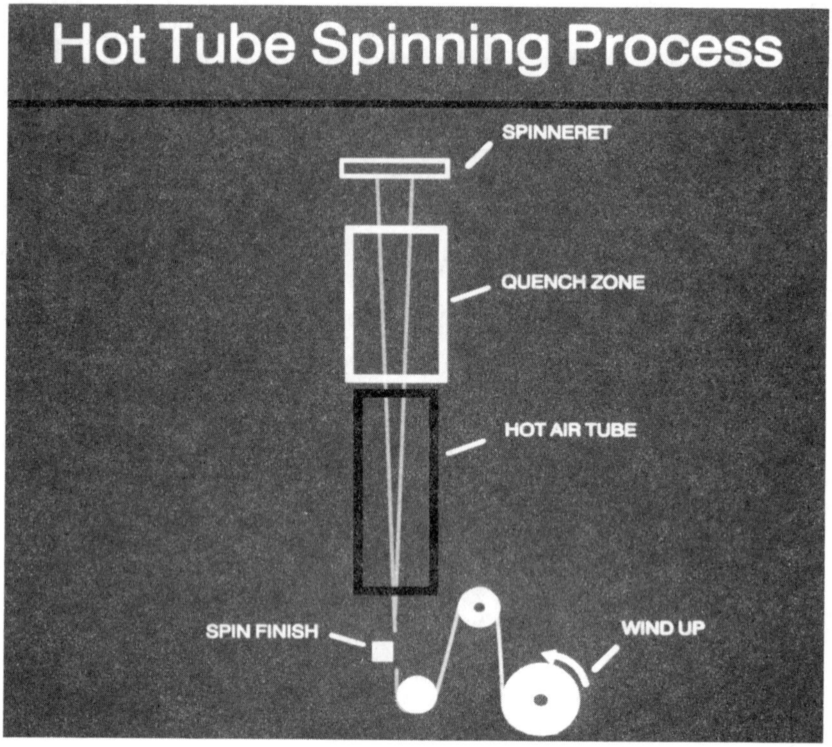
Hot Tube Spinning Process
SPINNERET
QUENCH ZONE
HOT AIR TUBE
SPIN FINISH
WIND UP

Figure 7.

Unfortunately, unlike wool or cotton yarns, the false twist polyester yarns of the time shrank considerably when subject to hot wet conditions, and this shrinkage was necessary to obtain the degree of bulk required to give the fabric or garment good aesthetic characteristics. The fabric had to be made oversize to accommodate the shrinkage, but most of the machinery used for knitting and weaving was originally intended for use with wool or cotton yarns. While it was possible on some machines to make the large allowances needed when knitting or weaving false twist textured yarns, it was an added complication, and one that could easily lead to errors, and non-uniform fabrics. If bulked filament yarns were to be successful in the outerwear trade, they needed to be non-torque so that they could be knitted in singles form, they had to have bulk without too much stretch, and if possible, they should lend themselves to package dyeing by a shorter route than the cumbersome muff dyeing procedure for false-twist nylon yarns.

The breakthrough came in the Macclesfield mill of Cheslene and Crepes Ltd. Here, Nova and Ruffini developed a process for bulking yarns which met all the requirements [22]. The technical advance was to allow a controlled relaxation of the yarn after texturing, and then to stabilise the yarn by a high temperature heat treatment. They did this by winding the false-twisted yarn at a slower speed than it was forwarded from the false-twist machine, so that a very soft package was produced. This package was then steamed in an autoclave at about 130°C to stabilise the bulk in its relaxed state. When applied to polyester yarns this process produced a low torque yarn with a low shrinkage compared to conventional false twist yarns [23]. In addition, the soft yarn package was the ideal density for even penetration of dye liquor, giving the possibility of package dyed yarns [24].

In 1958 Cheslene and Crepes approached ICI with a view to developing this new yarn, although several textile firms had already rejected it as being impracticable. ICI recognised the fact that the proces swas capable of development, and that it had advantages in the control it allowed over the final yarn properties [25]. The heat setting in the autoclave could be well controlled, and in addition it made the dye uptake of the yarn much less sensitive to variations in the false-twist process. However, there was still some residual problems with package to package dye uptake variations leading to stripiness in plain self coloured fabrics, and since jersey fabrics for outerwear were predominantly of this type, the new yarn appeared to have its main potential in the 20% of the market which was taken by Jacquard fabrics. It was launched by ICI under the name 'Crimplene' in 1959 [26] and had an enormous impact on the growth of polyester fibre.

Development work continued, but the problem of package to package dye variability persisted. ICI found that although there was no easy solution to stripiness, the effect was not so pronounced when the surface of the fabric was slightly broken up by deliberately knitting the fabric to give surface texture. In 1962 a breakthrough came with the development by ICI of two completely new fabric types which had a novel type of surface texture. Then, for the first time since ICI started work on the development of Crimplene, they were rewarded with a stroke of luck not of their own making. The fashion trend swung away from plain double jersey fabric to cloque and bourrelet fabrics having more surface appeal. Crimplene was ideal for this sort of fabric, and the 'Crimplene boom' was born [25], see Fig. 8.

The Crimplene process was well covered by patents, and could only be used with ICI's Terylene polyester yarns. Although other manufacturers of polyester yarns were quick to develop competitive yarns for the double jersey outerwear market, by using

Figure 8. Crimplene Garment

additional heaters on false-twist machines, these processes did not give such good quality yarns as the Crimplene process. Its share of the double jersey market rose from 3% in 1963 to 47% in 1968, in spite of the fact that it was a relatively expensive yarn [27].

The original patents held by ICI and Du Pont for polyester expired in 1964, which brought more competing bulked polyesters onto the market, but the boom in polyester double jersey fabrics, and Crimplene in particular, persisted through into the 1970s. A 60% return on capital was achieved by ICI during the mid-sixties from its Crimplene sales. A significant factor in this success can be attributed to the promotion being backed up by sound technical knowledge obtained from a continuous process of research and development [28].

The slow speed of bush false-twist crimping (and even slower pin-twisting) relative to the speed of spin-draw of flat yarn has not made it possible to produce a totally integrated spin-draw-false twist process and hence the continued importance of POY production. However gas jet texturising in a stuffer box is capable of high speed and in 1981 ICI launched their Mitrelle textured apparel yarns produced in a single stage process.

FABRICATION, DYEING AND FINISHING

Polyester 2GT is today used in practically every end use for fibres known to man. Sometimes it is used on its own, sometimes as a relatively insignificant component but more often as an important constituent, dependent on how cost-effective the final structure product properties are compared with those made from competitive fibre products. Each segment of each market will optimise the choice of initial fibre (continuous filament, flat or textured, 100% staple or staple blend spun), the method of this assembly into a structure, the dyeing route (if required to be dyed), and any surface or structure modification. If other fibres are incorporated into the structure, either as staple blends or as separate threadlines then dyeing and finishing routes will have to be a compromise between what would be normal for 100% polyester or 100% of the other fibre.

A comprehensive account of techniques used would therefore be a life-time's work and well outside the scope of this paper. Moreover the industry is in a continuous state of flux, with new products and technologies emerging, old techniques disappearing and acceptable chemical treatments being increasingly governed by toxilogical and environmental considerations. This section will therefore attempt to give a flavour of the enormous breadth of technologies involved by listing the more important ones, and then give an account of the major science developments concerned with the basic chemistry and physics of polyester fibre that have allowed this explosion of technologies to occur.

Fabric Manufacture[5]

The original piece of 100% polyester fabric was a woven square 10 in x 12 in of 96 denier fibre made as a panel in a substantially acetate warp in July 1946. The initial main problem with weaving was in the preparation of a processable warp, but this was quickly substantially overcome by the use of a size already developed for polyamide. Since that time:

(a) development of specialist multi-component spin finishes, to reduce static and friction and to aid cohesion

(b) if necessary, the augmentation of the finish with a cohesive size and
(c) improved yarn mechanical quality has led to the very low fault rates common today of $3/10^8$ metres of yarn and weaving efficiencies of 98%.

The first commercial sale of polyester in 1948 was 140 denier for brass bobbin yarn in lace-making. Product development quantities were also being sampled at the same time in weft knitting, warp knitting, braids, threads, nets, ropes and cordage. Tufted, flocked and non-woven fabrications followed later.

The key properties of polyester fibre that make it so attractive for volume fabric production are:

(1) Relatively low cost with high mechanical quality - the fibre produced not only has a cheap starting material but can also process it at maximum speeds.

(2) Satisfactorily high melting point (265°C) and T_g (115°C) - it can therefore be used without fear of melting in a large majority of end uses where fibres are required including ordinary apparel, domestic uses and many industrial applications. The T_g being above the temperature met in domestic washing machines, structure can be heat set into apparel fabrics without fear of them being destroyed at the first wash (crease resistance).

(3) Low sensitivity to moisture and hence the ability to go easily through the majority of finishing routes.

(4) Ability to easily alter fibre properties either by polymer or process variation to achieve maximum compatibility with other fibres in mixed fibre fabrics - especially with cotton in the largest volume outlet for polyester in apparel blends.

(5) High initial modulus which make it important for industrial structures such as tyre cords and conveyor belting where without brittle behaviour minimum elongation of the fabric is required for the in-use applied stresses.

Its major weaknesses are its poor elastic recovery from applied stress and its low flex fatigue life, both inferior to polyamides. This is because of lack of hydrogen bonding between polyester molecules, present in polyamides. Cost of dyeing is also more expensive than polyamide basically due to lack of chemical reactivity and the need to use carriers. However in spite of these deficiencies its spectrum of properties is perfectly suitable for many market segments and this has enabled polyester to gain the position of significance it enjoys.

In composite structures the original problems of separation of fibre from rubber or plastic matrix have long since been overcome by the development of purpose built 'glues', often based on epoxy chemistry, the exact formulations of which are closely guarded industrial secret.

Dyeing [6,29,30]

The optical appearance of a fabric can be significantly altered by properties of the basic fibre, such as its delustrant level, cross-section and textured format, and by the fabric construction itself and subsequent after treatments. However, the most important optical effect is to be able to easily colour the fabric.

When polyester was born it was born into a world that had a variety of dye-stuffs and dyeing procedures that were satisfactory for the fibres already in commercial production but none of which could acceptably dye polyester. Indeed at one stage the project was almost curtailed because of this problem - what was the use of an undyeable fibre particularly when there were several other alternative fibres awaiting scarce development resource including polyethylene, acrylics and peanut fibre (Ardil). Fortunately two acetate dyes (Dispersal Fast Scarlet B150 powder and Solacet Fast Orange 26 KS) were found to substantially dye polyester, wiser councils prevailed and ICI decided to concentrate on polyester as its major synthetic fibre product. Between 1945 and 1951 a major R&D effort on dyeing was mounted, extended to include textile firms, which was extremely successful. Not only were the fundamental principles of using carriers or/and excess pressure at dyeing established and the range of dyestuffs extended from disperse acetates to azo compounds but the applied techniques developed so that in 1951, technical handbooks were published on both dyeing and printing of polyester, immediately applicable in the trade.

The vast majority of polyester dyeing is carried out using disperse dyestuffs which are classified according to molecular weight. Occasionally other types are used including cationic if the basic polyester has been chemically modified with sulphuric acid groups, vat dyes and pigment dyes.

Disperse dyestuffs are generally hydrophobic aromatic molecules which exist as particles of about 1μ which are dispersed in water at around the 10 mg/l level. The continued stability of the dispersion during the dyeing process is important and has to be protected against a number of disrupting chemical and physical influences such as temperature, pH, agitation and carriers. This protection an either take the form of chemical modifications to the dye itself (inclusion of a low level of ionic species) or the use of surfactant dispersion aids.

Before the fabric is dyed it generally undergoes at least one pretreatment, a scour to remove the spinning finish oils and sizes (if used) that have been applied to aid fibre and fabric production. However other pretreatments are possible including heat setting (dry and by steam), alkali reduction, bleaching and optical brightening - dependent on the specific final fabric requirements. There are used both a variety of pretreatment equipment and more especially of dyeing equipment. The variety has stemmed both from the desire to perform different physical and chemical unit processes and from the machine manufacturers evolving their own preferred equipment.

Pretreatment washing machines for example can rely on a slow washing passage and a set of rollers through a scouring liquor, a perforated drum system when the liquid is forced onto the fabric as it passes over the drum, or a centrifugal batching machine where liquid is fed into the centre of a spinning drum containing the fabric fixed to the wall. However there are more specialist pieces of equipment such as pad-roll systems that prevent fabric creasing.

Dyeing machines are built either for batch or continuous operation. The more common batch machines for apparel operate above atmospheric pressure of steam, up to a maximum of 140°C. Below 110°C there is a danger of dimers and trimers (around 1.4% extractable in chloroform from fibre) precipitating in the dye bath. The advantage of operating at these high temperatures is that it is above the T_g of polyester and dyestuffs enter the fibre more quickly. There are a number of general principles to be observed in the design of the machines particularly as to how the dye liquor uniformly penetrates the fabric. Thus on the liquor side turbulence must be avoided by the use of

Dyeing Equipment c 1960

Figure 9. High-pressure Loose Stock Slubbing and Yarn Dyeing Machine (Longclose).

Such exotic fabrics using very fine fibres were not popular in Europe at that time, but in 1981 ICI launched a 0.8 denier per filament yarn known as Mitrelle [34] which was used to make high quality silk-like fabrics for women's wear.

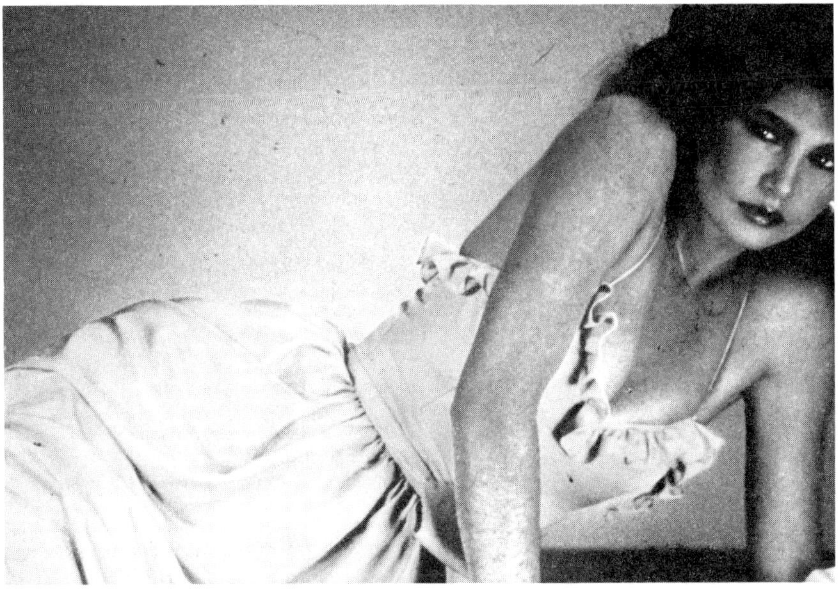

Figure 10. Mitrelle Silk-like Fabric

By the end of the 1980s the use of polyester microfibre had become well established, not only in apparel [35], but also for cleaning cloths [36] and household items such as towels [37], where the high surface area of such fine fibres is used to trap dirt or moisture to good effect.

CONCLUSIONS

Polyester is today the most widely used synthetic fibre in terms of volume. It owes this position to a combination of a low cost base with a number of inherent properties, such as its ability to blend well with natural fibres like cotton and wool. It was this feature above all, which enabled polyester to spearhead the revolutionary swing to 'easy-care' clothing in the 1950s. Other important properties which have ensured its success are its ability to be heat-set for permanent pleats in garments, and its resistance to sunlight through glass, which gives it great advantages as curtain material. Its high Young's modulus is particularly useful in many industrial applications, such as reinforcement for belting and tyres. However, none of this growth would have been possible without the timely development of the appropriate technology. Right at the beginning, polyester fibres might never have seen the light of day if Whinfield and Dickson had not discovered a way of purifying terephthalic acid so that a pure polymer could be produced. The continuing improvement in polymer purity achieved over the years facilitated further technical advances such as higher speed spinning and the production of microfibres. Another key enabling step was the development of suitable dyeing processes, without which polyester would not have been acceptable as a textile fibre. Finally, achieving reliable and economic production has needed engineering advances, some of which, like high speed winding machinery, took years to develop to the required degree of sophistication.

The invention of polyester was one of the great discoveries of history, one that now effects the daily life of the majority of the human race and in the future will do so increasingly more and more as polyester steadily overtakes cotton production.

REFERENCES

1. Textile Organon, June 1990.
2. D. A. Hounshell and J. K. Smith, Jr., Science and Corporate Strategy Du Pont R&D - 1902-1980, Cambridge, p. 409.
3. British Patent 578,079.
4. The Launching of a New Synthetic Fibre, ICI Terylene Polyester Fibre - A Historical Survey, ICI, 1954.
5. W. F. Osbourne, A History of the Early Development of 'Terylene' Polyester Fiber, ICI, 1960.
6. Chemiefasern/Textilindustrie, 40/92, Sept. 1990.
7. M. Grayson, ed., Kirk-Othmer Encyclopedia of Chemical Technology, 3rd ed., Vols. 11, 17, Wiley Interscience, 1980-82, Articles on glycols and phthalic acids.
8. K. Weissermel and H. J. Arpe (translated A. Mullen). Industrial Organic Chemistry. Verlay Chemie, 1978.
9. Internal ICI reports.
10. H. F. Mark, et al, eds, Encyclopedia of Polymer Science and Engineering, Vol. 12, John Wiley & Sons, Inc., 1988.
11. British Patent 1,296,242.
12. British Patent 719,062.
13. I. Marshall and A. B. Thompson, Proc. Roy. Soc. Long., A221, 541 (1954).
14. A. B. Thompson, J. Polymer. Sci., 34, 741 (1959).
15. U.S. Patent 3,361,859.
16. U.S. Patent 2,604,667.
17. W. A. B. Davidson, Modern Knitting Management, 7 (Oct. 1971).
18. Draw-Textured Yarn Technology, Monsanto Textile Co., 1974.
19. H. de Haas, Modern Textiles, 74 (Oct. 1971).
20. U.S. Patent 3,946,100.
21. Hosiery Times, 83 (April 1964).
22. British Patent 881,729.
23. British Patent 921,583.
24. C.S. Bruce, Modern Textiles Magazine, 47, 62 (Aug. 1966).
25. J. E. Gibbon, Hosiery Trade Journal, 110 (Sept. 1965).
26. C. M. Shorthouse, Textile Mercury and Argus, 142, 275 (Feb. 1962).
27. I. M. Law, Knitting International, 78 (June 1981).
28. ICI Ltd. Annual Reports, 1965-1970.
29. "The Dyeing of Polyester Fibres," 3rd ed., ICI Ltd. Dyestuffs Division, 1964.
30. Dyeing and Finishing of Polyester Fibres, BASF Manual.
31. S. H. Zeronium and M. J. Collins, Surface Modification of Polyester by Alkaline Treatments. Textile Progress, Vol. 20 No. 2, The Textile Institute, 1989.
32. J. D. Geerdes, Int. Fiber Journal, 5(3), 24 (1990).
33. High Performance Textiles, 6(4), 5 (1985).
34. Daily News Record, 20, 94 (May 1981).
35. Manufacturing Clothier, 70, 16 (1989).
36. Textile Horizons, 9, 8 (Nov. 1989).
37. Textile Horizons, 10, 10 (July 1990).

HISTORY OF POLY(PHENYLENE SULFIDE) FIBERS

JACK G. SCRUGGS AND MARVIN WISHMAN
Phillips Fibers Corporation, P. O. Box 66, Greenville, SC 29602

Introduction

The generic names of some of the man-made fibers, such as polyester, nylon, acrylic, olefin and spandex, have become so familiar that they are literally household words. Many of the trademarks associated with these fibers are also widely recognized by almost everyone, e.g., Hoechst Celanese "Trevira" polyester, Allied "Anso" nylon, Monsanto "Acrilan" acrylic, Phillips "Marvess" olefin and Dupont "Lycra" spandex fibers.

Official recognition of fiber generic names in the U.S. is obtained by petitioning the Federal Trade Commission (FTC) with evidence that the new fiber meets three criteria: (1) the chemical composition is radically different from other fibers resulting in distinctive physical properties of significance to the general public, (2) the fiber must be in active commercial use or such use must be immediately foreseen, and (3) the grant of a generic name must be of importance to the consuming public at large. In 1986 when the FTC approved the petition by Phillips Fibers Corporation for recognition of poly(phenylene sulfide) fiber, or PPS fiber, as representing a new generic class of fibers, no new generic class of fiber had been approved by the FTC since 1969. The rather descriptive name approved by the FTC for PPS fiber was "sulfar,"[1] defined as "a manufactured fiber in which the fiber-forming substance is a long chain synthetic polysulfide in which at least 85% of the sulfide (-S-) linkages are attached directly to two aromatic rings." This first commercial sulfar fiber was introduced commercially by Phillips Fibers Corporation in 1983 under the trademark Ryton*. Following is a history of the technical developments related to the development of PPS polymer and PPS fibers.

Development of PPS Polymer for Fibers

In 1897, Grenvesse[2] reported that the reaction of benzene and sulfur in the presence of aluminum chloride resulted in an amorphous, insoluble resin which was presumed by later investigators to have been PPS. This reaction had been described in some of the original work by Friedel and Crafts[3], for whom this type of reaction was named. However, later work by Cleary reported in 1984[4] concluded that the reaction produced diphenyl sulfide, thianthrene and phenylthiothianthrene but no PPS.

In 1948, Macallum[5] reported that he had successfully produced PPS by a melt polymerization process starting with a mixture of p-dichlorobenzene, sodium carbonate and sulfur in a sealed container at 300-340°C.

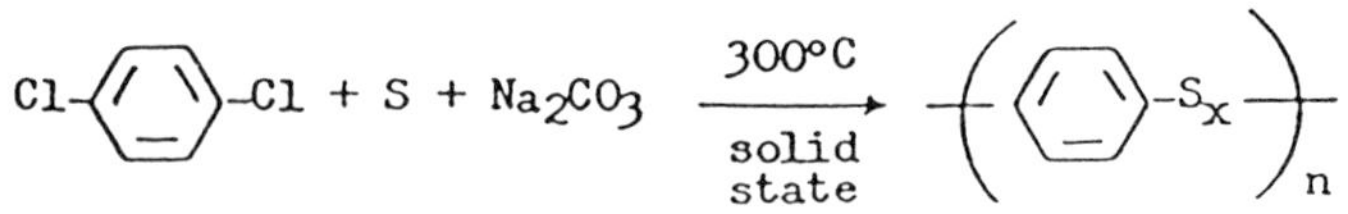

Variations in the Macallum process were studied by Lenz and coworkers at Dow Chemical[6,7] who produced PPS by the self-condensation of metal salts of a p-halothiophenol. However, removing the by-product copper bromide from the polymer proved to be difficult.

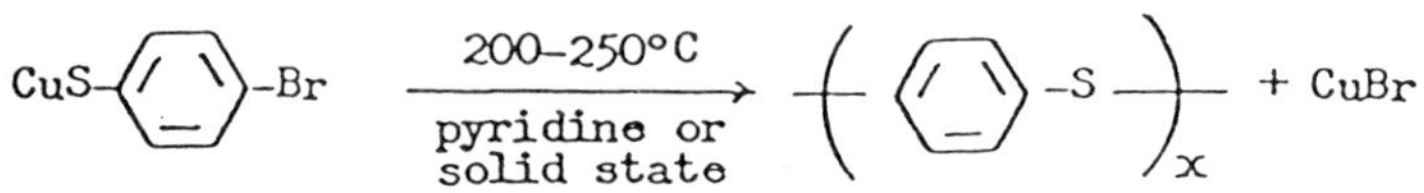

In 1967, a U.S. patent by Edmonds and Hill[8] was issued to Phillips Petroleum Company for a new process for producing arylene sulfide polymers. The process described in this patent is a polymerization reaction involving poly-halo-substituted aromatic compounds and alkali metal sulfides in a polar organic compound.

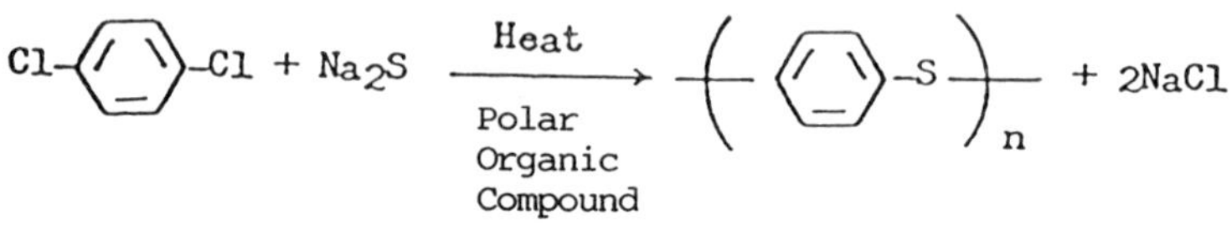

This process was based on readily available raw materials and proved to be commercially feasible. Phillips Petroleum Company began production of PPS in a semi-works facility and did extensive work to characterize the polymer and develop to applications for it. Commercial production of PPS began in 1973[9] with the product sold under the Phillips Petroleum Company trademark Ryton.

This first commercial PPS was a linear polymer of moderate molecular weight and mechanical properties.[10] It was useful as a curable coating resin and as a feedstock for the preparation of injection molding resins by an air curing process involving both chain extension and cross-linking reactions.[11] Excellent molding compounds were produced by compounding this cured resin with chopped glass fibers and minerals, but neither the relatively low molecular weight uncured polymer nor the cured polymer was satisfactory for producing good fibers.

Further research[12] led to a process for preparing a linear high molecular weight uncured polymer suitable for producing fibers by a melt spinning process. As polymerized, this polymer is a light tan, free-flowing powder. It is not soluble in any known solvent below 200 °C; above this temperature, it has limited solubility in some solvents such as aromatics, chloro-aromatics, and certain heterocyclic compounds.[13]

Differential scanning calorimetry (DSC) shows that this linear "virgin" PPS polymer has a melting endotherm at 278 °C. Polymer rapidly quenched from the melt shows a glass transition temperature (Tg) at about 85 °C and a crystallization endotherm (Tc) at 136 °C[14].

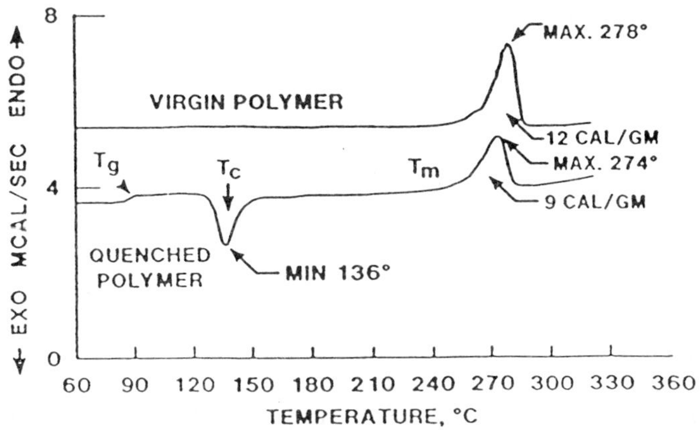

Figure 1. Differential Scanning Calorimetry of PPS Polymer

This process became the basis for producing fiber grade polymer for use by Phillips Fibers Corporation to produce PPS fibers.

Development of PPS Fibers

The characteristics of molded products made from PPS suggested that a potentially outstanding industrial fiber could be made from a suitable melt spinnable polymer. Among these characteristics were: good dimensional stability, inherent flame retardancy, high melting temperature, long term stability at elevated temperatures, outstanding resistance to most chemicals, good electrical properties, and ease of processing.

Fiber was produced from PPS polymer by melt spinning and was then drawn at elevated temperatures and converted into staple fiber using conventional crimping and cutting processes. The undrawn fiber was largely amorphous, as indicated by a pronounced DSC crystallization exotherm at 125°C. Drawn fiber showed no distinct Tg or crystallization exotherm, since it was crystallized during the drawing process.[14]

Phillips Fibers Corporation began development of a PPS fiber in the mid 1970's, including market evaluations to determine suitable applications for the fiber. In 1978, development quantities of a 4 dpf staple fiber were supplied to the Carter-Day Company for use in pulse jet bag filters for coal-fired boilers. For this application, the fiber must be able to resist degradation in a hot, acidic environment while being subjected to rather vigorous flexing. Carter-Day found that only Ryton PPS fiber and PTFE fiber would withstand these conditions, and Ryton PPS fiber was found to have better filtration characteristics and dimensional stability in the needled felt fabrics preferred for this application. Many other fibers were evaluated but all others failed the rigors of this test.[15]

Compared to today's Ryton PPS fibers, these early fibers were inferior in physical properties:

Table 1. Ryton PPS Fiber Properties

	1978 (Development)	1991 (Commercial)
Denier (dpf)	4.4	3.0
Tenacity (gpd)	1.8	3.5
Elongation (%)	32	35
Shrinkage (130 °C)	25	3

After proving the acceptability of PPS fiber in several small boiler operations, in 1979 Carter Day installed a large baghouse equipped with Ryton PPS bags on a coal-fired boiler at the Adolph Coors plant in Golden, Colorado. Performance of Ryton PPS fiber in this application has been excellent, as described in several technical presentations by Coors engineers.[16,17] Many of these original Ryton PPS bags lasted over six years in continuous service at operating temperatures in the range of 300-370 °F.[18]

Another application where Ryton PPS fiber filter fabrics have shown outstanding performance compared to other fibers is in bag filters for municipal waste incinerators.[19] Of major concern in the incineration of refuse are the products of oxidation of various sulfur compounds, i.e., sulfur dioxide and sulfur trioxide, which combine with the condensed water of combustion to form sulfurous and sulfuric acid, respectively. Laboratory tests on filter fabrics prepared from a variety of speciality fibers showed only PPS fabrics and PTFE fabrics were resistant to attack by a 2% solution of sulfuric acid at 350 °F.[20]

Phillips Fibers Corporation began commercial production of PPS fiber in 1983 with a 3 dpf staple fiber sold under the trademark Ryton. The principal application for this fiber is in filter fabrics for pulse jet baghouses installed on coal-fired boilers or waste incinerators. Other applications being evaluated include papermakers' felts, electrolysis membranes, composites, filter fabrics and electrical insulation. Its fire resistance (limiting oxygen index or LOI = 38) and its low smoke toxicity make it of interest for fabrics for the interior of aircraft.[21]

By producing a drapable woven fabric from two comingled multifilament yarns, such as carbon fiber and PPS fiber, followed by a molding step in which the thermoplastic yarn is melted and becomes the continuous phase, high performance composites can be produced.[22]

PPS Monofilaments

The good thermal stability and resistance of PPS to aqueous hydrolysis under either acid or alkaline conditions resulted in interest in the polymer for monofilaments for use in dryer fabrics for the paper industry. One U.S. patent[23] described monofilaments produced from blends of PPS with a minor amount of a halogenated polymer added to reduce brittleness and improve knot and loop strength. Fabrics made from these monofilaments were found to last longer than fabrics made from polyester or nylon monofilaments because of the superior chemical and temperature resistance of

PPS. Another U.S. patent[24] describes a similar monofilament made from a blend of PPS and up to 20% nylon 66 for "improved toughness." This monofilament is also for use in dryer fabrics for paper machines, again because of its thermal stability and hydrolysis resistance.

Recent Activity in PPS by Other Companies

Several Japanese companies have begun production of PPS, some with a U.S. partner involved. Phillips Petroleum Company and Toray Industries, Inc. formed a joint venture to produce PPS resins in Japan using Phillips Petroleum Company technology.[25] General Electric announced plans to "make and sell a family of PPS resins" using technology licensed from Toso Susteel[26] and sold under the trademark "Supec." Hoechst Celanese Corporation in a joint venture with Kureha Chemical Industry, Inc., under a license from Phillips Petroleum Company, announced plans to build a PPS plant in the U.S. scheduled for completion by 1993. Hoechst sells this polymer under the trademark "Fortron".[27] A U.S. patent issued to Kureha described a process for producing fibers from a polyarylene sulfide copolymer produced by using a mixture of para- and meta-dichlorobenzene as starting materials.[28] Tohpren, a joint venture company of Tonen Petrochemical and Toto Chemical, appointed Solvay to market their PPS resin in Europe. Both Teijin and Toyobo have announced plans to produce PPS fibers. Toyobo's fibers will be sold under the oxymoronic trademark "Procon." A recent U.S. patent issued to Asahi[29] describes a nonwoven filter fabric produced from microdenier PPS fibers and having superior heat resistance, chemical resistance and fire retardant properties. Dainippon Ink and Chemicals, Inc. has described in Japanese patents a process for producing a hollow fiber from PPS.[30]

Among European companies, patents issued to Bayer indicate a lot of activity in PPS polymer, fiber and monofilament.[31] Bayer recently announced[32] the startup of a plant in Antwerp with a capacity to produce 4,000 tons per year of PPS based on their own technology. This polymer is to be used for plastics, fibers and monofilaments. This same announcement stated that its fiber plant at Dornagen would produce staple fibers and monofilaments under the trademark "Tregalon" for hot gas/dust separation, wet and dry filtration, dryer felts for paper machines and other industrial applications.

If the number of companies now active in PPS polymer, fibers and monofilaments is any indication of what is to come, we should see a lot of new variants of PPS fibers for a lot of new applications within the next few years.

References

1. Federal Register, 51, No. 110, 20809 (June 9, 1986).
2. Grenvesse, P., Bull. Soc. Chim. France 17, 599 (1897).
3. Friedel, C., and Crafts, J. M., Ann. Chim, Phys. 14 (6), 433 (1888).
4. Chemical & Engineering News 62, 46 (Sept. 10, 1984).
5. Macallum, A. D., J. Org. Chem. 13, 154 (1948); U. S. Pats. 2,513,188 (June 27, 1950) and 2,538,941 (Jan. 23, 1951).
6. Lenz, R. W., and Handlovitz, C. E., J. Polym. Sci. 43, 167 (1960).
7. Lenz, R. W., Handlovitz, C. E., and Smith, H. A., J. Polym. Sci. 58, 351 (1962).
8. Edmonds, J. T., Jr., and Hill, H. W., Jr., U.S. Pat. 3,354,129 to Phillips Petroleum Co. (Nov. 21, 1967).

9. Hill, H. W., Jr., "Polyphenylene Sulfide: Stability and Long Term Behavior," in Durability of Macromolecular Materials (R. K. Eby, Ed.), ACS Symposium Series 95, p. 183, (presented at the 176th Meeting of the American Chemical Society, Sept. 11-14, 1978).
10. Hill, H. W., Jr., I.&E.C. Product Research & Development 18, 252 (Dec., 1979).
11. Hawkins, R. T., Macromolecules 9, No. 2, 189 (March-April, 1976).
12. Campbell, R. W., U.S. Pat. 3,919,177 to Phillips Petroleum Co. (Nov. 11, 1975).
13. Bailey, G. C., and Hill, H. W., Jr., "Polyphenylene Sulfide; A New Industrial Resin," in New Industrial Polymers (Rudolph D. Deanin, Ed.), ACS Symposium Series 4, p. 83, presented at the 167th Meeting of the American Chemical Society, April 1-2, 1974.
14. Scruggs, J. G., and Reed, J. O., "Polyphenylene Sulfide Fibers," Handbook of Fiber Science and Technology: Vol. III, High Technology Fibers, Part A, p. 335 (M. Lewin and J. Preston, Ed.), Marcel Dekker, Inc., New York and Basel (1985).
15. Wellan, W. G., "Ryton Felt for Flue Gas Applications," presented at the Clemson University Textile Conference on Developments in Filtration Technology, May 14-15, 1986.
16. Pearson, G. L., "Pulse-Jet Type Fabric Experience at Air to Cloth Ratio of 5 to 1 on a Boiler Firing Pulverized Coal," presented at the Third EPA Symposium on the Transfer and Utilization of Particulate Control Technology, March 9-12, 1981.
17. Pearson, G. L., and Capps, D. D., "Pulse-Jet Fabric Filter Experience Using Non-Glass Media at Air to Cloth Ratio of 5 to 1 on a Pulverized Coal Fired Boiler," presented at the Fourth EPA Symposium on the Transfer and Utilization of Particulate Control Technology, Oct. 11-15, 1982.
18. Grubb, W. T., "Evaluation of Acid-Resistant Synthetic Needled Felt Filter Media in Coal-Fired Boiler Pulse-Jet Baghouses," presented at the Sixth EPA Symposium on the Transfer and Utilization of Particulate Control Technology, March 1986.
19. Olofsson, H., and Barranger, C. B., "Bag Fabric and Fabric Finish Evaluation at a Municipal Waste Incineration Plant," presented at the Third Conference on Fabric Filter Technology for Coal-Fired Power Plants, Nov. 19-21, 1985.
20. Parker, R. B., "Comparisons of High Temperature Fibers in Hot Gas Filtration," presented at the Industrial Fabrics Association International Annual Convention, Oct. 22-25, 1989.
21. Nonwovens Report International, No. 217, 11 (April, 1989).
22. Modern Plastics International 17, No. 5, p. 6 (May, 1987).
23. Ballard, L., U.S. Pat. 4,610,916 to Shakespeare Co. (Sep. 9, 1986).
24. Baker, S. M., Best, F. B., and Bhatt, G. M., U.S. Pat. 4,755,420 to JWI Ltd. (Jul. 5, 1988).
25. Modern Plastics, 61, No. 2, 93 (Feb., 1984).
26. Chemical Week, 141, No. 22, 10 (Nov. 25, 1987).
27. Chemical & Engineering News, 68, No. 36, 13 (Sept. 3, 1990).
28. Konyama, T., Egawa, H., Ilzuka, Y., and Shiiki, Z., U.S. Pat. 4,816,335 to Kureha Kagaku Kogyo Kabushiki Kaisha (Mar. 28, 1989).
29. Ikeda, M., and Shima, T., U.S. Pat. 4,950,529 to Asahi Kasei Kabushiki Kaisha (Aug. 21, 1990).
30. "Microporous Hollow Fibres Containing Poly(para-phenylene sulfide)," Jap. Pat. 168446 (Sept. 22, 1982); "Fibre Production with Microporous Surface by Melt Spinning Poly(p-phenylene sulfide), Drawing and Heat Setting," Jap. Pat. 169722 (Sept. 30, 1982).
31. Idel, K., and Willenberg, B., U.S. Pat. 4,645,825 to Bayer Aktiengesselschaft (Feb. 24, 1987).
32. Nonwovens Industry, 21, No. 10, 24 (Oct., 1990).

HISTORY OF CARBON FIBERS

ROOP C. BANSAL
Department of Chemical Engineering & Technology, Panjab University, Chandigarh-160014, India
J. B. DONNET
CNRS Centre de Recherches sur la Physico-chimie des Surfaces Solides, F-60200 Mulhouse, France

INTRODUCTION

Aerospace structures need light weight but strong materials that can withstand heavy loadings with minimum deflection. The aerospace industry has, therefore, generated a considerable amount of interest for continued development of advanced materials. Carbon fibres are one such high performance material which has attracted world wide attention. Although their initial applications were confined to military aircraft and space fields, they are now being extensively considered important for nonmilitary applications were confined to military aircraft and space fields, they are now being extensively considered important for nonmilitary applications such as for commercial and civilian aircrafts and in recreational, industrial and transportation fields.

Carbon fibers are thin black filaments with diameter smaller than 10 m and containing about 99% chemically pure carbon. They have low density and thus possess the highest specific strength and specific modulus of all currently available engineering materials. They are elastic to failure at normal temperatures which renders them creep resistant and non-susceptible to fatigue. They are chemically inert except in strongly oxidising environments or in contact with some molten metals. Carbon fibers have exceptional thermophysical properties and excellent damping characteristics.

They can be used as yarn, felt, fabric or powder like short monofilaments to reinforce light weight materials, generally an epoxy resin and occasionally a polyester or a polyamide. For high temperature applications they are also being frequently used to reinforce certain pitch and other refractory materials which on pyrolysis produce a carbon matrix. In the composite, the carbon fiber contributes towards stiffness and strength and causes a large reduction in thermal expansion coefficient, a vast improvement in fatigue strength and a decrease in electrical resistivity. Thus a tremendous amount of research and developmental activity has been directed in the carbon fiber field.

HISTORY OF CARBON FIBER DEVELOPMENT

Carbon fibres had been known for nearly a century when Thomas Edison used them as filaments in incandescent lamps. These early carbon fibers were prepared by the carbonisation of bamboo fibres and later from regenerated cellulose. Soon after the development of tungsten filaments this use of carbon fibres was abandoned. Later carbon fibres were prepared from rayon to be used for thermal insulation but these fibers could not compete with glass fibres. These rayon-based carbon fibres had retained the fibril structure of rayon in the carbonised state but were isotropic as shown by X-ray examination of their preferred orientation.

The story of the development of high performance carbon fibers started in 1950s at Wright Patterson Air Force Base where the first major effort was undertaken to prepare high strength carbon fibres. Union Carbide [1] developed this process and was able to demonstrate that high modulus of carbon fibres can be achieved by hot stretching of the isotropic carbon fibre during the heat treatment at the graphitisation temperature. This process led to the preparation of high strength and high modulus carbon fibers. The carbon fibre types available commercially from Union Carbide were called Thornel-25, 40, and 50 indicating the modulus number in million psi. The tensile strength of these fibres was around 2000 MPa. This was the time when U.S. Airforce was thinking of replacing boron fibres for their new generation of military aeroplanes.

At about the same time Sindo in Japan [2] and Watt and coworkers in U.K. [3], working independently, successfully prepared carbon fibres from polyacrylonitrile (PAN). In 1963 L.N. Phillips referred Watt to "black orlon" which could be prepared from the commercial PAN fibres by heating at 200°C for several hours in air followed by pyrolysis in a flame. Watt and Johnson then began the first experiment with the Courtaulds PAN fibre courtelle which was aimed at bringing the exothermic oxidation reaction under control. The carbon fibre so produced when carbonised at 1000°C had a modulus of 150 GPa which could be further enhanced by heat treatment at 2500°C. An important development during this work which eventually led to the RAE British patent for producing high modulus carbon fibres was the understanding of the need to restrict the PAN fibres during the oxidation process. This early batch process was later developed into a continuous process and in 1966 was used on fully commercial lines at Morganite limited and Courtaulds limited.

The PAN process of carbon fibre production received the immediate attention of several manufacturers because it turned out to be more economical due to the less expensive precursor polymer and to the simple fabrication process. The yield of carbon fibre was almost double of its yield from rayon and the process did not entail the difficult and expensive high temperature stressing treatment.

The preferred orientation needed for the high performance carbon fibres is achieved, in case of the PAN precursor, by stretching the polymeric precursor fibre when it is still thermoplastic. This helps in compensating for the entropy shrinkage which occurs during subsequent transformation into the more stable step ladder polymer by heat treatment in air around 200°C. The carbonisation of the stabilised fibre between 1200-1400°C improves strength values to about 3000 MPa while a modulus of about 250 GPa is obtained. The post carbonisation heat treatment to graphitisation temperature improves the modulus of the carbon fibre to around 350 GPa although the strength values tend to decrease by this heat treatment (Fig. 1).

These developments and the realisation that carbon fibres may be used successfully to produce light weight automobile parts that reduce energy consumption gave incentive to investigators to search for new and low cost sources for carbon fibre production. Thus several types of precursor materials have been investigated and their feasibility for large scale production of carbon fibres assessed [11].

The first pitch based carbon fibres were prepared by Otani [4] by the pyrolysis of PVC pitch fibres. The pitch fibres were obtained by extruding molten pitch through a pin hole on the bottom of a cell used for melting pitch. The pitch filaments were extended and solidified as they flowed down through air and were wound on to bobbins at high speed. However, the carbon fibres obtained were isotropic and had poor mechanical properties. A considerable improvement in mechanical properties was observed by applying tensile load during the heat treatment of PVC pitch carbon fibres. A strength of 2.18 GPa and a young's modulus of 440

GPa were achieved by hot stretching of PVC carbon fibres above 2000°C. Petroleum pitch is the new material used for producing pitch based carbon fibres. Kueha chemical company of Japan built a commercial plant with a production capacity of 250 tons per year of carbon fibres and extended it later.

However, the hot stretching process for the preparation of carbon fibre is a difficult and costly process. Attempts were, therefore, made to produce highly oriented carbon fibres from other varieties of pitch materials. One such approach employed special pitches consisting of aromatic molecules with structures capable of forming anisotropic pitch fibres. These pitches were prepared [5] by thermal treatment of small polynuclear materials such as phenanthracene, benzopynene and chrysene or were obtained from by products of thermal cracking of petroleum fractions or of certain crudes above 700°C preferably between 1000-1200°C.

While measuring the viscosity of acenapthalene pitch between 400-500°C, Singer [6] observed that this "taffylike" stage exists for a considerable time before the material solidified so that it was possible to draw fibres from the mesophase. Thus the first mesophase fibres were drawn from acenapthalene pitch at 450°C by repeatedly dipping a quartz rod into the pitch as it gradually thickened. These investigations gave rise to the development of spinning equipment and also resulted in attempts to use low temperature melting pitches which could be produced in a separate operation and then spun at a temperature well below the rapid cooking temperature [7,8]. Union Carbide completed a new mesophase pitch based carbon fibre plant at Greenville, South Carolina which attained its maximum capacity of about 230 tons per year in March 1981. This plant is producing a variety of continuous and yarn of 200 and 4000 filaments with elastic modulus varying between 170 and 830 GPa.

Pitch fibres have a high degree of preferred orientation and maintain this orientation during the thermosetting step. The degree of preferred orientation increases with heat treatment temperature above 1000°C. After heat treatment above 3000°C the Young's modulus approaches a value of 880 GPa close to the graphite single crystal value of 1000 GPa [9]. Union Carbide (now Amoco) and Toray in Japan have produced pitch fibres in two forms: as fibre mat and as continuous filament yarn. The two types are primarily differentiated by the spinning technique but differ significantly in appearance, properties and cost potential. The carbon fibre mat which is designated as Thornel Type P mat was marketed by Union Carbide in 1973. SEM micrograph [10] of the product show that the filaments have random layered orientation and an average diameter of 10 μm. The curved nature of the filament results from the spinning process. The pitch based continuous carbon fibres (Thornel Type P continuous yarn) are round and have a diameter of 7.8 μm and a density of 1.95 g/cm^3. Their strength is lower than PAN based carbon fibres.

Recently carbon fibres have also been obtained by the pyrolysis of certain gases and vapours. They are formed on ceramic surfaces by the thermal decomposition of a mixture of hydrogen and a hydrocarbon such as benzene, methane, naphthalene at 1000-1300°C. The carbon fibres grow straight to short lengths up to 25 cm depending upon the length of the heating zone and have diameter varying between 3 and 100 μm. They differ in morphology and growth conditions from curved or bent vermicular carbon filaments of much smaller dimensions (several micrometer in length only) which grow on transition metal surfaces at much lower temperature (400°C). These vapour grown carbon fibres have a concentric layer structure of turbostratic carbon. These concentric sheets of carbon have a high degree of alignment of the graphitic basal planes along the fibre axis. Their strength and modulus are as high as that of commercial carbon fibres prepared from organic polymer precursors such as PAN, rayon and pitch. They have good electrical properties which can be further enhanced by intercalation.

CLASSIFICATION OF CARBON FIBRES

The classification of carbon fibres poses a very difficult problem [11] because new brands of carbon fibres with enhanced mechanical properties are being produced and put into the market rapidly. Figure 2 shows the range of strengths and modulii and the approximate ultimate strains for a variety of carbon fibres manufactured by different industries. The plot although not comprehensive is restricted to the higher - quality continuous tows available on spool. Attempts are being made to bring the properties of high strength PAN-based carbon fibres closer to the properties of high modulus mesophase pitch-based carbon fibres.

As the mechanical properties of carbon fibres which make them versatile materials can be enhanced by varying the process parameters such as the final heat treatment temperature and the nature of the precursor fibre material, they can be classified in a number of ways. The IUPAC international committee on the nomenclature and characterisation of Carbons and Graphite, after considerable discussion and debate has classified carbon fibres into five types on the basis of scientifically verifiable mechanical properties.

UHM Type	Carbon fibres with modulus greater than 500 GPa.
HM Type	Carbon fibres with modulus greater than 300 GPa and with a tensile strength-to-tensile modulus ratio of less than 1%.
IM Type	Carbon fibres with modulus up to 300 GPa and with a strength to modulus ratio above $1x10^{-2}$.
Low Modulus Type	Carbon fibres with modulus as low as 100 GPa and with low strength. These are carbon fibres that have an isotropic structure.
HT Type	Carbon fibres with tensile strength values greater than 3000 MPa and strength to stiffness ratio between 1.5 and $2x10^{-2}$.

PREPARATION OF CARBON FIBRES

Several organic polymeric materials such as rayon, PAN, aromatic heterocyclic polymers, linear polymers and pitch precursor fibres have been used as starting materials for the preparation of carbon fibres. However, for the large scale production of carbon fibres, three important starting materials used are PAN, rayon and pitch precursor fibres. At present the PAN based carbon fibres constitute the most important group which are highly desirable in high performance composites for aircraft, aerospace and other high technology applications. The pitch based carbon fibres are relatively cheaper and possess unique structural characteristics which have opened up a variety of new applications. Therefore, new pitch materials capable of producing better fibres and new processes to obtain better materials from existing pitch sources are in the process of development. In the case of rayon-based carbon fibres the yield is very low (15-20%) and the production of high performance materials requires a difficult and expensive high temperature stress treatment. Thus their production is very small.

The pyrolytic manufacture of carbon fibres from different precursors essentially involves similar steps although the details are difficult. Essentially they involve a stabilising heat treatment to prevent melting or fusion of the precursor fibre, a carbonising heat treatment to eliminate the non-carbon elements and a high temperature heat treatment at graphitisation temperatures to enhance the mechanical properties of the final carbon fibres. In order to produce carbon fibres with a high degree of preferred orientation along the fibre axis (which determines their mechanical properties), the precursor fibres are given a stretching treatment during one of the three stages of the pyrolysis process (Fig. 3). This orientation is generally carried out during the spinning stage in pitch fibres, during oxidative stabilisation step in PAN fibres and during the high temperature treatment in case of rayon fibres.

PREPARATION OF CARBON FIBRES FROM RAYON (CELLULOSE)

Cellulose precursor fibres were the first to be used as raw materials for the large scale production of carbon fibres. Natural cellulose fibres such as those of cotton and ramie are not used for the production of carbon fibres because they produce discontinuous filaments, have a low degree of orientation and are associated with impurity materials. The most commonly employed cellulose fibres for producing carbon fibres are textile grade rayon, regenerated cellulose and cuperammonium rayon because in these cases the basic composition, the degree of polymerisation and the spinning conditions can be controlled.

The pyrolysis is generally carried out at low temperature and in a reactive atmosphere such as oxygen, air, chlorine or HCl vapours or in the presence of flame retardants. These chemical compounds promote dehydration of cellulose and inhibit the formation of tar resulting in a greater yield of the carbon fibre. The pyrolysis involves four different stages, the major degradation occurring between 210-320°C resulting in a weight loss of about 75-90% depending upon the precursor fibre and the rate of heating.

In the first stage of pyrolysis up to 150°C the desportion of water amounting to about 12% takes place. This is followed by two processes; the first involving random dehydrogenation of hydroxyl group as structural water (150-240°C) producing dehydrocellulose which is the main source of carbon at high temperature. The second step involves the cleavage of C-O and C-C bond (240-400°C) and leads to the formation of levoglucosan which at higher temperatures produces unwanted tarry products. These condensation and dehydrogenation reactions break up each cellulose unit into a four carbon atom residue which polymerises to give a carbon polymer ultimately producing a graphite-like structure with a carbon yield of less than 20%. The carbon fibre yield has been found to be increased to 30-35% when pyrolysis is carried out in the presence of HCl vapours.

The ultimate yield of carbon fibre by rayon pyrolysis depends upon the nature of the precursor fibre, the rate of heating and the environment (atmosphere) in which the pyrolysis is carried out. The molecular weight of the cellulose precursor also plays a vital role as it determines the number of end groups available for initiation of the degradation process. A slow rate of heating results in a higher yield but is less economical and, therefore, the pyrolysis is carried out in a reactive atmosphere or in the presence of flame retardants. These chemical species change the course of the decomposition process so that it occurs at a lower temperature but at a much faster rate leading to higher carbon yield.

The carbonisation of cellulose fibres is carried out in the temperature range 1000-1500°C in an inert atmosphere usually for a few seconds. Carbonisation at higher temperatures under tension improves the mechanical properties. The carbonised fibres are subjected to high temperature heat treatment in the graphitisation temperature range (2500°C) under stress for a fraction of a second when the modulus improves further.

PREPARATION OF CARBON FIBRES FROM PAN

PAN fibres are the most widely used precursor for the large scale production of present day carbon fibres. This is due to the fact that they have high degree of molecular orientation, high melting point and produce a higher yield of carbon fibres. When subjected to a low temperature heat treatment, PAN fibres give rise to a thermally stable, highly oriented molecular structure which is not significantly disrupted during the carbonisation treatment at higher temperatures. The PAN fibres as precursor fibres usually have the following characteristics [11].

- Denier: 1.0-1.5
- Comonomer content: 2-10%
- Molecular weight: M_n $2.5x10^4$
  M_w $25x10^4$
- Crystalline content: 55%
- No surface finish

The comonomers usually present are methyl acrylate (MA) and methyl methacrylate (MMA) along with a small amount of a third comonomer which initiates the stabilisation reactions at lower temperatures. Methyl acrylate acts as a plasticizer to help in the spinning process and facilitates subsequent stretching of the fibre during the oxidative stabilisation process. The fibres are stretched about 2.5 times in a coagulation bath and then by about 14 times in steam at 100°C [12] after washing. This stretching aligns the molecular chains along the fibre axis and enhances the mechanical properties of the carbon fibres.

The oxidative stabilisation of PAN fibres is carried out between 200-300°C in air under a constant load. The chemistry of the stabilisation process has been examined using IR spectroscopy [13] thermal analysis (DSC) [14-16], gas chromatography [17,18] and mass spectrometry [19,20]. Although the exact nature of the process is still not clear but it is certain that oxidative stabilisation involves cyclisation, dehydrogenation, and oxidation reactions which transform the thermoplastic chain molecule into a thermally stable cyclised structure often referred to as a "ladder polymer". Several different structures have been proposed for this cyclised polymer,[11] but the view that the pendent nitrile groups in PAN cyclise to produce polyimine structures [11,21] appears to be more generally accepted. The formation of the stabilised polymer and the evolution of major volatiles such as CO_2, H_2O and HCN during the stabilisation process are shown in Fig. 4.

The properly stabilised polymer should contain between 8 and 12% oxygen which is present as C-OH and C=O groups and promotes crosslinking. It also helps in the aromatisation of the cyclised sequences by the elimination of water by reaction of oxygen with the backbone $-CH_2$ groups. An oxygen content of greater than 12% results in poor quality while an oxygen content less than 8% results in a low carbon yield due to excessive weight loss during the subsequent carbonisation process.

The oxidative stabilisation also causes shrinkage of the fibre which may amount to between 13-34%. According to Fitzer [21,22] this shrinkage has two constituents: the entropy shrinkage caused by molecular relaxation and the chemical shrinkage resulting from cyclisation and dehydrogenation reactions. A part of the chemical shrinkage may also be due to crosslinking reactions caused by interaction with the oxygen or the comonomer. The two contributions can be clearly distinguished by a minimum in the first derivation curve (Fig. 5). The entropy shrinkage is completed below 200°C while the chemical shrinkage starts thereafter and is influenced by the increase in heating rate and by the presence of the copolymer. At higher heating rates the chemical shrinkage starts later but takes place at an enhanced rate. This indicates the crosslinking reaction contribution to the basic shrinkage caused by cyclisation and dehydrogenation reactions. The entropy shrinkage can be minimised either by constraining the fibre longitudinally to constant length or by restraining under constant force during the stabilisation heat treatment. In a continuous process the stabilisation can be carried out by feeding and removing the fibre at the same speed from the heat treatment zone so that the fibres can undergo compensatory local shrinkage and extension during the treatment.

The stabilised PAN fibres are subjected to thermal pyrolysis in an inert atmosphere when the non-carbon elements are evolved as volatiles with a carbon fibre yield of about 50%. The pyrolysis is carried out in three stages. The first stage (300-600°C) involves chemical reactions which evolve H_2O, NH_3, HCN and CO_2 along with some hydrocarbons in the form of tar. The release of tar depends upon the degree of stabilisation of the PAN fibre while the evolution of volatiles depends upon the rate of heating which has to be kept low (~1.5° C min^{-1}). A faster release of volatiles damages the fibre while a larger formation of tar results in a low carbon yield. During this stage the hydroxyl groups initiate crosslinking condensation reactions which fix the structure of the polymer while the remaining linear segments either become cyclised or undergo chain scission reactions evolving the gaseous volatiles.

During the second stage of pyrolysis (600-1300°C) the cyclised structures undergo dehydrogenation reactions and start linking in the lateral directions bounded by nitrogen atoms. These fused, oriented heterocyclic molecules coalesce into graphite like ribbons and fibrils with the evolution of hydrogen and nitrogen. The carbon fibre yield which is about 50% after pyrolysis can be enhanced by carbonising in an atmosphere of HCl vapours. The presence of HCl vapours inhibits the formation of HCN by removing nitrogen as ammonia and also acts as a dehydrating agent.

The third stage of pyrolysis is carried out in the graphitising temperature range (1300-2500°C) when the structure leads to become more and more graphite like. Heat treatment in this step which involves only a few second does not result in any appreciable loss in weight but improves the orientation of the crystallites in the direction of the fibre axis. This results in enhancing the mechanical properties of the carbon fibres.

CARBON FIBRES FROM PITCH

Pitch is readily available in large quantities and is an important and low cost material for the large scale production of carbon fibres. However, the low cost of this material is partially offset by the purification and the spinning costs. Consequently at least ten Japanese companies are carrying out research and development work in this field. The poor mechanical properties of carbon fibres from isotropic pitch have been improved by developing certain type of mesophase pitches in which case the expensive high temperature stretching treatment can be avoided. The high degree of preferred orientation in case of mesophase pitch fibres is introduced into the fibre during a low temperature spinning process. These high performances mesophase pitch fibres are now available commercially.

The more important characteristics of a mesophase pitch suitable for producing high performance carbon fibres are

- it should be generally isotropic

- the mesophase should generally not develop but stay dormant during the spinning process
- the mesophase should develop during the heat treatment following spinning.

The preparation of such a mesophase pitch is a difficult process and is a part of the patented literature. Two procedures which are generally used for the preparation of such mesophase pitches are classified into hydrogenation and non hydrogenation methods.

In the hydrogenation process which was first developed at Kyushu Industrial Research Institute in Japan, the feed pitch after deashing and distillation is hydrogenated catalytically using donor solvents (e.g. tetrahydroquinone). The hydrogenated pitch is then filtered and heated when the low volatile contents are eliminated and the formation of mesophase takes place. The distilled low volatile contents can be recycled into the hydrogenation step. The mesophase pitch thus obtained is optically isotropic at spinning temperatures but it quickly orients itself under the heat treatment after spinning. In this method coal derived pitches are preferred to petroleum pitches [23].

In a non-hydrogenation process the feed pitch, with or without deashing and distillation is heated to transform into mesophase pitch containing about 70-80% mesophase. The mesophase pitch is slightly cooled and allowed to stand. Due to the difference in density and molecular weight, the mesophase tends to settle down at the bottom and can be separated by centrifuging.

It may, however, be mentioned that the structures of the final carbon fibre depends very much on the temperature of spinning and the type of the spinning process used. For example the radial type structure in mesophase pitch fibres results from a laminar flow of mesophase pitch through a spinneret and can be suppressed by changing a laminar flow to a turbulent flow. In fact spinnerets of different shapes produce carbon fibres with different structures (Fig. 6). The important parameters for spinning of mesophase pitch fibre are the spinning temperature, the mesophase pitch fibre are the spinning temperature, the mesophase content in the pitch, the filament speed and diameter, the length to diameter ratio of the spinneret and the draw ratio. Some of these important parameters used by Union Carbide and Kureha Company of Japan are recorded in Table I. By careful control of the various spinning parameters, the filament diameter can be reduced to as low as 708 μm by increasing the wind up rate.

The as-spun mesophase fibres are highly anisotropic and show elongated anisotropic domains. About 75% of the larger planes lie within an angle of ±15° of the fibre axis indicating a high degree of preferential alignment.

The mesophase pitch fibres are stabilised between 250-300°C in electrically heated air ovens and then pyrolysed by passing continuously through a series of heating zones until the final carbonisation temperature is attained. This carbonisation eliminates remaining heteroatoms as volatiles and helps in the development of a larger structure. When the carbon fibres are heat treated between 2500-3000°C, the Young's modulus approaches a value of 880 GPa close to the graphite single crystal value of 1000 GPa. This graphitisation step is optional and is carried out only to obtain UHM carbon fibres.

VAPOUR GROWN CARBON FIBRES: (VGCF)

These carbon fibres are of recent origin and are prepared by the seeding method. The substrate ceramic cylinder surface is seeded with ultrafine particles of iron with diameters between 100-500 Å which act as very effective catalysts for the formation of carbon fibres. Alternatively the fluid catalyst seeds are introduced in the form of floating iron particles. A high concentration of hydrocarbon -hydrogen mixture is then introduced into the reaction tube maintained at a temperature of about 1100°C. The hydrocarbon decomposes to produce carbon fibres which are formed in the central hot zone of the reaction tube and can be brought to the other end by gas flow.

X-ray diffraction studies indicate that VGCF have a turbostratic structure with an interlayer spacing of 3.48 Å. TEM and selected area diffraction (SAD) show that they have a high degree of preferred orientation along the fibre axis. Their high strength, high surface area and high electrical conductivity make them useful both in reinforcement and electrical applications.

STRUCTURE OF CARBON FIBRES

Several aspects of the structure of carbon fibres have been examined during the last more than two decades using X-ray diffraction, electron microscopy as well as optical microscopy. These techniques have provided useful information about the micro as well as macro structural features of carbon fibres. There is general agreement that carbon fibres are composed of two dimensional turbostratic graphite crystallites oriented preferentially parallel to the fibre axis. These techniques have also provided qualitative and quantitative information about several structural parameters such as crystallite size and orientation, stack height and width, the shape size and distribution of pores and the presence of a three dimensional order.

Several structural models have been proposed for carbon fibre. According to the microfibrillar model (Fig. 7) which was proposed by Ruland and Perret [24.25]. The basic unit of the structure consists of ribbon shaped monatomic layers of sp^2 type carbons with an average width of 6 nm and a length of several hundred nanometers. The ribbons exhibit irregular contours which may contain holes. These ribbons run parallel to form microfibrils which have a preferred orientation parallel to the fibre axis. The ribbons are stacked parallel to each other and pass smoothly from one turbostratic domain to another resulting in a turbostratic disorder. As the contours of the microfibrils are wrinkled, the space between two ribbons creates a number of needle like voids.

Diefendorf and Tokarsky [26], on the basis of TEM studies of high modulus HM carbon fibres prepared a wrinkled ribbon (Basket weave) model (Fig. 8) with ribbons varying in thickness between 13-30 layers and in width between 409 nm. The ribbons have almost zero amplitude and are essentially parallel to the fibre axis. In fact the structure could be described as a wrinkled sheet. However, Oberlin et al [27] based on their dark field examinations are of the view that a two dimensional model for the longitudinal and transverse structure on HM and HT PAN based carbon fibres could be better represented by Fig. 9.

Knibbs [28], while examining end-on-end longitudinal sections of different carbon fibre by optical microscopy, identified three different types of fibre structures which differ in the orientation of the graphitic crystallites within the transverse plane (Fig. 10). Structures (a) and (b) are two zone structures while (c) is a single zone structure. Structure (a) has an almost isotropic orientation in the central zone and circumferential orientation in the outer zone while structure (b) exhibits radial preferred orientation in the central zone and circumferential in the outer zone. The structure (c) however shows a circumferential orientation throughout the cross section. The existence of these three structures was found to depend only on the difference in the precursor material and the processing conditions. X-ray diffraction studies showed that the outer sheath has a thickness of 1.5 μm and is separated by a distinct annular region which was approximately 3 μm in diameter. In the case of carbon fibres graphitised at lower temperatures (<2000°C). The transition between the sheath and the core region was less distinct and the transition region occupied a larger proportion of the fibre volume.

This sheath-core heterogeneity can be attributed to the difference in the degree of stabilisation in the outer and the inner region of the carbon fibre. The fibre, after stabilisation and before high temperature treatment, has an almost completely stabilised outer region and only a partially stabilised inner region.

Johnson [29], on the basis of lattice fringe images of transverse sections of a number of different types of carbon fibres have proposed the three dimensional structure shown in Fig. 11. This model exhibits the difference in ordering and density of the carbon ribbons between the peripheral zone and also the spin of the fibres which is in agreement with the behaviour of carbon fibres in many processes.

SURFACE TREATMENT OF CARBON FIBRES

Carbon fibres when used without any surface treatment produce composites with poor interlaminar shear strength (ILSS). This has been attributed to weak adhesion and poor bonding between the fibre and the resin matrix. This interfacial bond can be enhanced either by enlarging the surface area, which provided more points of contact between the fibre and the matrix or by enhancing the physico-chemical interaction between the two. Thus a number of surface treatments have been developed which improve the fibre-matrix interfacial bond, although the exact nature of these surface treatments is a closely guarded secret. Therefore, whatever information is available has been the result of laboratory investigations by different groups of workers. The situation is complicated because virgin (untreated) carbon fibres are difficult to obtain from the manufacturer.

The surface treatment has to be adaptable to the matrix and several have been examined these treatments are collected in Fig. 12 and the average improvement in ILSS that can be achieved by different surface treatments are reported in Table II.

The exact mechanism of improvement of adhesion as a result of the surface treatment is unknown and is a matter of speculation. However, a considerable amount of evidence has accumulated that a direct chemical reaction between the functional groups on the carbon fibre surface and those in the resin takes place. More recent studies, the results have confirmed the primary importance of COOH groups and their contribution to adhesion. The desorption of surface oxides by heat treatment in argon or vacuum results in the elimination of CO_2 and CO as volatile products. CO_2 is formed by the decomposition of carboxylic groups while CO has its origin in the non-acidic groups. The gireus clearly shows a decrease in ILSS with progressive elimination of CO_2. Fitzer and coworkers [30,31], on the basis of very comprehensive studies on the functional group analysis and the ILSS measurements have suggested that the overall adhesion between carbon fibres and the epoxy matrix has two components. The basic adhesion which is measured after desorption or blocking of the surface functional groups is caused by the residual nitrogen content on the carbon fibre and is of the order of 60-80 MPa in HT and in the range of 20 MPa in HM carbon fibres. The other component, is chemical in nature and is caused by the creation of carbon-oxygen surface groups as a result of the surface treatment. The acidic oxygen groups interact with the amine groups of the hardener while the second amine group of the bifunctional hardener acts as a coupling agent by reacting with the epoxy group of the prepolymer. Alternatively, there can be a dipole-dipole interchange between oxidised fibres and the resins. In both cases the bonding between the cured matrix and the reinforcing fibre is enhanced causing an improvement in ILSS. It may be mentioned that the extent of improvement in adhesion on surface treatment also depends upon the composition and the structural perfections of the carbon layers on the fibre surface. Thus there is more improvement in adhesion on surface treatment of HM carbon fibres compared to that of HT carbon fibres although the final adhesion is higher in the case of HT carbon fibres.

Based on their measurement of polar and dispersive components of surface energy for a number of HM and HT PAN-based and mesophase pitch based carbon fibres before and after several treatments, Donnet and coworkers [32] have suggested that the improvement in adhesion between the fibre and the matrix is due not only to the creation of surface functionality but also to an increase in surface energy and surface roughness caused by the surface treatment.

PROPERTIES OF CARBON FIBRES

High tensile modulus and high tensile strength are the two most important mechanical properties which make carbon fibres as high performance materials. The earlier carbon fibres prepared from rayon, PAN or from pitches had poor mechanical properties. But in the later sixties it was realised that these mechanical properties could be improved by heat treatment at graphitisation temperatures after the preparation of the carbon fibres. It was also found that the tensile modulus of a carbon fibre was related to the degree of preferred orientation of the layer planes along the fibre axis. As carbon fibres have a highly anisotropic structure, a small misalignment of the layer planes can result in a marked decrease of the tensile modulus. Thus, a sample of carbon fibres which had basal planes of crystallites lying within $\pm 35^{\circ}$ to the fibre axis showed a modulus of 103 GPa while another samples with an orientation angle of $\pm 10^{\circ}$ showed a modulus of 410 GPa. Therefore, any treatment such as high temperature treatment or the application of stress during one of the processing stage which reduces the angle of layer plane preferred orientation enhances the tensile modulus. The effect of heat treatment was first studied by Shindo in Japan in 1961 obviously without achieving any preferred orientation but was improved later in U.K. in 1964 by Watt at the Royal Air Force base (Fig. 1). This was really the first time when high performance carbon fibres were obtained. Thus to produce carbon fibres with a high degree of preferred orientation along the fibre axis and consequently with a high modulus, the precursor fibres are given a stretching treatment during one of the stages of the pyrolysis process. This stretch treatment is usually carried out during the low temperature heat treatment in the case of PAN, during graphitisation in the case of rayon and during spinning of the pitch fibres (Fig. 3).

The elastic compliance of a single crystal graphite parallel to the basal plane is 0.98×10^{-12} Pa^{-1} [25]. Therefore, the theoretical maximum tensile modulus of graphite having perfectly aligned carbon fibres, a value corresponding to about 80% of the theoretical value for graphite. The tensile properties for a representative selection of commercially available carbon fibres are given in Table III. It is unfortunate that high modulus fibres have a very low strain to failure when a high strain to failure is essential for many CFRP applications. However, with improvements in the preparation of precursor fibres and the processing conditions significant improvements in the strain to failure with high strength and intermediate modulus fibre types have been obtained in the recent years. It is now possible to obtain strains to failure of around 2% with a commensurate tensile strength of between 5 and 7 GPa.

The tensile strength of a carbon fibre depends upon the nature of the precursor fibre, processing conditions such as heat treatment temperature, the degree of oxidative stabilisation and the presence of flaws and defects which include dislocations, stacking faults, metal particle inclusions, discontinuities and damages of the pyrolysing fibre. The tensile strength increases on heat treatment, reaches a maximum value at about 1500°C and decreases thereafter at higher temperatures. The tensile strength of a carbon fibre, in general, decreases with increase in tensile modulus. The corresponding ultimate tensile strain decreases from about 2.5% for the best of the high strength carbon fibres (tensile modulus 300 GPa) to about 0.25% for carbon fibres with the highest modulus (~800 GPa) (Table III).

CARBON FIBRE APPLICATIONS

Carbon fibres are versatile high technology materials because they have high specific strength and stiffness, light weight, low coefficient of thermal expansion and good thermal and electrical conductivity, good vibration damping, low abrasion and high fatigue resistance, biological compatibility and chemical inertness. They, however, have some shortcomings as well. They are brittle, have a low impact resistance and low break extension. They remain expensive materials and can therefore be used only when cost is not a major factor. However, the price of these materials is decreasing with increased production.

The two main sectors of carbon fibre applications are the high technology sector which includes aerospace and nuclear engineering and the general engineering and the transportation sector which include engineering components such as bearings, cams, fan blades and automobile bodies. However, the requirements of the two sectors are entirely different. The large scale use of carbon fibres in aircraft and aerospace is driven by maximum performance and fuel efficiency while the cost factor and production requirements are not critical. The use of carbon fibres in general engineering and surface transportation is dominated by cost constraints, high production rate requirements and generally less critical performance needs. This points to very different considerations regarding material forms, acceptable matrices, and potential manufacturing methods.

Carbon fibres are unidirectional reinforcements and can be arranged in such a way in the composite that it is stronger in the direction which must bear loads. Another arrangement of fibres can be used to prevent deflection on loading in one direction and not in the other. But in general the properly requirements for the processed bodies are becoming increasingly complex because they have to bear loads, resist deflection and give high performance in more than one direction and have to operate under varying physical, mechanical and chemical environments. Consequently the level as well as the combination of composite properties has to be more severe and complex. Thus the carbon fibers are arrayed in more than one direction in the composite to attain better efficiency. This selective reinforcement by carbon fibers permits designers to fashion articles suited to their needs. The designer may arrange more fibers in one direction than in the other or use two different types of carbon fibers, one known for its high strength to bear loads in one direction and the other for its high stiffness to resist deflection in the other direction.

It is now realised that the use of more than one reinforcement material in composites is going to be of importance since no one fiber can tackle all the technical and economic requirements of aerospace, transport, sports and other industries. Thus it is important to consider a more general composite behaviour rather than a single property such as strength of stiffness. This has led to the development of a variety of reinforcement fibers with different mechanical, physical and electrical properties. The structures constructed using Kevlar and carbon fibers simultaneously in the same composite have impact strengths better than those reinforced with carbon fibers alone and have better compressive strength than those reinforced with Kevlar fibers alone. Such composites in which more than one kind of reinforcements is used are called hybrids and can constitute carbon and glass fibers, carbon and boron fibers, carbon and Kevlar fibers and several other combinations. Hybridisation within a matrix can be achieved by using two or more fibers in the same fabric (intraply hybrid) or by mixing single fiber plies through the composite cross section (interply hybrid) Fig. 13.

The basic characteristics of importance in all applications of carbon fiber reinforced composites are high specific strength and stiffness together with lighter weight. Composites with properties superior to those of steel and lighter in weight by 40-70% have been realized. In addition several other physical properties of carbon fibre composite such as their thermal conductivity and corrosion resistance can be varied by a suitable choice of the matrix material and by varying the processing conditions for preparation of the carbon fibre - reinforced composite. Table IV gives the carbon fibre composite applications of the basis of their most significant physical properties in a particular application, but by no means does this imply that other physical properties do not play their role.

AEROSPACE APPLICATIONS

Aerospace is the largest market sector for CFRP and consumes about 60-70% of the current world consumption of about 8000 tons of carbon fibres. It is also likely to sustain a growth rate of about 10% per year during the next decade. The most important flight tested components which have performed very well are the air brakes on the Vulcan, a door strut on the Trident, a flap operating torque tube on the Andover and the rudder twin tabs on the jet Provost [33]. These components showed absence of corrosion and the absence of foreign object impact damage and apparent resistance to aircraft fluid in service. Figs. 14 and 15 illustrate some of the areas in which advanced carbon fibre composites are considered to be competitive with metals and where design specification on deflection limits, buckling and dynamic response can not be met without them. The largest piece of CFRP structure in service at the present time is the pay load bay door of NASA Shuttle which is 60 feet long and 1500 square feet in area and has made an estimated weight saving of 23% over the same structure in aluminum. Goodyear has developed CFRC airbrakes for F-15 fighter planes which weight about 30 Kg (against 100 Kg) and perform 3-5 times better than their steel counterparts.

The low coefficient of linear expansion of carbon fibres makes it possible to design structures with almost zero or very low linear or planar thermal expansion. This is a valuable characteristic for precision instrument components such as telescopes and aerospace antennas. British Aircraft Corporation (BAC) has developed radar antennas based on CFRP aluminum honeycomb construction capable of retaining dimensional stability under widely different environmental conditions.

SURFACE TRANSPORT APPLICATIONS

The use of CFRP in road and marine transport has been very slow because cost is an important factor in this sector. Two major load bearing components which are being examined are the drive shaft and the body spring. A CFRP drive shaft fabricated by wet filament binding technique has 50% weight compared to its steel counterpart with comparable flexural strength and stiffness. The body spring made of CFRP has a weight 25% less than its metal counterpart and shows better fatigue resistance. CFRP has also been used to effect weight reduction in surface vessels and hovercrafts and in making transportable bridges (Fig. 16). Ford Motor Company has demonstrated the possibility of using CFRP parts by replacing steel parts such as doors, bonnets, drive shafts, structural brackets, wheels and frames resulting in a weight reduction of 33% without compromising on strength.

SPORTS APPLICATIONS

The high specific strength and stiffness coupled with rapid damping and good fatigue resistance make CFRP as versatile material for certain sports an de=leisure time products such as golf clubs, fishing rods, ski poles, tennis and badminton rackets, hockey sticks, racing bike frames, vaulting poles, racing cars and bows and arrows. A Ford GT-40 sports car with body made from bundles of short carbon fibres applied as grid pattern over the hood, fender and roof which had its weight reduced by 40% has won the Lemans race several times.

METALLURGICAL APPLICATIONS

The excellent high temperature and mechanical properties of CFRC have been used for making molds for metallurgical hot forging processes in the superplastic stage and for hot molding in powder metallurgy [22]. One such multiple reusable mold for cobalt-based products has been found to be economical despite higher investment costs.

NUCLEAR FIELD APPLICATIONS

Chemical inertness of carbon fibres outweigh the cost disadvantage in the construction of a gas centrifuge for uranium enrichment. The GRP and metal react with uranium hexafluoride involved in the uranium enrichment and therefore, require a protective coating while no such coating is needed for carbon fibre composites. Carbon fibres in the form of mat, felt or cloth impregnated by a resin matrix can be used as protective lining for chemical plants and anticorrosion equipment including bulk liquid chemical transport.

BIOMEDICAL APPLICATIONS

The biocompatibility of carbon fibres with blood, soft tissues and bones makes them useful for surgical implants as replacement for damaged or absent tendons and for regenerating tissues. Carbon fibre components can be used as components of bone plates, hip join prostheses, ligaments and hydraulic motors for artificial heart implants. The main shin structure and the rigid keel of the foot (Fig. 17) has also been successfully made from CFRP. The use of CFRP reduces the weight of the limb to a level suitable for enfeebled geriatric patient and provides sufficient strength and durability to an active high weight amputee. This application is expected to expand to cover all levels of amputation from the hip downward.

GENERAL ENGINEERING APPLICATIONS

Carbon fibre composites are used to make several general engineering components such as bearings, gears, cams, fan blades, screws, nuts and bolts. They can also be obtained in the form of tubes, sleeves and other auxiliary aids in the field of apparatus construction. The textile machinery components such as picking sticks, reciprocating guide, bars, push rods and slays in weaving looms are also made from CFRP. The reciprocating guiding bars which reciprocates longitudinally at the rate of 750 cycles per minuet when made from CFRP has increased the speed by about 10% and has increased the lift of the bar by twice.

ELECTRICAL APPLICATIONS

Carbon fibres composites with good electrical conductivity can now be prepared and these can be used for electromagnetic interference shielding (EMI), lightening strike tolerant aircraft surfaces, electrostatic dissipation for the protection of micro circuiting in computer field and for making large conducting structures. The electrical conductivity of the composite can be controlled by proper selection of the carbon fibre and the matrix material to meet different applications.

Besides, these applications, there is no facet of human endeavour where carbon fibres do not have applications. The importance of carbon fibre reinforced composites can also be judged by the growth rate of world production of carbon fibres which has remained steady between 24-27% in the years between 1975 and 1983. The consumption of carbon fibres which has remained at about 50% of the normal production capacity has varied from 1200 tons in 1982 to 3150 tons in 1985. It has been predicted that the consumption of carbon fibre will exceed 8000 tons in the year 1991 (Fig. 18). The major consumer markets are the United States which consumes about 50% of the total world consumption followed by japan which consumes about 22%. However, the end-use distribution of carbon fibres for various applications is different in the two countries, while 67% of the carbon fibres consumed in the United States goes towards aerospace applications, 78% of the Japanese consumption is towards sports and leisure time applications (Fig. 19). Industrial use of carbon fibres is still of minor importance but it will increase with decrease in the price of carbon fibres and with their mass production.

It is worthwhile to mention that several cooperative ventures are underway between carbon fibre companies in Japan, the United States and Europe. Several companies are developing special divisions devoted to the manufacture of carbon fibres. New production facilities are being established in France, Germany, India and elsewhere in Europe. Governments of industrialised countries are also realising the importance of these high performance materials and are supporting and sponsoring civilian and military projects designed to develop these materials.

REFERENCES

1. R. Bacon and W. H. Smith, Proc. 2nd Conf. Ind. Carbon and Graphite, Soc. Chim. Ind. London 203 (1965).
2. A. Shindo, Report No. 317 Govt. Ind. Res. Instt. Osaka (1961).
3. W. Watt, L. N. Phillips and W. Johnson, Engineer (London), 221, 816 (1966).
4. S. Otani, Carbon, 3, 31 (1965); 5, 219 (1967).
5. British Patent 1,310,769 (March 21, 1973).
6. L. S. Singer, Carbon, 16, 409 (1978).
7. U.S. Patent 3,976,729 (1976).
8. U.S. Patent 4,026,788 (1977).
9. J. B. Barr, S. Chwastiak, R. Didchenko, I. C. Lewis, R. T. Lewis and L. S. Singer, Appl. Polymer Symposia No. 29, 161 (1976).
10. H. F. Volk, 4th London Intern. Carbon and Graphite Conf. Soc. Chem. Ind. London, 183 (1976).
11. J. B. Donnet and R. C. Bansal, "Carbon Fibers," Marcel Dekker Inc.: New York, 1990.
12. G. Dorey, Phys. Technol., 11, 56 (1980).
13. O. P. Bahl and L. M. Manocha, Carbon, 13, 297 (1975).
14. E. Fitzer and D. J. Müller, Carbon, 13, 63 (1975).
15. M. Minagawa and T. J. Iwamatru, Polym. Sci. Polym. Chem. Ed., 18, 481 (1980).
16. G. Radhakrishanan, T. Nagabushanam, K. J. Joseph and M. Santappa, Makromol. Chem., 150, 2923 (1979).
17. E. Urbas and E. J. Kullik, Chromatogr., 137, 210 (1977).
18. R. S. Lehrle, J. C. Robb and J. R. Suggote, Eur. Polym. J., 18, 443 (1982).
19. W. Watt, D. J. Johnson and E. Parker, Proc. Plastics and Polymer Conf. Supplement 6 Plastics Inst., London, 3 (1975).
20. Y. Suchiya and K. Sumi, J. Appl. Polym. Sci., 21, 975 (1977).
21. E. Fitzer, W. Frohs, and M. Heine, Carbon, 24, 387 (1986).
22. E. Fitzer and M. Heym, Chem. Ind., 16, 663 (1976).
23. R. C. Bansal and J. B. Donnet, in Comprehensive Polymer Science, Academic Press, 6, 501 (1989).
24. A. Fourdeaux, C. Henrinchx, R. Perret and W. Ruland, C.R. Acad. Sci., Paris (Series C), 269, 1597 (1969).
25. W. Ruland, Advanced Materials, 2, 528 (1990).
26. R. J. Diefendorf and E. Tokarsky, Polym. Eng. Sci., 15, 150 (1975).
27. M. Guigen, A. Oberlin and G. Desarmot, Fibre Sci. Technol., 20, 117 (1984).
28. R. H. Knibbs, J. Microsc., 94, 273 (1971).
29. S. C. Bennet and D. J. Johnson, Carbon, 17, 25 (1979).
30. E. Fitzer and R. Weiss, 16th Brenn. Conf. Carbon 1983 Ext. Abstracts and Program, p. 473.
31. E. Fitzer and H. Jaeger, 17th Brenn. Conf. Carbon 1985 Ext. Abstracts and Program, p. 317.
32. J. B. Donnet, M. Brendle, T. L. Dhami, and O. P. Bahl, Carbon, 24, 757 (1986).
33. D. K. Thomas, Plastics Rubber Intern., 8, 53 (1983).

Table I. Pitch Spinning Parameters used by Kureha and Union Carbide

	Kureha	Union Carbide
Spinning temperature (°C)	240-330	350-436
Spinneret diameter (nm)	0.1-0.9	0.07-0.38
Number of holes	30-120	40-1,000
Spinneret capillary length to diameter ratio	---	2-5
Filament diameter (microns)	8-30	8-50
Filament speed (m/min)	400-1,700	30-225
Draw ratio	400-5,000	3-1,700

Table II. Improvement in Composite ILSS by Surface Treatment of Carbon Fibres

Treatment	Improvement in ILSS (%)
Gaseous oxidation (air, ozone, RF plasma)	10-15
Liquid phase oxidation (HNO_3, NaClO, electrolytic)	100-200
Whiskerisation (Si_3N_4, TiO_4, SiC)	200-300
Pyrolytic carbon coating (CH_4, FeC, SiC)	60-100
Polymer grafting	80-100

Table III. Mechanical Properties of Carbon Fibres (Manufacturers' data)

Manufacturer	Fibre	E GPa	σ GPa	e%
PAN-based High Modulus (Low strain to failure)				
Celanese	Celion GY-70	517	1.86	0.4
Hercules	HM-S Magnamite	345	2.21	0.6
Hysol Grafil	Grafil HM	370	2.75	0.7
Toray	M-50	500	2.50	0.5
Courtaulds	HM	350	2.60	0.7
PAN-based Int. Modulus (Int. strain to failure)				
Celanese	Celion 1000	234	3.24	1.4
Hercules	IM-6	276	4.40	1.4
Hysol Grafil	Apollo IM 43-600	300	4.00	1.3
Toho Beslon	Sta-grade Besfight	240	3.73	1.6
Union Carbide/Amoco	Thornel 300	230	3.10	1.3
Toray	M 30	294	3.92	1.3
PAN-based HT (High strain to failure)				
Celanese	Celion ST	235	4.34	1.8
Hercules	AS-6	241	4.14	1.7
Hysol Grafil	Apollo HS 38-750	260	5.00	1.9
Toray	T 800	300	5.70	1.9
Toray	T 1000	294	7.06	2.4
Mesophase-pitch-based				
Union Carbide	Thornel	P-25	140	1.0
		P-55	380	0.5
		P-75	500	.04
		P-100	690	0.3
		P-120	820	0.2
Osaka Gas	Donacarbo	F-140	140	1.3
		F-600	600	0.5
E - Young's Modulus; σ - Tensile Strength; **e** - Strain to failure				

Table IV. Characteristics and Applications of Carbon Fibers

Physical strength, specific toughness, light weight	Aerospace, road and marine transport, sporting goods
High dimensional stability, low coefficient of thermal expansion, and low abrasion	Missiles, aircraft brakes, aerospace antenna and support structures, large telescopes, optical benches, waveguides for stable high-frequency (GHz) precision measurement frames
Good vibration damping, strength, and toughness	Audio equipment, loudspeakers for Hi-fi equipment, pickup arms, robot arms
Electrical conductivity	Automobile hoods, novel tooling, casings and bases for electronic equipments, EMI and RF shielding, brushes
Biological inertness and X-ray permeability	Medical applications in prostheses, surgery and X-ray equipment, implants, tendon/ligament repair
Fatigue resistance, self-lubrication, high damping	Textile machinery, general engineering
Chemical inertness, high corrosion resistance	Chemical industry; nuclear field; valves, seals, and pump components in process plants
Electromagnetic properties	Large generator retaining rings, radiological equipment

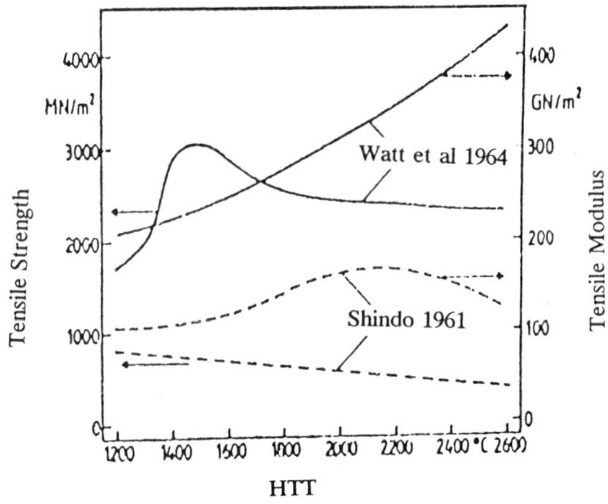

Figure 1. Variation of Tensile Strength and Tensile Modulus with Heat Treatment Temperature

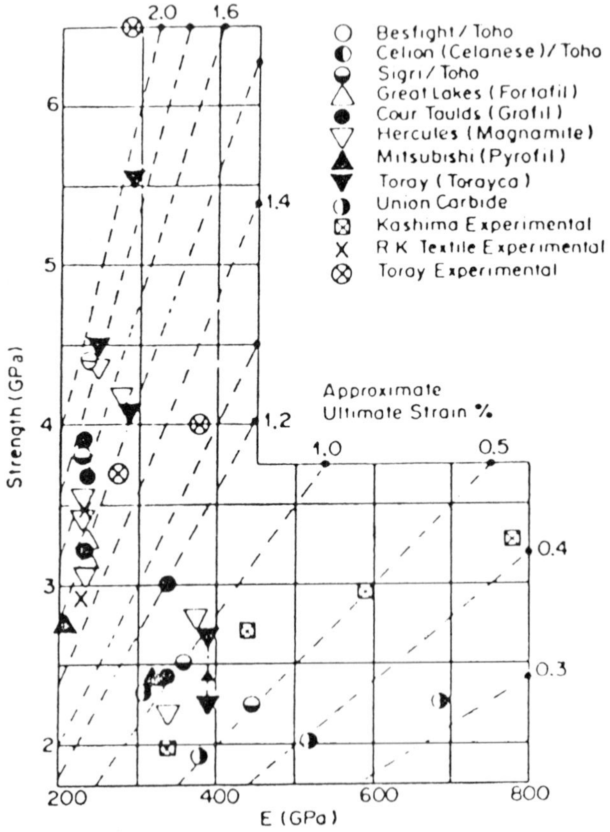

Figure 2. Tensile Strength and Tensile Modulus of Commercially Available Carbon Fibres

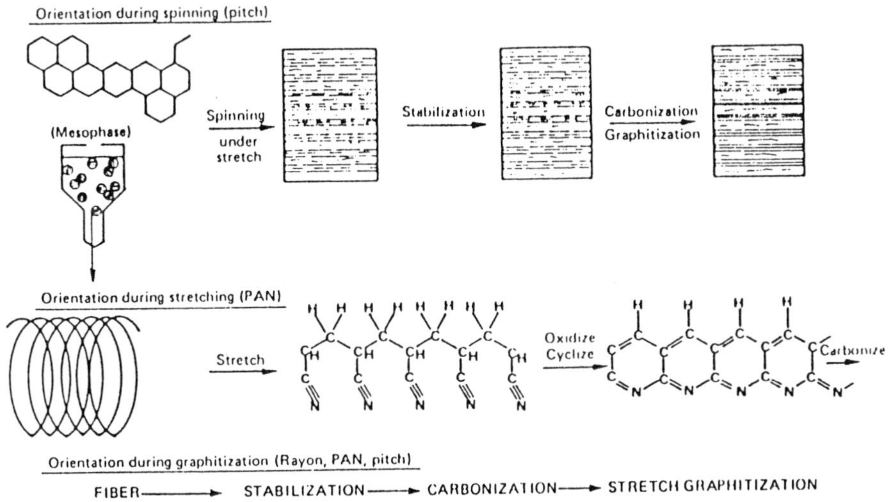

Figure 3. Creation of Preferred Orientation in Different Types of Carbon Fibres

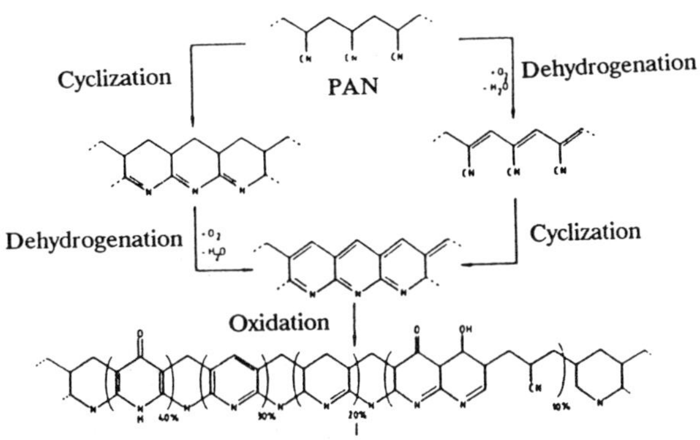

Figure 4. Chemistry of PAN Stabilisation

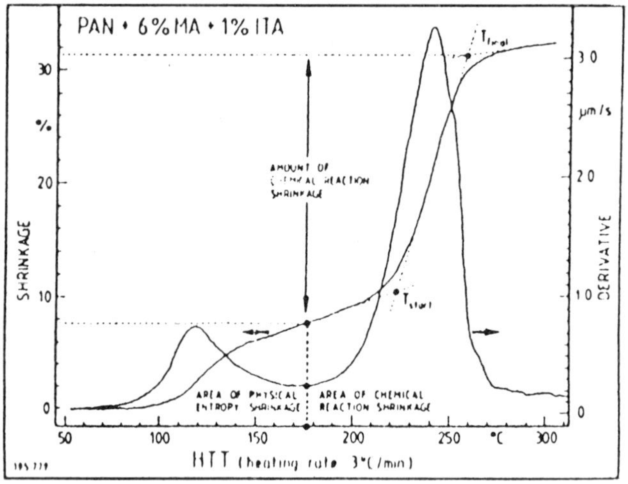

Figure 5. Shrinkage of PAN During Oxidative Stabilisation

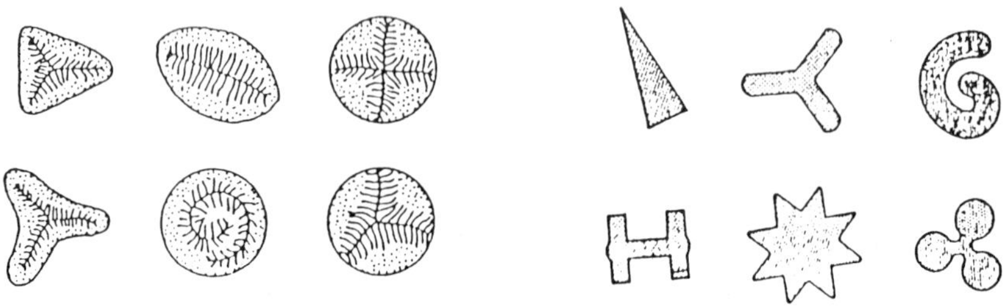

Figure 6. Microstructure of Carbon Fibres Using Spinneret of Different Shapes

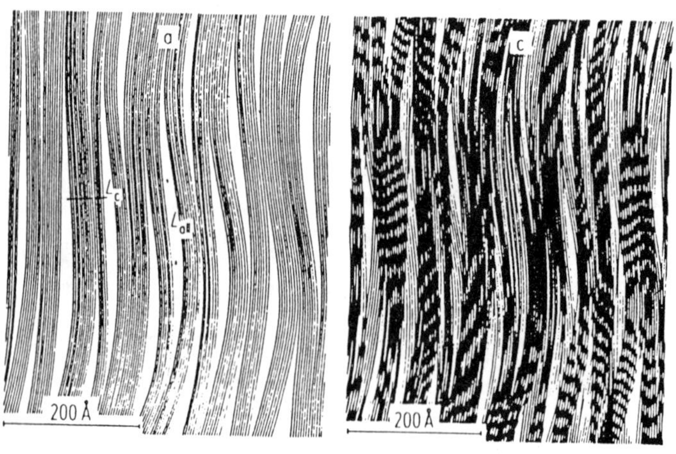

Figure 7. Micro Fibrillar Structure of Carbon Fibre

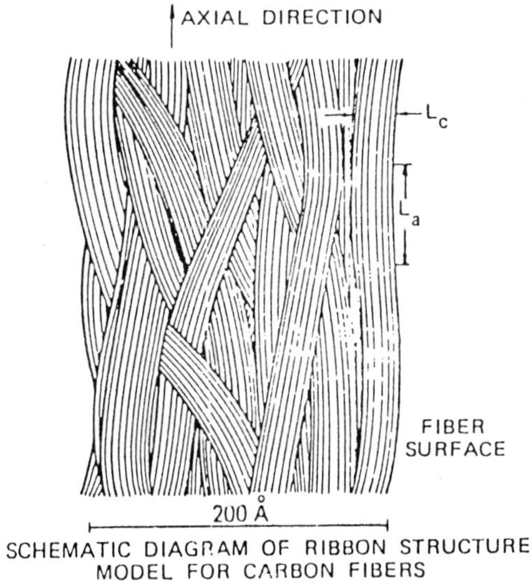

Figure 8. Basketweave Ribbon Structure of a Carbon Fibre

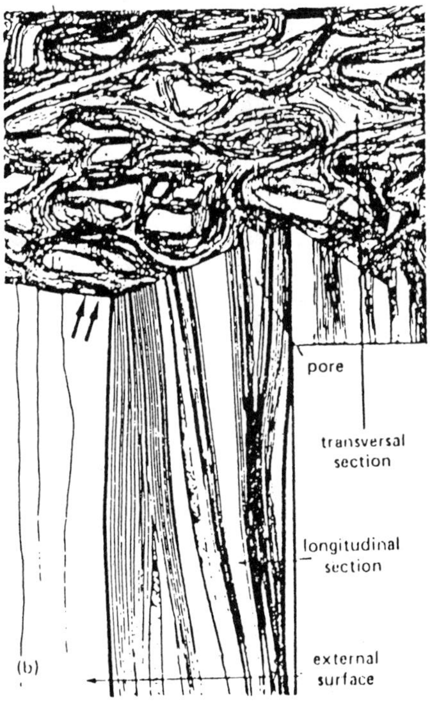

Figure 9. Two Dimensional Crystallite Interlinking Representation of Carbon Fibre Structure

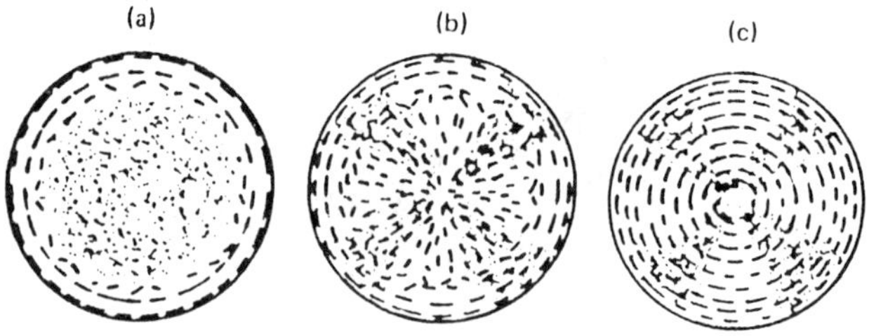

Figure 10. Schematic Representation of Carbon Fibre Structure (a) Isotropic Centre and Oriented Outer Skin (b) Different Orientation in the Centre and the Outer Layers (c) One Type of Preferred Orientation

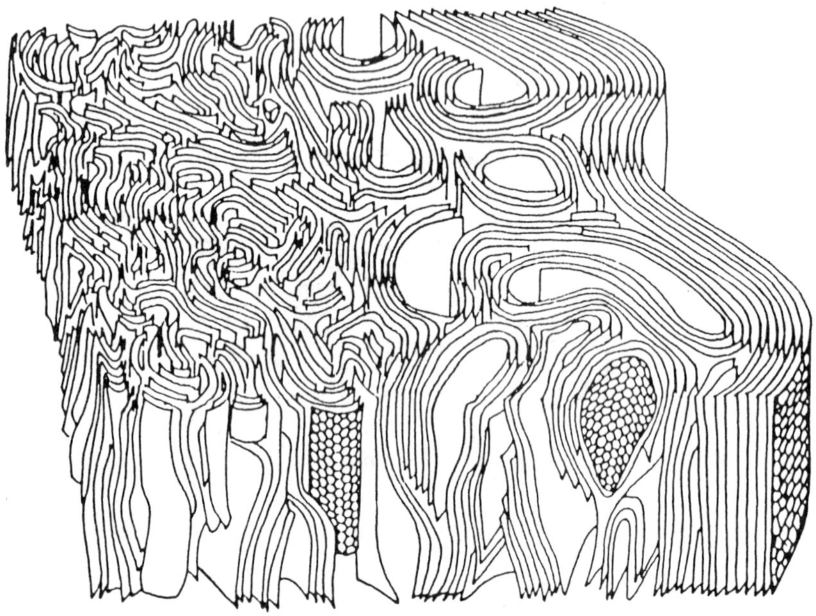

Figure 11. Three Dimensional Model of Carbon Structure

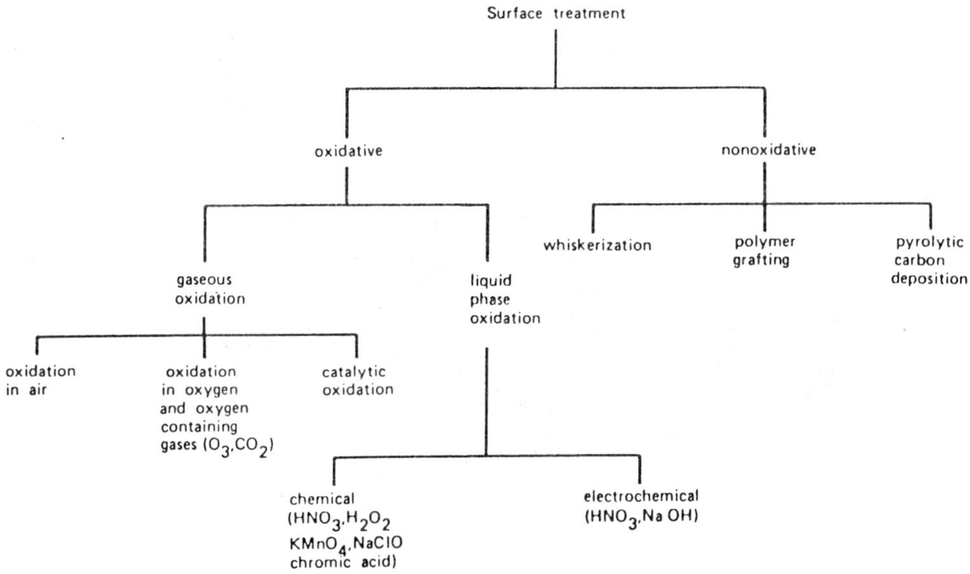

Figure 12. Various Types of Surface Treatment of Carbon Fibres

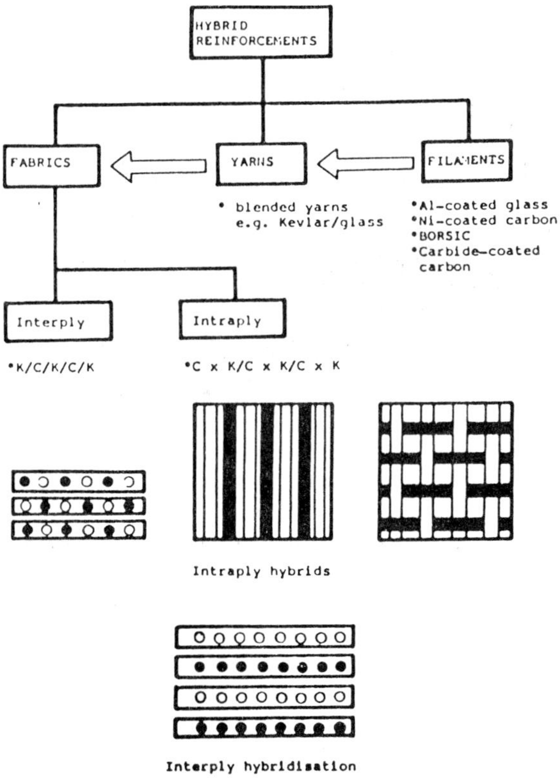

Figure 13. Hybrid Fabric Reinforcement

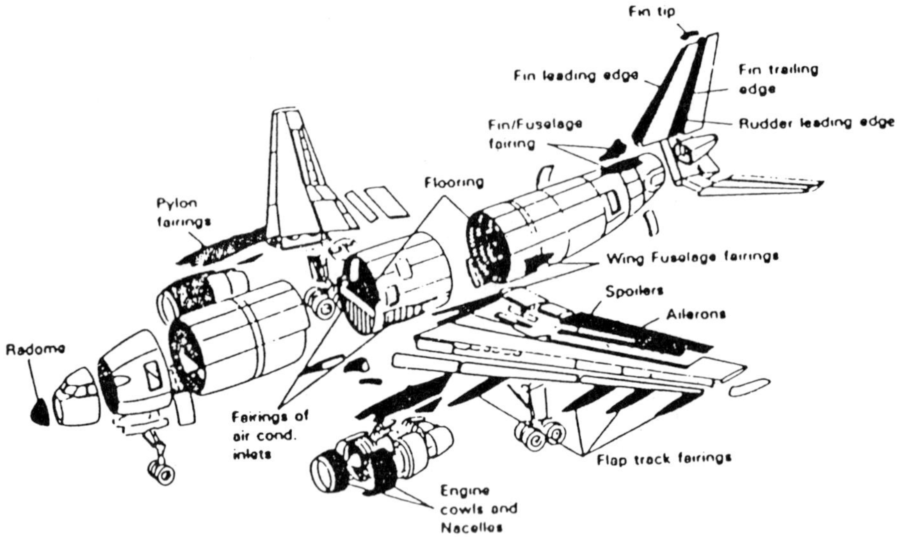

Figure 14. Aerospace Applications of Carbon Fibres

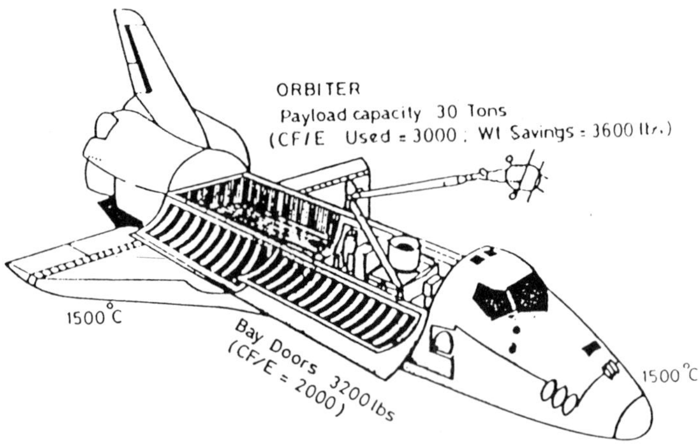

Figure 15. Aerospace Applications of Carbon Fibres in NASA Space Shuttle

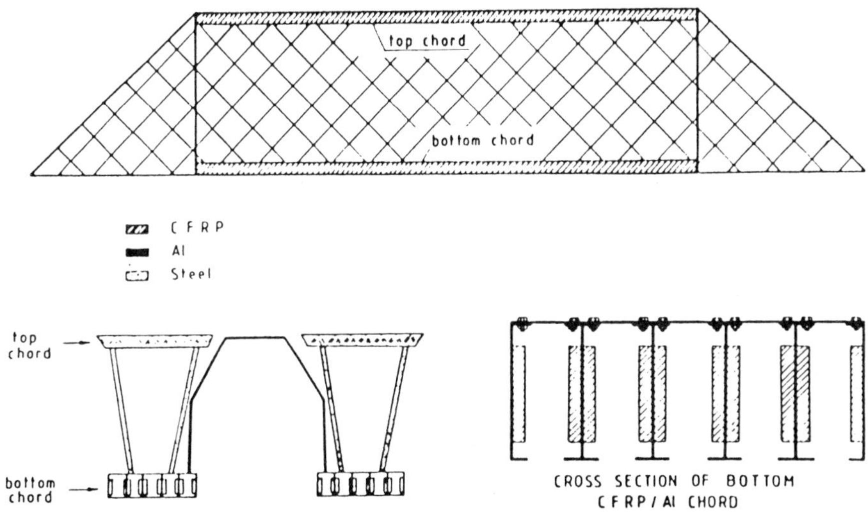

Figure 16. Transportable Bridge from CFRP and Aluminum

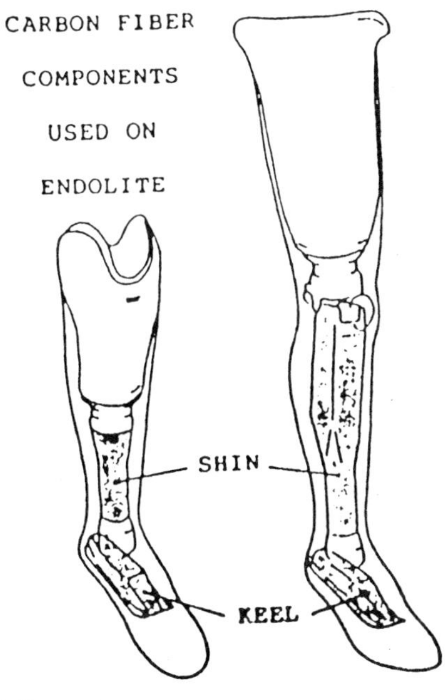

Figure 17. CFRP Endolite

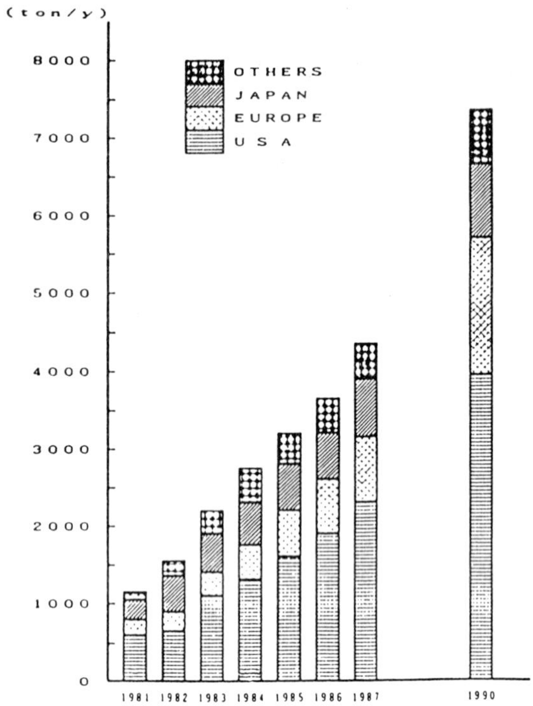

Figure 18. Consumption of PAN-based Carbon Fibres

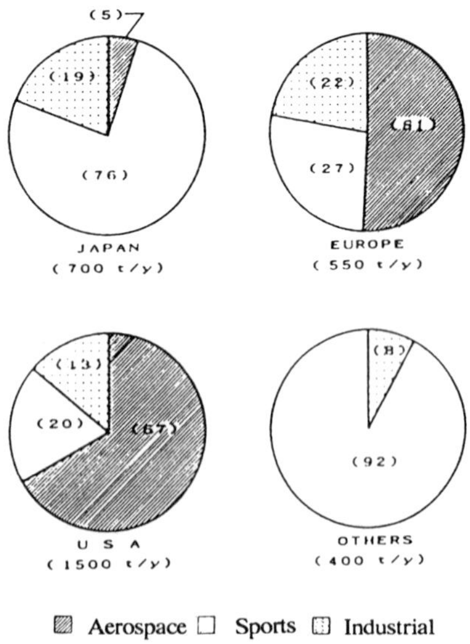

Figure 19. End Use Distribution of PAN-based Carbon Fibres

HISTORY OF MISCELLANEOUS MANMADE FIBERS

RAYMOND B. SEYMOUR
Department of Polymer Science, University of Southern Mississippi, Southern Station Box 10076, Hattiesburg, MS 39406-0076

The first manmade fibers, i.e., rayons were regenerated cellulose produced via the denitration of filaments of cellulose nitrate, the coagulation of extruded cuprammonium solutions of cellulose or the coagulation of aqueous solutions of cellulose xanthate. These commercial successes catalyzed comparable developments based on other natural sources, such as alginic acid in seaweed and the protein in milk, peanuts, corn and soybeans.

PROTEINACEOUS FIBERS (AZLON)

In 1860, E.C.S. Stanford obtained alginic acid by the acidification of alkaline solutions of brown seaweed but the first alginic fiber was not commercialized until green chromium alginate monofilaments were used by J. B. Speakman as camouflaging textiles. These and calcium alginate textiles are flame resistant and soluble and dilute alkaline solutions. These filaments are now used as draw threads for separating socks.

In 1897, A. Spittler produced artificial horn by molding casein. In 1904 H. Todtenhaupt extruded monofilaments from aqueous alkaline urea solutions of casein from milk. However, casein fibers were not produced commercially until A. Ferratti developed a commercial process which he patented in 1935.

Snia Viscose and Atlantic Research Associates produced these fibers under the trade names of Lanital, Fibrolane, and Aralac. The production of these manmade wool-like fibers has been discontinued. The trade names Merrinova and Caslen have been used for formaldehyde treated casein filaments.

Proteinaceous filaments, called Ardil have been produced in Scotland by the extrusion of alkaline solutions of peanut protein. Blends of Ardil, wool or cotton have been produced commercially, but their manufacture has been discontinued.

Chemists at the USDA Research Laboratory at Peoria produced monofilaments by the extrusion of alkaline solutions of zein which was extracted from corn by isopropyl alcohol. This filament was produced commercially by Virginia-Carolina Chemical Corp. under the trade name of Vicara in 1948. Other regional USDA Laboratories produced fibers from Keratin (hair), feathers, and blood plasma but the production of all azlon type fibers has been discontinued.

After their importation from the orient in 1804, soya beans became an important commercial crop in the USA. Protein was solvent-extracted from soya beans in the 1930's. Henry Ford championed the use of this protein for plastic production. Subsequently, techniques developed for spinning of other azlon filaments were applied to alkaline solutions of soya bean protein but the production of soya bean fibers has been discontinued.

HALOGEN CONTAINING FIBERS

Vinyl chloride (VCM) was described by V. Regnault and E. Baumann in 1838 and 1872, respectively. The insoluble mass obtained when Baumann exposed (VCM) to sunlight was reinvestigated by I. Ostromislensky in 1914 and this PVC was plasticized by W. Semon and coworkers in the early 1930's. PVC fibers were patented in 1913 and produced commercially in 1931. However, these relatively insoluble fibers were displaced by those extruded from solutions of a copolymer of VCM and vinyl acetate which was described by A. Doolittle and G. Powell in 1936.

E. Rugely, T. Field and J. Conlon patented Vinyon fibers which were based on a copolymer of VCM (88) and vinyl acetate (12). Vinyon has been used for filter cloths, fish lines and nets and as a thermally softened bonding agent for the production of bonded cellulosic fabrics.

Copolymers of VCM (60) and acrylonitrile (40) called Vinyon N which were introduced in 1942 have replaced Vinyon in many applications. R. Seymour produced monofilaments from a copolymer of vinyl chloride (90) and vinylidene (10) in 1938. This polymer was called Plivoic and/or Plioflex by Goodyear Tire and Rubber Co. Similar names were used by Goodyear for other copolymers of vinyl chloride, such as those in which the comonomer was diethyl maleate.

German chemists produced monofilaments, called Igelit MP and Mipolam from copolymers of VCM and methyl acrylate. German chemists also improved the solubility of PVC in acetone by postchlorination. Fibers from postchlorinated PVC were called PeCe fibers, Rhovyl, Fibracryl, Thermovyl, Isocyl, Retractyl, and Tevilon. Monofilaments produced from blends of aqueous PVC and polyvinyl alcohol called Polychlal have been produced in Japan.

J. Reilly and P. Wiley produced copolymers of vinylidiene chloride (90) and vinyl chloride (10) in the late 1930's. Solutions of this copolymer, which was called Saran, were used to produce monofilaments called Permalon by Pierce Plastics, Velan by Firestone, and Tygan by Fothergill and Harvey, Krehalon by Kuahi Kasei and Chlorene by LaSociete' Rhovyl. Saran monofilaments continue to be used for insect screens, filter cloths, and upholstery.

R. Plunkett discovered polytetrafluoroethylene (PTFE) (Teflon) when his assistant J. Rebock sawed open the "empty" cylinder of tetrafluoroethylene which contained a solid polymer. In spite of its insolubility, fibers have been spun by forcing an aqueous dispersion of PTFA through an orifice into hydrochloric acid and sintering and stretching the dried filaments. Teflon filaments are used as fibers in packing chemical pump shafts, filtration clothes, and water proof textiles.

OTHER MANMADE ORGANIC FIBERS

In spite of the fact that monomeric vinyl alcohol is an unstable tautomer of acetaldehyde, its polymer was produced in 1924 by the saponification of polyvinyl acetate. PVA fibers, which are available under the generic name of vinylon, were produced in 1951, in Japan, by coagulating the extrudate from an aqueous PVA solution in an aqueous sodium sulfate solution.

These water soluble filaments (Kuralon) are insolubilized by surface treatement with formaldehyde. Kuralon, which is also called Cremona, is used for surgical thread, separation of socks, and for reinforcing polyester plastics.

A silk-like filament has been produced in Japan by the extrusion of poly p-(β-hydroxyethyl) benzoic acid. Polyglycolides such as poly(hydroxyacetic acid) have been used as absorbable sutures. J. Economy produced phenolic fibers (Kynol) by extruding A-stage resole resins and then heat curing the filaments produced. These flame resistant filaments are produced commercially in Japan.

INORGANIC FIBERS

Glass Fiber (Pele's hair) were discovered centuries ago near volcanoes and glass filaments were used for decorative purposes by Venetian artisans for several centuries before fiberglass was produced commercially. A machine for the production of glass filaments was developed in UK in 1841 but this machine had little use. However, German technicians produced fiberglass, during World War I, by heating glass rods and winding the filaments on a man propelled bicycle wheel. E. Libbey and M. Owens exhibited clothing made from glass fibers at the Chicago fair in 1893.

In 1936, G. Slayter and J. Thomas patented the commercial production of glass fibers which they called Fiberglass. These filaments were produced by cooling molten glass extrudates by the Owens-Corning Fiberglass Corp. at Newark, OH, St. Helens and Glasgow, Scotland. These fibers are widely used for insulation and for reinforcing plastics. Haveg Industries produced silica textiles (Sil-Tip, Refrasil) by leaching fiberglass textiles with hydrochloric acid.

In 1911, E. Wintraub produced boron by the chemical vapor decomposition (CVD) reaction of hydrogen with boron trichloride and deposited this boron on a tungsten filament. This investigation was continued at Texaco Research Laboratories and these boron filaments were produced by AVCO. The principal use for boron filaments is as a reinforcement for epoxy resins in high performance composites.

When boron filaments with a Tungsten core are heated at about 560°C, they are oxidized. In 1965, J. Economy and R. Anderson heated this boron oxide in an ammonia atmosphere at about 1200°C and produced boron nitride, which is used as a reinforcement for aluminum.

In 1961, C. Potter and W. Lindenthal produced fused silica fibers which were used for tiles in space shuttles. In 1949, I. Harter, C. Horton and J. Christie produced heat-resistant fibers from molten kaolin.

In 1969, W. Ekdahl produced relatively long fibers by blowing a molten mass of alumina-silica and chromia. The trade name Nextal is used to describe alumina-silica-boria transparent fibers.

In 1968, B. Hamline produced zirconia fibers (Zircar) by the thermal decomposition of a mixture of zirconia and organic polymers. In 1972, chemists at ICI produced zirconia fibers and alumina fibers (Saffil) by post heating filaments extruded from viscous aqueous salt solutions.

In 1968, chemists at Thermokinetic Fibers and Carborundum deposited silicon carbide on carbon filaments by chemical vapor decomposition of hydrogen and chloroalkylsilanes. Short silicon carbide fibers (Nicalon) are also produced commercially by the pyrolysis of rice hulls in an inert or ammonia atmosphere.

HISTORY OF BLENDS OF FIBERS

IAN R. HARDIN
Department of Consumer Affairs, Auburn University, Auburn, Alabama 36849

Any discussion of the history of blends of fibers quite naturally starts with the beginnings of history itself, at least as far as we can know it. If we ask the question, how long have there been blends of fibers for textile uses, then we will almost certainly restate the question to be, how long have fibers been used by humans? Once the usefulness of fibers was discovered by the early humans, experimentation followed with the result that almost any possible blend was tried out. We tend to think of only a few fibers among the natural sources, but in fact there are probably hundreds of sources in nature. Only a few have proved to have the qualities of sufficient strength and overall availability to make them a part of our heritage. These are the fibers we find in historical fabrics.

A cursory review of the literature reveals numerous examples of fiber blends. Various prehistoric investigations have shown that blends of fibers were used for varying purposes. Investigations of Paracas textiles revealed use of llama and other hair fibers of different colors, indicating dyeing before yarn formation [1]. Wool and linen blends, usually a wool weft with linen warp, are found in Ptolemaic textiles [2], Coptic textiles [3], and in textiles of early New England [4]. The linen warp was used to provide strength while the wool filling maintained the aesthetic qualities of the wool for the fabric. These fabrics, often known as "linsey-woolsey", were finely woven materials with lengthwise threads of handspun linen. Handspun wool yarns were the weft or filling. Linsey-woolsey was closely woven for household as well as apparel uses [5]. These fabrics were said to be moth-resistant, presumably because the linen was present in large enough amounts to convince the moth larvae that the fabric was not a good food source.

There are at least two interesting historical references to linen-wool blends. Herodotus comments that Egyptians considered wool to be ritually unclean. On the other hand, the biblical book of Leviticus warns the believer, "neither shall a garment mingled with linen and woolen come upon thee" [6]. Nothing that follows explains this enjoinder, but commentators have speculated that linen and wool may have represented Egypt and Babylonia, respectively [7].

Other examples of blends in historical fabrics include silk and linen in Regensburg fabrics [2], silk and cotton in fabrics from Hyderabad [2], oriental yarns with silk and gold or silver intertwined [8], and central asian fabrics with wool filling and warps of either linen or hemp [8]. The Phoenicians [9] wove fine soft fabrics of silk mixed with cotton or linen. Both the Tunisians [10] and the Peruvians [9] made watertight tent fabrics of wool mixed with either camel hair or goat hair. In Nigeria we find fabrics constructed from mixtures of cotton and various bast fibers [11]. In India brocades of cotton and silk were woven with great skill.

For a few decades after 1720, the printing of all-cotton cloth was forbidden in England with the purpose of protecting the wool industry. This prohibition was circumvented by using the linen-cotton blend called fustian [12]. In France there were women's garments of silk warp and cotton weft, men's clothing with linen and cotton twills, linen-cotton dimity, and silk and wool damask for upholstery [12].

Why blend fibers? Blending of fibers is done for a variety of reasons, such as to produce new effects or corrections of deficits, to reduce costs, to achieve decoration effects, and to take advantage of the best qualities in each fiber [13]. No fiber can provide all of the properties that are needed under all conditions. A sole fiber may serve well under some conditions but fail under others. By choosing a combination of fibers whose properties complement one another, one can create a fabric whose performance matches the specified conditions.

On an operational level, the blend must have adequate properties for the spinner. The price of the blend must be accessible, it must meet the fabric requirements, and it must be versatile enough to meet the changes in industry. Factors to consider are the staple length of the fibers in the blend, the fiber diameter, and the extensibility of the fibers in the blend. The latter becomes important because a mismatch can create blend whose strength will actually be less than either of the components. The drafting and spinning properties are obviously important for the blend, and the dyeing characteristics will be of prime importance in the results that blending will achieve [14].

The act of blending fibers can be viewed as the engineering of fabrics. By carefully choosing fibers and the proportions that are used in blends, particular yarn and fabric properties can be engineered. For example, by using fibers at opposite ends of the moisture absorption scale, fabrics with specific moisture regain characteristics can be created. By blending fibers of different deniers, desired stiffness and drapability values can be obtained. A fabric of 4.5 denier fabric can be matched in stiffness by one made of a 50/50 blend with 6- and 3-denier fibers. The yarns in the latter contain 12 percent more fibers and thus will spin better [5].

Blends of man made fibers. Blends of man made fibers appeared on the market very shortly after the introduction of artificial silk (rayon) in 1892. The early blends were with the prevailing natural fibers, flax, silk, wool, and cotton. The latter blends, particularly, were usually dress materials of a rayon warp and cotton filling [15]. One of the attractions of such a blend was the resulting capability to achieve various cross-dyed effects.

With the introduction of acetate to the market in the 1920's the possibilities of blends included not only natural fibers but blends with rayon and even three-way blends with the natural fibers, and with rayon, itself. Blends of acetate and rayon gave potential for cross dyeing, with the acetate also lending considerable improvement in crease resistance to the rayon [16]. The effects were gained by cross dyeing 25 percent dope-dyed acetate with 20 percent natural acetate and 50 percent rayon to give a three color blend after the dyeing was done [16].

Acetate blends with wool reduced the shrinkage of the cloths, as well as providing the potential for the cross dyeing. The handle of the wool was maintained but the skin irritation sometimes caused by the wool was reduced significantly. Acetate and wool blends were used for tropical suitings, flannels, and blouse fabrics. The use of the acetate in the blends improved the pleating and permanent creases of the resulting fabric.

A 1942 paper on analysis of fiber mixtures [17] gave chemical methods to determine a number of different fiber mixtures, presumably those in actual use. Table I lists these fiber mixtures.

A justification given during the war time [15] for the blending of rayon was to develop new ideas for blends with wool. Part of the reasoning was that wool had to be shipped long distances and therefore more uses should be made of rayon staple with wool.

Several examples of blending of rayon during this period can be given. Staple viscose rayon in coarser deniers was blended with low crossbred wools, mohair, and goat hair [18]. Casein fibers were used with rayon and wool in felts. Blends of rayon staple were blended with casein in 50/50 mixtures to give fabrics with a "wooly handle". Rayolanda X, a rayon with a modified structure to give affinity for acid dyes, was used to give two-tone effects in blends. Dyeing effects were also the object of various three-way blends such as Fibro/Rayolanda X/wool and Fibro/casein/wool.

Other examples of blends with wool that appear in the literature in 1950 include wool and nylon, for exceptional wearing qualities, rayon or acetate with wool for piece dyeing to obtain a heather look, wool and Rayolanda X for the tone on tone effect, and wool and casein for a soft handle in the fabric [19]. Certainly one of the reasons for the increasing experimentation with blends of wool in the early fifties was the rising prices of wool and the attempt to retain wool's qualities in the yarn while reducing the percentage of wool. Blends such as 60/40 wool/rayon, 75/25 wool/nylon and 40/40/20 wool/acetate/nylon were commonly tried in hosiery [20]. An interesting note is that comments were made that at this time (1952), that the United Kingdom lagged behind the U.S. in the use of both 100 percent and blends of man-mades. The question was posed as to whether this was the result of advertising or of a "less susceptible" British public?

Blends with rayon staple were also popular. A mixture containing 15-20 percent acetate provided a heather effect through differential dyeing. Other common blends were rayon with casein fibers, and with flax, nylon, cotton, and wool [21].

Blends with synthetic fibers. The introduction of nylon to the market began a new era in textiles. Nylon was the first truly synthetic fiber, as opposed to fibers like rayon and acetate whose polymers come from a natural wood source. In 1929, W.H. Carothers, who had been hired away from Harvard University, initiated fundamental research that resulted in the development of what we now call nylon 6,6. This nylon is based on the polymer formed from the 6-carbon precursors of adipic acid and hexamethylenediamine.

After several years of laboratory work, pilot scale development, and mill trials, nylon fibers for commerce were produced in October of 1939. The first nylon stockings were then marketed in May, 1940 [22]. World War II quickly interrupted the supply of nylon to the consumer market but, following the cessation of hostilities, nylon reentered the consumer market and was an immediate success. It was recognized early on that the superior strength and abrasion properties of nylon made it a prime candidate to blend with weaker fibers and bolster their yarn properties [23].

Some examples from a 1952 paper on the benefits of blending with nylon are military half hose, constructed from 75 percent wool and 25 percent nylon, which lasted over five times as long as the equivalent all-wool product; twills of wool/nylon blends that gave the same performance as the all-wool counterpart but with half the weight; and flags of a 25/75 wool/nylon blend that withstood five to ten times the exposure as that of the all-wool flag [24].

However, although nylon was quite successful in blends with wool, there were serious problems with the properties of nylon/cotton blends. Blends of less than 50 percent nylon were actually weaker than all-cotton [24]. Nylon's modulus is lower than cotton's. When a yarn containing a blend of the two fibers is stretched, the nylon gives way and the load is increasingly borne by the cotton in the blend. Thus, in effect, the load is concentrated on the weaker cotton fibers and, since these fibers are less numerous than in a homogeneous yarn, the yarn breaks at a lower value.

One of the responses to this problem was the creation of DuPont 420 nylon, which was engineered to have a stress-strain curve very similar to that of cotton up to the breaking point [5]. Table II shows some results from a comparison of three twill fabrics, one of which is all-cotton, and the other two containing cotton fillings and 25 percent nylon in the warp [5].

Acrylic blends. The first commercial acrylic fiber in the United States was DuPont's Orlon, introduced in 1949. Production of continuous filament actually began in 1950, while staple fiber became available in 1952 [25]. Blends with acrylic fibers could be with other fibers or with modified acrylic fibers themselves. Hibulk acrylic fiber (which shrinks on heating) was blended with regular acrylic fiber and made into yarn. Heating made the Hibulk component shrink, thus producing a lofty yarn. Other blends of acrylic included the addition of nylon or polyester to improve wrinkle resistance in socks, the blending with polyester to give woven fabrics with improved wrinkle resistance, and blends with wool, angora and silk to reduce costs and to improve physical performance [5].

Modacrylic blends. Modacrylic fibers were introduced in 1948 by Union Carbide with Dynel staple becoming available in 1950 [25]. Blending with modacrylic fibers was usually done in order to take advantage of the inherent flame resistance of the modacrylic fibers and thence add flame resistance to the blend. For the modacrylic fabrics themselves, the addition of polyester improved the relatively low abrasion resistance of the fabric.

Polyester. Polyester fibers were invented in the United Kingdom and introduced into the United States by the DuPont company as Dacron in 1952 [26]. Two attributes of polyester fibers were recognized early: one was problems with dyeability and the other was the ability of polyester to impart crease resistance to fabrics. The latter property, along with the inherent strength of polyester, have been the reasons for polyester's status as the leading synthetic fiber and has led to polyester's use in blends with cotton and wool to provide wrinkle resistance, and to add strength and abrasion resistance to the blends.

The blending of polyester with wool has made possible light-weight fabrics with excellent crease retention and wrinkle resistance that retain the visual and tactile aesthetics of all-wool fabrics. In addition, the lower cost of these blends has kept men's suits made of these fabric in the affordable range. For blends of wool with polyester, the low modulus fiber was used at levels between 40-60 percent [26].

The development of wool/polyester blends for suiting fabric began early. A 1955 paper surveying blends of fibers [27] notes that polyester could be used "to advantage in clear-finish suitings". The recommendation was that a 65/35 polyester/wool blend be used, although a range of 40-70 percent polyester was acceptable. Fabrics containing lower or higher percentages of either fiber did not benefit from the blending.

The wool/polyester blend continued to develop through the 1960's to become a standard. The blends that had great impact were in the range of 50-65 percent polyester with 50-35 percent wool, and in 4-5 ounces per linear yard weights [28]. The blends have a wool-like feel and appearance, are durable and easy to care for, and reasonable in cost.

The greatest success in fiber blends is the union of cotton and polyester. These blends take advantage of polyester's relatively low cost, its ability to impart extra wrinkle resistance and strength, and its relatively good match with the stress-strain characteristics of cotton. The advantages of cotton in these blends are its aesthetic characteristics of handle and overall perceived comfort, cost and easy dyeability. The impetus to develop these blends is easy to identify - the desire to alleviate the long hours spent in home washing and ironing of all-cotton clothes.

An early conference on fiber blends in 1952 mentioned the ability of polyester to impart wrinkle resistance, but did not extend the reasoning to the concept of wash-and-wear [16]. By 1958, however, it was recognized that the ability to create such fabrics was perhaps the major advantage that polyester brought to blends of cotton. As an example, the American Home Laundry Manufacturers Association had wash-and-wear as its theme in 1957 [5]. Polyester has an advantage over nylon in these blends in that its initial modulus matches better with cotton than it does nylon. The blend levels of the high modulus staple fibers vary widely, from 25 percent to 80 percent. Generally lower blend levels are used for the heavier weight fabrics.

Beginning in the late fifties, interest grew in developing cotton fabrics that would give true wash/wear properties. The primary shortcoming of the fabrics that were developed was the inability to create fabrics that would retain creases once they were made into garments. The breakthrough that led eventually to the polyester/cotton blend fabrics of the late sixties was the development of a durable press finish that allowed deferred curing of the garment [27]. The garment was constructed and creases set in before the complete reaction of the finish with the cotton was accomplished. A garment that kept its creases after washing was the result.

Initially, this process was used on an all-cotton fabric and put into the market in 1963. Severe problems occurred, however, because the durable press (DP) finish caused the strength and abrasion resistance of the garment to drop significantly. The creases and the cuffs wore through very quickly. The solution was to apply the process to polyester/cotton blend fabrics to take advantage of the strength and inherent wrinkle resistance of the polyester [30]. In order to do this, improvements were made in the finish, in the method of blending the two fibers in fabric manufacturing, and in the garment pressing methods.

The results were spectacular. The use of polyester in blends became the rule rather than the exception [31]. The old patterns of laundry work within the home were changed drastically, and even the formulation of detergents and the design of automatic washing machines and driers were profoundly changed by the success of the DP-finished polyester/cotton blend fabrics.

The question of what blend level of polyester was needed to insure adequate wrinkle resistance became an important issue, particularly in the mid and latter part of the sixties. Some polyester fiber producers believed that 65 percent polyester was necessary to obtain satisfactory durable press performance [32], while other companies maintained that their research showed a 50 percent polyester content to be adequate [33]. The weight of the fabric was also a factor, with heavier weight fabrics needing less polyester than the lighter fabrics. The trend over the years has been to move toward more cotton in DP blends. This trend has been caused by both technological developments and by very successful campaigns from cotton marketing groups to convince consumers of the advantages from using more cotton. Today, men's dress shirts, the biggest single market for DP fabrics, is dominated by 60/40 cotton/polyester blends.

Spandex. Spandex fibers were introduced to the market in 1960. These are elastomeric polyurethane fibers with a block copolymer structure that allows them to have unique elongation properties without some of the susceptibility to oils, chlorine water and sunlight that are major drawbacks to fibers made of natural rubber. Because of their high elongation these fibers are rarely used except in blends. Blend levels range from 3-5 percent in filling stretch cotton (core spun) to over 40 percent in warp knit tricots. Foundation garments, for example, typically contain 20-30 percent spandex [34].

Although a few fibers in a small number of combinations dominate the market of blend fabrics, many different combinations and blend proportions have been tried. Some have established niches in the overall market. Table III, which primarily covers the 1960's and 1970's, lists many combinations that have been offered, some which have been successful and some which, like the blends discussed in the beginning of this review, are just a historical footnote.

References

1. K.A. Jakes, ACPTC Proceedings 44 (1989).
2. V. Thurston, Short History of Ancient Decorative Fabrics 6, 24, 46, Fariel Press Ltd, London, 1954.
3. D.L. Carroll, Looms and Textiles of the Copts 20-31, California Academy of Sciences, 1986.
4. N.D. Bogdanoff, Handwoven Textiles of Early New England 37, Stackpole Books, Harrisburg, 1975.
5. R.M. Hoffman and R.W. Peterson, J. Textile Institute 49, 418-434 (1958).
6. Leviticus 19:19, *The Holy Bible*.
7. C.S. Groves and C.F. Ward, J. Textile Institute 43, 626 (1952).
8. A. Geiger, A History of Textile Art, 245-255, W. S. Mancy and Son, Ltd., London. 1979.
9. P.A. Means, A Study of Peruvian Textiles, 39, Boston Museum of Fine Arts, Boston, 1932.
10. I. Reswick, Traditional Textiles of Tunisian and Related North African Weavings, 38, University of Washington Press, Seattle, 1985.
11. D. Idiens, Textile History, 11, 15 (1980).
12. F.H. Montgomery, Studies in Textile History, 217-222, ed. by V. Gervers, Royal Ontario Museum, Toronto, 1977.
13. J.B. Speakman, J. Textile Institute 49, 580 (1958).
14. H.R. Crone, J. Textile Institute 43, 556 (1952).
15. A.B. Shearer, J. Textile Institute-Proceedings 16, 146-148 (1925).
16. H. Wild, J. Textile Institute 37, 190-197 (1946).
17. F. Hawlett, M.J. Morley. and A.R. Urquhart, J. Textile Institute 42, T75-103 (1942).
18. W. Hardacre, J. Textile Institute 34, 71-77 (1943).
19. C.S. Whewell, J. Textile Institute 41, 219-228 (1950).
20. K.H. Parr, J. Textile Institute 43, 88-93 (1952).
21. R.S. Greenwood, J.Textile Institute 43, 511-515 (1952).
22. J.H. Saunders, Encyclopedia of Textiles, Fibers and Nonwoven Fabrics, 347-380, John Wiley and Sons, New York, 1984.
23. W. Hardacre J. Textile Institute 43, 80-87 (1952).
24. F. Millard and P. Thornton, J. Textile Institute 43, 413-449, (1952).
25. P.A. Hobson and M. Grayson, Encyclopedia of Textiles, Fibers and Nonwoven Fabrics, 1-31, John Wiley and Sons, New York, 1984.
26. G.W. Davis and E.S. Hill, Encyclopedia of Textiles, Fibers and Nonwoven Fabrics, 381-398, John Wiley and Sons, New York, 1984.
27. M.K. Ryan, Textile World, 236 (Feb. 1955).
28. Anon., Textile World, 8 (Dec. 1973).
29. H.B. Goldstein and J.M. May, Textile Research J. 34, 325-330, (1964).
30. Anon., Textile World, 79 (Dec. 1973).
31. D.A. Bowen, Textile World, 99 (Nov. 1973).
32. N.C. Heimhold, Textile World, 75 (Nov. 1966).

33. N.C. Heimhold, Textile World, 149 (Dec. 1966).
34. T. Peters, Encyclopedia of Textiles, Fibers and Nonwoven Fabrics, 136-151, John Wiley and Sons, New York, 1984.

TABLE I. Fiber mixtures.

Acetate-Vinyon	Nylon-Cotton
Cotton-Rayon	Nylon-Flax
Cotton-Rayon	Nylon-Rayon
Cotton-Silk	Nylon-Wool
Cotton-Asbestos	Nylon-Camel Hair
Rayon-Silk	Wool-Casein
Silk-Wool	Silk-Nylon

TABLE II. Comparison of twill fabrics.

Fabric Composition	Flex Abrasion	Breaking Strength(lb)	Tearing Resistance
Cotton	1378	191	9.0
Cotton and Regular nylon	6750	171	8.7
Cotton and Type 420 nylon	6750	242	11.4

Table III. Miscellaneous blends of fibers.

Fiber in the Blend	%	Year Cited	Type of Fabric
Metallic/Mylar		1955	
Acrilan/Cotton	50/50	1955	Wash wear
Rayon/Wool/ Acetate/Nylon	57/28/10/5	1955	Uniforms
Beaver/Dynel		1956	
Dacron/Flax	65/35	1956	Resort apparel
Bamboo/Cotton		1956	Home furnishings
Cotton/Rayon	66/33	1958	
Wool/Dynel	70/30	1959	Wash wear
Dacron/Orlon/ Wool/Other	47/28/19/5	1959	
Orlon/Cotton	80/20	1959	Underwear
Topel/Kodel	70/30	1959	Batiste & Poplin
Orlon/Triacetate	85/15	1959	
Wool/Orlon	50/50	1959	Mens suits & slacks
Triacetate/Acetate	50/50	1960	
Cotton/Acrylic	70/30	1960	Knitted Outerwear
Nylon/Acrylic	80/20	1960	Trimmings
Triacetate/Rayon		1960	Machine washable
Rayon/Acetate/Dynel		1960	
Acrilan/Rayon	50/50	1960	
Cotton/Nylon	60/40	1960	
Kodel/Rayon	50/50	1960	Wash wear
Dacron/Pima Cotton	65/35	1960	
Dacron/Cotton	65/35	1960	Linen Look
Triacetate/Cotton	50/50	1960	Suiting
Acrylic/Rayon		1960	
Cotton/Nylon/Rayon/ Jute/Taslan		1960	Upholstery
Cotton/Vycron/Bemberg	50/45/5	1960	Shantung
Vycron/Cotton/Dacron	45/55	1961	Sheer Casement
Acrylic/Dynel	60/40	1961	Rugs
Orlon/Rayon	50/50	1961	Lining
Kodel/Wool	60/40	1961	Double Knit
Orlon/Wool/Polyester	80/20	1962	Coats & Suits

Fiber in the Blend	%	Year Cited	Type of Fabric
Cotton/Rayon	60/40	1962	Sheets & Pillows
Orlon/Nylon	80/20	1962	Mens socks
Nylon/Cotton	50/50	1962	Stretch Sail Cloth Womens Sportswear
Cotton/Zantrel/Nylon	50/35/15	1963	Slacks
Dacron/Cotton	65/35	1963	Linen Look
Polypropylene/Rayon	50/50	1963	
Orlon/Rayon	50/50	1964	Slacks & Skirts
Nylon/Lycra	80/20	1964	Seamless Stockings
Polypropylene/Wool	50/50	1964	
Polyester/Asbestos			Heat Resistant
Triacetate/Cotton		1964	Diapers
Acrylic/Spandex	92/2	1964	Sweaters
Acrylic/Mohair	50/50	1964	Knit
Rayon/Polypropylene/ Silk		1964	Dresses & Skirts
Dacron/Cotton	65/35	1964	Corduroy
Acrylic/Polypropylene	70/30	1964	Hose
Nylon/Polyester	70/30	1965	Tire Cord
Rayon/Polypropylene		1965	Upholstery
Olefin/Rayon		1965	Shirting
Acrylic/Cotton		1965	Knitted Sleepwear
Triacetate/Nylon	75/25	1965	Tricot Knit
Dacron/Cotton	65/35	1965	Denim
Wool/Acetate/Flax	70/15/15	1965	Dress Fabric
Rayon/Acetate/Flax	85/8/7	1965	
Wool/Nylon	75/25	1965	
Nylon/Rayon/Acetate	40/30/30	1965	Stretch Fabric
Rayon/Nylon	65/35	1965	Satin Look
Wool/Nylon	65/35	1965	Textured Stretch Fabric
Wool/Nylon	78/22	1965	Dress Fabric
Fortrel/Combed Cotton	50/50	1966	No-Iron Sheets
Polyester/Cotton	35/65	1966	Shirts
Nylon/Polyester	67/34	1966	Tire Cord
Polypropylene/Wool	50/50	1966	Socks
Polyester/Cotton	50/50	1966	Jeans

Fiber in the Blend	%	Year Cited	Type of Fabric
Poltyester/Flax	67/33	1966	Hand Knitting & Hosiery
Cotton/Rayon/Linen	60/25/15	1966	Towels & Potholders
Nylon/Rayon	50/50	1967	Water Repellent
Dacron/Acrylic	55-45	1967	Slacks
Polyester/Linen	70/30	1968	Shantung Look
Acrilan/Lambs Wool/Kempy Wool		1968	Shetland Wool Look
Modacrylic/Fiberglass		1970	Flameproof fabric
Wool/Rayon	70/30	1970	
Triacetate/Nylon	85/15	1972	
Fiberglass/Acetate	60/40	1972	Draperies
Acrylic/Cotton	90/10	1973	Fake fur
Acrylic/Polyester	50/50	1973	Knit
Wool/Vinyon	60/40	1973	
Wool/Polyester	80/20	1974	Flame-Resistant Uniforms
Wool/Nomex	80/20	1974	Flame-Resistant Fabric
Ramie/Polyester	52/48	1977	
Cotton/Kevlar		1979	Twill Fabrics

SOME BACKGROUND FOR THE TEXTILE INDUSTRY IN THE 20TH CENTURY

CHARLES H. FISHER
Chemistry Department, Roanoke College, Salem, VA 24153

INTRODUCTION

Textiles, through the history of civilization, as yarns, cords and two-dimensional goods always have been encompassing an enormous range of properties and applications: from the softest and finest silk, cotton and Vicuna knittings and fabrics to the strongest and toughest cords for fishing nets, sail cloth and anchor ropes.

Yet, until the beginning of the 20th century, there were only two compositions which served as base materials for this amazingly wide variety of uses: polysaccharides in the form of seed (cotton) of bast (hemp) fibers and polypeptides in the form of silk and wool. All the processing was experience, skill and taste in the selection of the best fiber, or mixture thereof, pretreatment (loading of silk mercerization of cotton) and appropriate entanglement to form two-dimensional systems by weaving, knitting and tufting. It was all essentially mechanical; with the exception of dying and printing no chemistry was involved. There was some knowledge about the composition of the fibers but nothing about their structure. Nevertheless, innumerable exquisite objects of art were produced in the course of the centuries in the form of carpets for floors and walls, curtains, upholstery and beautiful garments for daily use and for ceremonial occasions.

All steps of preparation, processing and finishing, were done by hand until the middle of the 19th century when the advent of the steam engine compelled producers to mechanize such processes like spinning yarns from staple fiber, weaving, knitting or composing a fabric on fast moving machines. This conversion of a labor intensive home industry into a highly mechanized, continuous operation was one of the earliest sources for the industrial and social revolution.

THE MOLECULES

Cotton and wool are staple fibers (up to 20 cm long) which have to be "spun" into continuous yarns before being made into a fabric. In the domain of polypeptides silk was a continuous filament with all its advantages for the further processing. Could one not do the same thing for a polysaccharide like cotton? Cellulose is not soluble in any liquid, but certain derivatives are: the nitrate, the acetate and the xanthate.

CHAIN $CH_2{-}O{-}NO_2$

CHAIN $CH_2{-}O{-}C({=}C){-}CH_3$

CHAIN $CH_2{-}S({=}S){-}ONa$

These cellulose derivative solutions may be extruded and shaped into endless filaments; into filaments of what? Evidently of the molecules of the dissolved cellulose derivative. This focussed the situation one step further down to the fundamentals: namely to the ultimate structural unit of textile fibers to their molecule. Already early investigators[1] assumed that these fiber forming molecules have the shape of long chains with very high molecular weights - 100,000 and more - but it was H. Staudinger who in 1920 considered their existence as a fact and, a few years later, coined for them the term macromolecules. Two of the above cellulose derivatives - the nitrate and the xanthate are reconverted into cellulose. As a result, there existed at the beginning of this century two continuous polysaccharide fibers - the regenerates cellulose and the cellulose acetate.

The exhaustive characterization of a macromolecule is complex: there is first its molecular weight, or, since all actual samples are mixtures of many species - its average molecular weight. There exist rigorous methods to measure it: ultracentrifuge, osmometry, lightscattering and a simple, approximate method - viscometry.

Purely empirically it was observed that polysaccharides and polypeptides are being degraded by the action of chemicals by heat and irradiation.[2] It was correctly interpreted as a reduction of the molecular weight, corresponding to a decrease of monomer units in the chain, known as degree of polymerization (DP), until at the end a powderous mass was obtained which showed no fiber forming properties any more. Evidently "degradation: meant a shortening of the chains and favored a slippage of them under tension which occurred easier if the chains are short. Hence it became necessary to study experimentally the influence of DP (chain length or MW) on tensile strength. Such tests were made int he 1920s with Viscose Rayon and Cellulose Acetate. They have the characteristic form of a hyperbola shown in Fig. 1, where it can be seen that at (and below) a certain critical value of DP-DP_o - no strength exists at all; increase of DP is developing strength rather efficiently because it reduces the probability to find a large number of chain ends in a specific cross-section and eventually reaches a (mildly inclined) plateau indicating that there is no chain slippage anymore but that now, chains have to be broken, i.e., covalent bonds have to be severed.[3]

As a result the tensile strength of a fiber can be expressed by

$$T_n = T_{max} - A \cdot n \quad (1)$$

where T(max) is the largest possible strength - if there is no slippage but if only chains are broken. This maximum possible tensile strength is reduced proportional to the number n of chain ends. The constant A indicates the efficiency of a chain end on tensile strength. If A is large then the influence of a chain end is important i.e., there are no intermolecular forces which "camouflage" the chain end; this is, f.i. the case in Cellulose Acetate, Polystyrene or Polymethylmethacrylate. On the other hand, if there are strong intermolecular forces, like in polyamides or if the chain ends are "covered up" by crystallization or crosslinking then their influence - represented by A - is relatively small and even short chains are giving reasonable strength. Obviously the number n of chain ends is reciprocal to the (average) molecular weight M and equation (1) may be written as:

$$T_n = T_{max} - \frac{A}{M_n} \quad (2)$$

an expression which was published by Flory in 1945. Figure 1 shows curves of TS vs MW for cellulose and cellulose acetate where A is smaller for cellulose because of hydrogen bonding between the chains.

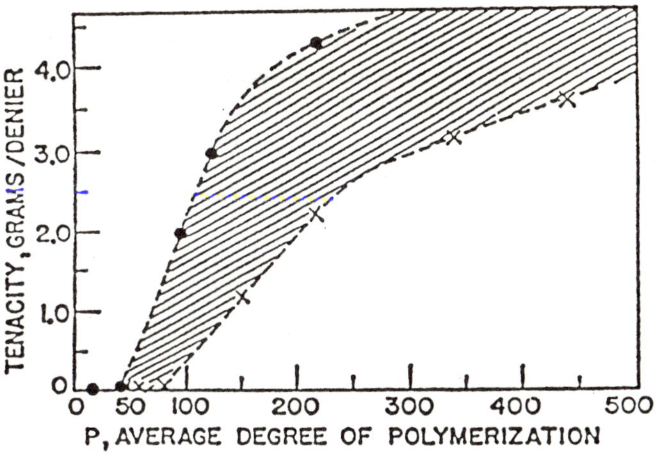

Figure 1. Effect of Average Degree of Polymerization on Tenacity of High Polymers.

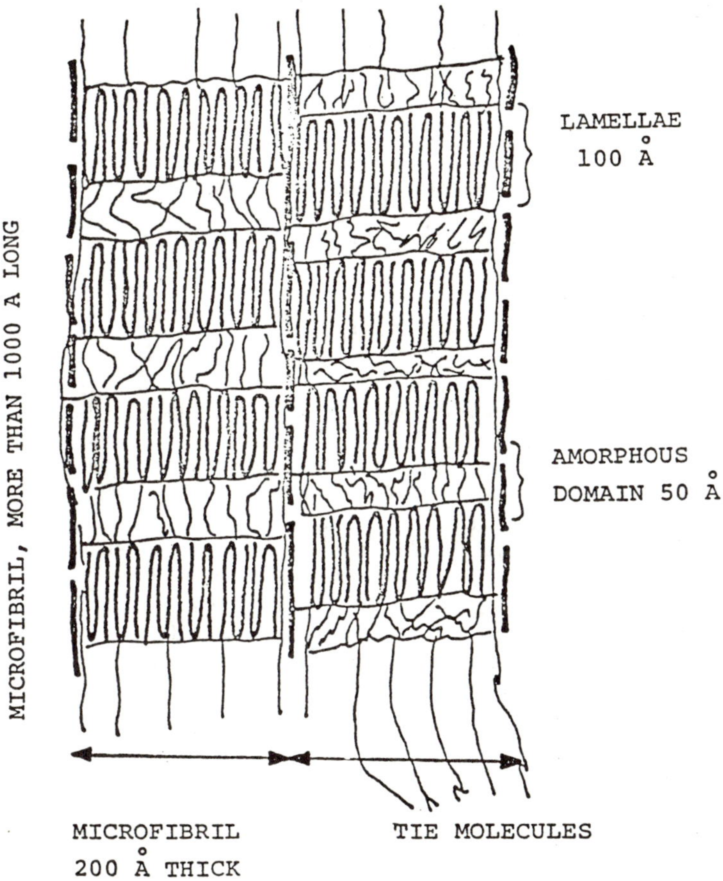

Figure 2. Structural Features of Fibers with Folded Chains

A few years after the advent of viscose and acetate rayon the first fully synthetic fiber - a polyamide - was invented and studied by W. H. Carothers.[5] As a result of all of these investigations (from 1910 to 1930) there emerged the concept of an ideal **fiber** former a linear macromolecule which might be described about as follows:

a) it should be capable to crystallize on drawing with a melting point between 210° and 250°C.

b) it should be insoluble in all organic liquids up to 150°C and resistant against acids and bases up to 100°C.

c) in the fully drawn state (draw ratio from 4-6) it should have a tensile modulus of 400-500 g/den* and a tensile strength of 8-10 g/den

d) at ambient conditions (60-70% relative humidity) it should have a moisture regain of 8-10%

e) it should have an elongation to break of 20-30% with an elastic part of 10%.

In spite of many synthetic polymers which were prepared and tested, there is none which completely fulfills all the above listed properties. However, there are numerous materials which are very satisfactory in certain respects and offer a wide range for maneuvering and experimentation. It is also possible to formulate a number of principles or recipes which will be useful for the search of better fiber formers.

Evidently strong intermolecular attraction - polar forces or hydrogen bonding - will favor the fulfillment of conditions [a], [b] and [c] but there are other structural features which are of importance. For instance, according to simple thermodynamics the melting point of a substance is given by

$$T_m = \frac{\Delta H_m}{\Delta S_m} \tag{3}$$

A large value of ΔH - the heat of melting - corresponds to strong intermolecular attraction and confirms their importance for high melting. However, at a given value of H, T_m still depends on S - the entropy of melting. The smaller it is, the higher will be Tm and the better will be the fulfillment of [b] and [c]. In fact, if the chain segments are difficulty mobile - i.e. the chains are rigid or stiff - then the configurational difference between the solid state and the melt will be small and T_m is high. Chains of this kind are going also to offer more resistance to gliding under dress since the segments have less mobility. Thus the quantities A in Equation (2) and S in Equation (3) are related to each other and indicate that chain rigidity is an important attribute of macromolecules; it leads to high melting and high strength. This was already qualitatively recognized by Lilienfeld for cellulose[6] and became later one of the most important principles for the search of better fiber formers.[7]

Altogether in the 1950s the following fundamental molecular properties could be considered to be good guidelines for the synthesis of improved fiber forming polymers:

- Strong intermolecular forces
- Crystallizability
- Capacity to be eventually crosslinked
- Segmental rigidity of the chains

With their aid numerous synthetic fibers were made, none of which covered all important characteristics but which, in their entirety, gave a new and extremely broad basis for the textile activities of this century. Even more than that, these principles also indicated how additional progress could be reasonably expected.

THE FIBERS

At the basis of all efforts to prepare a useful synthetic fiber is the "spinning process."

There exist presently three prevailing types of fluid polymeric compositions which are used in fiber production:

a) Solutions or melts of linear macromolecules which have "flexible" chains, i.e., long sequences of single -C-C-, -C-O-, or -C-N- bonds without bulky substituents or cyclic chain units. Examples of this type are polyolefins, polyvinyls, polyacrylics, aliphatic polyesters, polyamides and polyacetals. The chains of these materials possess a pronounced tendency to fold into lamellar structures having on the average a thickness of about 100 Å.

b) Solutions or melts of linear macromolecules which have more or less rigid chains, i.e., longer sequence of conjugated multiple bonds or of cycloaliphatic viz. aromatic rings. These chains have little or no folding character but tend to form under certain conditions (solvent, concentration, temperature) microgels or liquid crystals or supermolecular dimensions; examples are cellulose and its derivatives, aromatic polyesters, polyamides and more complicated polyconjugated heterocyclics.

c) Dispersions or emulsions of solid polymers in a polymeric solution, which serves as a viscous carrier during the spinning and stretching process. Examples are Teflon, PVC, Silica, PAN, and $PVCL_2$ in cellulosexanthate, polyvinylalcohol or proteins.

d) On top of these classical spinning modes there exists the conversion of films and membranes **to** fibers by slicing a thin film and drawing, orienting and twisting the resulting filaments.

Let us now discuss in some more detail the characteristic features of each of these modes which are summarized in Table 1.

Foldable Chain Spinning[8]

For the purpose of a simplified classification we shall report on this type of fiber formation in three subsections, namely:

a) semidrawing, relaxation and heat setting

b) drawing, essentially to the natural draw ratio plus heat setting, and

c) superdrawing and heat setting

It will readily be understood that there exist many "gray areas" between these types, i.e., conditions under which the specific characteristics operations overlap; they lead to products of intermediate structure and properties.

The structural features of such fibers have been developed with increasing clarity and reliability during the last ten years and are schematically presented in Fig. 2.

The fibers consist essentially of long (1000 Å and more) substantially parallelized microfibrils having average diameters of 200 Å or more--(This leads to the conclusion that the cross-section of a two denier fiber contains about 10,000 such microfibrils.)[8] Each of them is built up of layers of folded chain lamellae which extend throughout the microfibril and are about 100 Å thick. Between them there are somewhat thinner layers (30-60 Å) in which the segments of the polymer molecules are more or less random (amorphous domains). Along the length of the microfibrils there exist also bundles of extended chains which pass through many lamellae and many amorphorous domains (up to 20 or more) and provide for the firm axial structure of the system; they are called tie chains and also possess a substantial lateral order. Hence the "crystalline" portion of such fibers consist of crystalline folded chains and of crystalline extended chains and the "amorphous" portions consist of the disordered layers in the microfibrils between the folded lamellae. Most significant changes in the properties - as the results of stretching, relaxing and heat setting - are caused by manipulating a transition of chains from the folded crystalline type into the extended crystalline type.

Working with filament extensions in the neighborhood of slightly above the natural draw ratios (4-6 times) one arrives at the broad spectrum of commercial products from semidrawn to highly drawn yarns; this classical technique is of enormous importance for the man-made fiber industry and covers most important applications in hosiery, underwear, garments and coats. The technology of these processes is thoroughly developed leading to high levels of productivity (speed) and uniformity (quality). In each case a specific structural compromise is established between the three elements which are building up the body of the fiber as shown in Fig. 1 maneuvering particularly with the ratio of the two crystalline phases.

A few years ago a new attempt was initiated to reduce the degree of chain folding and to push into higher proportions of the molecules by applying a technique of superdrawing. First demonstrated by Pennings & Porter[9] with polyethylene successful experiments were soon carried out with several other foldable chains.

The process involves drawing first at normal speed (e.g. 50%/sec) to the natural draw ratio and then more slowly (e.g. 50%/min) and in the presence of a gel former at high viscosity to an extension 3-4 times higher, for instance 20-25 times the original length.

D. C. Prevorsek,[9] who has successfully developed the "spectra 900 fiber" for Allied Chemical has shown that tensile moduli up to almost 1000 g denier and strength values up to 40 g per denier can be obtained for polyethylene and there are indications that 6-, 66 nylon and polyester follow a similar pattern.

The observation on superdrawn fibers are still recent and their interpretation is only tentative, but the properties are such that further studies are obviously very attractive and a new family of hard and strong fibers could well result from such investigations. Even if they should only find practical applications in the fringes of the textile industry - in carpets, upholstery, curtains and the like - they could be very valuable as industrial yarns and cords in hoses, safety and transport belts, fishing lines and nets and for the construction of drying sieves and felts for paper machines.

It seems, therefore, that the art of foldable chain spinning has not only a promising future in improvements and expansions of conventional man-made fibers but may also have the chance to develop new fiber types utilizing old and well known polymers with a novel design of their supermolecular structure.

Spinning of Rigid Chains

There exist many linear macromolecules which have such rigid chains that there is little or no tendency for the formation of folded chain conformations. Consequently, their spinning differs considerably from that of foldable chains.

The oldest example for the spinning of relatively rigid chains are the Cuprammonium, Viscose and Acetate processes. Best investigated is the Viscose Rayon Process, i.e., the spinning of cellulose xanthate from strongly alkaline solutions into an acid bath. Commercial solutions containing 6-10% cellulose are clear, yellow, isotropic systems which have, at room temperature, viscosities between 100 and 300 poises depending on concentration, degree of xanthation and molecular weight of the cellulose. Upon entry into the spinning bath the gradual decomposition of the cellulose xanthate produces a highly swollen gel of moderately folded cellulose chains which are progressively oriented by shear and form essentially a mixture of extended chain crystallites separated by disordered or amorphous domains.[6]

Recently another family of rigid, non-folding chains have become known and available through the synthesis of wholly aromatic polyesters, polyamides and other polyconjugated heterocyclic species. Of particular importance is one specific member of this family because it has been thoroughly studied and shows the nematic crystal spinning in a particularly clear and impressive manner.[7] It is the poly-para-phenylemediaminterephthalate (PPT) of the schematic formula:

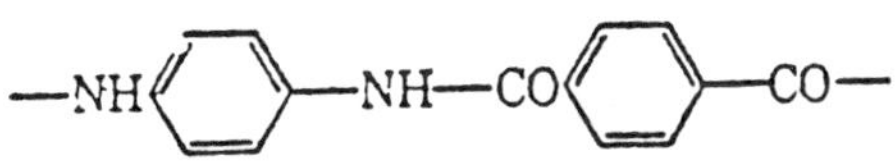

KEVLAR

The polymer is prepared by the interfacial polymerization of paraphenylene diamine and terephthalic acid dichloride and represents a hard yellowish powder with a melting point above 500°C. It is partly soluble in certain organic liquids such as mixtures of N-methyl-pyrrolidone and hexamethyl phosphamide but the only useful solvent is concentrated sulfuric acid (89% and higher) eventually with addition of fluoro-, chloro- or methane sulfuric acid. Depending on the conditions of the synthesis inherent viscosities up to 6.5 have been reached with a preferred range between 3.0 and 5.0. Spinning solutions containing about 20 weight percent of polymer may be prepared between 90 and 100°C; they have viscosities from 2000 to 3000 poises and are completely solid at room temperature. The most striking property of these solutions is their highly anistropic character which was found to be caused by the pronounced tendency to form nematic liquid crystals. Even very slow flow through a capillary (between 100° and 110°C), very slow stirring in a beaker or very weak shear between parallel glass plates produce strong optical birefringence, the character of which indicates the presence of a nematic mesophase.

These solutions are spun in the neighborhood of 100°C in cold water and the filaments are carefully washed and wound up on a bobbin. Fibers of this type were first made known by Du Pont under the laboratory code of fiber B and PRD-49.

The most striking fact of this technique is that very high moduli and tensile strengths are obtained with very small SSF's. This is being attributed to the fact that the partly charged, intrinsically rigid macromolecules have a strong tendency to aggregate spontaneously to nematic liquid crystal structures while the solution approaches the spinnerette. The viscosity of these spinning dopes is much higher - around 1000 poises between 80° and 90°C - than that of conventional spinning solutions in the wet spinning of rayon and in the wet or dry spinning of cellulose acetate and of acrylics, which is 100 to 200 poises at spinning temperatures.

It has been found that most linear macromolecules which contain rings, particularly aromatic, but in some cases also transcycloaliphatic, have a tendency to form nematic melts or spinning solutions and permit to get high tensile modulus and strength without large stretch ratios showing that there is a considerable amount of self-organization in the system even before it is extruded.

Emulsion and Suspension Spinning

Already many years ago it was shown that Teflon emulsions can be dispersed in a cellulose xanthate dope and spun into fine and uniform filaments. These filaments may be stretched at elevated temperatures and the cellulosic matrix can be removed by hydrolysis or pyrolysis. Strong, oriented Teflon yarns are obtained in this manner.

A similar procedure was used with dispersions of graphite and silica to give inorganic fibers which excel by their flame resistance and extreme heat stability.

In all these cases the spinning aid is destroyed or otherwise removed after fiber formation and a one component filament is obtained; there exist, however, several cases where an emulsion is spun from a viscous polymeric dope and a composite system is formed in which, after stretching, very fine elongated fibrils of the emulsified polymer are distributed in an axially oriented manner in the coagulated materials of the spin carrier. Fibers of this type are usually referred to as "Matrix" fibers; in most cases thin (0.1 to 1.0 microns) fibrils of a semicrystalline polymer are embedded in the amorphous and often crosslinked matrix of another polymer. A fiber of this characteristic composition and structure is Cordela; it consists of fine oriented semicrystalline fibrils of PVC embedded in a crosslinked matrix of polyvinylalcohol; its properties are very attractive because the very thin oriented PVC component contributes softness, reluctance to ignition, slow burning and good mechanical properties (moduli from 30-50 g/den, tensile values from 2.5-5.0 g/den with break elongations up to 50%) whereas the crosslinked PVA matrix contributes a favorable moisture metabolism, good dying characteristics and resistance against solvents.

Similar matrix fibers are now studied which contain polyvinylidene chloride, polyacrylonitrile and various vinyl-acrylic copolymers as fibril forming component and polyvinyls, cellulosics and proteins as amorphous, reticulated matrix.

Film to Fiber Technology[10]

The development of film to fiber technology actually started in the beginning of the 1960s with high expectations. In the following decades a large number of technologies in the manufacture of film tapes, split-film yarns, and film fibers were developed by inventors and companies throughout the world, including the best known industrial giants involved in the development of conventional spinnerette fiber manufacture. The initially exaggerated expectations have given way, in the meantime, to a more realistic judgment of the potential of this simplified technique for the manufacture of yarnlike and fibrous products. Film tapes, split-film yarns, and fibrillated film fibers have found their permanent place in a good number of end-use applications, such as in wide areas of the packaging field or in the manufacture of technical fabrics, household textiles, or carpets.

The first step in the manufacture of fibrous film products such as film tapes and film fibers is the film casting. It is understandable that the manufacture of such products is only economically feasible when the film processing is relatively simple and inexpensive. For this reason, film processing be melt extrusion, particularly of polyolefins, is widely applied. Other polymers calling for more elaborate film processing methods, such as casting from solutions, are only suited for film to fiber technologies when they yield special products for expensive end-use applications. Furthermore, uniaxial stretching methods are most generally applied in the manufacture of fibrous products from films.

CONCLUSIONS

A large number of individual steps lead from the fiber to the ultimate sellable goods of the textile industry and much important progress has been made during the last few decades. The advent of the stronger and tougher synthetic fibers made it possible to operate each process faster than before. Intrinsically all man-made synthetics are produced in the form of continuous, endless filaments. If one wants to blend them with each other or with cotton and wool they must be cut into staple fibers which have an exactly controlled length, and a certain crimp to increase the frictional adhesion to each other. These staple fibers are then converted into spun yarns either on one of the traditional spinning equipments or, more recently, on a variety of modern machines such as open end spinning and similar techniques. Many important improvements have also been made in the conversion of a filaments and yarns into two dimensional systems by weaving, knitting, tufting and related techniques. In all these cases the main purpose was the saving of energy, the speed of production and the quality of uniformity of the merchandise. Altogether the individual steps were essentially mechanic and electronic and provide a highly computerized and strictly controlled sequence of individual steps. There is not very much novel chemistry involved, except in dying, printing and finishing. However, the appearance of the synthetics, particularly of polyesters and polyamides has given rise to another new and very promising art - the technology of texturizing. The beneficial air and humidity metabolism of wool (and partly also of cotton) depends on its crimp, i.e., on the existence of a permanent randomly distorted pattern of segments which is not completely straightened and during the formulation of the yarns and the fabrics, straight, endless filaments do not have any crimp and, therefore, they are having a slick and uncomfortable feeling. But both polyamides and polyesters may be drawn and heat set. Thus, if a plurality of such filaments are exposed to rapid, random distortions and a subsequent heat shock - they are being texturized, i.e., they assume along their length irregular distortions and convolutions, which because of the heat setting process are as permanent as the crimp of wool.

<u>In review:</u> At the beginning of the last century - around 1800 - Textile Industry operated essentially with four species of fibers - two cellulosics: the staple fiber cotton and the nominally continuous filaments of coarse bast fibers, such as hemp, flax and others, and two proteinic species, the staple fiber wool and the luxurious continuous super filament, silk. At the end of the century the industry still worked, in general, with the same materials. Some important innovations had been made in such processes as sizing, dying and finishing and also, from the 1880s on, significant pioneering work on new fiber types had been begun. But, by and large, cotton, wool and silk dominated the main stream of commercial textile operation.

In sharp contradistinction, "our" century started with the same - small - number of raw materials and, already now, has at its disposal a very large number of useful fiber formers, encompassing almost any class of synthetic organic polymers: it is the "Century of 'fiber science and engineering'". And, even beyond that, we have developed principles which permit to improve logically beyond the presently existing limits in the factory.

Let us now, briefly, review, how the additional useful species came up during its decades. First to appear - in the 1920s - were two continuous filaments of cellulosic origin, two soluble derivatives - the xanthate and the acetate - which were spun from solution. The xanthate was "regenerated" back into cellulose, the acetate remained a cellulose ester in the solid state as fiber, film or plastics. Both newcomers were beautiful, lustrous, and uniform continuous filaments and were first introduced as "Kunstseid" - "artificial silk" - even though the macromolecule itself - cellulose - was taken from Nature and only appropriately manipulated to arrive at useful raw materials. These two "rayons" as they were alter called had for 20 years a very beneficial expansive influence on the textile industry, but they also had serious deficiencies: viscose rayon is very weak when wet and acetate rayon is extremely sensitive against many organic solvents. Thus this early enrichment of our textile fiber arsenal was welcome but not decisive.

The breakthrough came about 15 years later when three "fully synthetic" fibers appeared on the scene: the polyamides,[7] the polyesters and the polyacrylics. They added not only a large collection of individual fiber formers but also new processing steps like melt spinning, cold drawing and heat setting. The "nylons" initiated a new branch of very shear, strong and durable stockings and related knitted goods and invaded with great success the entire field of flat fabrics. The "acrylics" alone or in blends - proved to be real "hits" for knitted goods including socks, pullovers, sweaters and the like and the polyesters - the first fiber former with an aromatic component[7] - became the real workhorse for textile activities, not only because of the wide range of its uses, alone or in blends for ladies' and men's shirts, skirts, suits and sportswear but also in related applications like carpets and upholstery and even, very successfully as tire cords and as reinforcing fibers in thermoplastic compositions.

Within a few years, each of these classes broadened in a generation of fibers: to the acrylic were added the modacrylics having inferior mechanical properties but being less expensive and having better fire-retardant characteristics, the nylons branched out into other molecular compositions like 4, 10, 11 and 12 nylons together with an aromatic species made from meta-phenylene diamine and isophthalic acids - Nomex - which has an exceptionally high softening point, is high flame resistant, tough and stiff. The polyesters had already a very wide range of applications and started to invade the carpet field by special spinning and finishing steps.

Another, independent and significant addition to the synthetics of these years were the "Spandex" fibers; they are highly elastic, soft and low denier filaments - elongations up to 1000% - which consist of segmented polyurethanes and have eliminated the use of coarse natural and synthetic rubber threads in all higher quality textiles.

Something amounting to a revolution took place when, in the mid 1950s Giulio Natta succeeded to bring to the scene a low cost, polyolefinic fiber former.[11] With the aid of Ziegler catalyst he was able to synthesize a high melting (160° - 180°C) stereoregulated - isotactic - polypropylene which, because of its low cost, toughness and chemical inertness soon invaded certain sectors of textile operations, such as the manufacturing of floor and wall covers, upholstery, geo-textiles, ropes, fishing nets and also as base material for the surface of sport fields and tennis courts. The use of the new catalysts for the polymerization and copolymerization of olefins has become a very important branch of macromolecular science; at least 20 monographs have been written and over 1000 articles published since it began, some 30 years ago. Many problems still remain to be solved and this field is far from being at a close. A similar lively development is going on in the synthesis of rigid chain molecules which produce liquid crystalline solution and permit to manufacture heat stable filaments of high tensile modulus and strength and of remarkable resistance to chemicals and solvents.[7] Also this development is far from its end, and even though most of its products are not sued within the boundaries of textile operations it has greatly widened our horizons on the potential of additional textile materials.

Thus, as this century draws to a close we are looking back on a tremendous increase in available materials and processing abilities, much more than any four or five earlier centuries had ever provided for us. This, obviously leads to the questions: what about the future? Is the 21st Century going to be equally full of revolutions and expansive discoveries? Evidently there cannot be an intelligent general answer to this question but I would like to close this survey with the mentioning of two problems, the solution of which should not be too far away.

From the point of view of the ultimate consumer of textile goods one may characterize an "ideal" fiber - staple or texturized filament - as having characteristics within the following ranges:

- Melting or softening between 220° and 250°C
- Glass transition somewhat below room temperature
- Tensile modulus and tensile strength depending on draw ratio, from 200 to 300 and 6 to 8 g/den*
- Elongation to break - from 10 to 30%
- Elastic recovery from linear extension above 80%
- Moisture regain at 60-70% relative humidity between 6 and 12%
- Resistance against chemicals and solvents in the domain of wetting, dyeing and bleaching

Presently, there exist many materials which fulfill several of these conditions very well but there exists none as yet which fulfills them all - at least not at costs which would make it commercially attractive. However, silk, certain selected types of wool, high MW polyvinylalcohol and Qiana very close and there is good reason to believe that, eventually, another synthetic fiber will come even closer.

Another problem is the combination of rigidity and toughness, we are able to produce fibers with tensile moduli around 50 million psi* but their extensibility is less than one percent - they are brittle. In contradistinction hereto many metals - steel titanium and many others are having similar hardness but are deformable and even malleable. This is due to their larger interatomic forces and to their different type of deformation mechanism. Present attempts for progress are made through the synthesis of composites and it is reasonable to expect some progress is the not too distant future.

REFERENCES AND NOTES

1. A. Kekule; E. Fischer; B. 47, 1983 (1914); H. Staudinger; B. 53 (1920); K. Freudenberg; B. 54, 771 (1921); A. Polanyi; Naturwissensch. (1921).
2. E. Berl et al, Ber., 41, 1840 (1908); E. Ost, Z. S. Angew. Chem., 32, 66 (1919).
3. Comp - Meyer, K. H. and Mark, H., "Der Aufbau," Akad. Ver. Ges., Leipzig (1928); H. Mark, "Physik und Chemie der Zellulose, J. Springer: Berlin (1932).
4. Flory, P. J., JACS, 67, 817 (1945).
5. Compare: "Collected Papers of W. H. Carother," Interscience Pub.: New York (1940).
6. Leon Lilienfeld, German Patent 448,984 (1920).
7. Compare - Whinfield, R. E., Dickson, J. T., Tanner, D., Kwolek, S. L., Magat, E. E., Schaefgen, J. R. and others in "Condensation Polymers," edited by P. W. Morgan, Interscience: New York (1965); Comp - U.S. Patent 3,414,645 (1968); U.S. Patent 3,642,706 (1972); U.S. Patent 3,671,542 (1972) and others.
8. Kraessig, H. A., Mark. Rev., 12, 321 (1971).
9. Pennings, A. J., et al, Koll. Z. S., 205, 160 (1965); Porter, R. S., et al, J. Appl. Phys., 41, 4284 (1970); also several lectures of D. C. Prevorsek of Allied Chemical Corporation on Spectra-900 Fiber in 1986 and 1987.
10. Comp - Krawssig, H. A., Lenz, J. and Mark, H. F., "Film to Fiber Technology," Marcel Dekker: New York (1984).
11. Comp - Natta, G., J. Poly. Sci., 16, 134 (1955), also Raff, R. and Doak, K. W., "Crystalline Polyolefins," 2 Vols., Interscience Publishers: New York (1963 and 1964).

Note: * 1 g/den = 2000 psi = 14 x 106PA; 1 g/den = 0.9 g per dtex

Table 1. Modes of Fiber Formation

1. Spinning of foldable chains
   - a) Semidrawing, relaxing, setting
   - b) Drawing, essentially to the natural draw ratio and heat setting
   - c) Superdrawing of plasticized linear chains and heat setting under tension
2. Spinning of rigid chains
   - a) Coagulation of oriented filaments to a spinning gel
   - b) Nematic liquid crystal spinning
3. Emulsion spinning
   - a) with destruction of the matrix
   - b) with conservation and crosslinking of the matrix
4. Film to fiber technology

INDEX OF CONTRIBUTORS

SUBJECT INDEX